Summary of contents

Contents

Module 15   Statistics   498

Preface

It is now nearly 40 years since Ken Stroud first developed his approach to personalized learning with his classic text *Engineering Mathematics*, now in its sixth edition and having sold over half a million copies. That unique and hugely successful programmed learning style is exemplified in this text and I am delighted to have been asked to contribute to it. I have endeavoured to retain the very essence of his style, particularly the time-tested Stroud format with its close attention to technique development throughout. This style has, over the years, contributed significantly to the mathematical abilities of so many students all over the world.

Student readership

Over recent years there have been many developments in a wide range of university disciplines. This has led to an increase in the number of courses that require knowledge of mathematics to enable students to participate in their studies with confidence. Also, by widening access to these courses, many students arrive at university with a need to refresh and amplify the mathematical knowledge that they have previously acquired. This book was written with just those students in mind, starting as it does from the very beginnings of the subject. Indeed, the content of the book ranges from the earliest elements of arithmetic to differential and integral calculus. The material is presented in a manner that will be appreciated by those students requiring to review and extend their current mathematical abilities from a level of low confidence to one of confident proficiency.

Acknowledgements

This is a further opportunity that I have had to work on the Stroud books. It is as ever a challenge and an honour to be able to work with Ken Stroud's material. Ken had an understanding of his students and their learning and thinking processes which was second to none, and this is reflected in every page of this book. As always, my thanks go to the Stroud family for their continuing support for and encouragement of new projects and ideas which are allowing Ken's hugely successful teaching methodology to be offered to a whole new range of students. Finally, I should like to thank the entire production team at Palgrave Macmillan for all their care, and principally my editor Helen Bugler whose dedication and professionalism are an inspiration to all who work with her.

Huddersfield   Dexter J Booth
February 2009

How to use this book

This book contains nineteen **Modules**, each module consisting of a number of **Units**. In total there are 62 Units and each one has been written in a way that makes learning more effective and more interesting. It is like having a personal tutor because you proceed at your own rate of learning and any difficulties you may have are cleared before you have the chance to practise incorrect ideas or techniques.

You will find that each Unit is divided into numbered sections called **Frames**. When you start a Unit, begin at Frame 1. Read each frame carefully and carry out any instructions or exercise you are asked to do. In almost every frame, you are required to make a response of some kind, so have your pen and paper at the ready to test your understanding of the information in the frame. You can immediately compare your answer and how you arrived at it with the correct answer and the working given in the frame that follows. To obtain the greatest benefit, you are strongly advised to cover up the following frame until you have made your response. When a series of dots occurs, you are expected to supply the missing word, phrase, number or mathematical expression. At every stage you will be guided along the right path. There is no need to hurry: read the frames carefully and follow the directions exactly. In this way you must learn.

Each Module opens with a list of **Learning outcomes** that specify exactly what you will learn by studying the contents of the Module. The material is then presented in a number of short **Units**, each designed to be studied in a single sitting. At the end of each Unit there is a **Review summary** of the topics in the Unit. Next follows a **Review exercise** of questions that directly test your understanding of the Unit material and which comes complete with worked solutions. Finally, a **Review test** enables you to consolidate your learning of the Unit material. You are strongly recommended to study the material in each Unit in a single sitting so as to ensure that you cover a complete set of ideas without a break.

Each Module ends with a checklist of **Can You?** questions that matches the Learning outcomes at the beginning of the Module, and enables you to rate your success in having achieved them. If you feel sufficiently confident then tackle the short **Test exercise** that follows. Just like the Review tests, the Test exercise is set directly on what you have learned in the Module: the questions are straightforward and contain no tricks. To provide you with the necessary practice, a set of **Further problems** is also included: do as many of the these problems as you can. Remember that in mathematics, as in many other situations, practice makes perfect – or nearly so.

Useful background information

Symbols used in the text

$=$	is equal to	$\rightarrow$	tends to
$\approx$	is approximately equal to	$\neq$	is not equal to
$>$	is greater than	$\equiv$	is identical to
$\geq$	is greater than or equal to	$<$	is less than
$n!$	factorial $n = 1 \times 2 \times 3 \times \ldots \times n$	$\leq$	is less than or equal to
$|k|$	modulus of k, i.e. size of k	∞	infinity
	irrespective of sign	$\underset{n \to \infty}{Lim}$	limiting value as $n \to \infty$
$\sum$	summation		

Useful mathematical information

1 Algebraic identities

$$(a+b)^2 = a^2 + 2ab + b^2 \qquad (a+b)^3 = a^3 + 3a^2b + 3ab^2 + b^3$$

$$(a-b)^2 = a^2 - 2ab + b^2 \qquad (a-b)^3 = a^3 - 3a^2b + 3ab^2 - b^3$$

$$(a+b)^4 = a^4 + 4a^3b + 6a^2b^2 + 4ab^3 + b^4$$

$$(a-b)^4 = a^4 - 4a^3b + 6a^2b^2 - 4ab^3 + b^4$$

$$a^2 - b^2 = (a-b)(a+b) \qquad a^3 - b^3 = (a-b)(a^2 + ab + b^2)$$

$$a^3 + b^3 = (a+b)(a^2 - ab + b^2)$$

2 Trigonometrical identities

(a) $\sin^2\theta + \cos^2\theta = 1$; $\sec^2\theta = 1 + \tan^2\theta$; $\mathrm{cosec}^2\theta = 1 + \cot^2\theta$

(b) $\sin(A+B) = \sin A \cos B + \cos A \sin B$

$\sin(A-B) = \sin A \cos B - \cos A \sin B$

$\cos(A+B) = \cos A \cos B - \sin A \sin B$

$\cos(A-B) = \cos A \cos B + \sin A \sin B$

$$\tan(A+B) = \frac{\tan A + \tan B}{1 - \tan A \tan B}$$

$$\tan(A-B) = \frac{\tan A - \tan B}{1 + \tan A \tan B}$$

(c) Let $A = B = \theta$ $\therefore$ $\sin 2\theta = 2 \sin\theta \cos\theta$

$$\cos 2\theta = \cos^2\theta - \sin^2\theta = 1 - 2\sin^2\theta = 2\cos^2\theta - 1$$

$$\tan 2\theta = \frac{2\tan\theta}{1 - \tan^2\theta}$$

(d) Let $\theta = \dfrac{\phi}{2}$ $\therefore$ $\sin\phi = 2\sin\dfrac{\phi}{2}\cos\dfrac{\phi}{2}$

$$\cos\phi = \cos^2\dfrac{\phi}{2} - \sin^2\dfrac{\phi}{2} = 1 - 2\sin^2\dfrac{\phi}{2} = 2\cos^2\dfrac{\phi}{2} - 1$$

$$\tan\phi = \dfrac{2\tan\dfrac{\phi}{2}}{1 - 2\tan^2\dfrac{\phi}{2}}$$

(e) $\sin C + \sin D = 2\sin\dfrac{C+D}{2}\cos\dfrac{C-D}{2}$

$\sin C - \sin D = 2\cos\dfrac{C+D}{2}\sin\dfrac{C-D}{2}$

$\cos C + \cos D = 2\cos\dfrac{C+D}{2}\cos\dfrac{C-D}{2}$

$\cos D - \cos C = 2\sin\dfrac{C+D}{2}\sin\dfrac{C-D}{2}$

(f) $2\sin A\cos B = \sin(A+B) + \sin(A-B)$

$2\cos A\sin B = \sin(A+B) - \sin(A-B)$

$2\cos A\cos B = \cos(A+B) + \cos(A-B)$

$2\sin A\sin B = \cos(A-B) - \cos(A+B)$

(g) Negative angles: $\sin(-\theta) = -\sin\theta$

$\cos(-\theta) = \cos\theta$

$\tan(-\theta) = -\tan\theta$

(h) Angles having the same trigonometrical ratios:

(i) Same sine: θ and $(180° - \theta)$

(ii) Same cosine: θ and $(360° - \theta)$, i.e. $(-\theta)$

(iii) Same tangent: θ and $(180° + \theta)$

(i) $a\sin\theta + b\cos\theta = A\sin(\theta + \alpha)$

$a\sin\theta - b\cos\theta = A\sin(\theta - \alpha)$

$a\cos\theta + b\sin\theta = A\cos(\theta - \alpha)$

$a\cos\theta - b\sin\theta = A\cos(\theta + \alpha)$

where $\begin{cases} A = \sqrt{a^2 + b^2} \\ \alpha = \tan^{-1}\dfrac{b}{a} \quad (0° < \alpha < 90°) \end{cases}$

3 Standard curves

(a) *Straight line*

Slope, $m = \dfrac{dy}{dx} = \dfrac{y_2 - y_1}{x_2 - x_1}$

Angle between two lines, $\tan\theta = \dfrac{m_2 - m_1}{1 + m_1 m_2}$

For parallel lines, $m_2 = m_1$

For perpendicular lines, $m_1 m_2 = -1$

Equation of a straight line (slope $= m$)

(i) Intercept c on real y-axis: $y = mx + c$

(ii) Passing through (x_1, y_1): $y - y_1 = m(x - x_1)$

(iii) Joining (x_1, y_1) and (x_2, y_2): $\dfrac{y - y_1}{y_2 - y_1} = \dfrac{x - x_1}{x_2 - x_1}$

(b) *Circle*

Centre at origin, radius r: $\quad x^2 + y^2 = r^2$

Centre (h, k), radius r: $\quad (x - h)^2 + (y - k)^2 = r^2$

General equation: $\quad x^2 + y^2 + 2gx + 2fy + c = 0$

with centre $(-g, -f)$: radius $= \sqrt{g^2 + f^2 - c}$

Parametric equations: $x = r\cos\theta, \; y = r\sin\theta$

(c) *Parabola*

Vertex at origin, focus $(a, 0)$: $\quad y^2 = 4ax$

Parametric equations: $\quad x = at^2, \; y = 2at$

(d) *Ellipse*

Centre at origin, foci $\left(\pm\sqrt{a^2 + b^2}, 0\right)$: $\dfrac{x^2}{a^2} + \dfrac{y^2}{b^2} = 1$

where $a =$ semi-major axis, $b =$ semi-minor axis

Parametric equations: $x = a\cos\theta, \; y = b\sin\theta$

(e) *Hyperbola*

Centre at origin, foci $\left(\pm\sqrt{a^2 + b^2}, 0\right)$: $\dfrac{x^2}{a^2} - \dfrac{y^2}{b^2} = 1$

Parametric equations: $x = a\sec\theta, \; y = b\tan\theta$

Rectangular hyperbola:

Centre at origin, vertex $\pm\left(\dfrac{a}{\sqrt{2}}, \dfrac{a}{\sqrt{2}}\right)$: $xy = \dfrac{a^2}{2} = c^2$

$$\text{i.e. } xy = c^2 \qquad\qquad \text{where } c = \dfrac{a}{\sqrt{2}}$$

Parametric equations: $x = ct, \; y = c/t$

4 Laws of mathematics

(a) *Associative laws* – for addition and multiplication

$a + (b + c) = (a + b) + c$

$a(bc) = (ab)c$

(b) *Commutative laws* – for addition and multiplication

$a + b = b + a$

$ab = ba$

(c) *Distributive laws* – for multiplication and division

$a(b + c) = ab + ac$

$\dfrac{b + c}{a} = \dfrac{b}{a} + \dfrac{c}{a}$ (provided $a \neq 0$)

Arithmetic

Learning outcomes

When you have completed this Module you will be able to:

- Carry out the basic rules of arithmetic with integers
- Check the result of a calculation making use of rounding
- Write a natural number as a product of prime numbers
- Find the highest common factor and lowest common multiple of two natural numbers
- Manipulate fractions, ratios and percentages
- Manipulate decimal numbers
- Manipulate powers
- Use standard or preferred standard form and complete a calculation to the required level of accuracy
- Understand the construction of various number systems and convert from one number system to another.

Units

Types of numbers

1 The natural numbers

The first numbers we ever meet are the *whole numbers*, also called the *natural numbers*, and these are written down using *numerals*.

Numerals and place value

The *whole numbers* or *natural numbers* are written using the ten numerals 0, 1, ..., 9 where the position of a numeral dictates the value that it represents. For example:

246 stands for 2 hundreds and 4 tens and 6 units. That is $200 + 40 + 6$

Here the numerals 2, 4 and 6 are called the hundreds, tens and unit *coefficients* respectively. This is the place value principle.

Points on a line and order

The natural numbers can be represented by equally spaced points on a straight line where the first natural number is zero 0.

The natural numbers are ordered – they progress from small to large. As we move along the line from left to right the numbers increase as indicated by the arrow at the end of the line. On the line, numbers to the left of a given number are *less than* ($<$) the given number and numbers to the right are *greater than* ($>$) the given number. For example, $8 > 5$ because 8 is represented by a point on the line to the right of 5. Similarly, $3 < 6$ because 3 is to the left of 6.

Now move on to the next frame

2 The integers

If the straight line displaying the natural numbers is extended to the left we can plot equally spaced points to the left of zero.

These points represent *negative* numbers which are written as the natural number preceded by a minus sign, for example -4. These positive and negative whole numbers and zero are collectively called the *integers*. The notion of order still applies. For example, $-5 < 3$ and $-2 > -4$ because the point on the line representing -5 is to the *left* of the point representing 3. Similarly, -2 is to the *right* of -4.

The numbers $-10, 4, 0, -13$ are of a type called ....

You can check your answer in the next frame

3

They are integers. The natural numbers are all positive. Now try this:

Place the appropriate symbol $<$ or $>$ between each of the following pairs of numbers:

(a) -3 $\quad$ -6
(b) $\ 2$ $\quad$ -4
(c) -7 $\quad$ 12

Complete these and check your results in the next frame

4

(a) $\ -3 > -6$
(b) $\quad 2 > -4$
(c) $\ -7 < 12$

The reasons being:

(a) $-3 > -6$ because -3 is represented on the line to the *right* of -6
(b) $\ \ 2 > -4$ because 2 is represented on the line to the *right* of -4
(c) $-7 < 12$ because -7 is represented on the line to the *left* of 12

Now move on to the next frame

Brackets

5

Brackets should be used around negative numbers to separate the minus sign attached to the number from the arithmetic operation symbol. For example, $5 - -3$ should be written $5 - (-3)$ and 7×-2 should be written $7 \times (-2)$. *Never write two arithmetic operation symbols together without using brackets.*

Addition and subtraction

Adding two numbers gives their *sum* and subtracting two numbers gives their *difference*. For example, $6 + 2 = 8$. Adding moves to the right of the first number and subtracting moves to the left of the first number, so that $6 - 2 = 4$ and $4 - 6 = -2$:

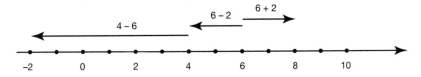

Adding a negative number is the same as subtracting its positive counterpart. For example $7 + (-2) = 7 - 2$. The result is 5. Subtracting a negative number is the same as adding its positive counterpart. For example $7 - (-2) = 7 + 2 = 9$.

So what is the value of:

(a) $8 + (-3)$
(b) $9 - (-6)$
(c) $(-4) + (-8)$
(d) $(-14) - (-7)$?

When you have finished these check your results with the next frame

6

(a)	5
(b)	15
(c)	−12
(d)	−7

Move now to Frame 7

7   **Multiplication and division**

Multiplying two numbers gives their *product* and dividing two numbers gives their *quotient*. Multiplying and dividing two positive or two negative numbers gives a positive number. For example:

$12 \times 2 = 24$ and $(-12) \times (-2) = 24$
$12 \div 2 = 6$ and $(-12) \div (-2) = 6$

Multiplying or dividing a positive number by a negative number gives a negative number. For example:

$12 \times (-2) = -24$, $(-12) \div 2 = -6$ and $8 \div (-4) = -2$

So what is the value of:

(a) $(-5) \times 3$
(b) $12 \div (-6)$
(c) $(-2) \times (-8)$
(d) $(-14) \div (-7)$?

When you have finished these check your results with the next frame

8

(a)	−15
(b)	−2
(c)	16
(d)	2

Move on to Frame 9

Brackets and precedence rules

9

Brackets and the precedence rules are used to remove ambiguity in a calculation. For example, $14 - 3 \times 4$ could be either:

$11 \times 4 = 44$ or $14 - 12 = 2$

depending on which operation is performed first.

To remove the ambiguity we rely on the precedence rules:

In any calculation involving all four arithmetic operations we proceed as follows:

(a) Working from the left evaluate divisions and multiplications as they are encountered;

this leaves a calculation involving just addition and subtraction.

(b) Working from the left evaluate additions and subtractions as they are encountered.

For example, to evaluate:

$4 + 5 \times 6 \div 2 - 12 \div 4 \times 2 - 1$

a first sweep from left to right produces:

$4 + 30 \div 2 - 3 \times 2 - 1$

a second sweep from left to right produces:

$4 + 15 - 6 - 1$

and a final sweep produces:

$19 - 7 = 12$

If the calculation contains brackets then these are evaluated first, so that:

$$(4 + 5 \times 6) \div 2 - 12 \div 4 \times 2 - 1 = 34 \div 2 - 6 - 1$$
$$= 17 - 7$$
$$= 10$$

This means that:

$$14 - 3 \times 4 = 14 - 12$$
$$= 2$$

because, reading from the left we multiply before we subtract. Brackets must be used to produce the alternative result:

$$(14 - 3) \times 4 = 11 \times 4$$
$$= 44$$

because the precedence rules state that brackets are evaluated first.

So that $34 + 10 \div (2 - 3) \times 5 = \ldots\ldots\ldots\ldots$

Result in the next frame

10

$$\boxed{-16}$$

Because

$$34 + 10 \div (2 - 3) \times 5 = 34 + 10 \div (-1) \times 5 \qquad \text{we evaluate the bracket first}$$
$$= 34 + (-10) \times 5 \qquad \text{by dividing}$$
$$= 34 + (-50) \qquad \text{by multiplying}$$
$$= 34 - 50 \qquad \text{finally we subtract}$$
$$= -16$$

Notice that when brackets are used we can omit the multiplication signs and replace the division sign by a line, so that:

$$5 \times (6 - 4) \text{ becomes } 5(6 - 4)$$

and

$$(25 - 10) \div 5 \text{ becomes } (25 - 10)/5 \text{ or } \frac{25 - 10}{5}$$

When evaluating expressions containing *nested* brackets the innermost brackets are evaluated first. For example:

$$3(4 - 2[5 - 1]) = 3(4 - 2 \times 4) \quad \text{evaluating the innermost bracket } [\ldots] \text{ first}$$
$$= 3(4 - 8) \qquad \text{multiplication before subtraction inside the } (\ldots) \text{ bracket}$$
$$= 3(-4) \qquad \text{subtraction completes the evaluation of the } (\ldots) \text{ bracket}$$
$$= -12 \qquad \text{multiplication completes the calculation}$$

so that $5 - \{8 + 7[4 - 1] - 9/3\} = \ldots\ldots\ldots\ldots$

Work this out, the result is in the following frame

11

$$\boxed{-21}$$

Because

$$5 - \{8 + 7[4 - 1] - 9/3\} = 5 - \{8 + 7 \times 3 - 9 \div 3\}$$
$$= 5 - \{8 + 21 - 3\}$$
$$= 5 - \{29 - 3\}$$
$$= 5 - 26$$
$$= -21$$

Now move to Frame 12

Basic laws of arithmetic 12

All the work that you have done so far has been done under the assumption that you know the rules that govern the use of arithmetic operations as, indeed, you no doubt do. However, there is a difference between knowing the rules innately and being consciously aware of them, so here they are. The four basic arithmetic operations are:

addition and subtraction

multiplication and division

where each pair may be regarded as consisting of 'opposites' – in each pair one operation is the reverse operation of the other.

1 Commutativity

Two integers can be added or multiplied in either order without affecting the result. For example:

$$5 + 8 = 8 + 5 = 13 \text{ and } 5 \times 8 = 8 \times 5 = 40$$

We say that addition and multiplication are commutative operations

The order in which two integers are subtracted or divided *does* affect the result. For example:

$$4 - 2 \neq 2 - 4 \text{ because } 4 - 2 = 2 \text{ and } 2 - 4 = -2$$

Notice that $\neq$ means *is not equal to*. Also

$$4 \div 2 \neq 2 \div 4$$

We say that subtraction and division are not commutative operations

2 Associativity

The way in which three or more integers are associated under addition or multiplication does not affect the result. For example:

$$3 + (4 + 5) = (3 + 4) + 5 = 3 + 4 + 5 = 12$$

and

$$3 \times (4 \times 5) = (3 \times 4) \times 5 = 3 \times 4 \times 5 = 60$$

We say that addition and multiplication are associative operations

The way in which three or more integers are associated under subtraction or division does affect the result. For example:

$$3 - (4 - 5) \neq (3 - 4) - 5 \text{ because}$$
$$3 - (4 - 5) = 3 - (-1) = 3 + 1 = 4 \text{ and } (3 - 4) - 5 = (-1) - 5 = -6$$

Also

$$24 \div (4 \div 2) \neq (24 \div 4) \div 2 \text{ because}$$
$$24 \div (4 \div 2) = 24 \div 2 = 12 \text{ and } (24 \div 4) \div 2 = 6 \div 2 = 3$$

We say that subtraction and division are not associative operations

3 Distributivity

Multiplication is distributed over addition and subtraction from both the left and the right. For example:

$$3 \times (4 + 5) = (3 \times 4) + (3 \times 5) = 27 \text{ and } (3 + 4) \times 5 = (3 \times 5) + (4 \times 5) = 35$$

$$3 \times (4 - 5) = (3 \times 4) - (3 \times 5) = -3 \text{ and } (3 - 4) \times 5 = (3 \times 5) - (4 \times 5) = -5$$

Division is distributed over addition and subtraction from the right but not from the left. For example:

$$(60 + 15) \div 5 = (60 \div 5) + (15 \div 5) \text{ because}$$

$$(60 + 15) \div 5 = 75 \div 5 = 15 \text{ and } (60 \div 5) + (15 \div 5) = 12 + 3 = 15$$

However, $60 \div (15 + 5) \neq (60 \div 15) + (60 \div 5)$ because

$$60 \div (15 + 5) = 60 \div 20 = 3 \text{ and } (60 \div 15) + (60 \div 5) = 4 + 12 = 16$$

Also:

$$(20 - 10) \div 5 = (20 \div 5) - (10 \div 5) \text{ because}$$

$$(20 - 10) \div 5 = 10 \div 5 = 2 \text{ and } (20 \div 5) - (10 \div 5) = 4 - 2 = 2$$

but $20 \div (10 - 5) \neq (20 \div 10) - (20 \div 5)$ because

$$20 \div (10 - 5) = 20 \div 5 = 4 \text{ and } (20 \div 10) - (20 \div 5) = 2 - 4 = -2$$

On now to Frame 13

13　Estimating

Arithmetic calculations are easily performed using a calculator. However, by pressing a wrong key, wrong answers can just as easily be produced. Every calculation made using a calculator should at least be checked for the reasonableness of the final result and this can be done by *estimating* the result using *rounding*. For example, using a calculator the sum $39 + 53$ is incorrectly found to be 62 if $39 + 23$ is entered by mistake. If, now, 39 is rounded up to 40, and 53 is rounded down to 50 the reasonableness of the calculator result can be simply checked by adding 40 to 50 to give 90. This indicates that the answer 62 is wrong and that the calculation should be done again. The correct answer 92 is then seen to be close to the approximation of 90.

Rounding

An integer can be rounded to the nearest 10 as follows:

If the number is less than halfway to the next multiple of 10 then the number is rounded *down* to the previous multiple of 10. For example, 53 is rounded down to 50.

If the number is more than halfway to the next multiple of 10 then the number is rounded *up* to the next multiple of 10. For example, 39 is rounded up to 40.

If the number is exactly halfway to the next multiple of 10 then the number is rounded *up*. For example, 35 is rounded up to 40.

This principle also applies when rounding to the nearest 100, 1000, 10 000 or more. For example, 349 rounds up to 350 to the nearest 10 but rounds down to 300 to the nearest 100, and 2501 rounds up to 3000 to the nearest 1000.

Try rounding each of the following to the nearest 10, 100 and 1000 respectively:

(a) 1846
(b) −638
(c) 445

Finish all three and check your results with the next frame

14

> (a) 1850, 1800, 2000
> (b) −640, −600, −1000
> (c) 450, 400, 0

Because

(a) 1846 is nearer to 1850 than to 1840, nearer to 1800 than to 1900 and nearer to 2000 than to 1000.

(b) −638 is nearer to −640 than to −630, nearer to −600 than to −700 and nearer to −1000 than to 0. The negative sign does not introduce any complications.

(c) 445 rounds to 450 because it is halfway to the next multiple of 10, 445 is nearer to 400 than to 500 and nearer to 0 than 1000.

How about estimating each of the following using rounding to the nearest 10:

(a) $18 \times 21 - 19 \div 11$
(b) $99 \div 101 - 49 \times 8$

Check your results in Frame 15

15

> (a) 398
> (b) −499

Because

(a) $18 \times 21 - 19 \div 11$ rounds to $20 \times 20 - 20 \div 10 = 398$
(b) $99 \div 101 - 49 \times 8$ rounds to $100 \div 100 - 50 \times 10 = -499$

At this point let us pause and summarize the main facts so far on types of numbers

Review summary   **Unit 1**

16

1 The integers consist of the positive and negative whole numbers and zero.

2 The integers are ordered so that they range from large negative to small negative through zero to small positive and then large positive. They are written using the ten numerals 0 to 9 according to the principle of place value where the place of a numeral in a number dictates the value it represents.

3 The integers can be represented by equally spaced points on a line.

4 The four arithmetic operations of addition, subtraction, multiplication and division obey specific precedence rules that govern the order in which they are to be executed:

In any calculation involving all four arithmetic operations we proceed as follows:

(a) working from the left evaluate divisions and multiplications as they are encountered.

This leaves an expression involving just addition and subtraction:

(b) working from the left evaluate additions and subtractions as they are encountered.

5 Multiplying or dividing two positive numbers or two negative numbers produces a positive number. Multiplying or dividing a positive number and a negative number produces a negative number.

6 Brackets are used to group numbers and operations together. In any arithmetic expression, the contents of brackets are evaluated first.

7 Integers can be rounded to the nearest 10, 100 etc. and the rounded values used as estimates for the result of a calculation.

 Review exercise   **Unit 1**

17

1 Place the appropriate symbol < or > between each of the following pairs of numbers:

(a) $-1 \quad -6$ (b) $5 \quad -29$ (c) $-14 \quad 7$

2 Find the value of each of the following:

(a) $16 - 12 \times 4 + 8 \div 2$

(b) $(16 - 12) \times (4 + 8) \div 2$

(c) $9 - 3(17 + 5[5 - 7])$

(d) $8(3[2 + 4] - 2[5 + 7])$

3 Show that:

(a) $6 - (3 - 2) \neq (6 - 3) - 2$

(b) $100 \div (10 \div 5) \neq (100 \div 10) \div 5$

(c) $24 \div (2 + 6) \neq (24 \div 2) + (24 \div 6)$

(d) $24 \div (2 - 6) \neq (24 \div 2) - (24 \div 6)$

4 Round each number to the nearest 10, 100 and 1000:

(a) 2562   (b) 1500   (c) −3451   (d) −14 525

Complete all four questions. Take your time, there is no need to rush.
If necessary, look back at the Unit.
The answers and working are in the next frame.

---

1 (a) $-1 > -6$ because -1 is represented on the line to the right of -6 **18**

(b) $5 > -29$ because 5 is represented on the line to the right of -29

(c) $-14 < 7$ because -14 is represented on the line to the left of 7

2 (a) $16 - 12 \times 4 + 8 \div 2 = 16 - 48 + 4 = 16 - 44 = -28$
divide and multiply before adding and subtracting

(b) $(16 - 12) \times (4 + 8) \div 2 = (4) \times (12) \div 2 = 4 \times 12 \div 2 = 4 \times 6 = 24$
brackets are evaluated first

(c) $9 - 3(17 + 5[5 - 7]) = 9 - 3(17 + 5[-2])$
$$= 9 - 3(17 - 10)$$
$$= 9 - 3(7)$$
$$= 9 - 21 = -12$$

(d) $8(3[2 + 4] - 2[5 + 7]) = 8(3 \times 6 - 2 \times 12)$
$$= 8(18 - 24)$$
$$= 8(-6) = -48$$

3 (a) Left-hand side (LHS) $= 6 - (3 - 2) = 6 - (1) = 5$
Right-hand side (RHS) $= (6 - 3) - 2 = (3) - 2 = 1 \neq$ LHS

(b) Left-hand side (LHS) $= 100 \div (10 \div 5) = 100 \div 2 = 50$
Right-hand side (RHS) $= (100 \div 10) \div 5 = 10 \div 5 = 2 \neq$ LHS

(c) Left-hand side (LHS) $= 24 \div (2 + 6) = 24 \div 8 = 3$
Right-hand side (RHS) $= (24 \div 2) + (24 \div 6) = 12 + 4 = 16 \neq$ LHS

(d) Left-hand side (LHS) $= 24 \div (2 - 6) = 24 \div (-4) = -6$
Right-hand side (RHS) $= (24 \div 2) - (24 \div 6) = 12 - 4 = 8 \neq$ LHS

4 (a) 2560, 2600, 3000

(b) 1500, 1500, 2000

(c) −3450, −3500, −3000

(d) −14 530, −14 500, −15 000

Now on to the Unit test

Review test   Unit 1

19   **1** Place the appropriate symbol < or > between each of the following pairs of
          numbers:
          (a) −3   − 2   (b) 8   − 13   (c) −25   0
      2 Find the value of each of the following:
          (a) $13 + 9 \div 3 - 2 \times 5$ (b) $(13 + 9) \div (3 - 2) \times 5$
      3 Round each number to the nearest 10, 100 and 1000:
          (a) 1354   (b) 2501   (c) −2452   (d) −23 625

Factors and prime numbers   Unit 2

1 Factors

Any pair of natural numbers are called *factors* of their product. For example,
the numbers 3 and 6 are factors of 18 because $3 \times 6 = 18$. These are not the
only factors of 18. The complete collection of factors of 18 is 1, 2, 3, 6, 9, 18
because

$$18 = 1 \times 18$$
$$= 2 \times 9$$
$$= 3 \times 6$$

So the factors of:

(a) 12

(b) 25

(c) 17   are ....

The results are in the next frame

2

| (a) 1, 2, 3, 4, 6, 12 |
| (b) 1, 5, 25 |
| (c) 1, 17 |

Because

(a) $12 = 1 \times 12 = 2 \times 6 = 3 \times 4$

(b) $25 = 1 \times 25 = 5 \times 5$

(c) $17 = 1 \times 17$

Now move to the next frame

Prime numbers

If a natural number has only two factors which are itself and the number 1, the number is called a *prime number*. The first six prime numbers are 2, 3, 5, 7, 11 and 13. The number 1 is *not* a prime number because it only has one factor, namely, itself.

Prime factorization

Every natural number can be written as a product involving only prime factors. For example, the number 126 has the factors 1, 2, 3, 6, 7, 9, 14, 18, 21, 42, 63 and 126, of which 2, 3 and 7 are prime numbers and 126 can be written as:

$$126 = 2 \times 3 \times 3 \times 7$$

To obtain this *prime factorization* the number is divided by successively increasing prime numbers thus:

$$
\begin{array}{r|l}
2 & 126 \\ \hline
3 & 63 \\ \hline
3 & 21 \\ \hline
7 & 7 \\ \hline
& 1
\end{array}
$$

so that $126 = 2 \times 3 \times 3 \times 7$

Notice that a prime factor may occur more than once in a prime factorization.

Now find the prime factorization of:

(a) 84
(b) 512

Work these two out and check the working in Frame 4

(a) $84 = 2 \times 2 \times 3 \times 7$
(b) $512 = 2 \times 2 \times 2 \times 2 \times 2 \times 2 \times 2 \times 2 \times 2$

Because

$$
\text{(a)} \quad
\begin{array}{r|l}
2 & 84 \\ \hline
2 & 42 \\ \hline
3 & 21 \\ \hline
7 & 7 \\ \hline
& 1
\end{array}
$$

so that $84 = 2 \times 2 \times 3 \times 7$

(b) The only prime factor of 512 is 2 which occurs 9 times. The prime factorization is:

$$512 = 2 \times 2 \times 2 \times 2 \times 2 \times 2 \times 2 \times 2 \times 2$$

Move to Frame 5

5   Highest common factor (HCF)

The *highest common factor* (HCF) of two natural numbers is the largest factor that they have in common. For example, the prime factorizations of 144 and 66 are:

$$144 = 2 \times 2 \times 2 \times 2 \times 3 \times 3$$
$$66 = 2 \qquad\qquad \times 3 \quad\; \times 11$$

Only the 2 and the 3 are common to both factorizations and so the highest factor that these two numbers have in common (HCF) is $2 \times 3 = 6$.

Lowest common multiple (LCM)

The smallest natural number that each one of a pair of natural numbers divides into a whole number of times is called their *lowest common multiple* (LCM). This is also found from the prime factorization of each of the two numbers. For example:

$$144 = 2 \times 2 \times 2 \times 2 \times 3 \times 3$$
$$66 = 2 \qquad\qquad \times 3 \quad\; \times 11$$
$$\text{LCM} = 2 \times 2 \times 2 \times 2 \times 3 \times 3 \times 11 = 1584$$

The HCF and LCM of 84 and 512 are . . . .

6

> HCF: 4
> LCM: 10 752

Because

84 and 512 have the prime factorizations:

$$84 = 2 \times 2 \qquad\qquad\qquad\qquad\qquad\quad \times 3 \times 7$$
$$512 = 2 \times 2 \times 2 \times 2 \times 2 \times 2 \times 2 \times 2 \times 2 \qquad\qquad \text{HCF} = 2 \times 2 = 4$$
$$\text{LCM} = 2 \times 2 \times 2 \times 2 \times 2 \times 2 \times 2 \times 2 \times 2 \times 3 \times 7 = 10\,752$$

At this point let us pause and summarize the main facts on factors and prime numbers

Review summary

Unit 2

7

1 A pair of natural numbers are called factors of their product.

2 If a natural number only has one and itself as factors it is called a prime number.

3 Every natural number can be written as a product of its prime factors, some of which may be repeated.

4 The highest common factor (HCF) is the highest factor that two natural numbers have in common.

5 The lowest common multiple (LCM) is the lowest natural number that two natural numbers will divide into a whole number of times.

 # Review exercise

Unit 2

8

1 Write each of the following as a product of prime factors:

   (a) 429   (b) 1820   (c) 2992   (d) 3185

2 Find the HCF and the LCM of each pair of numbers:

   (a) 63, 42   (b) 92, 34

Complete both questions. Work through Unit 2 again if you need to.
Don't rush. Take your time.
The answers and working are in the next frame.

9

1 (a)

$$
\begin{array}{r|r}
3 & 429 \\
11 & 143 \\
13 & 13 \\
\hline
 & 1
\end{array}
$$
$429 = 3 \times 11 \times 13$

(b)

$$
\begin{array}{r|r}
2 & 1820 \\
2 & 910 \\
5 & 455 \\
7 & 91 \\
13 & 13 \\
\hline
 & 1
\end{array}
$$
$1820 = 2 \times 2 \times 5 \times 7 \times 13$

▶

(c)

2	2992
2	1496
2	748
2	374
11	187
17	17
	1

$2992 = 2 \times 2 \times 2 \times 2 \times 11 \times 17$

(d)

5	3185
7	637
7	91
13	13
	1

$3185 = 5 \times 7 \times 7 \times 13$

2 (a) The prime factorizations of 63 and 42 are:

$$63 = \quad\quad 3 \times 3 \times 7$$
$$42 = 2 \times 3 \quad\quad \times 7 \quad\quad\quad \text{HCF } 3 \times 7 = 21$$
$$\text{LCM} = 2 \times 3 \times 3 \times 7 = 126$$

(b) The prime factorizations of 34 and 92 are:

$$34 = 2 \quad\quad \times 17$$
$$92 = 2 \times 2 \quad\quad \times 23 \quad\quad\quad \text{HCF } 2$$
$$\text{LCM} = 2 \times 2 \times 17 \times 23 = 1564$$

Now for the Review test

 # Review test   Unit 2

10

1 Write each of the following as a product of prime factors:

(a) 170   (b) 455   (c) 9075   (d) 1140

2 Find the HCF and the LCM of each pair of numbers:

(a) 84, 88   (b) 105, 66

Fractions, ratios and percentages   Unit 3

Division of integers

A fraction is a number which is represented by one integer – the *numerator* – divided by another integer – the *denominator* (or the *divisor*). For example, $\frac{3}{5}$ is a fraction with numerator 3 and denominator 5. Because fractions are written as one integer divided by another – a *ratio* – they are called *rational* numbers. Fractions are either *proper*, *improper* or *mixed*:

- in a proper fraction the numerator is less than the denominator, for example $\frac{4}{7}$

- in an improper fraction the numerator is greater than the denominator, for example $\frac{12}{5}$

- a mixed fraction is in the form of an integer and a fraction, for example $6\frac{2}{3}$

So that $-\dfrac{8}{11}$ is a .... fraction?

The answer is in the next frame

$\boxed{\text{Proper}}$

Fractions can be either positive or negative.

Now to the next frame

Multiplying fractions

Two fractions are multiplied by multiplying their respective numerators and denominators independently. For example:

$$\frac{2}{3} \times \frac{5}{7} = \frac{2 \times 5}{3 \times 7} = \frac{10}{21}$$

Try this one for yourself. $\dfrac{5}{9} \times \dfrac{2}{7} = $ ....

$\boxed{\dfrac{10}{63}}$

Because

$$\frac{5}{9} \times \frac{2}{7} = \frac{5 \times 2}{9 \times 7} = \frac{10}{63}$$

Correct? Then on to the next frame

5 | Of

The word 'of' when interposed between two fractions means multiply. For example:

Half of half a cake is one-quarter of a cake. That is

$$\frac{1}{2} \text{ of } \frac{1}{2} = \frac{1}{2} \times \frac{1}{2} = \frac{1 \times 1}{2 \times 2} = \frac{1}{4}$$

So that, for example:

$$\frac{1}{3} \text{ of } \frac{2}{5} = \frac{1}{3} \times \frac{2}{5} = \frac{1 \times 2}{3 \times 5} = \frac{2}{15}$$

So that $\frac{3}{8}$ of $\frac{5}{7}$ = ...

6

$$\boxed{\frac{15}{56}}$$

Because

$$\frac{3}{8} \text{ of } \frac{5}{7} = \frac{3}{8} \times \frac{5}{7} = \frac{3 \times 5}{8 \times 7} = \frac{15}{56}$$

On now to the next frame

7 | Equivalent fractions

Multiplying the numerator and denominator by the same number is equivalent to multiplying the fraction by unity, that is by 1:

$$\frac{4 \times 3}{5 \times 3} = \frac{4}{5} \times \frac{3}{3} = \frac{4}{5} \times 1 = \frac{4}{5}$$

Now, $\frac{4 \times 3}{5 \times 3} = \frac{12}{15}$ so that the fraction $\frac{4}{5}$ and the fraction $\frac{12}{15}$ *both represent the same number* and for this reason we call $\frac{4}{5}$ and $\frac{12}{15}$ *equivalent fractions*.

A second fraction, equivalent to a first fraction, can be found by multiplying the numerator and the denominator of the first fraction by the same number.

So that if we multiply the numerator and denominator of the fraction $\frac{7}{5}$ by 4 we obtain the equivalent fraction . . . .

Check your result in Frame 8

$$\boxed{\dfrac{28}{20}}$$

Because

$$\frac{7 \times 4}{5 \times 4} = \frac{28}{20}$$

We can reverse this process and find the equivalent fraction that has the smallest numerator by *cancelling out* common factors. This is known as reducing the fraction to its *lowest terms*. For example:

$\dfrac{16}{96}$ can be reduced to its lowest terms as follows:

$$\frac{16}{96} = \frac{4 \times 4}{24 \times 4} = \frac{4 \times \cancel{4}}{24 \times \cancel{4}} = \frac{4}{24}$$

by cancelling out the 4 in the numerator and the denominator

The fraction $\dfrac{4}{24}$ can also be reduced:

$$\frac{4}{24} = \frac{4}{6 \times 4} = \frac{\cancel{4}}{6 \times \cancel{4}} = \frac{1}{6}$$

Because $\dfrac{1}{6}$ cannot be reduced further we see that $\dfrac{16}{96}$ reduced to its lowest terms is $\dfrac{1}{6}$.

How about this one? The fraction $\dfrac{84}{108}$ reduced to its lowest terms is . . . .

Check with the next frame

$$\boxed{\dfrac{7}{9}}$$

Because

$$\frac{84}{108} = \frac{7 \times 3 \times 4}{9 \times 3 \times 4} = \frac{7 \times \cancel{3} \times \cancel{4}}{9 \times \cancel{3} \times \cancel{4}} = \frac{7}{9}$$

Now move on to the next frame

10 Dividing fractions

The expression $6 \div 3$ means the number of 3's in 6, which is 2. Similarly, the expression $1 \div \frac{1}{4}$ means the number of $\frac{1}{4}$'s in 1, which is, of course, 4. That is:

$$1 \div \frac{1}{4} = 4 = 1 \times \frac{4}{1}.$$ Notice how the numerator and the denominator of the divisor are switched and the division replaced by multiplication.

Two fractions are divided by switching the numerator and the denominator of the divisor and multiplying the result. For example:

$$\frac{2}{3} \div \frac{5}{7} = \frac{2}{3} \times \frac{7}{5} = \frac{14}{15}$$

So that $\dfrac{7}{13} \div \dfrac{3}{4} = $ ....

11

$$\boxed{\dfrac{28}{39}}$$

Because

$$\frac{7}{13} \div \frac{3}{4} = \frac{7}{13} \times \frac{4}{3} = \frac{28}{39}$$

In particular:

$$1 \div \frac{3}{5} = 1 \times \frac{5}{3} = \frac{5}{3}$$

The fraction $\dfrac{5}{3}$ is called the *reciprocal* of $\dfrac{3}{5}$

So that the reciprocal of $\dfrac{17}{4}$ is ....

12

$$\boxed{\dfrac{4}{17}}$$

Because

$$1 \div \frac{17}{4} = 1 \times \frac{4}{17} = \frac{4}{17}$$

And the reciprocal of -5 is ....

13

$$-\frac{1}{5}$$

Because

$$1 \div (-5) = 1 \div \left(-\frac{5}{1}\right) = 1 \times \left(-\frac{1}{5}\right) = -\frac{1}{5}$$

Move on to the next frame

Adding and subtracting fractions

14

Two fractions can only be added or subtracted immediately if they both possess the same denominator, in which case we add or subtract the numerators and divide by the common denominator. For example:

$$\frac{2}{7} + \frac{3}{7} = \frac{2+3}{7} = \frac{5}{7}$$

If they do not have the same denominator they must be rewritten in equivalent form so that they do have the same denominator – called the *common denominator*. For example:

$$\frac{2}{3} + \frac{1}{5} = \frac{10}{15} + \frac{3}{15} = \frac{10+3}{15} = \frac{13}{15}$$

The common denominator of the equivalent fractions is the LCM of the two original denominators. That is:

$$\frac{2}{3} + \frac{1}{5} = \frac{2 \times 5}{3 \times 5} + \frac{1 \times 3}{5 \times 3} = \frac{10}{15} + \frac{3}{15}$$ where 15 is the LCM of 3 and 5

So that $\dfrac{5}{9} + \dfrac{1}{6} = \ldots\ldots\ldots$

The result is in Frame 15

15

$$\frac{13}{18}$$

Because

The LCM of 9 and 6 is 18 so that $\dfrac{5}{9} + \dfrac{1}{6} = \dfrac{5 \times 2}{9 \times 2} + \dfrac{1 \times 3}{6 \times 3} = \dfrac{10}{18} + \dfrac{3}{18}$

$$= \frac{10+3}{18} = \frac{13}{18}$$

There's another one to try in the next frame

16

Now try $\dfrac{11}{15} - \dfrac{2}{3} = \ldots\ldots\ldots$

17

$$\boxed{\dfrac{1}{15}}$$

Because

$$\frac{11}{15} - \frac{2}{3} = \frac{11}{15} - \frac{2 \times 5}{3 \times 5} = \frac{11}{15} - \frac{10}{15}$$
$$= \frac{11 - 10}{15} = \frac{1}{15} \qquad \text{(15 is the LCM of 3 and 15)}$$

Correct? Then on to Frame 18

18 **Fractions on a calculator**

The $a^b/_c$ button on a calculator enables fractions to be entered and manipulated with the results given in fractional form. For example, to evaluate $\frac{2}{3} \times 1\frac{3}{4}$ using your calculator [*note*: your calculator may not produce the identical display in what follows]:

Enter the number 2
Press the $a^b/_c$ key
Enter the number 3

The display now reads 2 ⌐ 3 to represent $\frac{2}{3}$

Press the × key
Enter the number 1
Press the $a^b/_c$ key
Enter the number 3
Press the $a^b/_c$ key
Enter the number 4

The display now reads 1 ⌐ 3 ⌐ 4 to represent $1\frac{3}{4}$

Press the = key to display the result 1 ⌐ 1 ⌐ 6 $= 1\frac{1}{6}$, that is:

$$\tfrac{2}{3} \times 1\tfrac{3}{4} = \frac{2}{3} \times \frac{7}{4} = \frac{14}{12} = 1\tfrac{1}{6}$$

Now use your calculator to evaluate each of the following:

(a) $\frac{5}{7} + 3\frac{2}{3}$

(b) $\dfrac{8}{3} - \dfrac{5}{11}$

(c) $\dfrac{13}{5} \times \dfrac{4}{7} - \dfrac{2}{9}$

(d) $4\frac{1}{11} \div \left(-\dfrac{3}{5}\right) + \dfrac{1}{8}$

Check your answers in Frame 19

19

> (a) $4 \lrcorner 8 \lrcorner 21 = 4\frac{8}{21}$
>
> (b) $2 \lrcorner 7 \lrcorner 33 = 2\frac{7}{33}$
>
> (c) $1 \lrcorner 83 \lrcorner 315 = 1\frac{83}{315}$
>
> (d) $-6 \lrcorner 61 \lrcorner 88 = -6\frac{61}{88}$

In (d) enter the $\frac{3}{5}$ and then press the $+\!\!\!/\!\!\!-$ key.

On now to the next frame

Ratios

20

If a whole number is separated into a number of fractional parts where each fraction has the same denominator, the numerators of the fractions form a *ratio*. For example, if a quantity of brine in a tank contains $\frac{1}{3}$ salt and $\frac{2}{3}$ water, the salt and water are said to be in the ratio 'one-to-two' – written $1 : 2$.

What ratio do the components A, B and C form if a compound contains $\frac{3}{4}$ of A, $\frac{1}{6}$ of B and $\frac{1}{12}$ of C?

Take care here and check your results with Frame 21

$$9 : 2 : 1$$

21

Because the LCM of the denominators 4, 6 and 12 is 12, then:

$\frac{3}{4}$ of A is $\frac{9}{12}$ of A, $\frac{1}{6}$ of B is $\frac{2}{12}$ of B and the remaining $\frac{1}{12}$ is of C. This ensures that the components are in the ratio of their numerators. That is:

$9 : 2 : 1$

Notice that the sum of the numbers in the ratio is equal to the common denominator.

On now to the next frame

Percentages

22

A percentage is a fraction whose denominator is equal to 100. For example, if 5 out of 100 people are left-handed then the fraction of left-handers is $\frac{5}{100}$ which is written as 5%, that is 5 *per cent* (%).

So if 13 out of 100 cars on an assembly line are red, the percentage of red cars on the line is . . . .

23

13%

Because

The fraction of cars that are red is $\dfrac{13}{100}$ which is written as 13%.

Try this. What is the percentage of defective resistors in a batch of 25 if 12 of them are defective?

24

48%

Because

The fraction of defective resistors is $\dfrac{12}{25} = \dfrac{12 \times 4}{25 \times 4} = \dfrac{48}{100}$ which is written as 48%. Notice that this is the same as:

$$\left(\dfrac{12}{25} \times 100\right)\% = \left(\dfrac{12}{25} \times 25 \times 4\right)\% = (12 \times 4)\% = 48\%$$

A fraction can be converted to a percentage by multiplying the fraction by 100.

To find the percentage part of a quantity we multiply the quantity by the percentage written as a fraction. For example, 24% of 75 is:

$$24\% \text{ of } 75 = \dfrac{24}{100} \text{ of } 75 = \dfrac{24}{100} \times 75 = \dfrac{6 \times 4}{25 \times 4} \times 25 \times 3 = \dfrac{6 \times 4}{25 \times 4} \times 25 \times 3$$
$$= 6 \times 3 = 18$$

So that 8% of 25 is . . . .

Work it through and check your results with the next frame

25

2

Because

$$\dfrac{8}{100} \times 25 = \dfrac{2 \times 4}{25 \times 4} \times 25 = \dfrac{2 \times 4}{25 \times 4} \times 25 = 2$$

At this point let us pause and summarize the main facts on fractions, ratios and percentages

Review summary

Unit 3

26

1 A fraction is a number represented as one integer (the numerator) divided by another integer (the denominator or divisor).

2 The same number can be represented by different but equivalent fractions.

3 A fraction with no common factors other than unity in its numerator and denominator is said to be in its lowest terms.

4 Two fractions are multiplied by multiplying the numerators and denominators independently.

5 Two fractions can only be added or subtracted immediately when their denominators are equal.

6 A ratio consists of the numerators of fractions with identical denominators.

7 The numerator of a fraction whose denominator is 100 is called a percentage.

 # Review exercise

Unit 3

27

1 Reduce each of the following fractions to their lowest terms:

(a) $\dfrac{24}{30}$ (b) $\dfrac{72}{15}$ (c) $-\dfrac{52}{65}$ (d) $\dfrac{32}{8}$

2 Evaluate the following:

(a) $\dfrac{5}{9} \times \dfrac{2}{5}$ (b) $\dfrac{13}{25} \div \dfrac{2}{15}$ (c) $\dfrac{5}{9} + \dfrac{3}{14}$ (d) $\dfrac{3}{8} - \dfrac{2}{5}$

(e) $\dfrac{12}{7} \times \left(-\dfrac{3}{5}\right)$ (f) $\left(-\dfrac{3}{4}\right) \div \left(-\dfrac{12}{7}\right)$ (g) $\dfrac{19}{2} + \dfrac{7}{4}$ (h) $\dfrac{1}{4} - \dfrac{3}{8}$

3 Write the following proportions as ratios:

(a) $\dfrac{1}{2}$ of A, $\dfrac{2}{5}$ of B and $\dfrac{1}{10}$ of C

(b) $\dfrac{1}{3}$ of P, $\dfrac{1}{5}$ of Q, $\dfrac{1}{4}$ of R and the remainder S

4 Complete the following:

(a) $\dfrac{2}{5} = $ %   (b) 58% of 25 =

(c) $\dfrac{7}{12} = $ %   (d) 17% of 50 =

Complete the questions.
Look back at the Unit if necessary but don't rush.
The answers and working are in the next frame.

28

1 (a) $\dfrac{24}{30} = \dfrac{2 \times 2 \times 2 \times 3}{2 \times 3 \times 5} = \dfrac{2 \times 2}{5} = \dfrac{4}{5}$

(b) $\dfrac{72}{15} = \dfrac{2 \times 2 \times 2 \times 3 \times 3}{3 \times 5} = \dfrac{2 \times 2 \times 2 \times 3}{5} = \dfrac{24}{5}$

(c) $-\dfrac{52}{65} = -\dfrac{2 \times 2 \times 13}{5 \times 13} = -\dfrac{2 \times 2}{5} = -\dfrac{4}{5}$

(d) $\dfrac{32}{8} = \dfrac{2 \times 2 \times 2 \times 2 \times 2}{2 \times 2 \times 2} = 4$

2 (a) $\dfrac{5}{9} \times \dfrac{2}{5} = \dfrac{5 \times 2}{9 \times 5} = \dfrac{2}{9}$

(b) $\dfrac{13}{25} \div \dfrac{2}{15} = \dfrac{13}{25} \times \dfrac{15}{2} = \dfrac{13 \times 15}{25 \times 2} = \dfrac{13 \times 3 \times 5}{5 \times 5 \times 2} = \dfrac{39}{10}$

(c) $\dfrac{5}{9} + \dfrac{3}{14} = \dfrac{5 \times 14}{9 \times 14} + \dfrac{3 \times 9}{14 \times 9} = \dfrac{70}{126} + \dfrac{27}{126} = \dfrac{70 + 27}{126} = \dfrac{97}{126}$

(d) $\dfrac{3}{8} - \dfrac{2}{5} = \dfrac{3 \times 5}{8 \times 5} - \dfrac{2 \times 8}{5 \times 8} = \dfrac{15}{40} - \dfrac{16}{40} = \dfrac{15 - 16}{40} = -\dfrac{1}{40}$

(e) $\dfrac{12}{7} \times \left(-\dfrac{3}{5}\right) = \dfrac{12 \times (-3)}{7 \times 5} = \dfrac{-36}{35} = -\dfrac{36}{35}$

(f) $\left(-\dfrac{3}{4}\right) \div \left(-\dfrac{12}{7}\right) = \left(-\dfrac{3}{4}\right) \times \left(-\dfrac{7}{12}\right) = \dfrac{(-3) \times (-7)}{4 \times 12} = \dfrac{3 \times 7}{4 \times 3 \times 4} = \dfrac{7}{16}$

(g) $\dfrac{19}{2} + \dfrac{7}{4} = \dfrac{38}{4} + \dfrac{7}{4} = \dfrac{45}{4}$

(h) $\dfrac{1}{4} - \dfrac{3}{8} = \dfrac{2}{8} - \dfrac{3}{8} = -\dfrac{1}{8}$

3 (a) $\dfrac{1}{2}, \dfrac{2}{5}, \dfrac{1}{10} = \dfrac{5}{10}, \dfrac{4}{10}, \dfrac{1}{10}$ so ratio is $5 : 4 : 1$

(b) $\dfrac{1}{3}, \dfrac{1}{5}, \dfrac{1}{4} = \dfrac{20}{60}, \dfrac{12}{60}, \dfrac{15}{60}$ and $\dfrac{20}{60} + \dfrac{12}{60} + \dfrac{15}{60} = \dfrac{47}{60}$

so the fraction of S is $\dfrac{13}{60}$

so P, Q , R and S are in the ratio $20 : 12 : 15 : 13$

4 (a) $\dfrac{2}{5} = \dfrac{2 \times 20}{5 \times 20} = \dfrac{40}{100}$ that is 40% or $\dfrac{2}{5} = \dfrac{2}{5} \times 100\% = 40\%$

(b) $\dfrac{58}{100} \times 25 = \dfrac{58}{4} = \dfrac{29}{2} = 14\tfrac{1}{2}$

(c) $\dfrac{7}{12} = \dfrac{7}{12} \times 100\% = \dfrac{700}{12}\% = \dfrac{58 \times 12 + 4}{12}\% = 58\tfrac{4}{12}\% = 58\tfrac{1}{3}\%$

(d) $\dfrac{17}{100} \times 50 = \dfrac{17}{2} = 8\tfrac{1}{2}$

Now for the Review test

Review test

29

1 Reduce each of the following fractions to their lowest terms:

(a) $\dfrac{12}{18}$ (b) $\dfrac{144}{21}$ (c) $-\dfrac{49}{14}$ (d) $\dfrac{64}{4}$

2 Evaluate the following:

(a) $\dfrac{3}{7} \times \dfrac{2}{3}$ (b) $\dfrac{11}{30} \div \dfrac{5}{6}$ (c) $\dfrac{3}{7} + \dfrac{4}{13}$ (d) $\dfrac{5}{16} - \dfrac{4}{3}$

3 Write the following proportions as ratios:

(a) $\dfrac{1}{2}$ of A, $\dfrac{1}{5}$ of B and $\dfrac{3}{10}$ of C

(b) $\dfrac{1}{4}$ of P, $\dfrac{1}{3}$ of Q, $\dfrac{1}{5}$ of R and the remainder S

4 Complete the following:

(a) $\dfrac{4}{5} =$ %   (b) 48% of 50 =

(c) $\dfrac{9}{14} =$ %   (d) 15% of 25 =

Decimal numbers

Division of integers

1

If one integer is divided by a second integer that is not one of the first integer's factors the result will not be another integer. Instead, the result will lie between two integers. For example, using a calculator it is seen that:

$$25 \div 8 = 3 \cdot 125$$

which is a number greater than 3 but less than 4. As with integers, the position of a numeral within the number indicates its value. Here the number $3 \cdot 125$ represents

3 units + 1 tenth + 2 hundredths + 5 thousandths.

That is $3 + \dfrac{1}{10} + \dfrac{2}{100} + \dfrac{5}{1000}$

where the decimal point shows the separation of the units from the tenths. Numbers written in this format are called *decimal numbers*.

On to the next frame

2   Rounding

All the operations of arithmetic that we have used with the integers apply to decimal numbers. However, when performing calculations involving decimal numbers it is common for the end result to be a number with a large quantity of numerals after the decimal point. For example:

$$15 \cdot 11 \div 8 \cdot 92 = 1 \cdot 6939461883 \ldots$$

To make such numbers more manageable or more reasonable as the result of a calculation, they can be rounded either to a specified number of *significant figures* or to a specified number of *decimal places*.

Now to the next frame

3   Significant figures

Significant figures are counted from the first non-zero numeral encountered starting from the left of the number. When the required number of significant figures has been counted off, the remaining numerals are deleted with the following proviso:

If the first of a group of numerals to be deleted is a 5 or more, the last significant numeral is increased by 1. For example:

9·4534 to two significant figures is 9·5, to three significant figures is 9·45, and 0·001354 to two significant figures is 0·0014

Try this one for yourself. To four significant figures the number 18·7249 is ....

Check your result with the next frame

4

18·72

Because

The first numeral deleted is a 4 which is less than 5.

There is one further proviso. If the only numeral to be dropped is a 5 then the last numeral retained is rounded up. So that 12·235 to four significant figures (abbreviated to *sig fig*) is 12·24 and 3·465 to three sig fig is 3·47.

So 8·1265 to four sig fig is ....

Check with the next frame

$$8{\cdot}127$$

Because

The only numeral deleted is a 5 and the last numeral is rounded up.

Now on to the next frame

Decimal places

Decimal places are counted to the right of the decimal point and the same rules as for significant figures apply for rounding to a specified number of decimal places (abbreviated to *dp*). For example:

123·4467 to one decimal place is 123·4 and to two dp is 123·45

So, 47·0235 to three dp is . . . .

$$47{\cdot}024$$

Because

The only numeral dropped is a 5 and the last numeral retained is odd so is increased to the next even numeral.

Now move on to the next frame

Trailing zeros

Sometimes zeros must be inserted within a number to satisfy a condition for a specified number of either significant figures or decimal places. For example:

12 645 to two significant figures is 13 000, and 13·1 to three decimal places is 13·100.

These zeros are referred to as *trailing zeros*.

So that 1515 to two sig fig is . . . .

9

$$1500$$

And:

25·13 to four dp is . . . .

$$25{\cdot}1300$$

On to the next frame

11   Fractions as decimals

Because a fraction is one integer divided by another it can be represented in decimal form simply by executing the division. For example:

$$\frac{7}{4} = 7 \div 4 = 1{\cdot}75$$

So that the decimal form of $\frac{3}{8}$ is ....

12

$$\boxed{0{\cdot}375}$$

Because

$$\frac{3}{8} = 3 \div 8 = 0{\cdot}375$$

Now move on to the next frame

13   Decimals as fractions

A decimal can be represented as a fraction. For example:

$$1{\cdot}224 = \frac{1224}{1000} \text{ which in lowest terms is } \frac{153}{125}$$

So that 0·52 as a fraction in lowest terms is ....

14

$$\boxed{\frac{13}{25}}$$

Because

$$0{\cdot}52 = \frac{52}{100} = \frac{13}{25}$$

Now move on to the next frame

15   Unending decimals

Converting a fraction into its decimal form by performing the division always results in an infinite string of numerals after the decimal point. This string of numerals may contain an infinite sequence of zeros or it may contain an infinitely repeated pattern of numerals. A repeated pattern of numerals can be written in an abbreviated format. For example:

$$\frac{1}{3} = 1 \div 3 = 0{\cdot}3333\ldots$$

Here the pattern after the decimal point is of an infinite number of 3's. We abbreviate this by placing a dot over the first 3 to indicate the repetition, thus:

$$0{\cdot}3333\ldots = 0{\cdot}\dot{3} \quad \text{(described as zero point 3 recurring)}$$

For other fractions the repetition may consist of a sequence of numerals, in which case a dot is placed over the first and last numeral in the sequence. For example:

$$\frac{1}{7} = 0.142857142857142857\ldots = 0.\dot{1}4285\dot{7}$$

So that we write $\frac{2}{11} = 0.181818\ldots$ as ....

$$\boxed{0.\dot{1}\dot{8}}$$

16

Sometimes the repeating pattern is formed by an infinite sequence of zeros, in which case we simply omit them. For example:

$$\frac{1}{5} = 0.20000\ldots \text{ is written as } 0.2$$

Next frame

Unending decimals as fractions

17

Any decimal that displays an unending repeating pattern can be converted to its fractional form. For example:

To convert $0.181818\ldots = 0.\dot{1}\dot{8}$ to its fractional form we note that because there are two repeating numerals we multiply by 100 to give:

$$100 \times 0.\dot{1}\dot{8} = 18.\dot{1}\dot{8}$$

Subtracting $0.\dot{1}\dot{8}$ from both sides of this equation gives:

$$100 \times 0.\dot{1}\dot{8} - 0.\dot{1}\dot{8} = 18.\dot{1}\dot{8} - 0.\dot{1}\dot{8}$$

That is:

$$99 \times 0.\dot{1}\dot{8} = 18.0$$

This means that:

$$0.\dot{1}\dot{8} = \frac{18}{99} = \frac{2}{11}$$

Similarly, the fractional form of $2 \cdot 0\dot{3}1\dot{5}$ is found as follows:

$2 \cdot 0\dot{3}1\dot{5} = 2 \cdot 0 + 0 \cdot 0\dot{3}1\dot{5}$ and, because there are three repeating numerals:

$1000 \times 0 \cdot 0\dot{3}1\dot{5} = 31 \cdot 5\dot{3}1\dot{5}$

Subtracting $0 \cdot 0\dot{3}1\dot{5}$ from both sides of this equation gives:

$1000 \times 0 \cdot 0\dot{3}1\dot{5} - 0 \cdot 0\dot{3}1\dot{5} = 31 \cdot 5\dot{3}1\dot{5} - 0 \cdot 0\dot{3}1\dot{5} = 31 \cdot 5$

That is:

$999 \times 0 \cdot 0\dot{3}1\dot{5} = 31 \cdot 5$ so that $0 \cdot 0\dot{3}1\dot{5} = \dfrac{31 \cdot 5}{999} = \dfrac{315}{9990}$

This means that:

$2 \cdot 0\dot{3}1\dot{5} = 2 \cdot 0 + 0 \cdot 0\dot{3}1\dot{5} = 2 + \dfrac{315}{9990} = 2\frac{35}{1110} = 2\frac{7}{222}$

What are the fractional forms of $0 \cdot 2\dot{1}$ and $3 \cdot 2\dot{1}$?

The answers are in the next frame

18

$$\frac{7}{33} \text{ and } 3\frac{19}{90}$$

Because

$100 \times 0 \cdot \dot{2}\dot{1} = 21 \cdot \dot{2}\dot{1}$ so that $99 \times 0 \cdot \dot{2}\dot{1} = 21$

giving $0 \cdot \dot{2}\dot{1} = \dfrac{21}{99} = \dfrac{7}{33}$ and

$3 \cdot 2\dot{1} = 3 \cdot 2 + 0 \cdot 0\dot{1}$ and $10 \times 0 \cdot 0\dot{1} = 0 \cdot 1\dot{1}$ so that $9 \times 0 \cdot 0\dot{1} = 0 \cdot 1$ giving

$0 \cdot 0\dot{1} = \dfrac{0 \cdot 1}{9} = \dfrac{1}{90}$, hence $3 \cdot 2\dot{1} = \dfrac{32}{10} + \dfrac{1}{90} = \dfrac{289}{90} = 3\frac{19}{90}$

19   Rational, irrational and real numbers

A number that can be expressed as a fraction is called a *rational* number. An *irrational* number is one that *cannot* be expressed as a fraction and has a decimal form consisting of an infinite string of numerals that does not display a repeating pattern. As a consequence it is not possible either to write down the complete decimal form or to devise an abbreviated decimal format. Instead, we can only round them to a specified number of significant figures or decimal places. Alternatively, we may have a numeral representation for them, such as $\sqrt{2}$, e or π. The complete collection of rational and irrational numbers is called the collection of *real* numbers.

At this point let us pause and summarize the main facts so far on decimal numbers

Review summary **Unit 4**

1 Every fraction can be written as a decimal number by performing the division. **20**
2 The decimal number obtained will consist of an infinitely repeating pattern of numerals to the right of one of its digits.
3 Other decimals, with an infinite, non-repeating sequence of numerals after the decimal point are the irrational numbers.
4 A decimal number can be rounded to a specified number of significant figures (sig fig) by counting from the first non-zero numeral on the left.
5 A decimal number can be rounded to a specified number of decimal places (dp) by counting from the decimal point.

 # Review exercise **Unit 4**

1 Round each of the following decimal numbers, first to 3 significant figures and then to 2 decimal places: **21**

(a) 12·455 (b) 0·01356 (c) 0·1005 (d) 1344·555

2 Write each of the following in abbreviated form:
(a) 12·110110110... (b) 0·123123123...
(c) − 3·11111... (d) − 9360·936093609360...

3 Convert each of the following to decimal form to 3 decimal places:

(a) $\frac{3}{16}$ (b) $-\frac{5}{9}$

(c) $\frac{7}{6}$ (d) $-\frac{24}{11}$

4 Convert each of the following to fractional form in lowest terms:
(a) 0·6 (b) 1·$\dot{4}$
(c) 1·$\dot{2}\dot{4}$ (d) −7·3

Complete all four questions. Take your time, there is no need to rush.
If necessary, look back at the Unit.
The answers and working are in the next frame.

22

1 (a) 12·5, 12·46   (b) 0·0136, 0·01
   (c) 0·101, 0·10   (d) 1340, 1344·56

2 (a) 12·11̇0̇   (b) 0·12̇3̇
   (c) −3·1̇   (d) −9360·9̇36̇0̇

3 (a) $\dfrac{3}{16} = 0.1875 = 0.188$ to 3 dp

   (b) $-\dfrac{5}{9} = -0.555\ldots = -0.556$ to 3 dp

   (c) $\dfrac{7}{6} = 1.1666\ldots = 1.167$ to 3 dp

   (d) $-\dfrac{24}{11} = -2.1818\ldots = -2.182$ to 3 dp

4 (a) $0.6 = \dfrac{6}{10} = \dfrac{3}{5}$

   (b) $1.\dot4 = 1 + \dfrac{4}{9} = \dfrac{13}{9}$

   (c) $1.\dot2\dot4 = 1 + \dfrac{24}{99} = \dfrac{123}{99} = \dfrac{41}{33}$

   (d) $-7.3 = -\dfrac{73}{10}$

Now for the Review test

Review test

Unit 4

23

1 Round each of the following decimal numbers, first to 3 significant figures and then to 2 decimal places:
(a) 21·355   (b) 0·02456
(c) 0·3105   (d) 5134·555

2 Convert each of the following to decimal form to 3 decimal places:
(a) $\dfrac{4}{15}$ (b) $-\dfrac{7}{13}$ (c) $\dfrac{9}{5}$ (d) $-\dfrac{28}{13}$

3 Convert each of the following to fractional form in lowest terms:
(a) 0·8 (b) 2·8 (c) 3·3̇2̇ (d) −5·5

4 Write each of the following in abbreviated form:
(a) 1·010101…   (b) 9·2456456456…

Powers

Raising a number to a power

1

The arithmetic operation of raising a number to a *power* is devised from repetitive multiplication. For example:

$10 \times 10 \times 10 \times 10 = 10^4$ – the number 10 multiplied by itself 4 times

The power is also called an *index* and the number to be raised to the power is called the *base*. Here the number 4 is the power (index) and 10 is the base.

So $5 \times 5 \times 5 \times 5 \times 5 \times 5 = \ldots\ldots\ldots\ldots$ (in the form of 5 raised to a power)

Compare your answer with the next frame

5^6

2

Because the number 5 (the base) is multiplied by itself 6 times (the power or index).

Now to the next frame

The laws of powers

3

The laws of powers are contained within the following set of rules:

- *Power unity*
  Any number raised to the power 1 equals itself.

  $3^1 = 3$

So $99^1 = \ldots\ldots\ldots\ldots$

On to the next frame

99

4

Because any number raised to the power 1 equals itself.

▶

● *Multiplication of numbers and the addition of powers*
If two numbers are each written as a given base raised to some power then
the *product of the two numbers* is equal to the same base raised to the *sum of
the powers*. For example, $16 = 2^4$ and $8 = 2^3$ so:

$$16 \times 8 - 2^4 \times 2^3$$
$$= (2 \times 2 \times 2 \times 2) \times (2 \times 2 \times 2)$$
$$= 2 \times 2 \times 2 \times 2 \times 2 \times 2 \times 2$$
$$= 2^7$$
$$= 2^{4+3}$$
$$= 128$$

Multiplication requires powers to be added.

So $8^3 \times 8^5 = \ldots\ldots\ldots$ (in the form of 8 raised to a power)

Next frame

5

$$8^8$$

Because multiplication requires powers to be added.

Notice that we cannot combine different powers with different bases. For
example:

$2^2 \times 4^3$ cannot be written as 8^5

but we can combine different bases to the same power. For example:

$3^4 \times 5^4$ can be written as 15^4 because

$$3^4 \times 5^4 = (3 \times 3 \times 3 \times 3) \times (5 \times 5 \times 5 \times 5)$$
$$= 15 \times 15 \times 15 \times 15$$
$$= 15^4$$
$$= (3 \times 5)^4$$

So that $2^3 \times 4^3$ can be written as $\ldots\ldots\ldots$ (in the form of a number
                                                                                    raised to a power)

6

$$8^3$$

Next frame

● *Division of numbers and the subtraction of powers*

7

If two numbers are each written as a given base raised to some power then the *quotient of the two numbers* is equal to the same base raised to the *difference of the powers*. For example:

$$15\,625 \div 25 = 5^6 \div 5^2$$
$$= (5 \times 5 \times 5 \times 5 \times 5 \times 5) \div (5 \times 5)$$
$$= \frac{5 \times 5 \times 5 \times 5 \times 5 \times 5}{5 \times 5}$$
$$= 5 \times 5 \times 5 \times 5$$
$$= 5^4$$
$$= 5^{6-2}$$
$$= 625$$

Division requires powers to be subtracted.

So $12^7 \div 12^3 = \ldots\ldots\ldots\ldots$ (in the form of 12 raised to a power)

Check your result in the next frame

$$\boxed{12^4}$$

8

Because division requires the powers to be subtracted.

● *Power zero*

Any number raised to the power 0 equals unity. For example:

$$1 = 3^1 \div 3^1$$
$$= 3^{1-1}$$
$$= 3^0$$

So $193^0 = \ldots\ldots\ldots\ldots$

9

$$\boxed{1}$$

Because any number raised to the power 0 equals unity.

- *Negative powers*
  A number raised to a negative power denotes the reciprocal. For example:

$$6^{-2} = 6^{0-2}$$
$$= 6^0 \div 6^2 \qquad \text{subtraction of powers means division}$$
$$= 1 \div 6^2 \qquad \text{because } 6^0 = 1$$
$$= \frac{1}{6^2}$$

Also $6^{-1} = \dfrac{1}{6}$

A negative power denotes the reciprocal.

So $3^{-5} = \ldots\ldots\ldots\ldots$

10

$$\boxed{\dfrac{1}{3^5}}$$

Because

$$3^{-5} = 3^{0-5} = 3^0 \div 3^5 = \frac{1}{3^5}$$

A negative power denotes the reciprocal.

Now to the next frame

11

- *Multiplication of powers*
  If a number is written as a given base raised to some power then that number *raised to a further power* is equal to the base raised to the *product of the powers*. For example:

$$(25)^3 = \left(5^2\right)^3$$
$$= 5^2 \times 5^2 \times 5^2$$
$$= 5 \times 5 \times 5 \times 5 \times 5 \times 5$$
$$= 5^6$$
$$= 5^{2\times3}$$
$$= 15\,625 \qquad \text{Notice that } \left(5^2\right)^3 \neq 5^{2^3} \text{ because } 5^{2^3} = 5^8 = 390\,625.$$

Raising to a power requires powers to be multiplied.

So $\left(4^2\right)^4 = \ldots\ldots\ldots\ldots$ (in the form of 4 raised to a power)

$$4^8$$

12

Because raising to a power requires powers to be multiplied.

Now to the next frame.

Powers on a calculator

13

Powers on a calculator can be evaluated by using the x^y key. For example, enter the number 4, press the x^y key, enter the number 3 and press $=$. The result is 64 which is 4^3.

Try this one for yourself. To two decimal places, the value of $1 \cdot 3^{3 \cdot 4}$ is . . . .

The result is in the following frame

$$2 \cdot 44$$

14

Because

Enter the number $1 \cdot 3$
Press the x^y key
Enter the number $3 \cdot 4$
Press the $=$ key

The number displayed is $2 \cdot 44$ to 2 dp.

Now try this one using the calculator:

$8^{\frac{1}{3}} = $ . . . .   The 1/3 is a problem, use the $a^b/_c$ key.

Check your answer in the next frame

$$2$$

15

Because

Enter the number 8
Press the x^y key
Enter the number 1
Press the $a^b/_c$ key
Enter the number 3
Press $=$ the number 2 is displayed.

Now move on to the next frame

16   Fractional powers and roots

We have just seen that $8^{\frac{1}{3}} = 2$. We call $8^{\frac{1}{3}}$ the *third root* or, alternatively, the *cube root* of 8 because

$$\left(8^{\frac{1}{3}}\right)^3 = 8 \quad \text{the number 8 is the result of raising the 3rd root of 8 to the power 3}$$

Roots are denoted by such fractional powers. For example, the 5th root of 6 is given as $6^{\frac{1}{5}}$ because

$$\left(6^{\frac{1}{5}}\right)^5 = 6$$

and by using a calculator $6^{\frac{1}{5}}$ can be seen to be equal to 1·431 to 3 dp. Odd roots are unique in the real number system but even roots are not. For example, there are two 2nd roots – *square roots* – of 4, namely:

$$4^{\frac{1}{2}} = 2 \text{ and } 4^{\frac{1}{2}} = -2 \text{ because } 2 \times 2 = 4 \text{ and } (-2) \times (-2) = 4$$

Similarly:

$$81^{\frac{1}{4}} = \pm 3$$

Odd roots of negative numbers are themselves negative. For example:

$$(-32)^{\frac{1}{5}} = -2 \text{ because } \left[(-32)^{\frac{1}{5}}\right]^5 = (-2)^5 = -32$$

Even roots of negative numbers, however, pose a problem. For example, because

$$\left[(-1)^{\frac{1}{2}}\right]^2 = (-1)^1 = -1$$

we conclude that the square root of -1 is $(-1)^{\frac{1}{2}}$. Unfortunately, we cannot write this as a decimal number – we cannot find its decimal value because there is no decimal number which when multiplied by itself gives -1. We have to accept the fact that, at this stage, we cannot find the even roots of a negative number. This would be the subject matter for a book of more advanced mathematics.

Surds

An alternative notation for the square root of 4 is the surd notation $\sqrt{4}$ and, by convention, this is always taken to mean the positive square root. This notation can also be extended to other roots, for example, $\sqrt[7]{9}$ is an alternative notation for $9^{\frac{1}{7}}$.

Use your calculator to find the value of each of the following roots to 3 dp:

  (a) $16^{\frac{1}{7}}$ (b) $\sqrt{8}$ (c) $19^{\frac{1}{4}}$ (d) $\sqrt{-4}$

Answers in the next frame

17

> (a) 1·486   use the x^y key
> (b) 2·828   the positive value only
> (c) ±2·088   there are two values for even roots
> (d) We cannot find the square root of a negative number

On now to Frame 18

Multiplication and division by integer powers of 10

18

If a decimal number is multiplied by 10 raised to an integer power, the decimal point moves the integer number of places to the right if the integer is positive and to the left if the integer is negative. For example:

$1·2345 \times 10^3 = 1234·5$ (3 places to the right) and

$1·2345 \times 10^{-2} = 0·012345$ (2 places to the left).

Notice that, for example:

$1·2345 \div 10^3 = 1·2345 \times 10^{-3}$ and

$1·2345 \div 10^{-2} = 1·2345 \times 10^2$

So now try these:

(a) $0·012045 \times 10^4$

(b) $13·5074 \times 10^{-3}$

(c) $144·032 \div 10^5$

(d) $0·012045 \div 10^{-2}$

Work all four out and then check your results with the next frame

19

> (a) 120·45
> (b) 0·0135074
> (c) 0·00144032
> (d) 1·2045

Because

(a) multiplying by 10^4 moves the decimal point 4 places to the right

(b) multiplying by 10^{-3} moves the decimal point 3 places to the left

(c) $144·032 \div 10^5 = 144·032 \times 10^{-5}$ move the decimal point 5 places to the left

(d) $0·012045 \div 10^{-2} = 0·012045 \times 10^2$ move the decimal point 2 places to the right

Now move on to the next frame

20 Precedence rules

With the introduction of the arithmetic operation of raising to a power we need to amend our earlier precedence rules – *evaluating powers is performed before dividing and multiplying.* For example:

$$5(3 \times 4^2 \div 6 - 7) = 5(3 \times 16 \div 6 - 7)$$
$$= 5(48 \div 6 - 7)$$
$$= 5(8 - 7)$$
$$= 5$$

So that:

$$14 \div (125 \div 5^3 \times 4 + 3) = \ldots\ldots\ldots$$

Check your result in the next frame

21

$$\boxed{2}$$

Because

$$14 \div (125 \div 5^3 \times 4 + 3) = 14 \div (125 \div 125 \times 4 + 3)$$
$$= 14 \div (4 + 3)$$
$$= 2$$

22 Standard form

Any decimal number can be written as a decimal number greater than or equal to 1 and less than 10 (called the *mantissa*) multiplied by the number 10 raised to an appropriate power (the power being called the *exponent*). For example:

$$57{\cdot}3 = 5{\cdot}73 \times 10^1$$
$$423{\cdot}8 = 4{\cdot}238 \times 10^2$$
$$6042{\cdot}3 = 6{\cdot}0423 \times 10^3$$

and $$0{\cdot}267 = 2{\cdot}67 \div 10 = 2{\cdot}67 \times 10^{-1}$$
$$0{\cdot}000485 = 4{\cdot}85 \div 10^4 = 4{\cdot}85 \times 10^{-4} \text{ etc.}$$

So, written in standard form:

(a) $52\,674 = \ldots\ldots\ldots$ (c) $0{\cdot}0582 = \ldots\ldots\ldots$
(b) $0{\cdot}00723 = \ldots\ldots\ldots$ (d) $1\,523\,800 = \ldots\ldots\ldots$

23

(a) $5{\cdot}2674 \times 10^4$ (c) $5{\cdot}82 \times 10^{-2}$

(b) $7{\cdot}23 \times 10^{-3}$ (d) $1{\cdot}5238 \times 10^6$

Working in standard form

Numbers written in standard form can be multiplied or divided by multiplying or dividing the respective mantissas and adding or subtracting the respective exponents. For example:

$$0{\cdot}84 \times 23\,000 = (8{\cdot}4 \times 10^{-1}) \times (2{\cdot}3 \times 10^4)$$
$$= (8{\cdot}4 \times 2{\cdot}3) \times 10^{-1} \times 10^4$$
$$= 19{\cdot}32 \times 10^3$$
$$= 1{\cdot}932 \times 10^4$$

Another example:

$$175{\cdot}4 \div 6340 = (1{\cdot}754 \times 10^2) \div (6{\cdot}34 \times 10^3)$$
$$= (1{\cdot}754 \div 6{\cdot}34) \times 10^2 \div 10^3$$
$$= 0{\cdot}2767 \times 10^{-1}$$
$$= 2{\cdot}767 \times 10^{-2} \text{ to 4 sig fig}$$

Where the result obtained is not in standard form, the mantissa is written in standard number form and the necessary adjustment made to the exponent.

In the same way, then, giving the results in standard form to 4 dp:

(a) $472{\cdot}3 \times 0{\cdot}000564 = \ldots\ldots\ldots$
(b) $752\,000 \div 0{\cdot}862 = \ldots\ldots\ldots$

24

(a) $2{\cdot}6638 \times 10^{-1}$

(b) $8{\cdot}7239 \times 10^5$

Because

(a) $472{\cdot}3 \times 0{\cdot}000564 = (4{\cdot}723 \times 10^2) \times (5{\cdot}64 \times 10^{-4})$
$$= (4{\cdot}723 \times 5{\cdot}64) \times 10^2 \times 10^{-4}$$
$$= 26{\cdot}638 \times 10^{-2} = 2{\cdot}6638 \times 10^{-1}$$

(b) $752\,000 \div 0{\cdot}862 = (7{\cdot}52 \times 10^5) \div (8{\cdot}62 \times 10^{-1})$
$$= (7{\cdot}52 \div 8{\cdot}62) \times 10^5 \times 10^1$$
$$= 0{\cdot}87239 \times 10^6 = 8{\cdot}7239 \times 10^5$$

For *addition and subtraction in standard form* the approach is slightly different.

Example 1

$$4 \cdot 72 \times 10^3 + 3 \cdot 648 \times 10^4$$

Before these can be added, the powers of 10 must be made the same:

$$4 \cdot 72 \times 10^3 + 3 \cdot 648 \times 10^4 - 4 \cdot 72 \times 10^3 + 36 \cdot 48 \times 10^3$$
$$= (4 \cdot 72 + 36 \cdot 48) \times 10^3$$
$$= 41 \cdot 2 \times 10^3 = 4 \cdot 12 \times 10^4 \text{ in standard form}$$

Similarly in the next example.

Example 2

$$13 \cdot 26 \times 10^{-3} - 1 \cdot 13 \times 10^{-2}$$

Here again, the powers of 10 must be equalized:

$$13 \cdot 26 \times 10^{-3} - 1 \cdot 13 \times 10^{-2} = 1 \cdot 326 \times 10^{-2} - 1 \cdot 13 \times 10^{-2}$$
$$= (1 \cdot 326 - 1 \cdot 13) \times 10^{-2}$$
$$= 0 \cdot 196 \times 10^{-2} = 1 \cdot 96 \times 10^{-3} \text{ in standard form}$$

Using a calculator

Numbers given in standard form can be manipulated on a calculator by making use of the EXP key. For example, to enter the number $1 \cdot 234 \times 10^3$, enter 1·234 and then press the EXP key. The display then changes to:

$$1 \cdot 234 \quad 00$$

Now enter the power 3 and the display becomes:

$$1 \cdot 234 \quad 03$$

Manipulating numbers in this way produces a result that is in ordinary decimal format. If the answer is required in standard form then it will have to be converted by hand. For example, using the EXP key on a calculator to evaluate $(1 \cdot 234 \times 10^3) + (2 \cdot 6 \times 10^2)$ results in the display 1494 which is then converted by hand to $1 \cdot 494 \times 10^3$.

Therefore, working in standard form:

(a) $43 \cdot 6 \times 10^2 + 8 \cdot 12 \times 10^3 \quad = \dots\dots\dots\dots$
(b) $7 \cdot 84 \times 10^5 - 12 \cdot 36 \times 10^3 \quad = \dots\dots\dots\dots$
(c) $4 \cdot 25 \times 10^{-3} + 1 \cdot 74 \times 10^{-2} = \dots\dots\dots\dots$

(a) $1 \cdot 248 \times 10^4$
(b) $7 \cdot 7164 \times 10^5$
(c) $2 \cdot 165 \times 10^{-2}$

Preferred standard form

In the SI system of units, it is recommended that when a number is written in standard form, the power of 10 should be restricted to powers of 10^3, i.e. 10^3, 10^6, 10^{-3}, 10^{-6}, etc. Therefore in this *preferred standard form* up to three figures may appear in front of the decimal point.

In practice it is best to write the number first in standard form and to adjust the power of 10 to express this in preferred standard form.

Example 1

$\quad 5 \cdot 2746 \times 10^4$ in standard form

$\quad = 5 \cdot 2746 \times 10 \times 10^3$

$\quad = 52 \cdot 746 \times 10^3$ in preferred standard form

Example 2

$\quad 3 \cdot 472 \times 10^8$ in standard form

$\quad = 3 \cdot 472 \times 10^2 \times 10^6$

$\quad = 347 \cdot 2 \times 10^6$ in preferred standard form

Example 3

$\quad 3 \cdot 684 \times 10^{-2}$ in standard form

$\quad = 3 \cdot 684 \times 10 \times 10^{-3}$

$\quad = 36 \cdot 84 \times 10^{-3}$ in preferred standard form

So, rewriting the following in preferred standard form, we have

(a) $8 \cdot 236 \times 10^7 = \ldots\ldots\ldots$
(d) $6 \cdot 243 \times 10^5 = \ldots\ldots\ldots$
(b) $1 \cdot 624 \times 10^{-4} = \ldots\ldots\ldots$
(e) $3 \cdot 274 \times 10^{-2} = \ldots\ldots\ldots$
(c) $4 \cdot 827 \times 10^4 = \ldots\ldots\ldots$
(f) $5 \cdot 362 \times 10^{-7} = \ldots\ldots\ldots$

26

> (a) $82{\cdot}36 \times 10^6$ (d) $624{\cdot}3 \times 10^3$
> (b) $162{\cdot}4 \times 10^{-6}$ (e) $32{\cdot}74 \times 10^{-3}$
> (c) $48{\cdot}27 \times 10^3$ (f) $536{\cdot}2 \times 10^{-9}$

One final exercise on this piece of work:

Example 4

The product of $(4{\cdot}72 \times 10^2)$ and $(8{\cdot}36 \times 10^5)$

    (a) in standard form $=$ ....
    (b) in preferred standard form $=$ ....

27

> (a) $3{\cdot}9459 \times 10^8$
> (b) $394{\cdot}59 \times 10^6$

Because

    (a) $(4{\cdot}72 \times 10^2) \times (8{\cdot}36 \times 10^5) = (4{\cdot}72 \times 8{\cdot}36) \times 10^2 \times 10^5$
$$= 39{\cdot}459 \times 10^7$$
$$= 3{\cdot}9459 \times 10^8 \text{ in standard form}$$

    (b) $(4{\cdot}72 \times 10^2) \times (8{\cdot}36 \times 10^5) = 3{\cdot}9459 \times 10^2 \times 10^6$
$$= 394{\cdot}59 \times 10^6 \text{ in preferred standard form}$$

Now move on to the next frame

28 **Checking calculations**

When performing a calculation involving decimal numbers it is always a good idea to check that your result is reasonable and that an arithmetic blunder or an error in using the calculator has not been made. This can be done using standard form. For example:

$$59{\cdot}2347 \times 289{\cdot}053 = 5{\cdot}92347 \times 10^1 \times 2{\cdot}89053 \times 10^2$$
$$= 5{\cdot}92347 \times 2{\cdot}89053 \times 10^3$$

This product can then be estimated for reasonableness as:

    $6 \times 3 \times 1000 = 18\,000$ (see Frames $13-15$ of Unit 1)

The answer using the calculator is $17\,121{\cdot}968$ to three decimal places, which is $17\,000$ when rounded to the nearest 1000. This compares favourably with the estimated $18\,000$, indicating that the result obtained could be reasonably expected.

So, the estimated value of $800{\cdot}120 \times 0{\cdot}007953$ is ....

Check with the next frame

29

$$\boxed{6{\cdot}4}$$

Because

$$800{\cdot}120 \times 0{\cdot}007953 = 8{\cdot}00120 \times 10^2 \times 7{\cdot}953 \times 10^{-3}$$
$$= 8{\cdot}00120 \times 7{\cdot}9533 \times 10^{-1}$$

This product can then be estimated for reasonableness as:

$$8 \times 8 \div 10 = 6{\cdot}4$$

The exact answer is 6·36 to two decimal places.

Now move on to the next frame

Accuracy

30

Many calculations are made using numbers that have been obtained from measurements. Such numbers are only accurate to a given number of significant figures but using a calculator can produce a result that contains as many figures as its display will permit. Because any calculation involving measured values will not be accurate to *more significant figures than the least number of significant figures in any measurement*, we can justifiably round the result down to a more manageable number of significant figures.

For example:

The base length and height of a rectangle are measured as 114·8 mm and 18 mm respectively. The area of the rectangle is given as the product of these lengths. Using a calculator this product is 2066·4 mm². Because one of the lengths is only measured to 2 significant figures, the result cannot be accurate to more than 2 significant figures. It should therefore be read as 2100 mm².

Assuming the following contains numbers obtained by measurement, use a calculator to find the value to the correct level of accuracy:

$$19{\cdot}1 \times 0{\cdot}0053 \div 13{\cdot}345$$

31

$$\boxed{0{\cdot}0076}$$

Because

The calculator gives the result as 0·00758561 but because 0·0053 is only accurate to 2 significant figures the result cannot be accurate to more than 2 significant figures, namely 0·0076.

At this point let us pause and summarize the main facts so far on powers

Review summary   Unit 5

32
1   Powers are devised from repetitive multiplication of a given number.

2   Negative powers denote reciprocals and any number raised to the power 0 is unity.

3   Multiplication of a decimal number by 10 raised to an integer power moves the decimal point to the right if the power is positive and to the left if the power is negative.

4   A decimal number written in standard form is in the form of a mantissa (a number between 1 and 10 but excluding 10) multiplied by 10 raised to an integer power, the power being called the exponent.

5   Writing decimal numbers in standard form permits an estimation of the reasonableness of a calculation.

6   In preferred standard form the powers of 10 in the exponent are restricted to multiples of 3.

7   If numbers used in a calculation are obtained from measurement, the result of the calculation is a number accurate to no more than the least number of significant figures in any measurement.

 # Review exercise   Unit 5

33
1   Write each of the following as a number raised to a power:
(a) $5^8 \times 5^2$ (b) $6^4 \div 6^6$ (c) $\left(7^4\right)^3$ (d) $\left(19^{-8}\right)^0$

2   Find the value of each of the following to 3 dp:
(a) $16^{\frac{1}{4}}$ (b) $\sqrt[3]{3}$ (c) $(-8)^{\frac{1}{5}}$ (d) $(-7)^{\frac{1}{4}}$

3   Write each of the following as a single decimal number:
(a) $1{\cdot}0521 \times 10^3$ (b) $123{\cdot}456 \times 10^{-2}$
(c) $0{\cdot}0135 \div 10^{-3}$ (d) $165{\cdot}21 \div 10^4$

4   Write each of the following in standard form:
(a) $125{\cdot}87$ (b) $0{\cdot}0101$ (c) $1{\cdot}345$ (d) $10{\cdot}13$

5   Write each of the following in preferred standard form:
(a) $1{\cdot}3204 \times 10^5$ (b) $0{\cdot}0101$ (c) $1{\cdot}345$ (d) $9{\cdot}5032 \times 10^{-8}$

▷

6 In each of the following the numbers have been obtained by measurement. Evaluate each calculation to the appropriate level of accuracy:

(a) $13 \cdot 6 \div 0 \cdot 012 \times 7 \cdot 63 - 9015$

(b) $\dfrac{0 \cdot 003 \times 194}{13 \cdot 6}$

(c) $19 \cdot 3 \times 1 \cdot 04^{2 \cdot 00}$

(d) $\dfrac{18 \times 2 \cdot 1 - 3 \cdot 6 \times 0 \cdot 54}{8 \cdot 6 \times 2 \cdot 9 + 5 \cdot 7 \times 9 \cdot 2}$

Complete all of these questions.
Look back at the Unit if you need to.
You can check your answers and working in the next frame.

---

1 (a) $5^8 \times 5^2 = 5^{8+2} = 5^{10}$ (b) $6^4 \div 6^6 = 6^{4-6} = 6^{-2}$ (c) $\left(7^4\right)^3 = 7^{4 \times 3} = 7^{12}$

    (d) $\left(19^{-8}\right)^0 = 1$ as any number raised to the power 0 equals unity

34

2 (a) $16^{\frac{1}{4}} = \pm 2 \cdot 000$ (b) $\sqrt[3]{3} = 1 \cdot 442$

    (c) $(-8)^{\frac{1}{5}} = -1 \cdot 516$

    (d) $(-7)^{\frac{1}{4}}$ You cannot find the even root of a negative number

3 (a) $1 \cdot 0521 \times 10^3 = 1052 \cdot 1$ (b) $123 \cdot 456 \times 10^{-2} = 1 \cdot 23456$

    (c) $0 \cdot 0135 \div 10^{-3} = 0 \cdot 0135 \times 10^3 = 13 \cdot 5$

    (d) $165 \cdot 21 \div 10^4 = 165 \cdot 21 \times 10^{-4} = 0 \cdot 016521$

4 (a) $125 \cdot 87 = 1 \cdot 2587 \times 10^2$ (b) $0 \cdot 0101 = 1 \cdot 01 \times 10^{-2}$

    (c) $1 \cdot 345 = 1 \cdot 345 \times 10^0$ (d) $10 \cdot 13 = 1 \cdot 013 \times 10^1 = 1 \cdot 013 \times 10$

5 (a) $1 \cdot 3204 \times 10^5 = 132 \cdot 04 \times 10^3$ (b) $0 \cdot 0101 = 10 \cdot 1 \times 10^{-3}$

    (c) $1 \cdot 345 = 1 \cdot 345 \times 10^0$ (d) $9 \cdot 5032 \times 10^{-8} = 95 \cdot 032 \times 10^{-9}$

6 (a) $13 \cdot 6 \div 0 \cdot 012 \times 7 \cdot 63 - 9015 = -367 \cdot \dot{6} = -370$ to 2 sig fig

    (b) $\dfrac{0 \cdot 003 \times 194}{13 \cdot 6} = 0 \cdot 042794 \ldots = 0 \cdot 04$ to 1 sig fig

    (c) $19 \cdot 3 \times 1 \cdot 04^{2 \cdot 00} = 19 \cdot 3 \times 1 \cdot 0816 = 20 \cdot 87488 = 20 \cdot 9$ to 2 sig fig

    (d) $\dfrac{18 \times 2 \cdot 1 - 3 \cdot 6 \times 0 \cdot 54}{8 \cdot 6 \times 2 \cdot 9 + 5 \cdot 7 \times 9 \cdot 2} = \dfrac{35 \cdot 856}{77 \cdot 38} = 0 \cdot 46337554 \ldots = 0 \cdot 463$ to 3 sig fig

Now for the Review test

Review test

35

1   Write each of the following as a number raised to a power:

(a) $3^6 \times 3^3$ (b) $4^3 \div 2^5$ (c) $\left(9^2\right)^3$ (d) $\left(7^0\right)^{-8}$

2   Find the value of each of the following to 3 dp:

(a) $15^{\frac{1}{3}}$ (b) $\sqrt[5]{5}$ (c) $(-27)^{\frac{1}{3}}$ (d) $(-9)^{\frac{1}{2}}$

3   Write each of the following as a single decimal number:

(a) $3 \cdot 2044 \times 10^3$ (b) $16 \cdot 1105 \div 10^{-2}$

4   Write each of the following in standard form:

(a) $134 \cdot 65$ (b) $0 \cdot 002401$

5   Write each of the following in preferred standard form:

(a) $16 \cdot 1105 \div 10^{-2}$ (b) $9 \cdot 3304$

6   In each of the following the numbers have been obtained by measurement. Evaluate each calculation to the appropriate level of accuracy:

(a) $11 \cdot 4 \times 0 \cdot 0013 \div 5 \cdot 44 \times 8 \cdot 810$

(b) $\dfrac{1 \cdot 01 \div 0 \cdot 00335}{9 \cdot 12 \times 6 \cdot 342}$

Number systems

1   Denary (or decimal) system

This is our basic system in which quantities large or small can be represented by use of the symbols 0, 1, 2, 3, 4, 5, 6, 7, 8, 9 together with appropriate place values according to their positions.

For example	2	7	6	5	.	3	2	1_{10}
has place values	10^3	10^2	10^1	10^0		10^{-1}	10^{-2}	10^{-3}
	1000	100	10	1		$\dfrac{1}{10}$	$\dfrac{1}{100}$	$\dfrac{1}{1000}$

In this case, the place values are powers of 10, which gives the name *denary* (or *decimal*) to the system. The denary system is said to have a *base* of 10. You are, of course, perfectly familiar with this system of numbers, but it is included here as it leads on to other systems which have the same type of structure but which use different place values.

So let us move on to the next system

Binary system

2

This is widely used in all forms of switching applications. The only symbols used are 0 and 1 and the place values are powers of 2, i.e. the system has a base of 2.

For example $\begin{matrix} 1 & 0 & 1 & 1 & \cdot & 1 & 0 & 1_2 \end{matrix}$

has place values $\begin{matrix} 2^3 & 2^2 & 2^1 & 2^0 & & 2^{-1} & 2^{-2} & 2^{-3} \end{matrix}$

i.e. $\begin{matrix} 8 & 4 & 2 & 1 & & \dfrac{1}{2} & \dfrac{1}{4} & \dfrac{1}{8} \end{matrix}$

So: $\begin{matrix} 1 & 0 & 1 & 1 & \cdot & 1 & 0 & 1 \end{matrix}$ in the binary system

$= \quad 1 \times 8 \quad 0 \times 4 \quad 1 \times 2 \quad 1 \times 1 \quad 1 \times \frac{1}{2} \quad 0 \times \frac{1}{4} \quad 1 \times \frac{1}{8}$

$= \quad 8 \; + \; 0 \; + \; 2 \; + \; 1 \; + \; \frac{1}{2} \; + \; 0 \; + \; \frac{1}{8}$ in denary

$= \quad 11\frac{5}{8} = 11 \cdot 625$ in the denary system. Therefore $1011 \cdot 101_2 = 11 \cdot 625_{10}$

The small subscripts 2 and 10 indicate the bases of the two systems. In the same way, the denary equivalent of $1\ 1\ 0\ 1 \cdot 0\ 1\ 1_2$ is .... to 3 dp.

3

$$\boxed{13 \cdot 375_{10}}$$

Because

$$\begin{matrix} 1 & 1 & 0 & 1 & \cdot & 0 & 1 & 1_2 \end{matrix}$$
$$= \;\; 8 \; + \; 4 \; + \; 0 \; + \; 1 \; + \; 0 \; + \; \tfrac{1}{4} \; + \; \tfrac{1}{8}$$
$$= 13\tfrac{3}{8} = 13 \cdot 375_{10}$$

Octal system (base 8)

This system uses the symbols

0, 1, 2, 3, 4, 5, 6, 7

with place values that are powers of 8.

For example $\begin{matrix} 3 & 5 & 7 & \cdot & 3 & 2 & 1 \end{matrix}$ in the octal system

has place values $\begin{matrix} 8^2 & 8^1 & 8^0 & & 8^{-1} & 8^{-2} & 8^{-3} \end{matrix}$

i.e. $\begin{matrix} 64 & 8 & 1 & & \dfrac{1}{8} & \dfrac{1}{64} & \dfrac{1}{512} \end{matrix}$

So $\begin{matrix} 3 & 5 & 7 & \cdot & 3 & 2 & 1_8 \end{matrix}$

$= \quad 3 \times 64 \quad 5 \times 8 \quad 7 \times 1 \quad 3 \times \frac{1}{8} \quad 2 \times \frac{1}{64} \quad 1 \times \frac{1}{512}$

$= \quad 192 \;\; + \;\; 40 \;\; + \;\; 7 \;\; + \;\; \frac{3}{8} \;\; + \;\; \frac{1}{32} \;\; + \;\; \frac{1}{512}$

$= \quad 239\frac{209}{512} \;\; = 239 \cdot 4082_{10}$

That is

$$357 \cdot 321_8 = 239 \cdot 408_{10} \text{ to 3 dp}$$

As you see, the method is very much as before: the only change is in the base of the place values.

In the same way then, $263 \cdot 452_8$ expressed in denary form is .... to 3 dp.

4

$$179 \cdot 582_{10}$$

Because

$$2\ 6\ 3 \cdot 4\ 5\ 2_8$$
$$= 2 \times 8^2 + 6 \times 8^1 + 3 \times 8^0 + 4 \times 8^{-1} + 5 \times 8^{-2} + 2 \times 8^{-3}$$
$$= 2 \times 64 + 6 \times 8 + 3 \times 1 + 4 \times \tfrac{1}{8} + 5 \times \tfrac{1}{64} + 2 \times \tfrac{1}{512}$$
$$= 128 + 48 + 3 + \tfrac{1}{2} + \tfrac{5}{64} + \tfrac{1}{256}$$
$$= 179\tfrac{149}{256} = 179 \cdot 582_{10} \text{ to 3 dp}$$

Now we come to the duodecimal system, which has a base of 12.

So move on to the next frame

5   **Duodecimal system (base 12)**

With a base of 12, the units column needs to accommodate symbols up to 11 before any carryover to the second column occurs. Unfortunately, our denary symbols go up to only 9, so we have to invent two extra symbols to represent the values 10 and 11. Several suggestions for these have been voiced in the past, but we will adopt the symbols X and Λ for 10 and 11 respectively. The first of these calls to mind the Roman numeral for 10 and the Λ symbol may be regarded as the two strokes of 11 tilted together 1 1 to join at the top.

The duodecimal system, therefore, uses the symbols

$$0,\ 1,\ 2,\ 3,\ 4,\ 5,\ 6,\ 7,\ 8,\ 9,\ X,\ Λ$$

and has place values that are powers of 12.

For example	2	X	5	·	1	3	6_{12}
has place values	12^2	12^1	12^0		12^{-1}	12^{-2}	12^{-3}
i.e.	144	12	1		$\dfrac{1}{12}$	$\dfrac{1}{144}$	$\dfrac{1}{1728}$

So   2   X   5   ·   1   3 6_{12}
$$=\ 2 \times 144\ +\ 10 \times 12\ +\ 5 \times 1\ +\ 1 \times \tfrac{1}{12}\ +\ 3 \times \tfrac{1}{144}\ +\ 6 \times \tfrac{1}{1728}$$
$$= \ldots\ldots\ldots\ldots_{10} \text{ to 3 dp}$$

Finish it off

$$413 \cdot 108_{10}$$

Because

$$2 \text{ X } 5 \cdot 1 \text{ } 3 \text{ } 6_{12}$$
$$= 288 + 120 + 5 + \tfrac{1}{12} + \tfrac{1}{48} + \tfrac{1}{288} = 413 \tfrac{31}{288}$$

Therefore $2 \text{X} 5 \cdot 136_{12} = 413 \cdot 108_{10}$ to 3 dp.

Hexadecimal system (base 16)

This system has computer applications. The symbols here need to go up to an equivalent denary value of 15, so, after 9, letters of the alphabet are used as follows:

0, 1, 2, 3, 4, 5, 6, 7, 8, 9, A, B, C, D, E, F

The place values in this system are powers of 16.

For example	2	A	7	.	3	E	2_{16}
has place values	16^2	16^1	16^0		16^{-1}	16^{-2}	16^{-3}
i.e.	256	16	1		$\dfrac{1}{16}$	$\dfrac{1}{256}$	$\dfrac{1}{4096}$

Therefore $2 \text{ A } 7 \cdot 3 \text{ E } 2_{16} = \ldots\ldots\ldots$ to 3 dp.

$$679 \cdot 243_{10}$$

Here it is: $2 \text{ A } 7 \cdot 3 \text{ E } 2_{16}$

$$= 2 \times 256 + 10 \times 16 + 7 \times 1 + 3 \times \frac{1}{16} + 14 \times \frac{1}{256} + 2 \times \frac{1}{4096}$$
$$= 679 \tfrac{497}{2048} = 679 \cdot 243_{10} \text{ to 3 dp.}$$

And now, two more by way of practice.

Express the following in denary form:

(a) $3 \text{ } \Lambda \text{ } 4 \cdot 2 \text{ } 6 \text{ } 5_{12}$
(b) $3 \text{ C } 4 \cdot 2 \text{ } 1 \text{ } F_{16}$

Finish both of them and check the working with the next frame

8   Here they are:

(a)   3   Λ   4 ·   2   6 5_{12}

Place values   144   12   1 $\frac{1}{12}$ $\frac{1}{144}$ $\frac{1}{1728}$

So $3\,Λ\,4 \cdot 2\,6\,5_{12}$

$$= 3 \times 144 + 11 \times 12 + 4 \times 1 + 2 \times \frac{1}{12} + 6 \times \frac{1}{144} + 5 \times \frac{1}{1728}$$

$$= 432 + 132 + 4 + \frac{1}{6} + \frac{1}{24} + \frac{5}{1728}$$

$$= 568\tfrac{365}{1728} \qquad\qquad = 568 \cdot 211_{10}$$

Therefore $3\,Λ\,4 \cdot 265_{12} = 568 \cdot 211_{10}$ to 3 dp.

(b)   3   C   4 ·   2   1 F_{16}

Place values   256   16   1 $\frac{1}{16}$ $\frac{1}{256}$ $\frac{1}{4096}$

So $3\,C\,4 \cdot 2\,1\,F_{16}$

$$= 3 \times 256 + 12 \times 16 + 4 \times 1 + 2 \times \frac{1}{16} + 1 \times \frac{1}{256} + 15 \times \frac{1}{4096}$$

$$= 768 + 192 + 4 + \frac{1}{8} + \frac{1}{256} + \frac{15}{4096}$$

$$= 964\tfrac{543}{4096} \qquad\qquad = 964 \cdot 133_{10}$$

Therefore $3\,C\,4 \cdot 2\,1\,F_{16} = 964 \cdot 133_{10}$ to 3 dp.

9   ## An alternative method

So far, we have changed numbers in various bases into the equivalent denary numbers from first principles. Another way of arriving at the same results is by using the fact that two adjacent columns differ in place values by a factor which is the base of the particular system. An example will show the method.

Express the octal $357 \cdot 121_8$ in denary form.

First of all, we will attend to the whole-number part 357_8.

Starting at the left-hand end, multiply the first column by the base 8 and add the result to the entry in the next column (making 29).

```
        3              5              7
      × 8    →        24    →       232
      ───            ───            ───
       24  ┐          29            239
          ┘        × 8              ───
                    ───
                    232  ┘
```

Now repeat the process. Multiply the second column total by 8 and add the result to the next column. This gives 239 in the units column.

So $357_8 = 239_{10}$

Now we do much the same with the decimal part of the octal number

10

The decimal part is $0 \cdot 1\,2\,1_8$

$$
\begin{array}{ccccc}
0 & \cdot & 1 & 2 & 1 \\
 & & \times\ 8 & 8 & 80 \\ \hline
 & & 8 & 10 & 81 \\
 & & & \times\ 8 & \\ \hline
 & & & 80 & \\
\end{array}
$$

Starting from the left-hand column immediately following the decimal point, multiply by 8 and add the result to the next column. Repeat the process, finally getting a total of 81 in the end column.

But the place value of this column is . . . . to 4 dp.

11

$$8^{-3}$$

The denary value of $0 \cdot 121_8$ is 81×8^{-3} i.e. $81 \times \dfrac{1}{8^3} = \dfrac{81}{512} = 0 \cdot 1582_{10}$

Collecting the two partial results together, $357 \cdot 121_8 = 239 \cdot 1582_{10}$ to 4 dp.

In fact, we can set this out across the page to save space, thus:

$$
\begin{array}{cccccccc}
3 & 5 & 7 & \cdot & 1 & 2 & 1 \\
\times\ 8 & 24 & 232 & & \times\ 8 & 8 & 80 \\ \hline
24 & 29 & 239 & & 8 & 10 & 81 \\
 & \times\ 8 & & & & \times\ 8 & \\ \hline
 & 232 & & & & 80 & \\
\end{array}
$$

$81 \times \dfrac{1}{8^3} = \dfrac{81}{512} = 0 \cdot 1582_{10}$ Therefore $357 \cdot 121_8 = 239 \cdot 158_{10}$

Now you can set this one out in similar manner.

Express the duodecimal $245 \cdot 136_{12}$ in denary form.

Space out the duodecimal digits to give yourself room for the working:

   2 4 5 $\cdot$ 1 3 6_{12}

Then off you go. $245 \cdot 136_{12} = $ . . . . to 4 dp.

12

$$341 \cdot 1076_{10}$$

Here is the working as a check:

2	4	5 ·	1	3	6
× 12	24	336	× 12	12	180
24	28	341	12	15	186
	× 12			× 12	
	336			180	

Place value of last column is 12^{-3}, therefore

$$0 \cdot 136_{12} = 186 \times 12^{-3} = \frac{186}{1728} = 0 \cdot 1076_{10}$$

$$\text{So } 245 \cdot 136_{12} = 341 \cdot 1076_{10} \text{ to 4 dp.}$$

On to the next

13

Now for an easy one. Find the denary equivalent of the binary number $1\,1\,0\,1\,1 \cdot 1\,0\,1\,1_2$.

Setting it out in the same way, the result is .... to 4 dp.

14

$$27 \cdot 6875_{10}$$

1	1	0	1	1 ·	1	0	1	1
× 2	2	6	12	26	× 2	2	4	10
2	3	6	13	27	2	2	5	11
	× 2	× 2	× 2			× 2	× 2	
	6	12	26			4	10	

$$11 \times 2^{-4} = \frac{11}{16} = 0 \cdot 6875_{10}$$

Therefore $1\,1\,0\,1\,1 \cdot 1\,0\,1\,1_2 = 27 \cdot 6875_{10}$ to 4 dp.

And now a hexadecimal. Express $4\,C\,5 \cdot 2\,B\,8_{16}$ in denary form. Remember that C = 12 and B = 11. There are no snags.

$4\,C\,5 \cdot 2\,B\,8_{16} = \ldots\ldots\ldots$ to 4 dp.

15

$$1221 \cdot 1699_{10}$$

Here it is:

4	C	5 $\cdot$	2	B	8

$\times\,16 \longrightarrow 64 \longrightarrow 1216$ | $\times\,16 \longrightarrow 32 \longrightarrow 688$ Place value 16^{-3}

$$\underline{64}$$

$$\begin{array}{cccccc} 4 & C & 5\cdot & 2 & B & 8 \\ \times 16 & 64 & 1216 & \times 16 & 32 & 688 \\ \underline{64} & \underline{76} & \underline{1221} & \underline{32} & \underline{43} & \underline{696} \\ & \times 16 & & & \times 16 & \\ & \underline{456} & & & \underline{258} & \\ & 76 & & & 43 & \\ & \underline{1216} & & & \underline{688} & \end{array}$$

Place value 16^{-3}

So $\dfrac{696}{4096} = 0 \cdot 16992_{10}$

Therefore $4\,C\,5 \cdot 2\,B\,8_{16} = 1221 \cdot 1699_{10}$ to 4 dp.

They are all done in the same way.

Change of base from denary to a new base

A denary number in binary form

16

The simplest way to do this is by repeated division by 2 (the new base), noting the remainder at each stage. Continue dividing until a final zero quotient is obtained.

For example, to change 245_{10} to binary:

2	245_{10}	
2	122	—1
2	61	—0
2	30	—1
2	15	—0
2	7	—1
2	3	—1
2	1	—1
	0	—1

Now write all the remainders in the reverse order, i.e. from bottom to top.

Then $245_{10} = 11110101_2$

A denary number in octal form

The method here is exactly the same except that we divide repeatedly by 8 (the new base). So, without more ado, changing 524_{10} to octal gives
. . . .

17

$$\boxed{1014_8}$$

For:

8	524_{10}	
8	65	—4
8	8	—1
8	1	—0
	0	—1

As before, write the remainders in order, i.e. from bottom to top.

$\therefore\ 524_{10} = 1014_8$

A denary number in duodecimal form

Method as before, but this time we divide repeatedly by 12.

So $897_{10} = \dots\dots\dots$

18

$$\boxed{629_{12}}$$

Because

12	897_{10}	
12	74	—9
12	6	—2
	0	—6

$\therefore\ 897_{10} = 629_{12}$

Now move to the next frame

19   The method we have been using is quick and easy enough when the denary number to be changed is a whole number. When it contains a decimal part, we must look further.

A denary decimal to octal form

To change 0.526_{10} to octal form, we multiply the decimal repeatedly by the new base, in this case 8, but on the second and subsequent multiplication, we do not multiply the whole-number part of the previous product.

$$
\begin{array}{rcl}
0 & \cdot & 526_{10} \\
  &   & \underline{8} \\
4 & \cdot & 208 \\
  &   & \underline{8} \\
1 & \cdot & 664 \\
  &   & \underline{8} \\
5 & \cdot & 312 \\
  &   & \underline{8} \\
2 & \cdot & 496 \\
\end{array}
$$

Now multiply again by 8, but treat only the decimal part

and so on

Finally, we write the whole-number numerals downwards to form the required octal decimal.

Be careful *not* to include the zero unit digit in the original denary decimal. In fact, it may be safer simply to write the decimal as $\cdot 526_{10}$ in the working.

So $\hspace{4cm} 0.526_{10} = 0.4152_8$

Converting a denary decimal into any new base is done in the same way. If we express 0.306_{10} as a duodecimal, we get . . . .

Set it out in the same way: there are no snags

$$\boxed{0.3809_{12}}$$

20

$$
\begin{array}{rcl}
  & \cdot & 306_{10} \\
  &   & \underline{12} \\
3 & \cdot & 672 \\
  &   & \underline{12} \\
8 & \cdot & 064 \\
  &   & \underline{12} \\
0 & \cdot & 768 \\
  &   & \underline{12} \\
9 & \cdot & 216 \\
  &   & \underline{12} \\
2 & \cdot & 592 \\
\end{array}
$$

There is no carryover into the units column, so enter a zero.

etc.

$$\therefore\ 0.306_{10} = 0.3809_{12}$$

Now let us go one stage further – so on to the next frame

21

If the denary number consists of both a whole-number and a decimal part, the two parts are converted separately and united in the final result. The example will show the method.

Express 492.731_{10} in octal form:

Then $492.731_{10} = 754.5662_8$

In similar manner, 384.426_{10} expressed in duodecimals becomes ....

Set the working out in the same way

22

$$280.5142_{12}$$

Because we have:

$$\therefore \ 384.426_{10} = 280.5142_{12}$$

That is straightforward enough, so let us now move on to see a very helpful use of octals in the next frame.

Use of octals as an intermediate step   **23**

This gives us an easy way of converting denary numbers into binary and hexadecimal forms. As an example, note the following.

Express the denary number $348 \cdot 654_{10}$ in octal, binary and hexadecimal forms.

(a) First we change $348 \cdot 654_{10}$ into octal form by the usual method.

This gives $348 \cdot 654_{10} = \ldots\ldots\ldots\ldots$

$534 \cdot 517_8$ **24**

(b) Now we take this octal form and write the binary equivalent of each digit in groups of three binary digits, thus:

$$101 \quad 011 \quad 100 \quad \cdot \quad 101 \quad 001 \quad 111$$

Closing the groups up we have the binary equivalent of $534 \cdot 517_8$

i.e. $348 \cdot 654_{10} = 534 \cdot 517_8$
$$= 101011100 \cdot 101001111_2$$

(c) Then, starting from the decimal point and working in each direction, regroup the same binary digits in groups of four. This gives $\ldots\ldots\ldots\ldots$

$0001 \quad 0101 \quad 1100 \quad \cdot \quad 1010 \quad 0111 \quad 1000$ **25**

completing the group at either end by addition of extra zeros, as necessary.

(d) Now write the hexadecimal equivalent of each group of four binary digits, so that we have $\ldots\ldots\ldots\ldots$

$1 \quad 5 \quad (12) \quad \cdot \quad (10) \quad 7 \quad 8$ **26**

Replacing (12) and (10) with the corresponding hexadecimal symbols, C and A, this gives $1\,5\,C \cdot A\,7\,8_{16}$

So, collecting the partial results together:

$348 \cdot 654_{10} = 534 \cdot 517_8$
$$= 101011100 \cdot 101001111_2$$
$$= 15C \cdot A78_{16}$$

Next frame

We have worked through the previous example in some detail. In practice, the   **27**
method is more concise. Here is another example.

Change the denary number $428 \cdot 371_{10}$ into its octal, binary and hexadecimal forms.

(a) First of all, the octal equivalent of $428 \cdot 371_{10}$ is $\ldots\ldots\ldots\ldots$

28

$$654 \cdot 276_8$$

(b) The binary equivalent of each octal digit in groups of three is ....

29

$$110 \quad 101 \quad 100 \quad \cdot \quad 010 \quad 111 \quad 110_2$$

(c) Closing these up and rearranging in groups of four in each direction from the decimal point, we have ....

30

$$0001 \quad 1010 \quad 1100 \quad \cdot \quad 0101 \quad 1111_2$$

(d) The hexadecimal equivalent of each set of four binary digits then gives ....

31

$$1AC \cdot 5F_{16}$$

So $428 \cdot 371_{10} = 654 \cdot 276_8$

$$= 110101100 \cdot 010111110_2$$

$$= 1AC \cdot 5F_{16}$$

This is important, so let us do one more.

Convert $163 \cdot 245_{10}$ into octal, binary and hexadecimal forms.

You can do this one entirely on your own. Work right through it and then check with the results in the next frame

32

$$163 \cdot 245_{10} = 243 \cdot 175_8$$

$$= 010100011 \cdot 001111101_2$$

$$= 1010 \quad 0011 \quad \cdot \quad 0011 \quad 1110 \quad 1000_2$$

$$= A3 \cdot 3E8_{16}$$

And that is it.

On to the next frame

33 **Reverse method**

Of course, the method we have been using can be used in reverse, i.e. starting with a hexadecimal number, we can change it into groups of four binary digits, regroup these into groups of three digits from the decimal point, and convert these into the equivalent octal digits. Finally, the octal number can be converted into denary form by the usual method.

Here is one you can do with no trouble.

Express the hexadecimal number $4B2{\cdot}1A6_{16}$ in equivalent binary, octal and denary forms.

    (a) Rewrite $4B2{\cdot}1A6_{16}$ in groups of four binary digits.
    (b) Regroup into groups of three binary digits from the decimal point.
    (c) Express the octal equivalent of each group of three binary digits.
    (d) Finally convert the octal number into its denary equivalent.

Work right through it and then check with the solution in the next frame

---

34

$4B2{\cdot}1A6_{16} =$
(a)   0100 1011 0010 · 0001 1010 0110$_2$
(b)   010 010 110 010 · 000 110 100 110$_2$
(c)   2 2 6 2 · 0 6 4 6$_8$
(d) $1202{\cdot}103_{10}$

Now one more for good measure.

Express $2E3{\cdot}4D_{16}$ in binary, octal and denary forms.

Check results with the next frame

35

$$2E3{\cdot}4D_{16} = 0010\ 1110\ 0011 \cdot 0100\ 1101_2$$
$$= 001\ 011\ 100\ 011 \cdot 010\ 011\ 010_2$$
$$= 1\ \ 3\ \ 4\ \ 3 \ \cdot\ 2\ \ 3\ \ 2_8$$
$$= 739{\cdot}301_{10}$$

Let us pause and summarize the main facts so far on number systems

---

 # Review summary

Unit 6

36

1	*Denary (or decimal) system*	Base 10 Numerals	Place values – powers of 10 0, 1, 2, 3, 4, 5, 6, 7, 8, 9
2	*Binary system*	Base 2 Numerals	Place values – powers of 2 0, 1
3	*Octal system*	Base 8 Numerals	Place values – powers of 8 0, 1, 2, 3, 4, 5, 6, 7
4	*Duodecimal system*	Base 12 Numerals	Place values – powers of 12 0, 1, 2, 3, 4, 5, 6, 7, 8, 9, X, Λ (10, 11)
5	*Hexadecimal system*	Base 16 Numerals	Place values – powers of 16 0, 1, 2, 3, 4, 5, 6, 7, 8, 9, A, B, C, D, E, F (10, 11, 12, 13, 14, 15)

Review exercise   **Unit 6**

37

1 Express the following numbers in denary form:

   (a) $11001 \cdot 11_2$ (b) $776 \cdot 143_8$ (c) $4X9 \cdot 2\Lambda5_{12}$ (d) $6F8 \cdot 3D5_{16}$

2 Express the denary number $427 \cdot 362_{10}$ as a duodecimal number.

3 Convert 139_{10} to the equivalent octal, binary and hexadecimal forms.

Complete the questions. Take care, there is no need to rush.
If you need to, look back to the Unit.
The answers and working are in the next frame

38

1 (a)

$$11001 \cdot 11_2 = 25 \cdot 750_{10}$$

$$3 \times 2^{-2} = \frac{3}{4} = 0 \cdot 75_{10}$$

(b)

$$776 \cdot 143_8 = 510 \cdot 193_{10}$$

$$99 \times 8^{-3} = \frac{99}{512} = 0 \cdot 19336_{10}$$

(c)

$$4X9 \cdot 2\Lambda5_{12} = 705 \cdot 246_{10}$$

$$425 \times 12^{-3} = \frac{425}{1728} = 0 \cdot 2459_{10}$$

(d)

$$6F8 \cdot 3D5_{16} = 1784 \cdot 240_{10}$$

$$981 \times 16^{-3} = \frac{981}{4096} = 0 \cdot 2395_{10}$$

2 $427 \cdot 362_{10}$

12	427	
12	35	r 7
12	2	r 11 — Λ
	0	r 2

so $427_{10} = 2\Lambda 7_{12}$

$$
\begin{array}{r|r}
0 & \cdot\ 362 \\
 & 12 \\
\hline
4 & \cdot\ 344 \\
 & 12 \\
\hline
4 & \cdot\ 128 \\
 & 12 \\
\hline
1 & \cdot\ 536 \\
 & 12 \\
\hline
6 & \cdot\ 432 \\
\end{array}
$$

so $0 \cdot 362_{10} = 0 \cdot 4416_{12}$

$\therefore\ 427 \cdot 362_{10} = 2\Lambda 7 \cdot 4416_{12}$

3 139_{10}

8	139	
8	17	r 3
8	2	r 1
	0	r 2

$139_{10} = 213_8$

$139_{10} = 213_8$

$\qquad = 010\,001\,011_2$

$\qquad = 1000\,1011_2$

$\qquad = 8(11)_{16}$

$\qquad = 8B_{16}$

Now for the Review test

 # Review test

Unit 6

39

1   Express the following numbers in denary form:
   (a) $1011 \cdot 01_2$ (b) $456 \cdot 721_8$
   (c) $123 \cdot \Lambda 29_{12}$ (d) $CA1 \cdot B22_{16}$

2   Convert $15 \cdot 605_{10}$ to the equivalent octal, binary, duodecimal and hexadecimal forms.

40   You have now come to the end of this Module. A list of **Can You?** questions
follows for you to gauge your understanding of the arithmetic in this Module.
You will notice that these questions match the **Learning Outcomes** listed at
the beginning of the Module.

Now try the **Test exercise**. *Work through the questions at your own pace, there is
no need to hurry.* A set of **Further problems** provides valuable additional
practice.

 # Can You?

1   Checklist: Module 1

Check this list before and after you try the end of Module test.

On a scale of 1 to 5 how confident are you that you can:

- Carry out the basic rules of arithmetic with integers?
  Yes ☐ ☐ ☐ ☐ ☐ *No*

- Check the result of a calculation making use of rounding?
  Yes ☐ ☐ ☐ ☐ ☐ *No*

- Write a natural number as a product of prime numbers?
  Yes ☐ ☐ ☐ ☐ ☐ *No*

- Find the highest common factor and lowest common multiple of two
  natural numbers?
  Yes ☐ ☐ ☐ ☐ ☐ *No*

- Manipulate fractions, ratios and percentages?
  Yes ☐ ☐ ☐ ☐ ☐ *No*

- Manipulate decimal numbers?
  Yes ☐ ☐ ☐ ☐ ☐ *No*

- Manipulate powers?
  Yes ☐ ☐ ☐ ☐ ☐ *No*

- Use standard or preferred standard form and complete a calculation
  to the required level of accuracy?
  Yes ☐ ☐ ☐ ☐ ☐ *No*

- Understand the construction of various number systems and convert
  from one number system to another?
  Yes ☐ ☐ ☐ ☐ ☐ *No*

Test exercise 1

2

1 Place the appropriate symbol < or > between each of the following pairs of numbers:
(a) − 12 − 15   (b) 9 − 17   (c) − 11 10

2 Find the value of each of the following:
(a) $24 - 3 \times 4 + 28 \div 14$ (b) $(24 - 3) \times (4 + 28) \div 14$

3 Write each of the following as a product of prime factors:
(a) 156   (b) 546   (c) 1445   (d) 1485

4 Round each number to the nearest 10, 100 and 1000:
(a) 5045   (b) 1100   (c) − 1552   (d) − 4995

5 Find (i) the HCF and (ii) the LCM of:
(a) 1274 and 195   (b) 64 and 18

6 Reduce each of the following fractions to their lowest terms:
(a) $\frac{8}{14}$ (b) $\frac{162}{36}$ (c) $-\frac{279}{27}$ (d) $-\frac{81}{3}$

7 Evaluate each of the following, giving your answer as a fraction:
(a) $\frac{1}{3} + \frac{3}{5}$ (b) $\frac{2}{7} - \frac{1}{9}$ (c) $\frac{8}{3} \times \frac{6}{5}$ (d) $\frac{4}{5}$ *of* $\frac{2}{15}$
(e) $\frac{9}{2} \div \frac{3}{2}$ (f) $\frac{6}{7} - \frac{4}{5} \times \frac{3}{2} \div \frac{7}{5} + \frac{9}{4}$

8 In each of the following the proportions of a compound are given. Find the ratios of the components in each case:
(a) $\frac{3}{4}$ of A and $\frac{1}{4}$ of B
(b) $\frac{2}{3}$ of P, $\frac{1}{15}$ of Q and the remainder of R
(c) $\frac{1}{5}$ of R, $\frac{3}{5}$ of S, $\frac{1}{6}$ of T and the remainder of U

9 What is:
(a) $\frac{3}{5}$ as a percentage?
(b) 16% as a fraction in its lowest terms?
(c) 17·5% of £12·50?

10 Evaluate each of the following (i) to 4 sig fig and (ii) to 3 dp:
(a) $13·6 \times 25·8 \div 4·2$
(b) $13·6 \div 4·2 \times 25·8$
(c) $9·1(17·43 + 7·2(8·6 - 4·1^2 \times 3·1))$
(d) $-8·4((6·3 \times 9·1 + 2·2^{1·3}) - (4·1^{-3·1} \div 3·3^3 - 5·4))$

11 Convert each of the following to decimal form to 3 decimal places:
(a) $\frac{3}{17}$ (b) $-\frac{2}{15}$ (c) $\frac{17}{3}$ (d) $-\frac{24}{11}$

12 Write each of the following in abbreviated form:
(a) $6 \cdot 7777 \ldots$ (b) $0 \cdot 01001001001 \ldots$

13 Convert each of the following to fractional form in lowest terms:
(a) $0 \cdot 4$ (b) $3 \cdot 68$ (c) $1 \cdot \dot{4}$ (d) $-6 \cdot 1$

14 Write each of the following as a number raised to a power:
(a) $2^9 \times 2^2$ (b) $6^2 \div 5^2$ (c) $((-4)^4)^{-4}$ (d) $(3^{-5})^0$

15 Find the value of each of the following to 3 dp:
(a) $11^{\frac{1}{4}}$ (b) $\sqrt[7]{3}$ (c) $(-81)^{\frac{1}{5}}$ (d) $(-81)^{\frac{1}{4}}$

16 Express in standard form:
(a) $537 \cdot 6$ (b) $0 \cdot 364$ (c) 4902 (d) $0 \cdot 000125$

17 Convert to preferred standard form:
(a) $6 \cdot 147 \times 10^7$ (b) $2 \cdot 439 \times 10^{-4}$ (c) $5 \cdot 286 \times 10^5$ (d) $4 \cdot 371 \times 10^{-7}$

18 Determine the following product, giving the result in both standard form and preferred standard form:
$$(6 \cdot 43 \times 10^3)(7 \cdot 35 \times 10^4)$$

19 Each of the following contains numbers obtained by measurement. Evaluate each to the appropriate level of accuracy:
(a) $18 \cdot 4^{1 \cdot 6} \times 0 \cdot 01$ (b) $\dfrac{7 \cdot 632 \times 2 \cdot 14 - 8 \cdot 32 \div 1 \cdot 1}{16 \cdot 04}$

20 Express the following numbers in denary form:
(a) $1111 \cdot 11_2$ (b) $777 \cdot 701_8$ (c) $3\Lambda 3 \cdot 9\Lambda 1_{12}$ (d) $E02 \cdot FAB_{16}$

21 Convert $19 \cdot 872_{10}$ to the equivalent octal, binary, duodecimal and hexadecimal forms.

Further problems 1

3

1 Place the appropriate symbol $<$ or $>$ between each of the following pairs of numbers:
(a) $-4 \quad -11$ (b) $7 \quad -13$ (c) $-15 \quad 13$

2 Find the value of each of the following:
(a) $6 + 14 \div 7 - 2 \times 3$ (b) $(6 + 14) \div (7 - 2) \times 3$

3 Round each number to the nearest 10, 100 and 1000:
(a) 3505 (b) 500 (c) -2465 (d) -9005

4 Calculate each of the following to:
  (i) 5 sig fig;
  (ii) 4 dp;
  (iii) the required level of accuracy given that each number, other than those indicated in brackets, has been obtained by measurement.

  (a) $\dfrac{3\cdot21^{2\cdot33} + (5\cdot77 - 3\cdot11)}{8\cdot32 - 2\cdot64 \times \sqrt{2\cdot56}}$

  (b) $\dfrac{3\cdot142 \times 1\cdot95}{6}(3 \times 5\cdot44^2 + 1\cdot95^2)$　　(power 2, divisor 6, multiplier 3)

  (c) $\dfrac{3\cdot142 \times 1\cdot234}{12}(0\cdot424^2 + 0\cdot424 \times 0\cdot951 + 0\cdot951^2)$ (power 2, divisor 12)

  (d) $\sqrt{\dfrac{2 \times 0\cdot577}{3\cdot142 \times 2\cdot64} + \dfrac{2\cdot64^2}{3}}$　　(power 2, divisor 3, multiplier 2)

  (e) $\dfrac{3\cdot26 + \sqrt{12\cdot13}}{14\cdot192 - 2\cdot4 \times 1\cdot63^2}$　　(power 2)

  (f) $\dfrac{4\cdot62^2 - (7\cdot16 - 2\cdot35)}{2\cdot63 + 1\cdot89 \times \sqrt{73\cdot24}}$　　(power 2)

5 Find the prime factorization for each of the following:
  (a) 924　　(b) 825　　(c) 2310　　(d) 35 530

6 Find the HCF and LCM for each of the following pairs of numbers:
  (a) 9, 21　　(b) 15, 85　　(c) 66, 42　　(d) 64, 360

7 Reduce each of the following fractions to their lowest terms:
  (a) $\dfrac{6}{24}$　(b) $\dfrac{104}{48}$　(c) $-\dfrac{120}{15}$　(d) $-\dfrac{51}{7}$

8 Evaluate:

  (a) $\dfrac{9}{2} - \dfrac{4}{5} \div \left(\dfrac{2}{3}\right)^2 \times \dfrac{3}{11}$　　(b) $\dfrac{\dfrac{3}{4} + \dfrac{7}{5} \div \dfrac{2}{9} \times \dfrac{1}{3}}{\dfrac{7}{3} - \dfrac{11}{2} \times \dfrac{2}{5} + \dfrac{4}{9}}$

  (c) $\left(\dfrac{3}{4} + \dfrac{7}{5}\right)^2 \div \left(\dfrac{7}{3} - \dfrac{11}{5}\right)^2$　　(d) $\dfrac{\left(\dfrac{5}{2}\right)^3 - \dfrac{2}{9} \div \left(\dfrac{2}{3}\right)^2 \times \dfrac{3}{2}}{\dfrac{3}{11} + \left(\dfrac{11}{2} \times \dfrac{2}{5}\right)^2 - \dfrac{7}{5}}$

9 Express each of the following as a fraction in its lowest terms:
  (a) 36%　　(b) 17·5%　　(c) 8·7%　　(d) 72%

10 Express each of the following as a percentage accurate to 1 dp:
  (a) $\dfrac{4}{5}$　(b) $\dfrac{3}{11}$　(c) $\dfrac{2}{9}$　(d) $\dfrac{1}{7}$
  (e) $\dfrac{9}{19}$　(f) $\dfrac{13}{27}$　(g) $\dfrac{7}{101}$　(h) $\dfrac{199}{200}$

11 Find:
  (a) 16% *of* 125　　　　(b) 9·6% *of* 5·63
  (c) 13·5% *of* (−13·5)　　(d) 0·13% *of* 92·66

12 In each of the following the properties of a compound are given. In each case find $A : B : C$.

(a) $\frac{1}{5}$ of A, $\frac{2}{3}$ of B and the remainder of C;

(b) $\frac{3}{8}$ of A with B and C in the ratio $1 : 2$;

(c) A, B and C are mixed according to the ratios $A : B = 2 : 5$ and $B : C = 10 : 11$;

(d) A, B and C are mixed according to the ratios $A : B = 1 : 7$ and $B : C = 13 : 9$.

13 Write each of the following in abbreviated form:
(a) $8 \cdot 767676 \ldots$ (b) $212 \cdot 211211211 \ldots$

14 Convert each of the following to fractional form in lowest terms:
(a) $0 \cdot 12$ (b) $5 \cdot 25$ (c) $5 \cdot 30\dot{6}$ (d) $-9 \cdot 3$

15 Write each of the following as a number raised to a power:
(a) $8^4 \times 8^3$ (b) $2^9 \div 8^2$ (c) $(5^3)^5$ (d) $3^4 \div 9^2$

16 Find the value of each of the following to 3 dp:
(a) $17^{\frac{2}{5}}$ (b) $\sqrt[4]{13}$ (c) $(-5)^{\frac{2}{3}}$ (d) $\sqrt{(-5)^4}$

17 Convert each of the following to decimal form to 3 decimal places:
(a) $\frac{5}{21}$ (b) $-\frac{2}{17}$ (c) $\frac{8}{3}$ (d) $-\frac{32}{19}$

18 Express in standard form:
(a) $52 \cdot 876$ (b) $15\,243$ (c) $0 \cdot 08765$
(d) $0 \cdot 0000492$ (e) $436 \cdot 2$ (f) $0 \cdot 5728$

19 Rewrite in preferred standard form:
(a) 4285 (b) $0 \cdot 0169$ (c) $8 \cdot 526 \times 10^{-4}$
(d) $3 \cdot 629 \times 10^5$ (e) $1 \cdot 0073 \times 10^7$ (f) $5 \cdot 694 \times 10^8$

20 Evaluate the following, giving the result in standard form and in preferred standard form:
$$\frac{(4 \cdot 26 \times 10^4)(9 \cdot 38 \times 10^5)}{3 \cdot 179 \times 10^2}$$

21 Convert each of the following decimal numbers to (i) binary, (ii) octal, (iii) duodecimal and (iv) hexadecimal format:
(a) $1 \cdot 83$ (b) $3 \cdot 425 \times 10^2$

22 Convert each of the following octal numbers to (i) binary, (ii) decimal, (iii) duodecimal and (iv) hexadecimal format:
(a) $0 \cdot 577$ (b) 563

23 Convert each of the following duodecimal numbers to (i) binary, (ii) octal, (iii) decimal and (iv) hexadecimal format:
(a) $0 \cdot \Lambda X$ (b) $9\Lambda 1$

24 Convert each of the following binary numbers to (i) decimal, (ii) octal, (iii) duodecimal and (iv) hexadecimal format:

(a) 0·10011   (b) 111001100

25 Convert each of the following hexadecimal numbers to (i) binary, (ii) octal, (iii) duodecimal and (iv) decimal format:

(a) 0·F4B   (b) 3A5

Introduction to algebra

Learning outcomes

When you have completed this Module you will be able to:

- Use alphabetic symbols to supplement the numerals and to combine these symbols using all the operations of arithmetic
- Simplify algebraic expressions by collecting like terms and by abstracting common factors from similar terms
- Remove brackets and so obtain alternative algebraic expressions
- Manipulate expressions involving powers and logarithms
- Multiply two algebraic expressions together
- Manipulate logarithms both numerically and symbolically
- Manipulate algebraic fractions and divide one expression by another
- Factorize algebraic expressions using standard factorizations
- Factorize quadratic algebraic expressions

Units

Algebraic expressions

1

Think of a number
Add 15 to it
Double the result
Add this to the number you first thought of
Divide the result by 3
Take away the number you first thought of

The answer is 10

Why?

Check your answer in the next frame

Symbols other than numerals

2

A letter of the alphabet can be used to represent a number when the specific number is unknown and because the number is unknown (except, of course, to the person who thought of it) we shall represent the number by the letter a:

Think of a number	a
Add 15 to it	$a + 15$
Double the result	$2 \times (a + 15) = (2 \times a) + (2 \times 15)$
	$\qquad\qquad\qquad = (2 \times a) + 30$
Add the result to the number you first thought of	$a + (2 \times a) + 30 = (3 \times a) + 30$
Divide the result by 3	$[(3 \times a) + 30] \div 3 = a + 10$
Take away the number you first thought of	$a + 10 - a = 10$
The result is 10	

Next frame

This little puzzle has demonstrated how:

3

an unknown number can be represented by a letter of the alphabet which can then be manipulated just like an ordinary numeral within an arithmetic expression.

So that, for example:

$a + a + a + a = 4 \times a$
$3 \times a - a = 2 \times a$
$8 \times a \div a = 8$

and

$a \times a \times a \times a \times a = a^5$

▶

If *a* and *b* represent two unknown numbers then we speak of the:

sum of *a* and *b* $a + b$

difference of *a* and *b* $a - b$

product of *a* and *b* $a \times b$, *a.b* or simply *ab*

                                (*we can and do suppress the multiplication sign*)

quotient of *a* and *b* $a \div b$, *a/b* or $\dfrac{a}{b}$ provided $b \neq 0$

and

raising *a* to the power *b* a^b

Using letters and numerals in this way is referred to as *algebra*.

Now move to the next frame

---

4   **Constants**

In the puzzle of Frame 1 we saw how to use the letter *a* to represent an unknown number – we call such a symbol a *constant*.

In many other problems we require a symbolism that can be used to represent not just one number but any one of a collection of numbers. Central to this symbolism is the idea of a variable.

Next frame

---

5   **Variables**

We have seen that the operation of addition is commutative. That is, for example:

$2 + 3 = 3 + 2$

To describe this rule as applying to any pair of numbers and not just 2 and 3 we resort to the use of alphabetic characters *x* and *y* and write:

$x + y = y + x$

where *x* and *y* represent any two numbers. Used in this way, the letters *x* and *y* are referred to as *variables* because they each represent, not just one number, but any one of a collection of numbers.

So how would you write down the fact that multiplication is an associative operation? (Refer to Frame 12 of Module 1, Unit 1.)

You can check your answer in the next frame

6

$$x(yz) = (xy)z = xyz$$

where x, y and z represent numbers. Notice the suppression of the multiplication sign.

While it is not a hard and fast rule, it is generally accepted that letters from the beginning of the alphabet, i.e. a, b, c, d, ... are used to represent constants and letters from the end of the alphabet, i.e. ... v, w, x, y, z are used to represent variables. In any event, when a letter of the alphabet is used it should be made clear whether the letter stands for a constant or a variable.

Now move on to the next frame

Rules of algebra

7

The rules of arithmetic that we met in the previous Module for integers also apply to any type of number and we express this fact in the *rules of algebra* where we use variables rather than numerals as specific instances. The rules are:

Commutativity

Two numbers x and y can be added or multiplied in any order without affecting the result. That is:

$x + y = y + x$ and

$xy = yx$

Addition and multiplication are commutative operations

The order in which two numbers are subtracted or divided *does* affect the result. That is:

$x - y \neq y - x$ unless $x = y$ and

$x \div y \neq y \div x, \ \left(\dfrac{x}{y} \neq \dfrac{y}{x}\right)$ unless $x = y$ and neither equals 0

Subtraction and division are not commutative operations except in very special cases

Associativity

The way in which the numbers x, y and z are associated under addition or multiplication *does not* affect the result. That is:

$x + (y + z) = (x + y) + z = x + y + z$ and

$x(yz) = (xy)z = xyz$

Addition and multiplication are associative operations

The way in which the numbers are associated under subtraction or division *does* affect the result. That is:

$x - (y - z) \neq (x - y) - z$ unless $z = 0$ and

$x \div (y \div z) \neq (x \div y) \div z$ unless $z = 1$ and $y \neq 0$

Subtraction and division are not associative operations except in very special cases

Distributivity

Multiplication is distributed over addition and subtraction from both the left and the right. For example:

$$x(y + z) = xy + xz \text{ and } (x + y)z = xz + yz$$
$$x(y - z) = xy - xz \text{ and } (x - y)z = xz - yz$$

Division is distributed over addition and subtraction from the right but not from the left. For example:

$$(x + y) \div z = (x \div z) + (y \div z) \text{ but}$$
$$x \div (y + z) \neq (x \div y) + (x \div z)$$

that is:

$$\frac{x + y}{z} = \frac{x}{z} + \frac{y}{z} \text{ but } \frac{x}{y + z} \neq \frac{x}{y} + \frac{x}{z}$$

Take care here because it is a common mistake to get this wrong

Rules of precedence

The familiar rules of precedence continue to apply when algebraic expressions involving mixed operations are to be manipulated.

Next frame

8 Terms and coefficients

An algebraic expression consists of alphabetic characters and numerals linked together with the arithmetic operators. For example:

$$8x - 3xy$$

is an algebraic expression in the two variables x and y. Each component of this expression is called a *term* of the expression. Here there are two terms, namely:

the x term and the xy term.

The numerals in each term are called the *coefficients* of the respective terms. So that:

8 is the coefficient of the x term and -3 is the coefficient of the xy term.

Collecting like terms

Terms which have the same variables are called *like* terms and like terms can be collected together by addition or subtraction. For example:

$4x + 3y - 2z + 5y - 3x + 4z$ can be rearranged as $4x - 3x + 3y + 5y - 2z + 4z$ and simplified to:

$$x + 8y + 2z$$

Similarly, $4uv - 7uz - 6wz + 2uv + 3wz$ can be simplified to ....

Check your answer with the next frame

$$6uv - 7uz - 3wz$$

9

Next frame

Similar terms

10

In the algebraic expression:

$ab + bc$

both terms contain the letter b and for this reason these terms, though not like terms, are called *similar* terms. Common symbols such as this letter b are referred to as *common factors* and by using brackets these common factors can be *factored out*. For example, the common factor b in this expression can be factored out to give:

$ab + bc = b(a + c)$ This process is known as *factorization*.

Numerical factors are factored out in the same way. For example, in the algebraic expression:

$3pq - 3qr$

the terms are similar, both containing the letter q. They also have a common coefficient 3 and this, as well as the common letter q, can be factored out to give:

$3pq - 3qr = 3q(p - r)$

So the algebraic expression $9st - 3sv - 6sw$ simplifies to ....

The answer is in the next frame

$$3s(3t - v - 2w)$$

11

Because

$$9st - 3sv - 6sw = 3s \times 3t - 3s \times v - 3s \times 2w$$
$$= 3s(3t - v - 2w)$$

Next frame

12 Expanding brackets

Sometimes it will be desired to reverse the process of factorizing an expression by *removing* the brackets. This is done by:

(a) multiplying or dividing each term inside the bracket by the term outside the bracket, but

(b) if the term outside the bracket is negative then each term inside the bracket changes sign.

For example, the brackets in the expression:

$3x(y - 2z)$ are removed to give $3xy - 6xz$ and the brackets in the expression $-2y(2x - 4z)$ are removed to give $-4yx + 8yz$.

As a further example, the expression:

$\dfrac{y + x}{8x} - \dfrac{y - x}{4x}$ is an alternative form of $(y + x) \div 8x - (y - x) \div 4x$ and the brackets can be removed as follows:

$$\dfrac{y + x}{8x} - \dfrac{y - x}{4x} = \dfrac{y}{8x} + \dfrac{x}{8x} - \dfrac{y}{4x} + \dfrac{x}{4x}$$

$$= \dfrac{y}{8x} + \dfrac{1}{8} - \dfrac{y}{4x} + \dfrac{1}{4}$$

$$= \dfrac{3}{8} - \dfrac{y}{8x} \text{ which can be written as } \dfrac{1}{8}\left(3 - \dfrac{y}{x}\right) \text{ or as } \dfrac{1}{8x}(3x - y)$$

13 Nested brackets

Whenever an algebraic expression contains brackets nested within other brackets the innermost brackets are removed first. For example:

$$7(a - [4 - 5(b - 3a)]) = 7(a - [4 - 5b + 15a])$$
$$= 7(a - 4 + 5b - 15a)$$
$$= 7a - 28 + 35b - 105a$$
$$= 35b - 98a - 28$$

So that the algebraic expression $4(2x + 3[5 - 2(x - y)])$ becomes, after the removal of the brackets ....

Next frame

14

$$\boxed{24y - 16x + 60}$$

Because

$$4(2x + 3[5 - 2(x - y)]) = 4(2x + 3[5 - 2x + 2y])$$
$$= 4(2x + 15 - 6x + 6y)$$
$$= 8x + 60 - 24x + 24y$$
$$= 24y - 16x + 60$$

At this point let us pause and summarize the main facts so far on algebraic expressions

Review summary

15

1   Alphabetic characters can be used to represent numbers and then be subjected to the arithmetic operations in much the same way as numerals.

2   An alphabetic character that represents a single number is called a *constant*.

3   An alphabetic character that represents any one of a collection of numbers is called a *variable*.

4   Some algebraic expressions contain terms multiplied by numerical coefficients.

5   Like terms contain identical alphabetic characters.

6   Similar terms have some but not all alphabetic characters in common.

7   Similar terms can be factorized by identifying their common factors and using brackets.

 # Review exercise

16

1   Simplify each of the following by collecting like terms:
  (a) $4xy + 3xz - 6zy - 5zx + yx$
  (b) $-2a + 4ab + a - 4ba$
  (c) $3rst - 10str + 8ts - 5rt + 2st$
  (d) $2pq - 4pr + qr - 2rq + 3qp$
  (e) $5lmn - 6ml + 7lm + 8mnl - 4ln$

2   Simplify each of the following by collecting like terms and factorizing:
  (a) $4xy + 3xz - 6zy - 5zx + yx$
  (b) $3rst - 10str + 8ts - 5rt + 2st$
  (c) $2pq - 4pr + qr - 2rq + 3qp$
  (d) $5lmn - 6ml + 7lm + 8mnl - 4ln$

3   Expand the following and then refactorize where possible:
  (a) $8x(y - z) + 2y(7x + z)$ (b) $(3a - b)(b - 3a) + b^2$
  (c) $-3(w - 7[x - 8(3 - z)])$ (d) $\dfrac{2a - 3}{4b} + \dfrac{3a + 2}{6b}$

Complete the questions.
Look back at the Unit if necessary but don't rush.
The answers and working are in the next frame.

17

1 (a) $4xy + 3xz - 6zy - 5zx + yx = 4xy + xy + 3xz - 5xz - 6yz$
$$= 5xy - 2xz - 6yz$$

Notice that the characters are written in alphabetic order.

(b) $-2u + 4ab + a - 4ba = -2a + a + 4ab - 4ab$
$$= -a$$

(c) $3rst - 10str + 8ts - 5rt + 2st = 3rst - 10rst + 8st + 2st - 5rt$
$$= -7rst + 10st - 5rt$$

(d) $2pq - 4pr + qr - 2rq + 3qp = 2pq + 3pq - 4pr + qr - 2qr$
$$= 5pq - 4pr - qr$$

(e) $5lmn - 6ml + 7lm + 8mnl - 4ln = 5lmn + 8lmn - 6lm + 7lm - 4ln$
$$= 13lmn + lm - 4ln$$

2 (a) $4xy + 3xz - 6zy - 5zx + yx = 5xy - 2xz - 6yz$
$$= x(5y - 2z) - 6yz \text{ or}$$
$$= 5xy - 2z(x + 3y) \text{ or}$$
$$= y(5x - 6z) - 2xz$$

(b) $3rst - 10str + 8ts - 5rt + 2st = -7rst + 10st - 5rt$
$$= st(10 - 7r) - 5rt \text{ or}$$
$$= 10st - rt(7s + 5) \text{ or}$$
$$= t(10s - 7rs - 5r) = t(s[10 - 7r] - 5r)$$

(c) $2pq - 4pr + qr - 2rq + 3qp = 5pq - 4pr - qr$
$$= p(5q - 4r) - qr \text{ or}$$
$$= q(5p - r) - 4pr \text{ or}$$
$$= 5pq - r(4p + q)$$

(d) $5lmn - 6ml + 7lm + 8mnl - 4ln = 13lmn + lm - 4ln$
$$= l(13mn + m - 4n)$$
$$= l(m[13n + 1] - 4n) \text{ or}$$
$$= l(n[13m - 4] + m)$$

3 (a) $8x(y - z) + 2y(7x + z) = 8xy - 8xz + 14xy + 2yz$
$$= 22xy - 8xz + 2yz$$
$$= 2(x[11y - 4z] + yz)$$

(b) $(3a - b)(b - 3a) + b^2 = 3a(b - 3a) - b(b - 3a) + b^2$
$$= 3ab - 9a^2 - b^2 + 3ab + b^2$$
$$= 6ab - 9a^2$$
$$= 3a(2b - 3a)$$

(c) $-3(w - 7[x - 8(3 - z)]) = -3(w - 7[x - 24 + 8z])$
$$= -3(w - 7x + 168 - 56z)$$
$$= -3w + 21x - 504 + 168z$$

(d) $\dfrac{2a-3}{4b}+\dfrac{3a+2}{6b}=\dfrac{2a}{4b}-\dfrac{3}{4b}+\dfrac{3a}{6b}+\dfrac{2}{6b}$

$\qquad\qquad\qquad = \dfrac{a}{2b}-\dfrac{3}{4b}+\dfrac{a}{2b}+\dfrac{1}{3b}$

$\qquad\qquad\qquad = \dfrac{a}{b}-\dfrac{5}{12b}$

$\qquad\qquad\qquad = \dfrac{1}{12b}(12a-5)$

Now for the Review test

Review test
Unit 7

1 Simplify each of the following:

    (a) $3pq+5pr-2qr+qp-6rp$

    (b) $5l^2mn+2nl^2m-3mln^2+l^2nm+4n^2ml-nm^2$

    (c) $w^4 \div w^{-a} \times w^{-b}$

    (d) $\dfrac{(s^{\frac{1}{3}})^{\frac{3}{4}} \times (t^{\frac{1}{4}})^{-1} \div (s^{\frac{2}{7}})^{-\frac{7}{4}}}{(s^{-\frac{1}{4}})^{-1} \div (t^{\frac{1}{2}})^4}$

2 Remove the brackets in each of the following:

    (a) $-4x(2x-y)(3x+2y)$

    (b) $(a-2b)(2a-3b)(3a-4b)$

    (c) $-\{-2[x-3(y-4)]-5(z+6)\}$

Powers
Unit 8

Powers
1

The use of *powers* (also called *indices* or *exponents*) provides a convenient form of algebraic shorthand. Repeated factors of the same base, for example $a \times a \times a \times a$ can be written as a^4, where the number 4 indicates the number of factors multiplied together. In general, the product of n such factors a, where a and n are positive integers, is written a^n, where a is called the *base* and n is called the *index* or *exponent* or *power*. Any number multiplying a^n is called the *coefficient* (as described in Unit 7, Frame 8).

$$5a^3 \;\longleftarrow\text{index or exponent or power}$$

coefficient $\nearrow$ $\uparrow$
                  base

From the definitions above a number of rules of indices can immediately be established.

Rules of indices

1 $a^m \times a^n = a^{m+n}$
e.g. $a^5 \times a^2 = a^{5+2} = a^7$
2 $a^m \div a^n = a^{m-n}$
e.g. $a^5 \div a^2 = a^{5-2} = a^3$
3 $(a^m)^n = a^{mn}$
e.g. $(a^5)^2 = a^5 \times a^5 = a^{10}$

These three basic rules lead to a number of important results.

4 $a^0 = 1$ 	because $a^m \div a^n = a^{m-n}$ and also $a^m \div a^n = \dfrac{a^m}{a^n}$

Then if $n = m$, $a^{m-m} = a^0$ and $\dfrac{a^m}{a^m} = 1$. So $a^0 = 1$

5 $a^{-m} = \dfrac{1}{a^m}$ because $a^{-m} = \dfrac{a^{-m} \times a^m}{a^m} = \dfrac{a^0}{a^m} = \dfrac{1}{a^m}$. So $a^{-m} = \dfrac{1}{a^m}$

6 $a^{\frac{1}{m}} = \sqrt[m]{a}$ because $\left(a^{\frac{1}{m}}\right)^m = a^{\frac{m}{m}} = a^1 = a$. So $a^{\frac{1}{m}} = \sqrt[m]{a}$

From this it follows that $a^{\frac{n}{m}} = \sqrt[m]{a^n}$ or $\left(\sqrt[m]{a}\right)^n$.

Make a note of any of these results that you may be unsure about.

Then move on to the next frame

2

So we have:

(a) $a^m \times a^n = a^{m+n}$ 	(e) $a^{-m} = \dfrac{1}{a^m}$

(b) $a^m \div a^n = a^{m-n}$ 	(f) $a^{\frac{1}{m}} = \sqrt[m]{a}$

(c) $(a^m)^n = a^{mn}$ 	(g) $a^{\frac{n}{m}} = \left(\sqrt[m]{a}\right)^n$

(d) $a^0 = 1$ 	 	 	 	or $\sqrt[m]{a^n}$

Now try to apply the rules:

$$\dfrac{6x^{-4} \times 2x^3}{8x^{-3}} = \ldots\ldots\ldots\ldots$$

3

$$\boxed{\tfrac{3}{2}x^2}$$

Because $\dfrac{6x^{-4} \times 2x^3}{8x^{-3}} = \dfrac{12}{8} \cdot \dfrac{x^{-4+3}}{x^{-3}} = \dfrac{12}{8} \cdot \dfrac{x^{-1}}{x^{-3}} = \dfrac{3}{2}x^{-1+3} = \dfrac{3}{2}x^2$

That was easy enough. In the same way:

Simplify $E = (5x^2 y^{-\frac{3}{2}} z^{\frac{1}{4}})^2 \times (4x^4 y^2 z)^{-\frac{1}{2}}$

$E = \ldots\ldots\ldots\ldots$

$$\frac{25x^2}{2y^4}$$

$$E = 25x^4 y^{-3} z^{\frac{1}{2}} \times 4^{-\frac{1}{2}} x^{-2} y^{-1} z^{-\frac{1}{2}}$$

$$= 25x^4 y^{-3} z^{\frac{1}{2}} \times \frac{1}{2} x^{-2} y^{-1} z^{-\frac{1}{2}}$$

$$= \frac{25}{2} x^2 y^{-4} z^0 = \frac{25}{2} x^2 y^{-4} . 1 = \frac{25x^2}{2y^4}$$

And one more:

Simplify $F = \sqrt[3]{a^6 b^3} \div \sqrt{\frac{1}{9} a^4 b^6} \times \left(4\sqrt{a^6 b^2}\right)^{-\frac{1}{2}}$ giving the result without fractional indices.

$$F = \dots\dots\dots$$

$$\frac{3}{2ab^2 \sqrt{ab}}$$

$$F = a^2 b \div \frac{1}{3} a^2 b^3 \times \frac{1}{(4a^3 b)^{\frac{1}{2}}} = a^2 b \times \frac{3}{a^2 b^3} \times \frac{1}{2a^{\frac{3}{2}} b^{\frac{1}{2}}}$$

$$= \frac{3}{b^2} \times \frac{1}{2a^{\frac{3}{2}} b^{\frac{1}{2}}} = \frac{3}{b^2 . 2a(ab)^{\frac{1}{2}}} = \frac{3}{2ab^2 \sqrt{ab}}$$

Logarithms

Powers

Any real number can be written as another number raised to a power. For example:

$9 = 3^2$ and $27 = 3^3$

By writing numbers in the form of a number raised to a power some of the arithmetic operations can be performed in an alternative way. For example:

$$9 \times 27 = 3^2 \times 3^3$$
$$= 3^{2+3}$$
$$= 3^5$$
$$= 243$$

Here the process of multiplication is replaced by the process of relating numbers to powers and then adding the powers.

If there were a simple way of relating numbers such as 9 and 27 to powers of 3 and then relating powers of 3 to numbers such as 243, the process of multiplying two numbers could be converted to the simpler process of adding two powers. In the past a system based on this reasoning was created. It was done using tables that were constructed of numbers and their respective powers.

In this instance:

Number	*Power of 3*
1	0
3	1
9	2
27	3
81	4
243	5
. . .	. . .

They were not called tables of powers but tables of *logarithms*. Nowadays, calculators have superseded the use of these tables but the logarithm remains an essential concept.

Let's just formalize this

7 Logarithms

If a, b and c are three real numbers where:

$a = b^c$ and $b > 1$

the power c is called the *logarithm* of the number a to the base b and is written:

$c = \log_b a$ spoken as c *is the log of a to the base b*

For example, because

$25 = 5^2$

the power 2 is the logarithm of 25 to the base 5. That is:

$2 = \log_5 25$

So in each of the following what is the value of x, remembering that if $a = b^c$ then $c = \log_b a$?

(a) $x = \log_2 16$
(b) $4 = \log_x 81$
(c) $2 = \log_7 x$

The answers are in the next frame

8

$$\boxed{\begin{array}{l} \text{(a)} \ \ x = 4 \\ \text{(b)} \ \ x = 3 \\ \text{(c)} \ \ x = 49 \end{array}}$$

Because

(a) If $x = \log_2 16$ then $2^x = 16 = 2^4$ and so $x = 4$

(b) If $4 = \log_x 81$ then $x^4 = 81 = 3^4$ and so $x = 3$

(c) If $2 = \log_7 x$ then $7^2 = x = 49$

Move on to the next frame

Rules of logarithms

9

Since logarithms are powers, the rules that govern the manipulation of logarithms closely follow the rules of powers.

(a) If $x = a^b$ so that $b = \log_a x$ and
$\quad\quad y = a^c$ so that $c = \log_a y$ then:

$xy = a^b a^c = a^{b+c}$ hence $\log_a xy = b + c = \log_a x + \log_a y$. That is:

$\log_a xy = \log_a x + \log_a y$
The log of a product equals the sum of the logs

(b) Similarly $x \div y = a^b \div a^c = a^{b-c}$ so that $\log_a(x \div y) = b - c = \log_a x - \log_a y$
That is:

$\log_a(x \div y) = \log_a x - \log_a y$
The log of a quotient equals the difference of the logs

(c) Because $x^n = (a^b)^n = a^{bn}$, $\log_a(x^n) = bn = n \log_a x$. That is:

$\log_a(x^n) = n \log_a x$
The log of a number raised to a power is the product of the power and the log of the number

The following important results are also obtained from these rules:

(d) $\log_a 1 = 0$ because, from the laws of powers $a^0 = 1$. Therefore, from the definition of a logarithm $\log_a 1 = 0$

(e) $\log_a a = 1$ because $a^1 = a$ so that $\log_a a = 1$

(f) $\log_a a^x = x$ because $\log_a a^x = x \log_a a = x.1$ so that $\log_a a^x = x$

(g) $a^{\log_a x} = x$ because if we take the log of the left-hand side of this equation:

$\log_a a^{\log_a x} = \log_a x \log_a a = \log_a x$ so that $a^{\log_a x} = x$

(h) $\log_a b = \dfrac{1}{\log_b a}$ because, if $\log_b a = c$ then $b^c = a$ and so $b = \sqrt[c]{a} = a^{\frac{1}{c}}$

Hence, $\log_a b = \dfrac{1}{c} = \dfrac{1}{\log_b a}$. That is $\log_a b = \dfrac{1}{\log_b a}$

▷

So, cover up the results above and complete the following

(a) $\log_a(x \times y) = \ldots\ldots\ldots$ (e) $\log_a a = \ldots\ldots\ldots$

(b) $\log_a(x \div y) = \ldots\ldots\ldots$ (f) $\log_a a^x = \ldots\ldots\ldots$

(c) $\log_a(x^n) = \ldots\ldots\ldots$ (g) $a^{\log_a x} = \ldots\ldots\ldots$

(d) $\log_a 1 = \ldots\ldots\ldots$ (h) $\dfrac{1}{\log_b a} = \ldots\ldots\ldots$

10

(a) $\log_a x + \log_a y$	(e) 1
(b) $\log_a x - \log_a y$	(f) x
(c) $n \log_a x$	(g) x
(d) 0	(h) $\log_a b$

Now try it with numbers. Complete the following:

(a) $\log_a(6{\cdot}788 \times 1{\cdot}043) = \ldots\ldots\ldots$ (e) $\log_7 7 = \ldots\ldots\ldots$

(b) $\log_a(19{\cdot}112 \div 0{\cdot}054) = \ldots\ldots\ldots$ (f) $\log_3 27 = \ldots\ldots\ldots$

(c) $\log_a(5{\cdot}889^{1{\cdot}2}) = \ldots\ldots\ldots$ (g) $12^{\log_{12} 4} = \ldots\ldots\ldots$

(d) $\log_8 1 = \ldots\ldots\ldots$ (h) $\dfrac{1}{\log_3 4} = \ldots\ldots\ldots$

11

(a) $\log_a 6{\cdot}788 + \log_a 1{\cdot}043$	(e) 1
(b) $\log_a 19{\cdot}112 - \log_a 0{\cdot}054$	(f) 3
(c) $12 \log_a 5{\cdot}889$	(g) 4
(d) 0	(h) $\log_4 3$

Next frame

12   **Base 10 and base *e***

On a typical calculator there are buttons that provide access to logarithms to two different bases, namely 10 and the exponential number $e = 2{\cdot}71828\ldots$.

Logarithms to base 10 were commonly used in conjunction with tables for arithmetic calculations – they are called *common logarithms* and are written without indicating the base. For example:

$\log_{10} 1{\cdot}2345$ is normally written simply as $\log 1{\cdot}2345$

You will see it on your calculator as .

The logarithms to base *e* are called *natural logarithms* and are important for their mathematical properties. These also are written in an alternative form:

$\log_e 1{\cdot}2345$ is written as $\ln 1{\cdot}2345$

You will see it on your calculator as .

So, use your calculator and complete the following (to 3 dp):

(a) $\log 5{\cdot}321 = \ldots\ldots\ldots$ (e) $\ln 13{\cdot}45 = \ldots\ldots\ldots$

(b) $\log 0{\cdot}278 = \ldots\ldots\ldots$ (f) $\ln 0{\cdot}278 = \ldots\ldots\ldots$

(c) $\log 1 = \ldots\ldots\ldots$ (g) $\ln 0{\cdot}00001 = \ldots\ldots\ldots$

(d) $\log(-1{\cdot}005) = \ldots\ldots\ldots$ (h) $\ln(-0{\cdot}001) = \ldots\ldots\ldots$

The answers are in the next frame

13

(a) $0{\cdot}726$	(e) $2{\cdot}599$
(b) $-0{\cdot}556$	(f) $-1{\cdot}280$
(c) 0	(g) $-11{\cdot}513$
(d) E	(h) E

Notice that for any base the:

logarithm of 1 is zero
logarithm of 0 is not defined
logarithm of a number greater than 1 is positive
logarithm of a number between 0 and 1 is negative
logarithm of a negative number cannot be evaluated as a real number.

Move to the next frame

Change of base

14

In the previous two frames you saw that $\log 0{\cdot}278 \neq \ln 0{\cdot}278$, i.e. logarithms with different bases have different values. The different values are, however, related to each other as can be seen from the following:

Let $a = b^c$ so that $c = \log_b a$ and let $x = a^d$ so that $d = \log_a x$. Now:

$x = a^d = (b^c)^d = b^{cd}$ so that $cd = \log_b x$. That is:

$\log_b a \log_a x = \log_b x$

This is the change of base formula which relates the logarithms of a number relative to two different bases. For example:

$\log_e 0{\cdot}278 = -1{\cdot}280$ to 3 dp and
$\log_e 10 \times \log_{10} 0{\cdot}278 = 2{\cdot}303 \times (-0{\cdot}556) = -1{\cdot}280$ which confirms that:

$\log_e 10 \log_{10} 0{\cdot}278 = \log_e 0{\cdot}278$

Now, use your calculator to complete each of the following (to 3 dp):

(a) $\log_2 3{\cdot}66 = \ldots\ldots\ldots$ (c) $\log_{9{\cdot}9} 6{\cdot}35 = \ldots\ldots\ldots$

(b) $\log_{3{\cdot}4} 0{\cdot}293 = \ldots\ldots\ldots$ (d) $\log_{7{\cdot}34} 7{\cdot}34 = \ldots\ldots\ldots$

15

| (a)   1·872 | (b)   −1·003 | (c)   0·806 | (d)   1 |

Because

(a) $(\log_{10} 2) \times (\log_2 3{\cdot}66) = \log_{10} 3{\cdot}66$ so that

$$\log_2 3{\cdot}66 = \frac{\log_{10} 3{\cdot}66}{\log_{10} 2} = \frac{0{\cdot}563\ldots}{0{\cdot}301\ldots} = 1{\cdot}872$$

(b) $(\log_{10} 3{\cdot}4) \times (\log_{3{\cdot}4} 0{\cdot}293) = \log_{10} 0{\cdot}293$ so that

$$\log_{3{\cdot}4} 0{\cdot}293 = \frac{\log_{10} 0{\cdot}293}{\log_{10} 3{\cdot}4} = \frac{-0{\cdot}533\ldots}{0{\cdot}531\ldots} = -1{\cdot}003$$

(c) $(\log_{10} 9{\cdot}9) \times (\log_{9{\cdot}9} 6{\cdot}35) = \log_{10} 6{\cdot}35$ so that

$$\log_{9{\cdot}9} 6{\cdot}35 = \frac{\log_{10} 6{\cdot}35}{\log_{10} 9{\cdot}9} = \frac{0{\cdot}802\ldots}{0{\cdot}995\ldots} = 0{\cdot}806$$

(d) $\log_{7{\cdot}34} 7{\cdot}34 = 1$ because for any base a, $\log_a a = 1$.

16   Logarithmic equations

The following four examples serve to show you how logarithmic expressions and equations can be manipulated.

Example 1

Simplify the following:

$$\log_a x^2 + 3\log_a x - 2\log_a 4x$$

Solution

$$\log_a x^2 + 3\log_a x - 2\log_a 4x = \log_a x^2 + \log_a x^3 - \log_a(4x)^2$$
$$= \log_a \left(\frac{x^2 x^3}{16x^2}\right)$$
$$= \log_a \left(\frac{x^3}{16}\right)$$

Example 2

Solve the following for x:

$$2\log_a x - \log_a(x-1) = \log_a(x-2)$$

Solution

$$\text{LHS} = 2\log_a x - \log_a(x-1)$$
$$= \log_a x^2 - \log_a(x-1)$$
$$= \log_a \left(\frac{x^2}{x-1}\right)$$
$$= \log_a(x-2) \text{ so that } \frac{x^2}{x-1} = x-2. \text{ That is:}$$

$$x^2 = (x-2)(x-1) = x^2 - 3x + 2 \text{ so that } -3x+2 = 0 \text{ giving } x = \frac{2}{3}$$

Example 3

Find y in terms of x:

$$5\log_a y - 2\log_a(x+4) = 2\log_a y + \log_a x$$

Solution

$$5\log_a y - 2\log_a(x+4) = 2\log_a y + \log_a x \text{ so that}$$
$$5\log_a y - 2\log_a y = \log_a x + 2\log_a(x+4) \text{ that is}$$
$$\log_a y^5 - \log_a y^2 = \log_a x + \log_a(x+4)^2 \text{ that is}$$
$$\log_a\left(\frac{y^5}{y^2}\right) = \log_a y^3 = \log_a x(x+4)^2 \text{ so that } y^3 = x(x+4)^2 \text{ hence}$$
$$y = \sqrt[3]{x(x+4)^2}$$

Example 4

For what values of x is $\log_a(x-3)$ valid?

Solution

$\log_a(x-3)$ is valid for $x - 3 > 0$, that is $x > 3$

Now you try some

1 Simplify $2\log_a x - 3\log_a 2x + \log_a x^2$

17

2 Solve the following for x:
$$4\log_a \sqrt{x} - \log_a 3x = \log_a x^{-2}$$

3 Find y in terms of x where:
$$2\log_a y - 3\log_a(x^2) = \log_a \sqrt{y} + \log_a x$$

Next frame for the answers

1 $\quad 2\log_a x - 3\log_a 2x + \log_a x^2 = \log_a x^2 - \log_a(2x)^3 + \log_a x^2$

18

$$= \log_a\left(\frac{x^2 x^2}{8x^3}\right)$$
$$= \log_a\left(\frac{x}{8}\right)$$

2 $\quad \text{LHS} = 4\log_a \sqrt{x} - \log_a 3x$

$$= \log_a\left(\sqrt{x}\right)^4 - \log_a 3x$$
$$= \log_a x^2 - \log_a 3x$$
$$= \log_a\left(\frac{x^2}{3x}\right)$$
$$= \log_a\left(\frac{x}{3}\right)$$
$$= \log_a x^{-2} \qquad \text{the right-hand side of the equation.}$$

So that:
$$x^{-2} = \frac{x}{3}, \text{ that is } x^3 = 3 \text{ giving } x = \sqrt[3]{3}$$

3 $2\log_a y - 3\log_a(x^2) = \log_a \sqrt{y} + \log_a x$, that is

$\log_a y^2 - \log_a(x^2)^3 = \log_a y^{\frac{1}{2}} + \log_a x$ so that

$\log_a y^2 - \log_a y^{\frac{1}{2}} = \log_a(x^2)^3 + \log_a x$ giving

$\log_a \dfrac{y^2}{y^{\frac{1}{2}}} = \log_a x^6.x$. Consequently $y^{\frac{3}{2}} - x^7$ and so $y - \sqrt[3]{x^{14}}$

At this point let us pause and summarize the main facts so far on powers and logarithms

---

 # Review summary   Unit 8

19 **1** Rules of powers:

(a) $a^m \times a^n = a^{m+n}$ (b) $a^{-m} = \dfrac{1}{a^m}$

(c) $a^m \div a^n = a^{m-n}$ (d) $a^{\frac{1}{m}} = \sqrt[m]{a}$

(e) $(a^m)^n = a^{mn}$ (f) $a^{\frac{n}{m}} = \sqrt[m]{a^n}$ or $a^{\frac{n}{m}} = (\sqrt[m]{a})^n$

(g) $a^0 = 1$

2 Rules of logarithms:

(a) $\log_a xy = \log_a x + \log_a y$ *The log of a product equals the sum of the logs*

(b) $\log_a(x \div y) = \log_a x - \log_a y$ *The log of a quotient equals the difference of the logs*

(c) $\log_a(x^n) = n\log_a x$ *The log of a number raised to a power is the product of the power and the log of the number*

(d) $\log_a 1 = 0$

(e) $\log_a a = 1$ and $\log_a a^x = x$

(f) $a^{\log_a x} = x$

(g) $\log_a b = \dfrac{1}{\log_b a}$

3 Logarithms to base 10 are called *common logarithms* and are written as $\log x$.

4 Logarithms to base $e = 2{\cdot}71828\ldots$ are called *natural logarithms* and are written as $\ln x$.

Review exercise

Unit 8

1 Simplify each of the following:

20

(a) $a^6 \times a^5$

(b) $x^7 \div x^3$

(c) $(w^2)^m \div (w^m)^3$

(d) $s^3 \div t^{-4} \times (s^{-3}t^{-2})^3$

(e) $\dfrac{8x^{-3} \times 3x^2}{6x^{-4}}$

(f) $(4a^3b^{-1}c)^2 \times (a^{-2}b^4c^{-2})^{\frac{1}{2}} \div \left[64(a^6b^4c^2)^{-\frac{1}{2}}\right]$

(g) $\sqrt[3]{8a^3b^6} \div \sqrt{\dfrac{1}{25}a^4b^7} \times \left(16\sqrt{a^4b^6}\right)^{-\frac{1}{2}}$

2 Express the following without logs:

(a) $\log K = \log P - \log T + 1{\cdot}3\log V$ (b) $\ln A = \ln P + rn$

3 Rewrite $R = r\sqrt{\dfrac{f+P}{f-P}}$ in log form.

4 Evaluate by calculator or by change of base where necessary (to 3 dp):

(a) $\log 5{\cdot}324$ (b) $\ln 0{\cdot}0023$ (c) $\log_4 1{\cdot}2$

Complete all four questions. Take your time, there is no need to rush.
If necessary, look back at the Unit.
The answers and working are in the next frame.

---

1 (a) $a^6 \times a^5 = a^{6+5} = a^{11}$

21

(b) $x^7 \div x^3 = x^{7-3} = x^4$

(c) $(w^2)^m \div (w^m)^3 = w^{2m} \div w^{3m} = w^{2m} \times w^{-3m} = w^{-m}$

(d) $s^3 \div t^{-4} \times (s^{-3}t^{-2})^3 = s^3 \times t^4 \times s^{-9}t^{-6} = s^{-6}t^{-2}$

(e) $\dfrac{8x^{-3} \times 3x^2}{6x^{-4}} = \dfrac{24x^{-1}}{6x^{-4}} = 4x^3$

(f) $(4a^3b^{-1}c)^2 \times (a^{-2}b^4c^{-2})^{\frac{1}{2}} \div 64(a^6b^4c^2)^{-\frac{1}{2}}$

$= (16a^6b^{-2}c^2) \times (a^{-1}b^2c^{-1}) \div 64(a^{-3}b^{-2}c^{-1})$

$= (16a^6b^{-2}c^2) \times (a^{-1}b^2c^{-1}) \times 64^{-1}(a^3b^2c^1)$

$= \dfrac{a^8b^2c^2}{4}$

(g) $\sqrt[3]{8a^3b^6} \div \sqrt{\dfrac{1}{25}a^4b^7} \times \left(16\sqrt{a^4b^6}\right)^{-\frac{1}{2}} = (2ab^2) \div \dfrac{a^2b^{\frac{7}{2}}}{5} \times \left(4ab^{\frac{3}{2}}\right)^{-1}$

$= (2ab^2) \times \dfrac{5}{a^2b^{\frac{7}{2}}} \times \dfrac{1}{4ab^{\frac{3}{2}}}$

$= \dfrac{5ab^2}{2a^2b^{\frac{7}{2}}ab^{\frac{3}{2}}} = \dfrac{5}{2a^2b^3}$

2 (a) $K = \dfrac{PV^{1\cdot3}}{T}$ (b) $A = Pe^{rn}$

3 $\log R = \log r + \dfrac{1}{2}(\log(f+P) - \log(f-P))$

4 (a) 0·726   (b) −6·075   (c) 0·132

Now for the Review test

 # Review test   Unit 8

22

1 Evaluate by calculator or by change of base where necessary (to 3 dp):
   (a) $\log 0\cdot0101$
   (b) $\ln 3\cdot47$
   (c) $\log_2 3\cdot16$

2 Express $\log F = \log G + \log m - \log\left(\dfrac{1}{M}\right) - 2\log r$ without logs.

3 Rewrite $T = 2\pi\sqrt{\dfrac{l}{g}}$ in log form.

Algebraic multiplication and division   Unit 9

1 **Multiplication of algebraic expressions of a single variable**

Example 1

$$(x+2)(x+3) = x(x+3) + 2(x+3)$$
$$= x^2 + 3x + 2x + 6$$
$$= x^2 + 5x + 6$$

Now a slightly harder one

Example 2

2

$(2x + 5)(x^2 + 3x + 4)$

Each term in the second expression is to be multiplied by $2x$ and then by 5 and the results added together, so we set it out thus:

$$x^2 + 3x + 4$$
$$2x + 5$$

Multiply throughout by $2x$ $2x^3 + \ \ 6x^2 + \ \ 8x$
Multiply by 5 $5x^2 + 15x + 20$

Add the two lines $2x^3 + 11x^2 + 23x + 20$

So $(2x + 5)(x^2 + 3x + 4) = 2x^3 + 11x^2 + 23x + 20$

Be sure to keep the same powers of the variable in the same column.

Next frame

Now look at this one.

3

Example 3

Determine $(2x + 6)(4x^3 - 5x - 7)$

You will notice that the second expression is a cubic (highest power x^3), but that there is no term in x^2. In this case, we insert $0x^2$ in the working to keep the columns complete, that is:

$$4x^3 + 0x^2 - 5x - 7$$
$$2x \ + 6$$

which gives ....

Finish it

$$8x^4 + 24x^3 - 10x^2 - 44x - 42$$

4

Here it is set out: $4x^3 + 0x^2 - 5x - 7$
                            $2x + 6$

$$8x^4 + \ \ 0x^3 - 10x^2 - 14x$$
$$24x^3 + \ \ 0x^2 - 30x - 42$$

$$8x^4 + 24x^3 - 10x^2 - 44x - 42$$

They are all done in the same way, so here is one more for practice.

Example 4

Determine the product $(3x - 5)(2x^3 - 4x^2 + 8)$

You can do that without any trouble. The product is ....

5

$$6x^4 - 22x^3 + 20x^2 + 24x - 40$$

All very straightforward:

$$2x^3 - 4x^2 + 0x + 8$$
$$3x - 5$$

$$6x^4 - 12x^3 + 0x^2 + 24x$$
$$- 10x^3 + 20x^2 + 0x - 40$$

$$6x^4 - 22x^3 + 20x^2 + 24x - 40$$

Fractions

6   **Algebraic fractions**

A numerical fraction is represented by one integer divided by another. Division of symbols follows the same rules to create *algebraic fractions*. For example:

$5 \div 3$ can be written as the fraction $\dfrac{5}{3}$ so $a \div b$ can be written as $\dfrac{a}{b}$

Addition and subtraction

The addition and subtraction of algebraic fractions follow the same rules as the addition and subtraction of numerical fractions – the operations can only be performed when the denominators are the same. For example, just as:

$$\frac{4}{5} + \frac{3}{7} = \frac{4 \times 7}{5 \times 7} + \frac{3 \times 5}{7 \times 5} = \frac{4 \times 7 + 3 \times 5}{5 \times 7}$$

$$= \frac{43}{35} \qquad \text{(where 35 is the LCM of 5 and 7)}$$

so:

$$\frac{a}{b} + \frac{c}{d} = \frac{a \times d}{b \times d} + \frac{c \times b}{d \times b} = \frac{ad + cb}{bd} \text{ provided } b \neq 0 \text{ and } d \neq 0$$

(where bd is the LCM of b and d)

So that:

$$\frac{a}{b} - \frac{c}{d^2} + \frac{d}{a} = \dots\dots\dots$$

Answer in the next frame

7

$$\frac{a^2d^2 - abc + bd^3}{abd^2}$$

Because

$$\frac{a}{b} - \frac{c}{d^2} + \frac{d}{a} = \frac{aad^2}{bad^2} - \frac{cab}{d^2ab} + \frac{dd^2b}{ad^2b}$$

$$= \frac{a^2d^2 - abc + bd^3}{abd^2}$$

where abd^2 is the LCM of a, b and d^2.

On now to the next frame

8

In the same way

$$\frac{2}{x+1} + \frac{4}{x+2} = \frac{2(x+2) + 4(x+1)}{(x+1)(x+2)}$$

$$= \frac{2x + 4 + 4x + 4}{(x+1)(x+2)}$$

$$= \frac{6x + 8}{x^2 + 3x + 2}$$

Multiplication and division

9

Fractions are multiplied by multiplying their numerators and denominators separately. For example, just as:

$$\frac{5}{4} \times \frac{3}{7} = \frac{5 \times 3}{4 \times 7} = \frac{15}{28}$$

so:

$$\frac{a}{b} \times \frac{c}{d} = \frac{ac}{bd}$$

The *reciprocal* of a number is unity divided by the number. For example, the reciprocal of a is $1/a$ and the reciprocal of $\frac{a}{b}$ is:

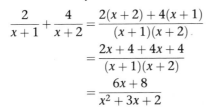

$$\frac{1}{a/b} = 1 \div \frac{a}{b} = 1 \times \frac{b}{a} = \frac{b}{a} \qquad \text{\textit{the numerator and denominator in the divisor are interchanged}}$$

To divide by an algebraic fraction we multiply by its reciprocal. For example:

$$\frac{a}{b} \div \frac{c}{d} = \frac{a}{b} \times \frac{d}{c} = \frac{ad}{bc}$$

So that $\dfrac{2a}{3b} \div \dfrac{a^2b}{6} = \ldots\ldots\ldots\ldots$

Check with the next frame

10

$$\frac{4}{ab^2}$$

Because

$$\frac{2a}{3b} \div \frac{a^2b}{6} = \frac{2a}{3b} \times \frac{6}{a^2b} = \frac{4}{ab^2}$$

Try another one:

$$\frac{2a}{3b} \div \frac{a^2b}{6} \times \frac{ab}{2} = \dots\dots\dots$$

The answer is in the next frame

11

$$\frac{2}{b}$$

Because

$$\frac{2a}{3b} \div \frac{a^2b}{6} \times \frac{ab}{2} = \frac{2a}{3b} \times \frac{6}{a^2b} \times \frac{ab}{2} = \frac{4}{ab^2} \times \frac{ab}{2} = \frac{2}{b}$$

Remember that by the rules of precedence we work through the expression from the left to the right so we perform the division before we multiply. If we were to multiply before dividing in the above expression we should obtain:

$$\frac{2a}{3b} \div \frac{a^2b}{6} \times \frac{ab}{2} = \frac{2a}{3b} \div \frac{a^3b^2}{12}$$
$$= \frac{2a}{3b} \times \frac{12}{a^3b^2}$$
$$= \frac{24a}{3a^3b^3} = \frac{8}{a^2b^3}$$

and this would be wrong.

In the following frames we extend this idea

Division of one expression by another

Let us consider $(12x^3 - 2x^2 - 3x + 28) \div (3x + 4)$. The result of this division is called the *quotient* of the two expressions and we find the quotient by setting out the division in the same way as we do for the long division of numbers:

12

$$3x + 4 \overline{\smash{\big)}\, 12x^3 \ - \ 2x^2 \ - 3x \ + 28}$$

To make $12x^3$, $3x$ must be multiplied by $4x^2$, so we insert this as the first term in the quotient, multiply the divisor $(3x + 4)$ by $4x^2$, and subtract this from the first two terms:

$$
\begin{array}{r}
4x^2 \\
3x + 4 \overline{\smash{\big)}\, 12x^3 \ - \ 2x^2 \ - 3x \ + 28} \\
12x^3 \ + 16x^2 \\
\hline
- 18x^2 \ - 3x
\end{array}
$$

Bring down the next term $(-3x)$ and repeat the process

To make $-18x^2$, $3x$ must be multiplied by $-6x$, so do this and subtract as before, not forgetting to enter the $-6x$ in the quotient. Do this and we get . . . .

13

$$
\begin{array}{r}
4x^2 \ - \ 6x \\
3x + 4 \overline{\smash{\big)}\, 12x^3 \ - \ 2x^2 \ - \ 3x \ + 28} \\
12x^3 \ + 16x^2 \\
\hline
- 18x^2 \ - \ 3x \\
- 18x^2 \ - 24x \\
\hline
21x
\end{array}
$$

Now bring down the next term and continue in the same way and finish it off.

So $(12x^3 - 2x^2 - 3x + 28) \div (3x + 4) = \ldots \ldots \ldots$

14

$$\boxed{4x^2 - 6x + 7}$$

As before, if an expression has a power missing, insert the power with zero coefficient. Now you can determine $(4x^3 + 13x + 33) \div (2x + 3)$

Set it out as before and check the result with the next frame

15

$$\boxed{2x^2 - 3x + 11}$$

Here it is:

$$
\begin{array}{r}
2x^2 \ - 3x \ +11 \\
2x+3 \overline{\smash{\big)}\ 4x^3 \ - 0x^2 + 13x + 33} \\
\underline{4x^3 \ + 6x^2} \\
-6x^2 + 13x \\
\underline{-6x^2 - 9x} \\
22x + 33 \\
\underline{22x + 33} \\
\cdot \qquad \cdot
\end{array}
$$

So $(4x^3 + 13x + 33) \div (2x + 3) = 2x^2 - 3x + 11$

And one more. Determine $(6x^3 - 7x^2 + 1) \div (3x + 1)$

Setting out as before, the quotient is . . . .

16

$$\boxed{2x^2 - 3x + 1}$$

After inserting the x term with zero coefficient, the rest is straightforward.

At this point let us pause and summarize the main facts so far for algebraic multiplication and division

 # Review summary   Unit 9

17   1   *Multiplication of algebraic expressions*
Two algebraic expressions are multiplied together by successively multiplying the second expression by each term of the first expression.

Long division of algebraic expressions
Two algebraic expressions are divided by setting out the division in the same way as we do for the long division of numbers.

2   The manipulation of algebraic fractions follows identical principles as those for arithmetic fractions.

3   Only fractions with identical denominators can be immediately added or subtracted.

4   Two fractions are multiplied by multiplying their respective numerators and denominators separately.

5   Two fractions are divided by multiplying the numerator fraction by the reciprocal of the divisor fraction.

Review exercise

1 Perform the following multiplications and simplify your results:

18

(a) $(2a + 4b)(a - 3b)$ (b) $(8x - 4)(4x^2 - 3x + 2)$

(c) $(9s^2 + 3)(s^2 - 4)$ (d) $(2x + 3)(5x^3 + 3x - 4)$

2 Simplify each of the following into a single algebraic fraction:

(a) $\dfrac{ab}{c} + \dfrac{cb}{a}$ (b) $\dfrac{ab}{c} - 1$ (c) $\left(\dfrac{ab}{c} + \dfrac{ac}{b}\right) + \dfrac{bc}{a}$

3 Perform the following divisions:

(a) $(x^2 + 5x - 6) \div (x - 1)$ (b) $(x^2 - x - 2) \div (x + 1)$

(c) $(12x^3 - 11x^2 - 25) \div (3x - 5)$ (d) $\dfrac{a^3 + 8b^3}{a + 2b}$

Complete the questions. Take one step at a time, there is no need to rush.
If you need to, look back at the Unit.
The answers and working are in the next frame.

---

1 (a) $(2a + 4b)(a - 3b) = 2a(a - 3b) + 4b(a - 3b) = 2a^2 - 2ab - 12b^2$

19

(b) $(8x - 4)(4x^2 - 3x + 2) = 8x(4x^2 - 3x + 2) - 4(4x^2 - 3x + 2)$

$$= 32x^3 - 24x^2 + 16x - 16x^2 + 12x - 8$$

$$= 32x^3 - 40x^2 + 28x - 8$$

(c) $(9s^2 + 3)(s^2 - 4) = 9s^2(s^2 - 4) + 3(s^2 - 4)$

$$= 9s^4 - 36s^2 + 3s^2 - 12$$

$$= 9s^4 - 33s^2 - 12$$

(d) $(2x + 3)(5x^3 + 3x - 4) = 2x(5x^3 + 3x - 4) + 3(5x^3 + 3x - 4)$

$$= 10x^4 + 6x^2 - 8x + 15x^3 + 9x - 12$$

$$= 10x^4 + 15x^3 + 6x^2 + x - 12$$

2 (a) $\dfrac{ab}{c} + \dfrac{cb}{a} = \dfrac{aab}{ac} + \dfrac{cbc}{ac} = \dfrac{b(a^2 + c^2)}{ac}$

(b) $\dfrac{ab}{c} - 1 = \dfrac{ab}{c} - \dfrac{c}{c} = \dfrac{ab - c}{c}$

(c) $\left(\dfrac{ab}{c} + \dfrac{ac}{b}\right) + \dfrac{bc}{a} = \dfrac{ab}{c} + \dfrac{ac}{b} + \dfrac{bc}{a} = \dfrac{a^2b^2 + a^2c^2 + b^2c^2}{abc}$

3 (a)

$$(x^2 + 5x - 6) \div (x - 1) = x - 1 \overline{\smash{\big)}\ x^2 + 5x - 6} \quad \begin{array}{r} x + 6 \\ \hline \end{array}$$

$$\underline{x^2 - x}$$

$$6x - 6$$

$$\underline{6x - 6}$$

$$\bullet \quad \bullet$$

(b)

$$(x^2 - x - 2) \div (x + 1) = x + 1 \quad \begin{array}{r} x - 2 \\ \hline x^2 - x - 2 \\ x^2 + x \\ \hline -2x - 2 \\ -2x - 2 \\ \hline \quad \bullet \quad \bullet \end{array}$$

(c)

$$(12x^3 - 11x^2 - 25) \div (3x - 5) = 3x - 5 \quad \begin{array}{r} 4x^2 + 3x + 5 \\ \hline 12x^3 - 11x^2 + 0x - 25 \\ 12x^3 - 20x^2 \\ \hline 9x^2 + 0x \\ 9x^2 - 15x \\ \hline 15x - 25 \\ 15x - 25 \\ \hline \quad \bullet \quad \bullet \end{array}$$

(d)

$$\frac{a^3 + 8b^3}{a + 2b} = a + 2b \quad \begin{array}{r} a^2 - 2ab + 4b^2 \\ \hline a^3 \qquad\qquad + 8b^3 \\ a^3 + 2a^2b \\ \hline -2a^2b \\ -2a^2b - 4ab^2 \\ \hline 4ab^2 + 8b^3 \\ 4ab^2 + 8b^3 \\ \hline \quad \bullet \quad \bullet \end{array}$$

Now for the Review test

Review test

Unit 9

20

1 Perform the following multiplications and simplify your results:
   (a) $(n^2 + 2n - 3)(4n + 5)$ (b) $(v^3 - v^2 - 2)(1 - 3v + 2v^2)$

2 Simplify each of the following:
   (a) $\dfrac{p}{q^3} \div \dfrac{q}{p^3}$ (b) $\dfrac{a^2b}{2c} \times \dfrac{ac^2}{2b} \div \dfrac{b^2c}{2a}$

3 Perform the following divisions:
   (a) $(2y^2 - y - 10) \div (y + 2)$ (b) $\dfrac{q^3 - 8}{q - 2}$

   (c) $\dfrac{2r^3 + 5r^2 - 4r - 3}{r^2 + 2r - 3}$

Factorization of algebraic expressions

Unit 10

1

An algebraic fraction can often be simplified by writing the numerator and denominator in terms of their factors and cancelling where possible.

For example $\dfrac{25ab^2 - 15a^2b}{40ab^2 - 24a^2b} = \dfrac{5ab(5b - 3a)}{8ab(5b - 3a)} = \dfrac{5}{8}$

This is an obvious example, but there are many uses for factorization of algebraic expressions in advanced processes.

Common factors

The simplest form of factorization is the extraction of highest common factors (HCF) from an expression. For example, $(10x + 8)$ can clearly be written $2(5x + 4)$.

Similarly with $(35x^2y^2 - 10xy^3)$:

the HCF of the coefficients 35 and 10 is 5

the HCF of the powers of x is x

the HCF of the powers of y is y^2

So $(35x^2y^2 - 10xy^3) = 5xy^2(7x - 2y)$

In the same way: (a) $8x^4y^3 + 6x^3y^2 = \ldots\ldots\ldots$

and (b) $15a^3b - 9a^2b^2 = \ldots\ldots\ldots$

2

> (a) $2x^3y^2(4xy + 3)$
>
> (b) $3a^2b(5a - 3b)$

Common factors by grouping

Four-termed expressions can sometimes be factorized by grouping into two binomial expressions and extracting common factors from each.

For example: $2ac + 6bc + ad + 3bd$

$= (2ac + 6bc) + (ad + 3bd) = 2c(a + 3b) + d(a + 3b)$

$= (a + 3b)(2c + d)$

Similarly: $x^3 - 4x^2y + xy^2 - 4y^3 = \ldots\ldots\ldots$

3

$$(x - 4y)(x^2 + y^2)$$

Because

$$x^3 - 4x^2y + xy^2 - 4y^3 = (x^3 - 4x^2y) + (xy^2 - 4y^3)$$
$$= x^2(x - 4y) + y^2(x - 4y) = (x - 4y)(x^2 + y^2)$$

In some cases it might be necessary to rearrange the order of the original four terms. For example:

$$12x^2 - y^2 + 3x - 4xy^2 = 12x^2 + 3x - y^2 - 4xy^2$$
$$= (12x^2 + 3x) - (y^2 + 4xy^2) = 3x(4x + 1) - y^2(1 + 4x)$$
$$= (4x + 1)(3x - y^2)$$

Likewise, $20x^2 - 3y^2 + 4xy^2 - 15x = \ldots\ldots\ldots\ldots$

4

$$(4x - 3)(5x + y^2)$$

Rearranging terms:

$$(20x^2 - 15x) + (4xy^2 - 3y^2) = 5x(4x - 3) + y^2(4x - 3) = (4x - 3)(5x + y^2)$$

5 Useful products of two simple factors

A number of standard results are well worth remembering for the products of simple factors of the form $(a + b)$ and $(a - b)$. These are:

(a) $(a + b)^2 = (a + b)(a + b) = a^2 + ab + ba + b^2$

 i.e. $(a + b)^2 = a^2 + 2ab + b^2$

(b) $(a - b)^2 = (a - b)(a - b) = a^2 - ab - ba + b^2$

 i.e. $(a - b)^2 = a^2 - 2ab + b^2$

(c) $(a - b)(a + b) = a^2 + ab - ba - b^2$

 i.e. $(a - b)(a + b) = a^2 - b^2$ *the difference of two squares*

For our immediate purpose, these results can be used in reverse:

$$a^2 + 2ab + b^2 = (a + b)^2$$
$$a^2 - 2ab + b^2 = (a - b)^2$$
$$a^2 - b^2 = (a - b)(a + b)$$

If an expression can be seen to be one of these forms, its factors can be obtained at once.

These expressions that involve the variables raised to the power 2 are examples of what are called *quadratic* expressions. If a quadratic expression can be seen to be one of these forms, its factors can be obtained at once.

On to the next frame

Remember

$$a^2 + 2ab + b^2 = (a + b)^2$$
$$a^2 - 2ab + b^2 = (a - b)^2$$
$$a^2 - b^2 = (a - b)(a + b)$$

Example 1

$$x^2 + 10x + 25 = (x)^2 + 2(x)(5) + (5)^2, \text{ like } a^2 + 2ab + b^2$$
$$= (x + 5)^2$$
So $x^2 + 10x + 25 = (x + 5)^2$

Example 2

$$4a^2 - 12a + 9 = (2a)^2 - 2(2a)(3) + (3)^2, \text{ like } a^2 - 2ab + b^2$$
$$= (2a - 3)^2$$
So $4a^2 - 12a + 9 = (2a - 3)^2$

Example 3

$$25x^2 - 16y^2 = (5x)^2 - (4y)^2$$
$$= (5x - 4y)(5x + 4y)$$
So $25x^2 - 16y^2 = (5x - 4y)(5x + 4y)$

Now can you factorize the following:

(a) $16x^2 + 40xy + 25y^2 = \ldots\ldots\ldots$
(b) $9x^2 - 12xy + 4y^2 = \ldots\ldots\ldots$
(c) $(2x + 3y)^2 - (x - 4y)^2 = \ldots\ldots\ldots$

7

> (a) $(4x + 5y)^2$
> (b) $(3x - 2y)^2$
> (c) $(x + 7y)(3x - y)$

Quadratic expressions as the product of two simple factors

1 $(x + g)(x + k) = x^2 + (g + k)x + gk$

The coefficient of the middle term is the sum of the two constants g and k and the last term is the product of g and k.

2 $(x - g)(x - k) = x^2 - (g + k)x + gk$

The coefficient of the middle term is minus the sum of the two constants g and k and the last term is the product of g and k.

3 $(x + g)(x - k) = x^2 + (g - k)x - gk$

The coefficient of the middle term is the difference of the two constants g and k and the last term is minus the product of g and k.

Now let's try some specific types of quadratic

8 ## Factorization of a quadratic expression $ax^2 + bx + c$ when $a = 1$

If $a = 1$, the quadratic expression is similar to those you have just considered, that is $x^2 + bx + c$. From rules **1–3** in the previous frame you can see that the values of f_1 and f_2 in $(x + f_1)$ and $(x + f_2)$, the factors of the quadratic expression, will depend upon the signs of b and c. Notice that b, c, f_1 and f_2 can be positive or negative. Notice that:

If c is positive   (a) f_1 and f_2 are factors of c and both have the sign of b

                     (b) the sum of f_1 and f_2 is b

If c is negative   (a) f_1 and f_2 are factors of c and have opposite signs, the numerically larger having the sign of b

                     (b) the difference between f_1 and f_2 is b

There are examples of this in the next frame

9 *Example 1*

$x^2 + 5x + 6$

(a) Possible factors of 6 are (1, 6) and (2, 3), so $(\pm 1, \pm 6)$ and $(\pm 2, \pm 3)$ are possible choices for f_1 and f_2.
(b) c is positive so the required factors add up to b, that is 5.
(c) c is positive so the required factors have the sign of b, that is positive, therefore $(2, 3)$.

$$\text{So } x^2 + 5x + 6 = (x + 2)(x + 3)$$

▶

Example 2

$x^2 - 9x + 20$

(a) Possible factors of 20 are $(1, 20)$, $(2, 10)$ and $(4, 5)$, so $(\pm 1, \pm 20)$, $(\pm 2, \pm 10)$ and $(\pm 4, \pm 5)$ are possible choices for f_1 and f_2.

(b) c is positive so the required factors add up to b, that is -9.

(c) c is positive so the required factors have the sign of b, that is negative, therefore $(-4, -5)$.

$$\text{So } x^2 - 9x + 20 = (x - 4)(x - 5)$$

Example 3

$x^2 + 3x - 10$

(a) Possible factors of 10 are $(1, 10)$ and $(2, 5)$, so $(\pm 1, \pm 10)$ and $(\pm 2, \pm 5)$ are possible choices for f_1 and f_2.

(b) c is negative so the required factors differ by b, that is 3.

(c) c is negative so the required factors differ in sign, the numerically larger having the sign of b, that is positive, therefore $(-2, 5)$.

$$\text{So } x^2 + 3x - 10 = (x - 2)(x + 5)$$

Example 4

$x^2 - 2x - 24$

(a) Possible factors of 24 are $(1, 24)$, $(2, 12)$, $(3, 8)$ and $(4, 6)$, so $(\pm 1, \pm 24)$, $(\pm 2, \pm 12)$, $(\pm 3, \pm 8)$ and $(\pm 4, \pm 6)$ are possible choices for f_1 and f_2.

(b) c is negative so the required factors differ by b, that is -2.

(c) c is negative so the required factors differ in sign, the numerically larger having the sign of b, that is negative, therefore $(4, -6)$.

$$\text{So } x^2 - 2x - 24 = (x + 4)(x - 6)$$

Now, here is a short exercise for practice. Factorize each of the following into two linear factors:

(a) $x^2 + 7x + 12$ (d) $x^2 + 2x - 24$

(b) $x^2 - 11x + 28$ (e) $x^2 - 2x - 35$

(c) $x^2 - 3x - 18$ (f) $x^2 - 10x + 16$

Finish all six and then check with the next frame

10

(a) $(x + 3)(x + 4)$	(d) $(x + 6)(x - 4)$
(b) $(x - 4)(x - 7)$	(e) $(x - 7)(x + 5)$
(c) $(x - 6)(x + 3)$	(f) $(x - 2)(x - 8)$

11 Factorization of a quadratic expression $ax^2 + bx + c$ when $a \neq 1$

If $a \neq 1$, the factorization is slightly more complicated, but still based on the same considerations as for the simpler examples already discussed.

To factorize such an expression into its linear factors, if they exist, we carry out the following steps.

(a) We obtain $|ac|$, i.e. the numerical value of the product ac ignoring the sign of the product.

(b) We write down all the possible pairs of factors of $|ac|$.

(c) (i) *If c is positive,* we select the two factors of $|ac|$ whose sum is equal to $|b|$: both of these factors have the same sign as b.

   (ii) *If c is negative,* we select the two factors of $|ac|$ which differ by the value of $|b|$: the numerically larger of these two factors has the same sign as that of b and the other factor has the opposite sign.

   (iii) In each case, denote the two factors so obtained by f_1 and f_2.

(d) Then $ax^2 + bx + c$ is now written $ax^2 + f_1x + f_2x + c$ and this is factorized by finding common factors by grouping – as in the previous work.

Example 1

To factorize $6x^2 + 11x + 3$ $(ax^2 + bx + c)$

In this case, $a = 6$; $b = 11$; $c = 3$. Therefore $|ac| = 18$

Possible factors of $18 = (1, 18)$, $(2, 9)$ and $(3, 6)$

c is positive. So required factors, f_1 and f_2, add up to ....

12

$$\boxed{|b|, \text{ i.e. } 11}$$

So required factors are $(2, 9)$.

c is positive. Both factors have the same sign as b, i.e. positive.

So $f_1 = 2$; $f_2 = 9$; and $6x^2 + 11x + 3 = 6x^2 + 2x + 9x + 3$

which factorizes by grouping into ....

13

$$\boxed{(2x + 3)(3x + 1)}$$

Because

$$6x^2 + 11x + 3 = 6x^2 + 2x + 9x + 3$$
$$= (6x^2 + 9x) + (2x + 3)$$
$$= 3x(2x + 3) + 1(2x + 3) = (2x + 3)(3x + 1)$$

Now this one.

Example 2

To factorize $3x^2 - 14x + 8$ $(ax^2 + bx + c)$

    $a = 3; b = -14; c = 8; |ac| = 24$

Possible factors of 24 =(1, 24), (2, 12), (3, 8) and (4, 6)

c is positive. So required factors total $|b|$, i.e. 14. Therefore (2, 12)

c is positive. So factors have same sign as b, i.e. negative, $f_1 = -2; f_2 = -12$

$$\text{So } 3x^2 - 14x + 8 = 3x^2 - 2x - 12x + 8$$

$$= \dots\dots\dots\dots$$

Finish it off

$(x - 4)(3x - 2)$

14

Because

$$3x^2 - 2x - 12x + 8 = (3x^2 - 12x) - (2x - 8)$$
$$= 3x(x - 4) - 2(x - 4)$$
$$3x^2 - 14x + 8 = (x - 4)(3x - 2)$$

And finally, this one.

Example 3

To factorize $8x^2 + 18x - 5$ $(ax^2 + bx + c)$

Follow the routine as before and all will be well

    So $8x^2 + 18x - 5 = \dots\dots\dots\dots$

$(2x + 5)(4x - 1)$

15

In this case, $a = 8; b = 18; c = -5; |ac| = 40$

Possible factors of 40 = (1, 40), (2, 20), (4, 10) and (5, 8)

c is negative. So required factors differ by $|b|$, i.e. 18. Therefore (2, 20)

c is negative. So numerically larger factor has sign of b, i.e. positive.

c is negative. So signs of f_1 and f_2 are different, $f_1 = 20; f_2 = -2$

$$\text{So } 8x^2 + 18x - 5 = 8x^2 + 20x - 2x - 5$$
$$= 4x(2x + 5) - 1(2x + 5)$$
$$= (2x + 5)(4x - 1)$$

Next frame

16 Test for simple factors

Some quadratic equations are not capable of being written as the product of *simple factors* – that is, factors where all the coefficients are integers. To save time and effort, a quick test can be applied before the previous routine is put into action.

To determine whether $ax^2 + bx + c$ can be factorized into two simple factors, first evaluate the expression $(b^2 - 4ac)$.

If $(b^2 - 4ac)$ *is a perfect square*, that is it can be written as k^2 for some integer k, $ax^2 + bx + c$ can be factorized into two simple factors.

If $(b^2 - 4ac)$ *is not a perfect square*, no simple factors of $ax^2 + bx + c$ exist.

Example 1

$3x^2 - 4x + 5 \qquad a = 3; b = -4; c = 5$

$b^2 - 4ac = 16 - 4 \times 3 \times 5 = 16 - 60 = -44$ (not a perfect square)

There are no simple factors of $3x^2 - 4x + 5$

Now test in the same way:

Example 2

$2x^2 + 5x - 3 \qquad a = 2; b = 5; c = -3$

$b^2 - 4ac = 25 - 4 \times 2 \times (-3) = 25 + 24 = 49 = 7^2$ (perfect square)

$2x^2 + 5x - 3$ can be factorized into simple factors.

Now as an exercise, determine whether or not each of the following could be expressed as the product of two simple factors:

(a) $4x^2 + 3x - 4$ (c) $3x^2 + x - 4$
(b) $6x^2 + 7x + 2$ (d) $7x^2 - 3x - 5$

17

(a) No simple factors	(c) Simple factors
(b) Simple factors	(d) No simple factors

Now we can link this test with the previous work. Work through the following short exercise: it makes useful revision.

Test whether each of the following could be expressed as the product of two simple factors and, where possible, determine those factors:

(a) $2x^2 + 7x + 3$ (c) $7x^2 - 5x - 4$
(b) $5x^2 - 4x + 6$ (d) $8x^2 + 2x - 3$

Check the results with the next frame

18

> (a) $(2x + 1)(x + 3)$ (c) No simple factors
> (b) No simple factors   (d) $(2x - 1)(4x + 3)$

Here is the working:

(a) $2x^2 + 7x + 3$ $a = 2;\ b = 7;\ c = 3$

$b^2 - 4ac = 49 - 4 \times 2 \times 3 = 49 - 24 = 25 = 5^2$. Factors exist.

$|ac| = 6$; possible factors of 6 are $(1, 6)$ and $(2, 3)$

c is positive. Factors add up to 7, i.e.$(1, 6)$

Both factors have the same sign as b, i.e. positive.

So $f_1 = 1$ and $f_2 = 6$

$2x^2 + 7x + 3 = 2x^2 + x + 6x + 3$

$$= (2x^2 + x) + (6x + 3) = x(2x + 1) + 3(2x + 1)$$

$$= (2x + 1)(x + 3)$$

(b) $5x^2 - 4x + 6$ $a = 5;\ b = -4;\ c = 6$

$b^2 - 4ac = 16 - 4 \times 5 \times 6 = 16 - 120 = -104$. Not a complete square.

Therefore, no simple factors exist.

(c) $7x^2 - 5x - 4$ $a = 7;\ b = -5;\ c = -4$

$b^2 - 4ac = 25 - 4 \times 7 \times (-4) = 25 + 112 = 137$. Not a complete square.

Therefore, no simple factors exist.

(d) $8x^2 + 2x - 3$ $a = 8;\ b = 2;\ c = -3$

$b^2 - 4ac = 4 - 4 \times 8 \times (-3) = 4 + 96 = 100 = 10^2$. Factors exist.

$|ac| = 24$; possible factors of 24 are $(1, 24)$, $(2, 12)$, $(3, 8)$ and $(4, 6)$

c is negative. Factors differ by $|b|$, i.e. 2. So $(4, 6)$

f_1 and f_2 of opposite signs. Larger factor has the same sign as b, i.e. positive. $f_1 = 6;\ f_2 = -4$.

$8x^2 + 2x - 3 = 8x^2 + 6x - 4x - 3$

$$= (8x^2 - 4x) + (6x - 3) = 4x(2x - 1) + 3(2x - 1)$$

$$\text{So } 8x^2 + 2x - 3 = (2x - 1)(4x + 3)$$

At this point let us pause and summarize the main facts so far on factorization of algebraic expressions

Review summary   Unit 10

19   **1** *Factorization of algebraic expressions*
            (a) Common factors of binomial expressions.
            (b) Common factors of expressions by grouping.
         Useful standard factors:
         $$a^2 + 2ab + b^2 = (a+b)^2$$
         $$a^2 - 2ab + b^2 = (a-b)^2$$
         $$a^2 - b^2 = (a-b)(a+b) \quad \text{(Difference of two squares)}$$

   2 *Factorization of quadratic expressions of the form* $ax^2 + bx + c$.
         Test for possibility of simple factors: $(b^2 - 4ac)$ *is a complete square.*
         Determination of factors of $ax^2 + bx + c$:
            (a) Evaluate $|ac|$.
            (b) Write down all possible factors of $|ac|$.
            (c) *If c is positive*, select two factors of $|ac|$ with sum equal to $|b|$.
                If c is positive, both factors have the same sign as b.
            (d) *If c is negative*, select two factors of $|ac|$ that differ by $|b|$.
                If c is negative, the factors have opposite signs, the numerically larger
                having the same sign as b.
            (e) Let the required two factors be f_1 and f_2.
                Then $ax^2 + bx + c = ax^2 + f_1 x + f_2 x + c$ and factorize this by the method
                of common factors by grouping.

 # Review exercise   Unit 10

20   **1** Factorize the following:
            (a) $18xy^3 - 8x^3y$ (b) $x^3 - 6x^2y - 2xy + 12y^2$
            (c) $16x^2 - 24xy - 18x + 27y$ (d) $(x - 2y)^2 - (2x - y)^2$
            (e) $x^2 + 7x - 30$ (f) $4x^2 - 36$
            (g) $x^2 + 10x + 25$ (h) $3x^2 - 11x - 4$

            Complete the questions, working back over the previous frames if you need to.
            Don't rush, take your time.
            The answers and working are in the next frame.

1 (a) $18xy^3 - 8x^3y = 2xy(9y^2 - 4x^2)$

$\qquad\qquad = 2xy(3y - 2x)(3y + 2x)$

(b) $x^3 - 6x^2y - 2xy + 12y^2 = x^2(x - 6y) - 2y(x - 6y)$

$\qquad\qquad\qquad\qquad = (x^2 - 2y)(x - 6y)$

(c) $16x^2 - 24xy - 18x + 27y = (16x^2 - 24xy) - (18x - 27y)$

$\qquad\qquad\qquad\qquad = 8x(2x - 3y) - 9(2x - 3y)$

$\qquad\qquad\qquad\qquad = (8x - 9)(2x - 3y)$

(d) $(x - 2y)^2 - (2x - y)^2 = x^2 - 4xy + 4y^2 - 4x^2 + 4xy - y^2$

$\qquad\qquad\qquad\qquad = 3y^2 - 3x^2$

$\qquad\qquad\qquad\qquad = 3(y^2 - x^2)$

$\qquad\qquad\qquad\qquad = 3(y - x)(y + x)$

(e) $x^2 + 7x - 30 = x^2 + (10 - 3)x + (10) \times (-3)$

$\qquad\qquad\qquad = (x + 10)(x - 3)$

(f) $4x^2 - 36 = (2x)^2 - (6)^2$

$\qquad\qquad = (2x - 6)(2x + 6)$

(g) $x^2 + 10x + 25 = x^2 + (2 \times 5)x + 5^2$

$\qquad\qquad\qquad = (x + 5)^2$

(h) $3x^2 - 11x - 4 = 3x(x - 4) + (x - 4)$

$\qquad\qquad\qquad = (3x + 1)(x - 4)$

Now for the Review test

21

 # Review test

Unit 10

1 Factorize the following:

(a) $18x^2y - 12xy^2$

(b) $x^3 + 4x^2y - 3xy^2 - 12y^3$

(c) $4(x - y)^2 - (x - 3y)^2$

(d) $12x^2 - 25x + 12$

22

You have now come to the end of this Module. A list of **Can You?** questions follows for you to gauge your understanding of the material in the Module. You will notice that these questions match the **Learning outcomes** listed at the beginning of the Module. Now try the **Test exercise**. *Work through the questions at your own pace, there is no need to hurry.* A set of **Further problems** provides additional valuable practice.

23

Can You?

1   Checklist: Module 2

Check this list before and after you try the end of Module test.

On a scale of 1 to 5 how confident are you that you can:

- Use alphabetic symbols to supplement the numerals and to combine these symbols using all the operations of arithmetic?
  Yes ☐ ☐ ☐ ☐ ☐   *No*

- Simplify algebraic expressions by collecting like terms and by abstracting common factors from similar terms?
  Yes ☐ ☐ ☐ ☐ ☐   *No*

- Remove brackets and so obtain alternative algebraic expressions?
  Yes ☐ ☐ ☐ ☐ ☐   *No*

- Manipulate expressions involving powers and logarithms?
  Yes ☐ ☐ ☐ ☐ ☐   *No*

- Multiply two algebraic expressions together?
  Yes ☐ ☐ ☐ ☐ ☐   *No*

- Manipulate logarithms both numerically and symbolically?
  Yes ☐ ☐ ☐ ☐ ☐   *No*

- Manipulate algebraic fractions and divide one expression by another?
  Yes ☐ ☐ ☐ ☐ ☐   *No*

- Factorize algebraic expressions using standard factorizations?
  Yes ☐ ☐ ☐ ☐ ☐   *No*

- Factorize quadratic algebraic expressions?
  Yes ☐ ☐ ☐ ☐ ☐   *No*

 # Test exercise 2

2   **1** Simplify each of the following:

   (a) $2ab - 4ac + ba - 2cb + 3ba$

   (b) $3x^2yz - zx^2y + 4yxz^2 - 2x^2zy + 3z^2yx - 3zy^2$

   (c) $c^p \times c^{-q} \div c^{-2}$

   (d) $\dfrac{\left(x^{\frac{1}{2}}\right)^{-\frac{2}{3}} \div \left(y^{\frac{3}{4}}\right)^2 \times \left(x^{\frac{3}{5}}\right)^{-\frac{5}{3}}}{\left(x^{\frac{1}{4}}\right)^{-1} \times \left(y^{\frac{1}{3}}\right)^6}$

2 Remove the brackets in each of the following:

(a) $2f(3g - 4h)(g + 2h)$

(b) $(5x - 6y)(2x + 6y)(5x - y)$

(c) $4\{3[p - 2(q - 3)] - 3(r - 2)\}$

3 Evaluate by calculator or by change of base where necessary:

(a) $\log 0{\cdot}0270$ (b) $\ln 47{\cdot}89$ (c) $\log_7 126{\cdot}4$

4 Rewrite the following in log form:

(a) $V = \dfrac{\pi h}{4}(D - h)(D + h)$ (b) $P = \dfrac{1}{16}(2d - 1)^2 N\sqrt{S}$

5 Express the following without logarithms:

(a) $\log x = \log P + 2\log Q - \log K - 3$

(b) $\log R = 1 + \dfrac{1}{3}\log M + 3\log S$

(c) $\ln P = \dfrac{1}{2}\ln(Q + 1) - 3\ln R + 2$

6 Perform the following multiplications and simplify your results:

(a) $(3x - 1)(x^2 - x - 1)$

(b) $(a^2 + 2a + 2)(3a^2 + 4a + 4)$

7 Simplify each of the following:

(a) $\dfrac{x^3}{y^2} \div \dfrac{x}{y^3}$

(b) $\dfrac{ab}{c} \div \dfrac{ac}{b} \times \dfrac{bc}{a}$

8 Perform the following divisions:

(a) $(x^2 + 2x - 3) \div (x - 1)$

(b) $\dfrac{n^3 + 27}{n + 3}$

(c) $\dfrac{3a^3 + 2a^2 + 1}{a + 1}$

9 Factorize the following:

(a) $36x^3y^2 - 8x^2y$ (e) $x^2 + 10x + 24$

(b) $x^3 + 3x^2y + 2xy^2 + 6y^3$ (f) $x^2 - 10x + 16$

(c) $4x^2 + 12x + 9$ (g) $x^2 - 5x - 36$

(d) $(3x + 4y)^2 - (2x - y)^2$ (h) $6x^2 + 5x - 6$

Further problems 2

3

1 Determine the following:

(a) $(2x^2 + 5x - 3)(4x - 7)$

(b) $(4x^2 - 7x + 3)(5x + 6)$

(c) $(5x^2 - 3x - 4)(3x - 5)$

(d) $(6x^3 - 5x^2 - 14x + 12) \div (2x - 3)$

(e) $(15x^3 + 46x^2 - 49) \div (5x + 7)$

(f) $(18x^3 + 13x + 14) \div (3x + 2)$

2 Simplify the following, giving the result without fractional indices:

$$(x^2 - 1)^2 \times \sqrt{x + 1} \div (x - 1)^{\frac{3}{2}}$$

3 Simplify:

(a) $\sqrt{a^{\frac{7}{3}}b^5c^{\frac{2}{3}}} \div \sqrt[3]{a^{\frac{1}{2}}b^3c^{-1}}$

(b) $\sqrt[3]{x^9 y^{\frac{1}{3}} z^{\frac{1}{2}}} \times y^{\frac{8}{9}} \times (2^{-8} x^6 y^2 z^{\frac{1}{3}})^{-\frac{1}{2}}$

(c) $(6x^3 y^{\frac{5}{2}} z^{\frac{1}{4}})^2 \div (9x^6 y^4 z^3)^{\frac{1}{2}}$

(d) $(x^2 - y^2)^{\frac{1}{2}} \times (x - y)^{\frac{3}{2}} \times (x + y)^{-\frac{1}{2}}$

4 Evaluate:

(a) $\log 0{\cdot}008472$ (b) $\ln 25{\cdot}47$ (c) $\log_8 387{\cdot}5$

5 Express in log form:

(a) $f = \dfrac{1}{\pi d \sqrt{LC}}$ (b) $K = \dfrac{a^3 \times \sqrt{b}}{c^{\frac{1}{6}} d^{\frac{2}{5}}}$

6 Rewrite the following without logarithms:

(a) $\log W = 2(\log A + \log w) - (\log 32 + 2\log \pi + 2\log r + \log c)$

(b) $\log S = \log K - \log 2 + 2\log \pi + 2\log n + \log y + \log r + 2\log L$
$$- 2\log h - \log g$$

(c) $\ln I = \ln(2V) - \{\ln(KR + r) - \ln K + KL\}$

7 Factorize the following:

(a) $15x^2 y^2 + 20xy^3$

(b) $14a^3 b - 12a^2 b^2$

(c) $2x^2 + 3xy - 10x - 15y$

(d) $4xy - 7y^2 - 12x + 21y$

(e) $15x^2 + 8y + 20xy + 6x$

(f) $6xy - 20 + 15x - 8y$

(g) $9x^2 + 24xy + 16y^2$

(h) $16x^2 - 40xy + 25y^2$

(i) $25x^3 y^4 - 16xy^2$

(j) $(2x + 5y)^2 - (x - 3y)^2$

8 Find simple factors of the following where possible:

(a) $5x^2 + 13x + 6$

(b) $2x^2 - 11x + 12$

(c) $6x^2 - 5x - 6$

(d) $3x^2 + 7x - 4$

(e) $5x^2 - 19x + 12$

(f) $4x^2 - 6x + 9$

(g) $6x^2 - 5x - 7$

(h) $9x^2 - 18x + 8$

(i) $10x^2 + 11x - 6$

(j) $15x^2 - 19x + 6$

(k) $8x^2 + 2x - 15$

Expressions and equations

Learning outcomes

When you have completed this Module you will be able to:

- Numerically evaluate an algebraic expression by substituting numbers for variables
- Recognize the different types of equation
- Evaluate an independent variable
- Change the subject of an equation by transposition
- Evaluate polynomial expressions by 'nesting'
- Use the remainder and factor theorems to factorize polynomials
- Factorize fourth-order polynomials

Units

Expressions and equations   Unit 11

Evaluating expressions   **1**

When numerical values are assigned to the variables and constants in an expression, the expression itself assumes a numerical value that is obtained by following the usual precedence rules. This process is known as *evaluating* the expression. For example, if $l = 2$ and $g = 9.81$ then the expression:

$$2\pi\sqrt{\frac{l}{g}}$$

is evaluated as:

$$2\pi\sqrt{\frac{2}{9.81}} = 2.84 \text{ to 2 dp where } \pi = 3.14159\ldots$$

So let's look at three examples:

Example 1

If $V = \dfrac{\pi h}{6}(3R^2 + h^2)$, determine the value of V when $h = 2.85$, $R = 6.24$ and $\pi = 3.142$.

Substituting the given values:

$$V = \frac{3.142 \times 2.85}{6}(3 \times 6.24^2 + 2.85^2)$$

$$= \frac{3.142 \times 2.85}{6}(3 \times 38.938 + 8.123)$$

$$= \ldots\ldots\ldots\ldots$$

Finish it off

$$\boxed{V = 186.46}$$

2

Example 2

If $R = \dfrac{R_1 R_2}{R_1 + R_2}$, evaluate R when $R_1 = 276$ and $R_2 = 145$.

That is easy enough, $R = \ldots\ldots\ldots\ldots$

3

$$R = 95{\cdot}06$$

Now let us deal with a more interesting one.

Example 3

If $V = \dfrac{\pi b}{12}(D^2 + Dd + d^2)$ evaluate V to 3 sig fig when $b = 1{\cdot}46$, $D = 0{\cdot}864$, $d = 0{\cdot}517$ and $\pi = 3{\cdot}142$.

Substitute the values in the expressions and then apply the rules carefully. Take your time with the working: there are no prizes for speed!

$V = \ldots\ldots\ldots\ldots$

4

$$V = 0{\cdot}558 \text{ to 3 sig fig}$$

Here it is:

$$V = \frac{3{\cdot}142 \times 1{\cdot}46}{12}(0{\cdot}864^2 + 0{\cdot}864 \times 0{\cdot}517 + 0{\cdot}517^2)$$

$$= \frac{3{\cdot}142 \times 1{\cdot}46}{12}(0{\cdot}7465 + 0{\cdot}864 \times 0{\cdot}517 + 0{\cdot}2673)$$

$$= \frac{3{\cdot}142 \times 1{\cdot}46}{12}(0{\cdot}7465 + 0{\cdot}4467 + 0{\cdot}2673)$$

$$= \frac{3{\cdot}142 \times 1{\cdot}46}{12}(1{\cdot}4605) = 0{\cdot}5583\ldots$$

$\therefore V = 0{\cdot}558$ to 3 sig fig

5 **Equations**

Because different values of the variables and constants produce different values for the expression, we assign these expression values to another variable and so form an *equation*. For example, the equation:

$r = 2s^3 + 3t$

states that the variable r can be assigned values by successively assigning values to s and to t, each time evaluating $2s^3 + 3t$. The variable r is called the *dependent* variable and *subject* of the equation whose value *depends* on the values of the *independent* variables s and t.

An *equation* is a statement of the equality of two expressions but there are different types of equation:

Conditional equation

A *conditional equation*, usually just called an *equation*, is true only for certain values of the symbols involved. For example, the equation:

$x^2 = 4$

is an equation that is only true for each of the two values $x = +2$ and $x = -2$.

Identity

An *identity* is a statement of equality of two expressions that is true for all values of the symbols for which both expressions are defined. For example, the equation:

$$2(5 - x) \equiv 10 - 2x$$

is true no matter what value is chosen for x – it is an *identity*. The expression on the left is not just equal to the expression on the right, it is *equivalent* to it – one expression is an alternative form of the other. Hence the symbol $\equiv$ which stands for 'is equivalent to'.

Defining equation

A *defining equation* is a statement of equality that defines an expression. For example:

$$a^2 \triangleq a \times a$$

Here the symbolism a^2 is defined to mean $a \times a$ where $\triangleq$ means 'is defined to be'.

Assigning equation

An *assigning equation* is a statement of equality that assigns a specific value to a variable. For example:

$$p := 4$$

Here, the value 4 is assigned to the variable p.

Formula

A *formula* is a statement of equality that expresses a mathematical fact where all the variables, dependent and independent, are well-defined. For example, the equation:

$$A = \pi r^2$$

expresses the fact that the area A of a circle of radius r is given as πr^2.

The uses of $\equiv$, $\triangleq$ and $:=$ as connectives are often substituted by the $=$ sign. While it is not strictly correct to do so, it is acceptable.

So what type of equation is each of the following?

(a) $l = \dfrac{T^2 g}{4\pi^2}$ where T is the periodic time and l the length of a simple pendulum and where g is the acceleration due to gravity

(b) $v = 23{\cdot}4$

(c) $4n = 4 \times n$

(d) $x^2 - 2x = 0$

(e) $\dfrac{r^3 - s^3}{r - s} = r^2 + rs + r^2$ where $r \neq s$

The answers are in the next frame

6

> (a) Formula
> (b) Assigning equation
> (c) Defining equation
> (d) Conditional equation
> (e) Identity

Because

    (a) It is a statement of a mathematical fact that relates the values of the variables T and l and the constant g where T, l and g represent well-defined entities.

    (b) It assigns the value $23 \cdot 4$ to the variable v.

    (c) It defines the notation $4n$ whereby the multiplication sign is omitted.

    (d) It is only true for certain values of the variables, namely $x = 0$ and $x = 2$.

    (e) The left-hand side is an alternative form of the right-hand side, as can be seen by performing the division. Notice that the expression on the left is not defined when $r = s$ whereas the expression on the right is. We say that the equality only holds true for numerical values when *both* expressions are defined.

Now to the next frame

7   **Evaluating independent variables**

Sometimes, the numerical values assigned to the variables and constants in a formula include a value for the dependent variable and exclude a value of one of the independent variables. In this case the exercise is to find the corresponding value of the independent variable. For example, given that:

$$T = 2\pi\sqrt{\frac{l}{g}} \text{ where } \pi = 3 \cdot 14 \text{ and } g = 9 \cdot 81$$

what is the length l that corresponds to $T = 1 \cdot 03$? That is, given:

$$1 \cdot 03 = 6 \cdot 28\sqrt{\frac{l}{9 \cdot 81}}$$

find l. We do this by isolating l on one side of the equation.

So we first divide both sides by $6 \cdot 28$ to give: $\dfrac{1 \cdot 03}{6 \cdot 28} = \sqrt{\dfrac{l}{9 \cdot 81}}$

An equation is like a balance, so if any arithmetic operation is performed on one side of the equation the identical operation must be performed on the other side to maintain the balance.

Square both sides to give: $\left(\dfrac{1 \cdot 03}{6 \cdot 28}\right)^2 = \dfrac{l}{9 \cdot 81}$ and now multiply both sides by 9·81 to give:

$$9 \cdot 81 \left(\frac{1 \cdot 03}{6 \cdot 28}\right)^2 = l = 0 \cdot 264 \text{ to 3 sig fig}$$

So that if:

$$I = \frac{nE}{R + nr}$$

and $n = 6$, $E = 2 \cdot 01$, $R = 12$ and $I = 0 \cdot 98$, the corresponding value of r is . . . .

Next frame

$$\boxed{r = 0 \cdot 051}$$

8

Because

Given that $0 \cdot 98 = \dfrac{6 \times 2 \cdot 01}{12 + 6r} = \dfrac{12 \cdot 06}{12 + 6r}$ we see, by taking the reciprocal of each side, that:

$$\frac{1}{0 \cdot 98} = \frac{12 + 6r}{12 \cdot 06} \text{ and hence } \frac{12 \cdot 06}{0 \cdot 98} = 12 + 6r$$

after multiplying both sides by 12·06. Subtracting 12 from both sides yields:

$$\frac{12 \cdot 06}{0 \cdot 98} - 12 = 6r$$

and, dividing both sides by 6 gives:

$$\frac{1}{6} \left(\frac{12 \cdot 06}{0 \cdot 98} - 12\right) = r, \text{ giving } r = 0 \cdot 051 \text{ to 2 sig fig}$$

By this process of arithmetic manipulation the independent variable r in the original equation has been *transposed* to become the dependent variable of a new equation, so enabling its value to be found.

You will often encounter the need to transpose a variable in an equation so it is essential that you acquire the ability to do so. Furthermore, you will also need to transpose variables to obtain a new equation rather than just to find the numerical value of the transposed variable as you have done so far. In what follows we shall consider the transposition of variables algebraically rather then arithmetically.

9 **Transposition of formulas**

The formula for the period of oscillation, T seconds, of a pendulum is given by:

$$T = 2\pi\sqrt{\frac{l}{g}}$$

where l is the length of the pendulum measured in metres, g is the gravitational constant ($9\cdot81$ m s^{-2}) and $\pi = 3\cdot142$ to 4 sig fig. The single symbol on the left-hand side (LHS) of the formula – the dependent variable – is often referred to as the *subject of the formula*. We say that *T is given in terms of l*. What we now require is a new formula where l is the subject. That is, where l is given in terms of T. To effect this transposition, keep in mind the following:

The formula is an equation, or balance. Whatever is done to one side must be done to the other.

To remove a symbol from the right-hand side (RHS) we carry out the opposite operation to that which the symbol is at present involved in. The 'opposites' are – *addition* and *subtraction*, *multiplication* and *division*, *powers* and *roots*.

In this case we start with:

$$T = 2\pi\sqrt{\frac{l}{g}}$$

To isolate l we start by removing the 2π by dividing both sides by 2π. This gives:

$$\frac{T}{2\pi} = \sqrt{\frac{l}{g}}$$

We next remove the square root sign by squaring both sides to give:

$$\frac{T^2}{4\pi^2} = \frac{l}{g}$$

Next we remove the g on the RHS by multiplying both sides by g to give:

$$\frac{gT^2}{4\pi^2} = l$$

Finally, we interchange sides to give:

$$l = \frac{gT^2}{4\pi^2}$$

because it is more usual to have the subject of the formula on the LHS.

Now try a few examples

Example 1

Transpose the formula $v = u + at$ to make a the subject.

(a) Isolate the term involving a by subtracting u from both sides:

$$v - u = u + at - u = at$$

(b) Isolate the a by dividing both sides by t:

$$\frac{v - u}{t} = \frac{at}{t} = a$$

(c) Finally, write the transposed equation with the new subject on the LHS,

$$a = \frac{v - u}{t}$$

Let's just try another

Example 2

Transpose the formula $a = \dfrac{2(ut - s)}{t^2}$ to make u the subject.

(a) u is part of the numerator on the RHS. Therefore first multiply both sides by t^2:

$$at^2 = 2(ut - s)$$

(b) We can now multiply out the bracket:

$$at^2 = 2ut - 2s$$

(c) Now we isolate the term containing u by adding $2s$ to each side:

$$at^2 + 2s = 2ut$$

(d) u is here multiplied by $2t$, therefore we divide each side by $2t$:

$$\frac{at^2 + 2s}{2t} = u$$

(e) Finally, write the transposed formula with the new subject on the LHS,

i.e. $u = \dfrac{at^2 + 2s}{2t}$

Apply the procedure carefully and take one step at a time.

Example 3

Transpose the formula $d = 2\sqrt{h(2r - h)}$ to make r the subject.

(a) First we divide both sides by 2:

$$\frac{d}{2} = \sqrt{h(2r - h)}$$

(b) To open up the expression under the square root sign, we . . . .

12

square both sides

So $\dfrac{d^2}{4} = h(2r - h)$

(c) At present, the bracket expression is multiplied by h. Therefore, we
. . . .

13

divide both sides by h

So $\dfrac{d^2}{4h} = 2r - h$

(d) Next, we . . . .

14

add h to both sides

So $\dfrac{d^2}{4h} + h = 2r$

Finish it off

15

$$r = \frac{1}{2}\left\{\frac{d^2}{4h} + h\right\}$$

Of course, this could be written in a different form:

$$r = \frac{1}{2}\left\{\frac{d^2}{4h} + h\right\} = \frac{1}{2}\left\{\frac{d^2 + 4h^2}{4h}\right\} = \frac{d^2 + 4h^2}{8h}$$

All these forms are equivalent to each other.
Now, this one.

Example 4

Transpose $V = \dfrac{\pi h(3R^2 + h^2)}{6}$ to make R the subject.

First locate the symbol R in its present position and then take the necessary
steps to isolate it. Do one step at a time.

$R = $ . . . .

16

$$R = \sqrt{\frac{2V}{\pi h} - \frac{h^2}{3}}$$

Because

$$V = \frac{\pi h(3R^2 + h^2)}{6}$$

$$6V = \pi h(3R^2 + h^2)$$

$$\frac{6V}{\pi h} = 3R^2 + h^2$$

$$\frac{6V}{\pi h} - h^2 = 3R^2 \qquad \text{So } \frac{2V}{\pi h} - \frac{h^2}{3} = R^2$$

Therefore $\sqrt{\frac{2V}{\pi h} - \frac{h^2}{3}} = R,$ $\qquad R = \sqrt{\frac{2V}{\pi h} - \frac{h^2}{3}}$

Example 5

This one is slightly different.

Transpose the formula $n = \dfrac{IR}{E - Ir}$ to make I the subject.

In this case, you will see that the symbol I occurs twice on the RHS. Our first step, therefore, is to move the denominator completely by multiplying both sides by $(E - Ir)$:

$$n(E - Ir) = IR$$

Then we can free the I on the LHS by multiplying out the bracket:

$$nE - nIr = IR$$

Now we collect up the two terms containing I on to the RHS:

$$nE = IR + nIr$$
$$= I(R + nr)$$

So $I = \dfrac{nE}{R + nr}$

Move on to the next frame

Example 6

Here is one more, worked in very much the same way as the previous example, so you will have no trouble.

Transpose the formula $\dfrac{R}{r} = \sqrt{\dfrac{f + P}{f - P}}$ to make f the subject.

Work right through it, using the rules and methods of the previous examples.

$$f = \dots\dots\dots$$

19

$$f = \frac{(R^2 + r^2)P}{R^2 - r^2}$$

Here it is:

$$\frac{R^2}{r^2} = \frac{f + P}{f - P}$$

$$\frac{R^2}{r^2}(f - P) = f + P$$

$$R^2(f - P) = r^2(f + P)$$

$$R^2 f - R^2 P = r^2 f + r^2 P$$

$$R^2 f - r^2 f = R^2 P + r^2 P$$

$$f(R^2 - r^2) = P(R^2 + r^2)$$

$$\text{So } f = \frac{(R^2 + r^2)P}{R^2 - r^2}$$

At this point let us pause and summarize the main facts so far on expressions and equations

 # Review summary   Unit 11

20

1 An algebraic expression is evaluated by substituting numbers for the variables and constants in the expression and then using the arithmetic precedence rules.

2 Values so obtained can be assigned to a variable to form an equation. This variable is called the *subject* of the equation.

3 The subject of an equation is called the *dependent variable* and the variables within the expression are called the *independent variables*.

4 *There is more than one type of equation:*

   Conditional equation
   Identity
   Defining equation
   Assigning equation
   Formula

5 Any one of the independent variables in a formula can be made the subject of a new formula obtained by transposing it with the dependent variable of the original formula.

6 Transposition is effected by performing identical arithmetical operations on both sides of the equation.

Review exercise

1 Evaluate each of the following to 3 sig fig:

21

(a) $I = \dfrac{nE}{R + nr}$ where $n = 4$, $E = 1{\cdot}08$, $R = 5$ and $r = 0{\cdot}04$

(b) $A = P\left(1 + \dfrac{r}{100}\right)^{n}$ where $P = 285{\cdot}79$, $r = 5{\cdot}25$ and $n = 12$

(c) $P = A\dfrac{(nv/u)^{\frac{3}{2}}}{1 + (nv/u)^{3}}$ where $A = 40$, $u = 30$, $n = 2{\cdot}5$ and $v = 42{\cdot}75$

2 Transpose the formula $f = \dfrac{S(M - m)}{M + m}$ to make m the subject.

Complete both questions. Take your time, there is no need to rush.
If necessary, refer back to the Unit.
The answers and working are in the next frame.

22

1 (a) $I = \dfrac{4 \times 1{\cdot}08}{5 + 4 \times 0{\cdot}04} = \dfrac{4{\cdot}32}{5 + 0{\cdot}16} = 0{\cdot}837$

(b) $A = 285{\cdot}79\left(1 + \dfrac{5{\cdot}25}{100}\right)^{12} = 285{\cdot}79(1{\cdot}0525)^{12}$

$= 285{\cdot}79 \times 1{\cdot}84784\ldots = 528$

(c) $P = 40\dfrac{(2{\cdot}5 \times 42{\cdot}75 \div 30)^{\frac{3}{2}}}{1 + (2{\cdot}5 \times 42{\cdot}75 \div 30)^{3}} = 40\dfrac{3{\cdot}5625^{\frac{3}{2}}}{1 + 3{\cdot}5625^{3}}$

$= 40\dfrac{6{\cdot}7240\ldots}{46{\cdot}2131\ldots} = 5{\cdot}82$

2 $f = \dfrac{S(M - m)}{M + m}$ so $f(M + m) = S(M - m)$ thus $fM + fm = SM - Sm$, that is

$fm + Sm = SM - fM$. Factorizing yields $m(f + S) = M(S - f)$ giving

$m = \dfrac{M(S - f)}{f + S}$.

Now for the Review test

Review test   Unit 11

23

1 Given $P = A\left(1 + \dfrac{r}{100}\right)^n$ find P to 2 dp given that $A = 12\,345\cdot66$, $r = 4\cdot65$
   and $n = 6\frac{255}{365}$.

2 Given that $T = 2\pi\sqrt{\dfrac{l^2 + 4t^2}{3g(r - t)}}$ find:

   (a) l in terms of T, t, r and g
   (b) r in terms of T, t, l and g.

Evaluating expressions   Unit 12

1 Systems

A *system* is a process that is capable of accepting an *input*, *processing* the input and producing an *output*:

We can use this idea of a system to describe the way we evaluate an algebraic expression. For example, given the expression:

  $3x - 4$

we evaluate it for $x = 5$, say, by multiplying 5 by 3 and then subtracting 4 to obtain the value 11; we *process* the *input* 5 to produce the *output* 11:

If we use the letter x to denote the input and the letter f to denote the process we denote the output as:

  $f(x)$, that is 'f acting on x'

where the process f, represented by the box in the diagram, is:

  multiply x by 3 and then subtract 4

How the evaluation is actually done, whether mentally, by pen and paper or by using a calculator is not important. What is important is that the prescription for evaluating it is given by the expression $3x - 4$ and that we can represent the *process* of executing this prescription by the label f.

The advantage of this notion is that it permits us to tabulate the results of evaluation in a meaningful way. For example, if:

$$f(x) = 3x - 4$$

then:

$$f(5) = 15 - 4 = 11$$

and in this way the corresponding values of the two variables are recorded.

So that, if $f(x) = 4x^3 - \dfrac{6}{2x}$ then:

(a) $f(3) = \ldots\ldots\ldots$

(b) $f(-4) = \ldots\ldots\ldots$

(c) $f(2/5) = \ldots\ldots\ldots$

(d) $f(-3\cdot24) = \ldots\ldots\ldots$ (to 5 sig fig)

Answers are in the next frame

2

> (a) $f(3) = 107$
>
> (b) $f(-4) = -255\cdot25$
>
> (c) $f(2/5) = -7\cdot244$
>
> (d) $f(-3\cdot24) = -135\cdot12$

Because

(a) $f(3) = 4 \times 3^3 - \dfrac{6}{2 \times 3} = 108 - 1 = 107$

(b) $f(-4) = 4 \times (-4)^3 - \dfrac{6}{2 \times (-4)} = -256 + 0\cdot75 = -255\cdot25$

(c) $f(2/5) = 4 \times (2/5)^3 - \dfrac{6}{2 \times (2/5)} = 0\cdot256 - 7\cdot5 = -7\cdot244$

(d) $f(-3\cdot24) = 4 \times (-3\cdot24)^3 - \dfrac{6}{2 \times (-3\cdot24)} = -136\cdot05 + 0\cdot92593 = -135\cdot12$

Polynomial equations

Polynomial expressions

3

A *polynomial* in x is an expression involving powers of x, normally arranged in descending (or sometimes ascending) powers. The degree of the polynomial is given by the highest power of x occurring in the expression, for example:

$$5x^4 + 7x^3 + 3x - 4 \qquad \text{is a polynomial of the 4th degree}$$

$$\text{and} \quad 2x^3 + 4x^2 - 2x + 7 \qquad \text{is a polynomial of the 3rd degree.}$$

Polynomials of low degree often have alternative names:

$2x - 3$ is a polynomial of the 1st degree – or a *linear* expression.

$3x^2 + 4x + 2$ is a polynomial of the 2nd degree – or a *quadratic* expression.

A polynomial of the 3rd degree is often referred to as a *cubic* expression.

A polynomial of the 4th degree is often referred to as a *quartic* expression.

Evaluation of polynomials

If $f(x) = 3x^4 - 5x^3 + 7x^2 - 4x + 2$, then evaluating $f(3)$ would involve finding the values of each term before finally totalling up the five individual values. This would mean recording the partial values – with the danger of including errors in the process.

This can be avoided by using the method known as *nesting* – so move to the next frame to see what it entails.

---

4   Evaluation of a polynomial by nesting

Consider the polynomial $f(x) = 5x^3 + 2x^2 - 3x + 6$. To express this in *nested* form, write down the coefficient and one factor x from the first term and add on the coefficient of the next term:

i.e. $5x + 2$

Enclose these in brackets, multiply by x and add on the next coefficient:

i.e. $(5x + 2)x - 3$

Repeat the process: enclose the whole of this in square brackets, multiply by x and add on the next coefficient:

i.e. $[(5x + 2)x - 3]x + 6$

So $f(x) = 5x^3 + 2x^2 - 3x + 6$

$\qquad = [(5x + 2)x - 3]x + 6$ in nested form.

Starting with the innermost brackets, we can now substitute the given value of x and carry on in a linear fashion. No recording is required:

$f(4) = [(22)4 - 3]4 + 6$

$\qquad = [85]4 + 6 = 346 \qquad$ So $f(4) = 346$

Note: The working has been set out here purely by way of explanation. Normally it would be carried out mentally.

So, in the same way, $f(2) = \ldots\ldots\ldots$ and $f(-1) = \ldots\ldots\ldots$

5

$$48,\ 6$$

Notes: (a) The terms of the polynomial must be arranged in descending order of powers.

(b) If any power is missing from the polynomial, it must be included with a zero coefficient before nesting is carried out.

Therefore, if $f(x) = 3x^4 + 2x^2 - 4x + 5$

(a) $f(x)$ in nested form = ...

(b) $f(2)$ = ...

6

(a) $f(x) = \{[(3x + 0)x + 2]x - 4\}x + 5$

(b) $f(2) = 53$

On to the next frame

7

Now a short exercise. In each of the following cases, express the polynomial in nested form and evaluate the function for the given value of x:

(a) $f(x) = 4x^3 + 3x^2 + 2x - 4$ $[x = 2]$

(b) $f(x) = 2x^4 + x^3 - 3x^2 + 5x - 6$ $[x = 3]$

(c) $f(x) = x^4 - 3x^3 + 2x - 3$ $[x = 5]$

(d) $f(x) = 2x^4 - 5x^3 - 3x^2 + 4$ $[x = 4]$

Results in the next frame

8

(a) $[(4x + 3)x + 2]x - 4$ $f(2) = \ \ 44$

(b) $\{[(2x + 1)x - 3]x + 5\}x - 6$ $f(3) = 171$

(c) $\{[(x - 3)x + 0]x + 2\}x - 3$ $f(5) = 257$

(d) $\{[(2x - 5)x - 3]x + 0\}x + 4$ $f(4) = 148$

This method for evaluating polynomials will be most useful in the following work, so let us now move on to the next topic.

Remainder theorem

9

The *remainder theorem* states that if a polynomial $f(x)$ is divided by $(x - a)$, the quotient will be a polynomial $g(x)$ of one degree less than the degree of $f(x)$, together with a remainder R still to be divided by $(x - a)$. That is:

$$\frac{f(x)}{x - a} = g(x) + \frac{R}{x - a}$$

$$\text{So } f(x) = (x - a).g(x) + R$$

When $x = a$ $f(a) = 0.g(a) + R$ i.e. $R = f(a)$

That is:

If $f(x)$ were to be divided by $(x - a)$, the remainder would be $f(a)$.

So, $(x^3 + 3x^2 - 13x - 10) \div (x - 3)$ would give a remainder
$R = f(3) = \dots\dots\dots\dots$

10

$$\boxed{5}$$

Because $f(x) = x^3 + 3x^2 - 13x - 10 = [(x + 3)x - 13]x - 10$

  so $f(3) = 5$

We can verify this by actually performing the long division:

$(x^3 + 3x^2 - 13x - 10) \div (x - 3) = \dots\dots\dots\dots$

11

$$\boxed{x^2 + 6x + 5 \text{ with remainder } 5}$$

Here it is:

$$
\begin{array}{r}
x^2 + 6x + 5 \\
x - 3 \overline{\big)\, x^3 + 3x^2 - 13x - 10} \\
\underline{x^3 - 3x^2} \\
6x^2 - 13x \\
\underline{6x^2 - 18x} \\
5x - 10 \\
\underline{5x - 15} \\
5 \leftarrow \text{Remainder}
\end{array}
$$

Now as an exercise, apply the remainder theorem to determine the remainder in each of the following cases.

(a) $(5x^3 - 4x^2 - 3x + 6) \div (x - 2)$

(b) $(4x^3 - 3x^2 + 5x - 3) \div (x - 4)$

(c) $(x^3 - 2x^2 - 3x + 5) \div (x - 5)$

(d) $(2x^3 + 3x^2 - x + 4) \div (x + 2)$

(e) $(3x^3 - 11x^2 + 10x - 12) \div (x - 3)$

Finish all five and then check with the next frame

12

(a)　24	(d) 2
(b) 225	(e) 0
(c)　65	

Factor theorem

If $f(x)$ is a polynomial and substituting $x = a$ gives a remainder of zero, i.e. $f(a) = 0$, then $(x - a)$ is a factor of $f(x)$.

For example, if $f(x) = x^3 + 2x^2 - 14x + 12 = [(x + 2)x - 14]x + 12$ and we substitute $x = 2$, $f(2) = 0$, so that division of $f(x)$ by $(x - 2)$ gives a zero remainder, i.e. $(x - 2)$ is a factor of $f(x)$. The remaining factor can be found by long division of $f(x)$ by $(x - 2)$.

$f(x) = (x - 2)(\ldots\ldots\ldots\ldots)$

13

$$x^2 + 4x - 6$$

So $f(x) = (x - 2)(x^2 + 4x - 6)$

The quadratic factor so obtained can sometimes be factorized further into two simple factors, so we apply the $(b^2 - 4ac)$ test – which we have used before. In this particular case $(b^2 - 4ac) = \ldots\ldots\ldots\ldots$

14

$$40$$

because $(b^2 - 4ac) = 16 - [4 \times 1 \times (-6)] = 16 + 24 = 40$. This is not a perfect square, so no simple factors exist. Therefore, we cannot factorize further.

So $f(x) = (x - 2)(x^2 + 4x - 6)$

As an example, test whether $(x - 3)$ is a factor of $f(x) = x^3 - 5x^2 - 2x + 24$ and, if so, determine the remaining factor (or factors).

$$f(x) = x^3 - 5x^2 - 2x + 24 = [(x - 5)x - 2]x + 24$$

$$\therefore \quad f(3) = \ldots\ldots\ldots\ldots$$

15

$$0$$

There is no remainder. So $(x-3)$ is a factor of $f(x)$. Long division now gives the remaining quadratic factor, so that $f(x) = (x-3)(\ldots\ldots\ldots)$

16

$$x^2 - 2x - 8$$

$$
\begin{array}{r}
x^2 - 2x - 8 \\
x-3\ \overline{\smash{\big)}\ x^3 - 5x^2 - 2x + 24} \\
\underline{x^3 - 3x^2} \\
-2x^2 - 2x \\
\underline{-2x^2 + 6x} \\
-8x + 24 \\
\underline{-8x + 24} \\
\bullet \quad \bullet
\end{array}
$$

$$f(x) = (x-3)(x^2 - 2x - 8)$$

Now test whether $x^2 - 2x - 8$ can be factorized further.

$$b^2 - 4ac = \ldots\ldots\ldots$$

17

$$36,\ \text{i.e. } 6^2$$

$(b^2 - 4ac) = 6^2$. Therefore there are simple factors of $x^2 - 2x - 8$. We have previously seen how to factorize a quadratic expression when such factors exist and, in this case:

$$x^2 - 2x - 8 = (\ldots\ldots\ldots)(\ldots\ldots\ldots)$$

18

$$(x-4)(x+2)$$

Collecting our results together:

$$
\begin{aligned}
f(x) &= x^3 - 5x^2 - 2x + 24 \\
&= (x-3)(x^2 - 2x - 8) \\
&= (x-3)(x-4)(x+2)
\end{aligned}
$$

And now another example. Show that $(x-4)$ is a factor of $f(x) = x^3 - 6x^2 - 7x + 60$ and, as far as possible, factorize $f(x)$ into simple factors.

Work through it just as before and then check with the next frame

$$f(x) = (x-4)(x+3)(x-5)$$

19

Here it is: $f(x) = x^3 - 6x^2 - 7x + 60 = [(x-6)x - 7]x + 60$
$\qquad f(4) = 0$, so $(x-4)$ is a factor of $f(x)$.

$$
\begin{array}{r}
x^2 - 2x - 15 \\
x - 4 \enclose{longdiv}{x^3 - 6x^2 - 7x + 60} \\
\underline{x^3 - 4x^2} \\
-2x^2 - 7x \\
\underline{-2x^2 + 8x} \\
-15x + 60 \\
\underline{-15x + 60} \\
\bullet \qquad \bullet
\end{array}
$$

$f(x) = (x-4)(x^2 - 2x - 15)$

Now we attend to the quadratic factor. $(b^2 - 4ac) = 64$, i.e. 8^2. This is a complete square. Simple factors exist.

$\qquad x^2 - 2x - 15 = (x+3)(x-5)$
$\quad$ so $f(x) = (x-4)(x+3)(x-5)$

But how do we proceed if we are not given the first factor? We will attend to that in the next frame.

If we are not given the first factor, we proceed as follows:

20

(a) We write the cubic function in nested form.
(b) By trial and error, we substitute values of x, e.g. $x = 1$, $x = -1$, $x = 2$, $x = -2$ etc. until we find a substitution $x = k$ that gives a zero remainder. Then $(x - k)$ is a factor of $f(x)$.

After that, of course, we can continue as in the previous examples.

So, to factorize $f(x) = x^3 + 5x^2 - 2x - 24$ as far as possible, we first write $f(x)$ in nested form, i.e. . . . .

$$f(x) = [(x+5)x - 2]x - 24$$

21

Now substitute values $x = k$ for x until $f(k) = 0$:

$\qquad f(1) = -20 \qquad\qquad$ so $(x-1)$ is not a factor
$\qquad f(-1) = -18 \qquad\quad$ so $(x+1)$ is not a factor
$\qquad f(2) = 0 \qquad\qquad\quad$ so $(x-2)$ *is a factor of* $f(x)$

Now you can work through the rest of it, giving finally that

$\qquad f(x) = $ . . . .

22

$$f(x) = (x-2)(x+3)(x+4)$$

Because the long division gives $f(x) = (x-2)(x^2 + 7x + 12)$ and factorizing the quadratic expression finally gives

$f(x) = (x-2)(x+3)(x+4)$

And now one more, entirely on your own:

Factorize $f(x) = 2x^3 - 9x^2 + 7x + 6$

There are no snags. Just take your time. Work through the same steps as before and you get

$f(x) = \ldots\ldots\ldots$

23

$$f(x) = (x-2)(x-3)(2x+1)$$

$$f(x) = 2x^3 - 9x^2 + 7x + 6 = [(2x-9)x + 7]x + 6$$

$x = 2$ is the first substitution to give $f(x) = 0$. So $(x-2)$ is a factor.
Long division then leads to $f(x) = (x-2)(2x^2 - 5x - 3)$.
$(b^2 - 4ac) = 49$, i.e. 7^2, showing that simple factors exist for the quadratic.
In fact $2x^2 - 5x - 3 = (x-3)(2x+1)$
    so $f(x) = (x-2)(x-3)(2x+1)$

Factorization of fourth-order polynomials

24

The same method can be applied to polynomials of the fourth degree or order, provided that the given function has at least one simple factor.
    For example, to factorize $f(x) = 2x^4 - x^3 - 8x^2 + x + 6$:

In nested form, $f(x) = \{[(2x-1)x - 8]x + 1\}x + 6$

$f(1) = 0$, so $(x-1)$ *is* a factor.

$$
\begin{array}{r}
2x^3 + x^2 - 7x - 6 \\
x-1 \overline{\smash{\big)}\, 2x^4 - x^3 - 8x^2 + x + 6} \\
\underline{2x^4 - 2x^3} \\
x^3 - 8x^2 \\
\underline{x^3 - x^2} \\
-7x^2 + x \\
\underline{-7x^2 + 7x} \\
-6x + 6 \\
\underline{-6x + 6} \\
\bullet \quad \bullet
\end{array}
$$

So $f(x) = 2x^4 - x^3 - 8x^2 + x + 6$

$\qquad = (x - 1)(2x^3 + x^2 - 7x - 6) = (x - 1).g(x)$

Now we can proceed to factorize $g(x) = (2x^3 + x^2 - 7x - 6)$ as we did with previous cubics:

$$g(x) = [(2x + 1)x - 7]x - 6$$

$g(1) = -10 \qquad$ so $(x - 1)$ is not a factor of $g(x)$

$g(-1) = 0 \qquad$ so $(x + 1)$ *is* a factor of $g(x)$

Long division shows that $g(x) = (x + 1)(2x^2 - x - 6)$

so $f(x) = (x - 1)(x + 1)(2x^2 - x - 6)$

Attending to the quadratic $(2x^2 - x - 6)$:

$(b^2 - 4ac) = 1 + 48 = 49 = 7^2 \quad$ There are simple factors.

In fact $2x^2 - x - 6 = (2x + 3)(x - 2)$

Finally, then, $f(x) = (x - 1)(x + 1)(x - 2)(2x + 3)$

On to the next frame for another example

Factorize $f(x) = x^4 + x^3 - 9x^2 + x + 10$.

25

First, in nested form, $f(x) = \ldots\ldots\ldots\ldots$

$$\boxed{f(x) = \{[(x + 1)x - 9]x + 1\}x + 10}$$

26

Now we substitute $x = 1, -1, 2, \ldots$ from which we get

$f(1) = 4 \qquad$ so $(x - 1)$ is not a factor

$f(-1) = 0 \qquad$ so $(x + 1)$ *is* a factor

$$
\begin{array}{r}
x^3 + 0x^2 - 9x + 10 \\
x + 1 \overline{\smash{)}\; x^4 + x^3 - 9x^2 + x + 10} \\
\underline{x^4 + x^3 \phantom{{}- 9x^2 + x + 10}} \\
\bullet \quad \bullet \quad - 9x^2 + x \\
\underline{- 9x^2 - 9x } \\
10x + 10 \\
\underline{10x + 10} \\
\bullet \quad \bullet
\end{array}
$$

So $f(x) = (x + 1)(x^3 + 0x^2 - 9x + 10) = (x + 1).g(x)$

Then in nested form, $g(x) = \ldots\ldots\ldots\ldots$

27

$$g(x) = [(x+0)x - 9]x + 10$$

Now we hunt for factors by substituting $x = 1, -1, 2\ldots$ in $g(x)$

$g(1) = \ldots\ldots\ldots\ldots; g(-1) = \ldots\ldots\ldots\ldots; g(2) = \ldots\ldots\ldots\ldots$

28

$$g(1) = 2; \ g(-1) = 18; \ g(2) = 0$$

$g(2) = 0$ so $5(x-2)$ *is a factor of* $g(x)$

Long division $(x^3 + 0x^2 - 9x + 10) \div (x - 2)$ gives the quotient $\ldots\ldots\ldots\ldots$

29

$$x^2 + 2x - 5$$

$f(x) = (x+1)(x-2)(x^2 + 2x - 5)$, so we finally test the quadratic factor for simple factors and finish it off.

There are no linear factors of the quadratic

  so $f(x) = (x+1)(x-2)(x^2 + 2x - 5)$

30

One stage of long division can be avoided if we can find two factors of the original polynomial.

For example, factorize $f(x) = 2x^4 - 5x^3 - 15x^2 + 10x + 8$.

In nested form, $f(x)\ldots\ldots\ldots\ldots$

31

$$f(x) = \{[(2x - 5)x - 15]x + 10\}x + 8$$

$f(1) = 0$ so $(x - 1)$ *is a factor of* $f(x)$
$f(-1) = -10$ so $(x + 1)$ *is not a factor of* $f(x)$
$f(2) = -40$ so $(x - 2)$ *is not a factor of* $f(x)$
$f(-2) = \ldots\ldots\ldots\ldots$

32

$$f(-2) = 0 \ \text{ so } \ (x + 2) \text{ is a factor of } f(x)$$

$f(x) = (x - 1)(x + 2)(ax^2 + bx + c)$
    $= (x^2 + x - 2)(ax^2 + bx + c)$

We can now find the quadratic factor by dividing $f(x)$ by $(x^2 + x - 2)$:

$$
\begin{array}{r}
2x^2 \quad - \quad 7x \ - 4 \\[2pt]
x^2 + x - 2 \,\big)\, \overline{2x^4 - 5x^3 - 15x^2 + 10x + 8} \\[2pt]
2x^4 + 2x^3 - 4x^2 \\[2pt]
\hline
-7x^3 - 11x^2 + 10x \\[2pt]
-7x^3 - 7x^2 + 14x \\[2pt]
\hline
-4x^2 - 4x + 8 \\[2pt]
-4x^2 - 4x + 8 \\[2pt]
\hline
\bullet \quad \bullet \quad \bullet
\end{array}
$$

So $f(x) = (x - 1)(x + 2)(2x^2 - 7x - 4)$

Finally test the quadratic factor for simple factors and finish it off.

$f(x) = \ldots\ldots\ldots$

$$\boxed{f(x) = (x - 1)(x + 2)(x - 4)(2x + 1)}$$

33

For the quadratic, $(2x^2 - 7x - 4)$, $b^2 - 4ac = 81 = 9^2$ so factors exist.

In fact, $2x^2 - 7x - 4 = (x - 4)(2x + 1)$

so $f(x) = (x - 1)(x + 2)(x - 4)(2x + 1)$

Next frame

Now one further example that you can do on your own. It is similar to the previous one.

34

Factorize $f(x) = 2x^4 - 3x^3 - 14x^2 + 33x - 18$.

Follow the usual steps and you will have no trouble.

$f(x) = \ldots\ldots\ldots$

35

$$f(x) = (x-1)(x-2)(x+3)(2x-3)$$

Here is the working:

$$f(x) = 2x^4 - 3x^3 - 14x^2 + 33x - 18$$
$$= \{[(2x-3)x - 14]x + 33\}x - 18$$

$f(1) = 0$ so $(x-1)$ *is a factor of* $f(x)$

$f(-1) = -60$ so $(x+1)$ is not a factor of $f(x)$

$f(2) = 0$ so $(x-2)$ *is a factor of* $f(x)$

So $f(x) = (x-1)(x-2)(ax^2 + bx + c)$
$$= (x^2 - 3x + 2)(ax^2 + bx + c)$$

$$
\begin{array}{r}
2x^2 + 3x - 9 \\
x^2 - 3x + 2 \overline{\smash{\big)}\ 2x^4 - 3x^3 - 14x^2 + 33x - 18} \\
\underline{2x^4 - 6x^3 + 4x^2} \\
3x^3 - 18x^2 + 33x \\
\underline{3x^3 - 9x^2 + 6x} \\
-9x^2 + 27x - 18 \\
\underline{-9x^2 + 27x - 18} \\
\bullet \qquad \bullet \qquad \bullet
\end{array}
$$

So $f(x) = (x-1)(x-2)(2x^2 + 3x - 9)$

For $2x^2 + 3x - 9, (b^2 - 4ac) = 81 = 9^2$ Simple factors exist.

So $f(x) = (x-1)(x-2)(x+3)(2x-3)$

At this point let us pause and summarize the main facts so far on the evaluation process

 # Review summary   **Unit 12**

36

1 The process whereby we evaluate an algebraic expression can be described as a system f which accepts input x, processes the input and produces an output $f(x)$.

2 *Remainder theorem*
   If a polynomial $f(x)$ is divided by $(x-a)$, the quotient will be a polynomial $g(x)$ of one degree less than that of $f(x)$, together with a remainder R such that $R = f(a)$.

3 *Factor theorem*
   If $f(x)$ is a polynomial and substituting $x = a$ gives a remainder of zero, i.e. $f(a) = 0$, then $(x-a)$ is a factor of $f(x)$.

Review exercise   **Unit 12**

1 Rewrite $f(x) = 6x^3 - 5x^2 + 4x - 3$ in nested form and determine the value of $f(2)$.

2 Without dividing in full, determine the remainder of
$(x^4 - 2x^3 + 3x^2 - 4) \div (x - 2)$.

3 Factorize $6x^4 + x^3 - 25x^2 - 4x + 4$.

	37

Complete all of these questions, looking back at the Unit if you need to.
You can check your answers and working in the next frame.

1 $f(x) = 6x^3 - 5x^2 + 4x - 3 = (((6x - 5)x + 0)x + 4)x - 3$ so that
$f(2) = (((12 - 5)2 + 0)2 + 4)2 - 3 = 61$

2 $f(x) = x^4 - 2x^3 + 3x^2 - 4 = (((x - 2)x + 3)x + 0)x - 4$ so that
$f(2) = (((2 - 2)2 + 3)2 + 0)2 - 4 = 8 = $ remainder

3 $f(x) = 6x^4 + x^3 - 25x^2 - 4x + 4 = (((6x + 1)x - 25)x - 4)x + 4$
$f(2) = 0$ so $x - 2$ *is a factor*, $f(-2) = 0$ so $x + 2$ *is a factor.*
$(x - 2)(x + 2) = x^2 - 4$ and:

38

$$
\begin{array}{r}
6x^2 \ + \ x \ \ - 1 \\
x^2 - 4\ \overline{\big)\ 6x^4\ + x^3\ - 25x^2\ - 4x\ \ + 4} \\
6x^4 \qquad\quad - 24x^2 \\
\overline{\ \ x^3\ -\ \ x^2} \\
x^3 \qquad\ \ - 4x \\
\overline{\ \ -\ \ x^2 \qquad\ + 4} \\
-\ \ x^2 \qquad\ + 4 \\
\overline{\ \ \bullet \qquad\qquad \bullet}
\end{array}
$$

Furthermore, $6x^2 + x - 1 = (2x + 1)(3x - 1)$ so that:

$$6x^4 + x^3 - 25x^2 - 4x + 4 = (x + 2)(x - 2)(2x + 1)(3x - 1)$$

Now for the Review test

 # Review test   **Unit 12**

1 Write $f(x) = 7x^3 - 6x^2 + 4x + 1$ in nested form and find the value of $f(-2)$.

2 Without dividing in full, determine the remainder of:
$(4x^4 + 3x^3 - 2x^2 - x + 7) \div (x + 3)$.

3 Factorize $6x^4 + 5x^3 - 39x^2 + 4x + 12$.

	39

40 You have now come to the end of this Module. A list of **Can You?** questions follows for you to gauge your understanding of the material in the Module. You will notice that these questions match the **Learning outcomes** listed at the beginning of the Module. Now try the **Test exercise.** *Work through the questions at your own pace, there is no need to hurry.* A set of **Further problems** provides additional valuable practice.

 # Can You?

1   Checklist: Module 3

Check this list before and after you try the end of Module test.

On a scale of 1 to 5 how confident are you that you can:

- Numerically evaluate an algebraic expression by substituting numbers for variables?
  Yes ☐ ☐ ☐ ☐ ☐ *No*

- Recognize the different types of equation?
  Yes ☐ ☐ ☐ ☐ ☐ *No*

- Evaluate an independent variable?
  Yes ☐ ☐ ☐ ☐ ☐ *No*

- Change the subject of an equation by transposition?
  Yes ☐ ☐ ☐ ☐ ☐ *No*

- Evaluate polynomial expressions by 'nesting'?
  Yes ☐ ☐ ☐ ☐ ☐ *No*

- Use the remainder and factor theorems to factorize polynomials?
  Yes ☐ ☐ ☐ ☐ ☐ *No*

- Factorize fourth-order polynomials?
  Yes ☐ ☐ ☐ ☐ ☐ *No*

Test exercise 3

2

1 Evaluate to 3 sig fig:

$$\sqrt{\frac{2V}{\pi h} - \frac{h^2}{3}} \quad \text{where}$$

(a) $V = 23 \cdot 05$ and $h = 2 \cdot 69$

(b) $V = 85 \cdot 67$ and $h = 5 \cdot 44$.

2 Given $q = \sqrt{(5p^2 - 1)^2 - 2}$ find p in terms of q.

3 Express each of the following functions in nested form and determine the value of the function for the stated value of x:

(a) $f(x) = 2x^3 - 3x^2 + 5x - 4$ $\quad f(3) = \ldots\ldots\ldots$

(b) $f(x) = 4x^3 + 2x^2 - 7x - 2$ $\quad f(2) = \ldots\ldots\ldots$

(c) $f(x) = 3x^3 - 4x^2 + 8$ $\quad f(4) = \ldots\ldots\ldots$

(d) $f(x) = 2x^3 + x - 5$ $\quad f(5) = \ldots\ldots\ldots$

4 Determine the remainder that would occur if $(4x^3 - 5x^2 + 7x - 3)$ were divided by $(x - 3)$

5 Test whether $(x - 2)$ is a factor of $f(x) = 2x^3 + 2x^2 - 17x + 10$. If so, factorize $f(x)$ as far as possible.

6 Rewrite $f(x) = 2x^3 + 7x^2 - 14x - 40$ as the product of three linear factors.

7 Express the fourth-order polynomial $f(x) = 2x^4 - 7x^3 - 2x^2 + 13x + 6$ as the product of four linear factors.

 # Further problems 3

3

1 Evaluate the following, giving the results to 4 sig fig:

(a) $K = 14 \cdot 26 - 6 \cdot 38 + \sqrt{136 \cdot 5} \div (8 \cdot 72 + 4 \cdot 63)$

(b) $P = (21 \cdot 26 + 3 \cdot 74) \div 1 \cdot 24 + 4 \cdot 18^2 \times 6 \cdot 32$

(c) $Q = \dfrac{3 \cdot 26 + \sqrt{12 \cdot 13}}{14 \cdot 192 - 2 \cdot 4 \times 1 \cdot 63^2}$

2 Transpose each of the following formulas to make the symbol in the square brackets the subject:

(a) $V = IR$ $\qquad\qquad$ $[I]$

(b) $v = u + at$ $\qquad\qquad$ $[u]$

(c) $s = ut + \frac{1}{2}at^2$ $\qquad\qquad$ $[u]$

(d) $f = \dfrac{1}{2\pi\sqrt{LC}}$ $\qquad\qquad$ $[L]$

(e) $P = \dfrac{S(C - F)}{C}$ [C]

(f) $S = \sqrt{\dfrac{3D(L - D)}{8}}$ [L]

(g) $T = \dfrac{M - m}{1 + Mm}$ [M]

(h) $A = \pi r \sqrt{r^2 + h^2}$ [h]

(i) $V = \dfrac{\pi h}{6}(3R^2 + h^2)$ [R]

3 Rewrite the following in nested form and, in each case, determine the value of the function for the value of x stated:

(a) $f(x) = 5x^3 - 4x^2 + 3x - 12$ $[x = 2]$

(b) $f(x) = 3x^3 - 2x^2 - 5x + 3$ $[x = 4]$

(c) $f(x) = 4x^3 + x^2 - 6x + 2$ $[x = -3]$

(d) $f(x) = 2x^4 - 4x^3 + 2x^2 - 3x + 6$ $[x = 3]$

(e) $f(x) = x^4 - 5x^3 + 3x - 8$ $[x = 6]$

4 Without dividing in full, determine the remainder in each of the following cases:

(a) $(5x^3 + 4x^2 - 6x + 3) \div (x - 4)$

(b) $(3x^3 - 5x^2 + 3x - 4) \div (x - 5)$

(c) $(4x^3 + x^2 - 7x + 2) \div (x + 3)$

(d) $(2x^3 + 3x^2 - 4x + 5) \div (x + 4)$

(e) $(3x^4 - 2x^3 - 10x - 5) \div (x - 4)$

5 Factorize the following cubics as completely as possible:

(a) $x^3 + 6x^2 + 5x - 12$ (f) $4x^3 - 39x + 35$

(b) $2x^3 + 9x^2 - 11x - 30$ (g) $2x^3 - x^2 - 10x + 8$

(c) $3x^3 - 4x^2 - 28x - 16$ (h) $15x^3 + 53x^2 + 8x - 48$

(d) $3x^3 - x^2 + x + 5$ (i) $x^3 + x^2 - 2x - 8$

(e) $6x^3 - 5x^2 - 34x + 40$ (j) $6x^3 + 37x^2 + 67x + 30$

6 Factorize the following fourth-order polynomials, expressing the results as the products of linear factors where possible:

(a) $f(x) = 2x^4 - 5x^3 - 15x^2 + 10x + 8$

(b) $f(x) = 3x^4 - 7x^3 - 25x^2 + 63x - 18$

(c) $f(x) = 4x^4 - 4x^3 - 35x^2 + 45x + 18$

(d) $f(x) = x^4 + 2x^3 - 6x^2 - 2x + 5$

(e) $f(x) = 6x^4 - 11x^3 - 35x^2 + 34x + 24$

(f) $f(x) = 2x^4 + 5x^3 - 20x^2 - 20x + 48$

Graphs

Learning outcomes

When you have completed this Module you will be able to:

- Construct a collection of ordered pairs of numbers from an equation
- Plot points associated with ordered pairs of numbers against Cartesian axes and generate graphs
- Appreciate the existence of asymptotes to curves and discontinuities
- Use an electronic spreadsheet to draw Cartesian graphs of equations
- Describe regions of the x–y plane that are represented by inequalities
- Draw graphs of and algebraically manipulate the absolute value or modulus function

Units

Graphs of equations   **Unit 13**

1 Ordered pairs of numbers

Consider the equation:

$$y = x^2$$

where, you will recall, x is referred to as the independent variable and y is the dependent variable. Evaluating such an equation enables a collection of ordered pairs of numbers to be constructed. For example, if we select $x = 2$ the corresponding value of the dependent variable y is found to be $2^2 = 4$. From these two values the ordered pair of numbers (2, 4) can be constructed. It is called an *ordered* pair because the first number of the pair is always the value of the independent variable (here x) and the second number is the corresponding value of the dependent variable (here y).

So the collection of ordered pairs constructed from $y = x^2$ using successive integer values of x from -5 to 5 is ....

See the next frame

2

> (−5, 25), (−4, 16), (−3, 9), (−2, 4), (−1, 1),
> (0, 0), (1, 1), (2, 4), (3, 9), (4, 16), (5, 25)

Cartesian axes

We can *plot* the ordered pair of numbers (2, 4) on a graph. On a sheet of graph paper, we draw two straight lines perpendicular to each other. On each line we mark off the integers so that the two lines intersect at their common zero points. We can then plot the ordered pair of numbers (2, 4) as a point in the plane referenced against the integers on the two lines thus:

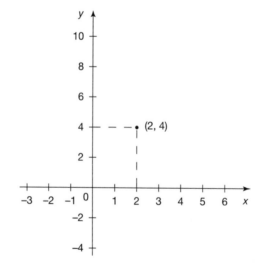

The arrangement of numbered lines is called the *Cartesian coordinate frame* and each line is called an *axis* (plural *axes*). The horizontal line is always taken to be the independent variable axis (here the *x*-axis) and the vertical line the dependent variable axis (here the *y*-axis). Notice that the scales of each axis need not be the same – the scales are chosen to make optimum use of the sheet of graph paper.

Now you try a couple of points. Plot the points (1, 1) and (−3, 9) against the same axes.

The result is in the next frame

3

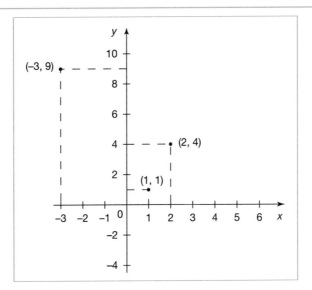

Drawing a graph

For an equation in a single independent variable, we can construct a collection of ordered pairs. If we plot each pair as a point in the same Cartesian coordinate frame we obtain a collection of isolated points.

On a sheet of graph paper plot the points (−5, 25), (−4, 16), (−3, 9), (−2, 4), (−1, 1), (0, 0), (1, 1), (2, 4), (3, 9), (4, 16) and (5, 25) obtained from the equation $y = x^2$.

Check your graph with the next frame

4

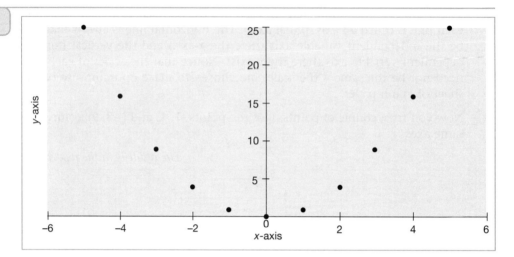

We have taken only integers as our x values. We could have taken any value of x. In fact, there is an infinite number of possible choices of x values and, hence, points. If we were able to plot this infinity of all possible points they would merge together to form a continuous line known as the *graph of the equation*. In practice, what we do is to plot a collection of isolated points and then join them up with a continuous line. For example, if we were to plot all the ordered pairs of numbers that could be constructed from the equation:

$$y = x^2 \text{ for } -5 \leq x \leq 5$$

we would end up with the shape given below:

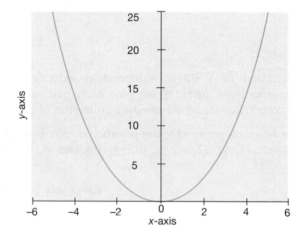

This is the graph of the equation $y = x^2$ for $-5 \leq x \leq 5$. We call the shape a *parabola*.

So what is the shape of the graph of $y = x + 1$ where $-4 \leq x \leq 4$? [Plot the graph using integer values of x.]

The answer is in the next frame

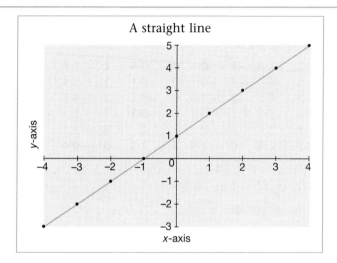

A straight line

Because, if we construct the following table:

x-value	−4	−3	−2	−1	0	1	2	3	4
y-value	−3	−2	−1	0	1	2	3	4	5

this gives rise to the following ordered pairs: $(-4, -3)$, $(-3, -2)$, $(-2, -1)$, $(-1, 0)$, $(0, 1)$, $(1, 2)$, $(2, 3)$, $(3, 4)$ and $(4, 5)$ which can be plotted as shown.

The plot we have obtained is the plot of just those ordered pairs of numbers that we have constructed, joined together by a continuous line.

Next frame

6

Try another one. Construct a table of *y*-values and then the graph of $y = x^3 - 2x^2 - x + 2$ with $-2 \leq x \leq 3$ at intervals of 0·5.

Check the next frame for the answer

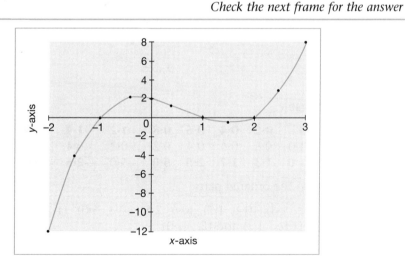

Because not all equations are polynomial equations we shall construct the ordered pairs by making use of a table rather than by *nesting* as we did in Module 3. Here is the table:

x	−2	−1·5	−1	−0·5	0	0·5	1	1·5	2	2·5	3
x^3	−8	−3·4	−1	−0·1	0	0·1	1	3·4	8	15·6	27
$-2x^2$	−8	−4·5	−2	−0·5	0	−0·5	−2	−4·5	−8	−12·5	−18
$-x$	2	1·5	1	0·5	0	−0·5	−1	−1·5	−2	−2·5	−3
	2	2	2	2	2	2	2	2	2	2	2
y	−12	−4·4	0	1·9	2	1·1	0	−0·6	0	2·6	8

Pairs: (−2, −12), (−1·5, −4·4), (−1, 0), (−0·5, 1·9), (0, 2), (0·5, 1·1), (1, 0), (1·5, −0·6), (2, 0), (2·5, 2·6), (3, 8)

How about the graph of:

$$y = \frac{1}{1-x} \text{ for } 0 \le x \le 2?$$

Select values of x ranging from 0 to 2 with intervals of 0·2 and take care here with the values that are near to 1.

The answer is in the next frame

8

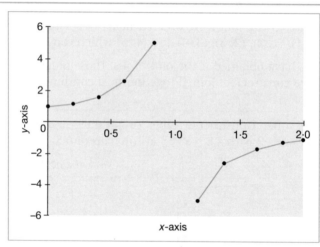

Here is the table:

x	0	0·2	0·4	0·6	0·8	1·2	1·4	1·6	1·8	2·0
$1-x$	1·0	0·8	0·6	0·4	0·2	−0·2	−0·4	−0·6	−0·8	−1·0
y	1·0	1·3	1·7	2·5	5·0	−5·0	−2·5	−1·7	−1·3	−1·0

giving rise to the ordered pairs:

(0, 1·0), (0·2, 1·3), (0·4, 1·7), (0·6, 2·5), (0·8, 5·0), (1·2, −5·0), (1·4, −2·5), (1·6, −1·7), (1·8, −1·3) and (2, −1·0)

Notice that we cannot find a value for y when $x = 1$ because then $1 - x = 0$ and we cannot divide by zero. The graph above can be improved upon by plotting more points – see below:

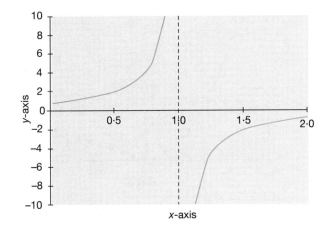

As x approaches the value 1 from the left, the graph rises and approaches (but never crosses) the vertical line through $x = 1$. Also as x approaches the value 1 from the right, the graph falls and approaches (but never crosses) the same vertical line. The vertical line through $x = 1$ is called a vertical *asymptote* to the graph. Not all asymptotes are vertical or even straight lines. Indeed, whenever a curve approaches a second curve without actually meeting or crossing it, the second curve is called an asymptote to the first curve.

Now, try finding the graph of the following:

$$y = \begin{cases} x^2 \text{ for } -5 \leq x < 0 \\ x \text{ for } \qquad x \geq 0 \end{cases} \quad -5 \leq x \leq 5 \text{ with intervals of } 0.5$$

Check the next frame

11

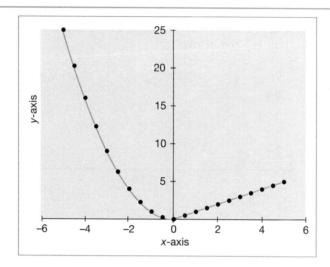

Because the equation:

$$y = \begin{cases} x^2 \text{ for } -5 \le x < 0 \\ x \text{ for } \qquad x \ge 0 \end{cases}$$

means that if x is chosen so that its value lies between -5 and 0 then $y = x^2$ and that part of the graph is the parabola shape. If x is greater than or equal to zero then $y = x$ and that part of the graph is the straight line. Notice that not all equations are of the simple form $y = $ *some expression in x*.

And finally, how about the graph of:

$$y = \begin{cases} 1 & \text{for } x \le 2 \\ -1 & \text{for } x > 2 \end{cases} \quad \text{for } -1 \le x \le 4$$

Next frame

10

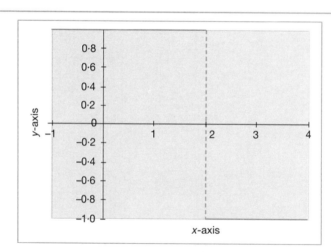

Because the equation:

$$y = \begin{cases} 1 & \text{for } x \le 2 \\ -1 & \text{for } x > 2 \end{cases}$$

means that no matter what the value is that is assigned to x, if it is less than or equal to 2 the value of y is 1. If the value of x is greater than 2 the corresponding value of y is -1. Notice the gap between the two continuous straight lines. This is called a *discontinuity* – not all equations produce smooth continuous shapes so *care must be taken when joining points together with a continuous line*. The dashed line joining the two end points of the straight lines is only there as a visual guide, *it is not part of the graph*.

At this point let us pause and summarize the main facts so far on graphs of equations

 # Review summary   Unit 13

1   Ordered pairs of numbers can be generated from an equation involving a **11** single independent variable.

2   Ordered pairs of numbers generated from an equation can be plotted against a Cartesian coordinate frame.

3   The graph of the equation is produced when the plotted points are joined by smooth curves.

4   Some equations have graphs that are given specific names, such as straight lines and parabolas.

5   Not all equations are of the simple form $y = some\ expression\ in\ x$.

6   Not all graphs are smooth, unbroken lines. Some graphs consist of breaks called discontinuities.

Review exercise   **Unit 13**

12   **1**   Given the equation:

$$x^2 + y^3 = 1$$

(a) Transpose the equation to make y the subject of the transposed equation.

(b) Construct ordered pairs of numbers corresponding to the integer values of x where $-5 \le x \le 5$.

(c) Plot the ordered pairs of numbers on a Cartesian graph and join the points plotted with a continuous curve.

2   Plot the graph of:

(a) $y = x^2 + \dfrac{1}{x}$ for $-3 \le x \le 3$ with intervals of 0.5.

(b) $y = \begin{cases} x^2 + x + 1 : x \le 1 \\ 3 - x \quad\;\; : x > 1 \end{cases}$ for $-3 \le x \le 4$ with intervals of 0.5.

Complete the questions. Take one step at a time, there is no need to rush.
If you need to, look back at the Unit.
The answers and working are in the next frame.

13   **1**   (a) $y = (1 - x^2)^{\frac{1}{3}}$

(b) $(-5, -2.9), (-4, -2.5), (-3, -2), (-2, -1.4), (-1, 0), (0, 1), (1, 0), (2, -1.4),$
$(3, -2), (4, -2.5), (5, -2.9)$

(c)

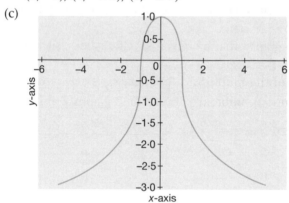

2 (a)

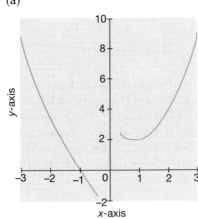

(b)

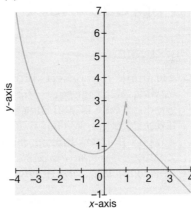

Now for the Review test

 # Review test

Unit 13

1 Given the equation:

14

$$y^2 - x^2 = 1$$

(a) Transpose the equation to make y the subject of the transposed equation.

(b) Construct ordered pairs of numbers corresponding to the even integer values of x where $-10 \leq x \leq 10$.

(c) Plot the ordered pairs of numbers on a Cartesian graph and join the points plotted with a continuous curve.

2 Plot the graph of:

(a) $y = \dfrac{1}{x^2}$ for $-3 \leq x \leq 3$ with intervals of 0·5

(b) $y = \begin{Bmatrix} -x & : x \leq 0 \\ x & : x > 0 \end{Bmatrix}$ for $-3 \leq x \leq 4$ with intervals of 0·5

Using a spreadsheet

Unit 14

1 Spreadsheets

Electronic spreadsheets provide extensive graphing capabilities and their use is widespread. Because *Microsoft* products are the most widely used software products on a PC, all the examples will be based on the *Microsoft* spreadsheet *Excel 2002* for *Windows*. It is expected that all later versions of *Excel* will support the handling characteristics of earlier versions with only a few minor changes. *If you have access to a computer algebra package you might ask your tutor how to use it to draw graphs.*

The features displayed are common to many spreadsheets but there may be minor differences in style between different products.

2 Rows and columns

Every electronic spreadsheet consists of a collection of cells arranged in a regular array of columns and rows. To enable the identification of individual cells each cell has an address given by the *column label* followed by the *row label*. In an *Excel* spreadsheet the columns are labelled alphabetically from **A** onwards and the rows are numbered from **1** onwards.

So that the cell with address **F123** is on the .... column of the .... row.

Check with the next frame

3

> **F** (6th) column of the **123**rd row

Because the address consists of the column label **F** (6th letter of the alphabet) followed by the row number **123**.

```
Microsoft Excel - Book1
File  Edit  View  Insert  Format  Tools  Data  Window  Help

⊢123              fx
        A      B      C      D      E      F      G
120
121
122
123                                          [    ]
124
125
```

At any time one particular cell boundary is highlighted with a *cursor* and this cell is called the *active cell*. An alternative cell can become the active cell by pressing the cursor movement keys on the keyboard (←, ↑, → and ↓) or, alternatively, by pointing at a particular cell with the *mouse pointer* () and then clicking the mouse button. Try it.

Next frame

Text and number entry

4

Every cell on the spreadsheet is capable of having numbers or text entered into it via the keyboard. Make the cell with the address **B10** the active cell and type in the text:

Text

and then press Enter (↵). Now make cell **C15** the active cell and type in the number 12 followed by Enter.

Next frame

5   Formulas

As well as text and numbers, each cell is capable of containing a formula. In an *Excel* spreadsheet every formula begins with the = (equals) sign when it is being entered at the keyboard. Move the cursor to cell **C16** and enter at the keyboard:

   $= 3*C15$

followed by Enter. The * represents multiplication ($\times$) and the formula states that the contents of cell **C16** will be three times the contents of cell **C15**. You will notice that the number displayed in cell **C16** is three times the number displayed in **C15**. Now change the number in **C15** by overwriting to see the effect. The number in **C16** also changes to reflect the change in the number in **C15**.

Next frame

6   Clearing entries

To clear an entry, point and click at the cell to be cleared. This makes it the active cell. Click the **Edit** command on the Command Bar to reveal a drop-down menu.

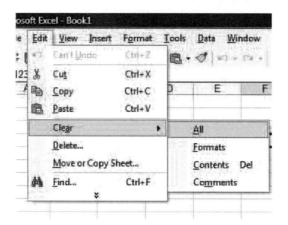

Select from this menu the option **Clear** to reveal a further drop-down menu. In this second menu select **All** and the cell contents are then cleared. Now make sure that all entries on the spreadsheet have been cleared because we want to use the spreadsheet to construct a graph.

Let's now put all this together in the next frame

Construction of a Cartesian graph

7

Follow these instructions to plot the graph of $y = (x - 2)^3$:

Enter the number -1 in **A1**
Highlight the cells **A1** to **A21** by pointing at **A1**, holding down the mouse button, dragging the pointer to **A21** and then releasing the mouse button (all the cells from **A2** to **A21** turn black to indicate that they have been selected)

Select the **Edit-Fill-Series** commands from the Command Bar and then in the *Series* window change the **Step value** from 1 to 0·3 and Click the **OK** button:

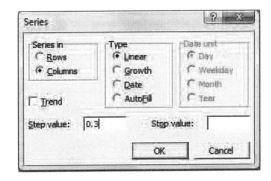

Cells **A2** to **A21** fill with single place decimals ranging from -1 to $+5$ with step value intervals of 0·3. These are the *x*-values, where $-1 \leq x \leq 5$.

In cell **B1**, type the formula $=(A1-2)\verb|^|3$ and then press Enter (the symbol $\wedge$ represents raising to a power). The number -27 appears in cell **B1** – that is $(-1 - 2)^3 = -27$ where -1 is the content of **A1**.

Activate cell **B1** and select the **Edit-Copy** commands
Highlight cells **B2** to **B21** and select **Edit-Paste**

Cells **B2** to **B21** fill with numbers, each being the number in the adjoining cell minus 2, all raised to the power 3; you have just copied the formula in **B1** into the cells **B2** to **B21**. These are the corresponding *y*-values.

Highlight the two columns of numbers – cells **A1:B21**
Click the *Chart Wizard* button

The *Chart Wizard Step 1 of 4* window has appeared, requesting a choice of chart type:

Click **XY (Scatter)** to reveal a further choice of types of XY Scatter charts:

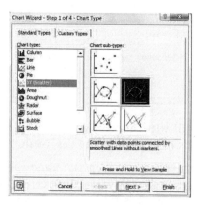

Click the type in the top right-hand corner to select it (*Scatter with data points connected by smooth lines without markers*). Press the **Next** button in this window to reveal the *Chart Wizard Step 2 of 4* window.

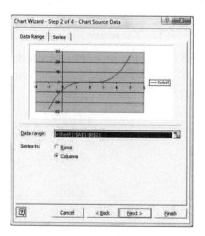

Click the **Next** button to reveal the *Chart Wizard Step 3 of 4* window:

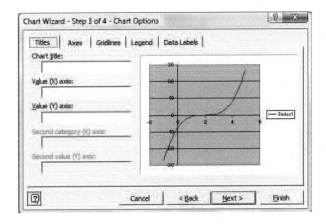

In the *Chart Wizard Step 3 of 4 window:*

> Click the *Legend* tab
> Clear the tick in the **Show legend** square (just Click the square)
> Click the *Titles* tab
> Enter in the **Value (X) Axis**
>> x-axis
> Enter in the **Value (Y) Axis**
>> y-axis
> Click the **Next** button to reveal the *Chart Wizard Step 4 of 4* window

In the *Chart Wizard Step 4 of 4* window ensure that the lower radio button is selected (it will have a black spot inside it when it is selected):

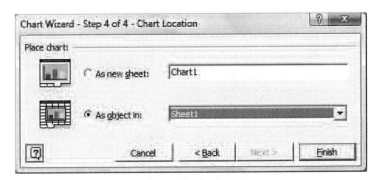

Click the **Finish** button to reveal the chart:

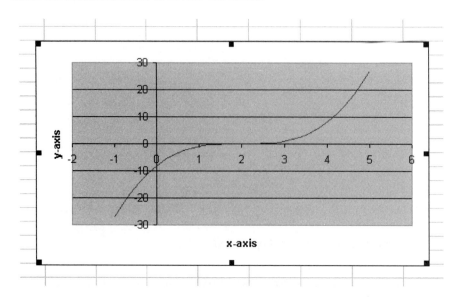

The small black squares (called *Handles*) around the Chart can be used to change the size of the Chart. Place the cursor over a Handle, hold down the mouse button and drag the mouse. Try it:

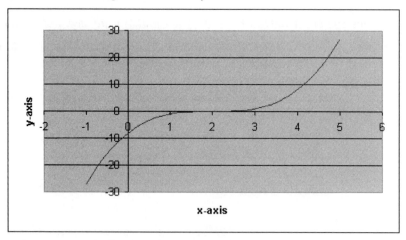

Now produce the graphs of the following equations. All you need to do is to change the formula in cell **B1** by activating it and then overtyping. Copy the new formula in **B1** down the **B** column and the graph will automatically update itself (*you do not have to clear the old graph, just change the formula*):

(a) $y = x^2 - 5x + 6$ Use * for multiplication so that $5x$ is entered as $5*x$
(b) $y = x^2 - 6x + 9$
(c) $y = x^2 - x + 1$
(d) $y = x^3 - 6x^2 + 11x - 6$

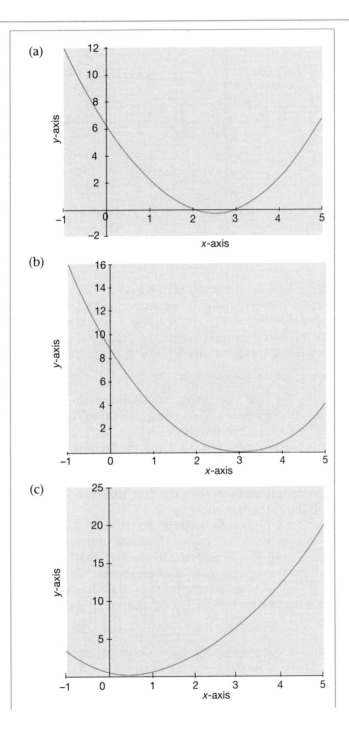

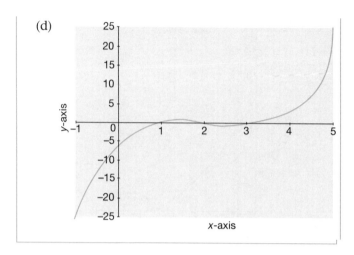

Because

(a) $y = x^2 - 5x + 6$ The formula in **B1** is =A1^2−5*A1+6. This shape is the parabola – every quadratic equation has a graph in the shape of a parabola. Notice that $y = 0$ at those points where the curve coincides with the x-axis, namely when $x = 2$ and $x = 3$. Also, because we can factorize the quadratic (see Module 2, Unit 10):

$$y = x^2 - 5x + 6 = (x - 2)(x - 3)$$

we can see that the graph demonstrates the fact that $y = 0$ when $x - 2 = 0$ or when $x - 3 = 0$ (see graph (a) in the box above).

(b) $y = x^2 - 6x + 9$ The formula in **B1** is =A1^2−6*A1+9. Notice that $y = 0$ at just one point when $x = 3$. Here, the factorization of the quadratic is:

$$y = x^2 - 6x + 9 = (x - 3)(x - 3)$$

so the graph demonstrates the fact that $y = 0$ only when $x = 3$ (see graph (b) in the box above).

(c) $y = x^2 - x + 1$ The formula in **B1** is =A1^2−A1+1. Notice that the curve never touches or crosses the x-axis so that there is no value of x for which $y = 0$. This is reflected in the fact that we cannot factorize the quadratic. (see graph (c) in the box above).

(d) $y = x^3 - 6x^2 + 11x - 6$ The formula in **B1** is =A1^3−6*A1^2+11*A1−6. Notice that $y = 0$ when $x = 1$, $x = 2$ and $x = 3$. Also, the cubic factorizes as:

$$y = x^3 - 6x^2 + 11x - 6 = (x - 1)(x - 2)(x - 3)$$

Again, we can see that the graph demonstrates the fact that $y = 0$ when $(x - 1) = 0$ or when $(x - 2) = 0$ or when $(x - 3) = 0$ (see graph (d) in the box above).

Let's try two graphs that we have already plotted manually. Repeat the same procedure by simply changing the formula in cell **B1** and then copying it into cells **B2** to **B21** to plot:

(a) $y = \dfrac{1}{1-x}$

(b) $y = \begin{cases} 1 & \text{for } x \leq 2 \\ -1 & \text{for } x > 2 \end{cases}$

You will have to give the second one some thought as to how you are going to enter the formula for the second equation.

Check your results in the next frame

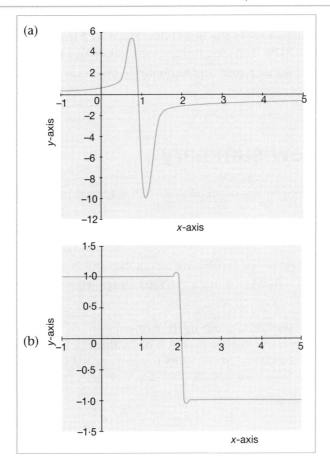

9

Because:

(a) $y = \dfrac{1}{1-x}$ The formula in **B1** is $= 1/(1-A1)$. Notice that the asymptotic behaviour is hidden by the joining of the two inside ends of the graph. See graph (a) in the box above. We can overcome this problem:

> Make cell **A8** the active cell
> On the Command Bar select **Insert-Rows**

An empty row appears above cell **A8** and the stray line on the graph disappears. (Clear the empty row using **Edit-Delete-Entire row**.)

(b) Here there are two formulas. The first formula in **B1** is $= 1$ copied down to cell **B11** and the second formula in **B12** is $= -1$ copied down to **B21**.

Notice that the two sides of the graph are joined together when they should not be – see graph (b) in the box above. Make **A12** the active cell and insert a row to remove the stray line just as you did in part (a).

At this point let us pause and summarize the main facts so far on using a spreadsheet

 # Review summary   Unit 14

10

1 A spreadsheet consists of an array of cells arranged in regular columns and rows.

2 Each cell has an address consisting of the column letter followed by the row number.

3 Each cell is capable of containing text, a number or a formula.

4 Cell entries are cleared by using the **Edit-Clear-All** sequence of commands.

5 To construct a graph:

> Enter the range of x-values in the first column
> Enter the corresponding collection of y-values in the second column
> Use the *Chart Wizard* to construct the graph using the *XY (Scatter)* option.

Review exercise

1 For x in the range $-2 \leq x \leq 6$ with a step value of $0\cdot4$ use a spreadsheet to draw   **11**
the graphs of:

(a) $y = x^2 - 3x + 2$

(b) $y = x^2 - 1$

(c) $y = 4x^2 - 3x + 25$

(d) $y = -x^3 + 6x^2 - 8x$

Complete the question. Take your time, there is no need to rush.
If necessary, refer back to the Unit.
The answers and working are in the next frame.

1 (a)   **12**

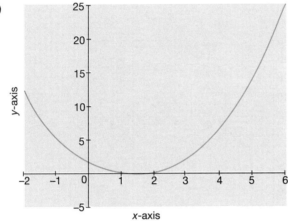

(b)

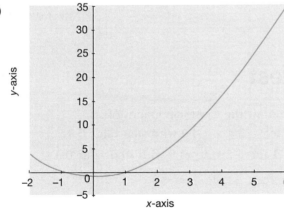

(c)

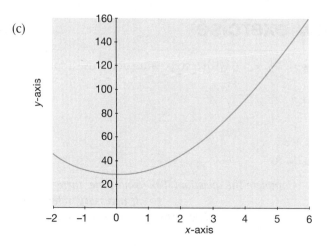

(d)

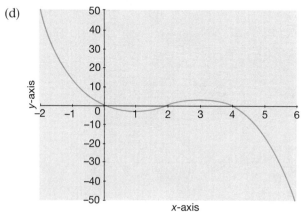

Now for the Review test

 # Review test   Unit 14

13   **1** Use a spreadsheet to draw the Cartesian graphs of:
       (a) $y = 2x^2 - 7x - 4$ for $-2 \leq x \leq 4.9$ with step value 0.3
       (b) $y = x^3 - x^2 + x - 1$ for $-2 \leq x \leq 4.9$ with step value 0.3

Inequalities

Less than or greater than

You are familiar with the use of the two inequality symbols $>$ and $<$. For example, $3 < 5$ and $-2 > -4$. We can also use them in algebraic expressions. For example:

$$y > x$$

The inequality tells us that whatever value is chosen for x the corresponding value for y is greater. Obviously there is an infinity of y-values greater than the chosen value for x, so the plot of an inequality produces an area rather than a line. For example, if we were to plot the graph of $y = x^2$ for $1 \leq x \leq 25$ we would obtain the graph shown below:

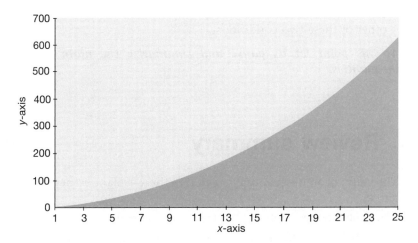

The graph is in the form of the curve $y = x^2$ acting as a separator for two different regions. The region above the line $y = x^2$ represents the plot of $y > x^2$ for $1 \leq x \leq 25$ because for every point in this region the y-value is greater than the square of the corresponding x-value. Similarly, the region below the line represents the plot of $y < x^2$ because for every point in this region the y-value is less than the square of the corresponding x-value.

So, without actually plotting it, what would be the region of the x–y plane that corresponds to the inequality $y < x^3 - 2x^2$?

Check the next frame for the answer

2

> The region of the graph below the curve $y = x^3 - 2x^2$

Because

For every point in this region the y-value is less than the corresponding x-value.

How about $y \geq 2x - 1$? The region of the x–y plane that this describes is . . . .

3

> The region of the graph on or above the line $y = 2x - 1$

Because

The inequality $y \geq 2x - 1$ means one of two conditions. Namely, $y > 2x - 1$ *or* $y = 2x - 1$ so that every point on or above the line satisfies one or the other of these two conditions.

At this point let us pause and summarize the main facts so far on inequalities.

 # Review summary   **Unit 15**

4

1 The graph of an inequality is a region of the x–y plane rather than a line or a curve.
2 Points above the graph of $y = f(x)$ are in the region described by $y > f(x)$.
3 Points below the graph of $y = f(x)$ are in the region described by $y < f(x)$.

 # Review exercise   **Unit 15**

5

1 Describe the regions of the x–y plane defined by each of the following inequalities:

   (a) $y < 3x - 4$
   (b) $y > -2x^2 + 1$
   (c) $y \leq x^2 - 3x + 2$

> *Complete the questions, working back over the previous frames if you need to.*
> *Don't rush, take your time.*
> *The answers and working are in the next frame.*

1 (a) The region below the line $y = 3x - 4$.

(b) The region above the line $y = -2x^2 + 1$.

(c) The region below and on the line $y = x^2 - 3x + 2$.

6

Now for the Review test

Review test
Unit 15

1 Describe the region of the x–y plane that corresponds to each of the following:

(a) $y > -x$ (b) $y \leq x - 3x^3$ (c) $x^2 + y^2 \leq 1$

7

Absolute values
Unit 16

Modulus
1

When numbers are plotted on a straight line the distance of a given number from zero is called the **absolute value** or **modulus** of that number. So the absolute value of -5 is 5 because it is 5 units distant from 0 and the absolute value of 3 is 3 because it is 3 units distant from 0.

So the absolute value of:

(a) -2.41 is ....

(b) 13.6 is ....

Answers in the next frame

2

(a) 2.41
(b) 13.6

Because

-2.41 is 2.41 units distant from 0 and 13.6 is 13.6 units distant from 0.

And the modulus of:

(a) -2.41 is ....

(b) 13.6 is ....

Answers in the next frame

3

> (a) 2.41
> (b) 13.6

Because

The modulus is just another name for the absolute value.

The absolute value of a number is denoted by placing the number between two vertical lines thus $|\ldots|$. For example, $|-7.35|$ represents the absolute value of -7.35. The absolute value 7.35 of the negative number -7.35 is obtained arithmetically by multiplying the negative number by -1.

We can use the same notation algebraically. For example, the equation $y = |x|$, (read as '$y = \text{mod } x$') means that y is equal to the absolute value of x. Where:

> $|x| = x$ if $x \geq 0$ and
> $|x| = -x$ if $x < 0$

so that:

> $|6| = 6$ because $6 > 0$ and
> $|-4| = -(-4) = 4$ because $-4 < 0$.

The absolute values of other expressions can similarly be found, for example

> $|x + 5| = x + 5$ if $x + 5 \geq 0$, that is $x \geq -5$ and
> $|x + 5| = -(x + 5)$ if $x + 5 < 0$, that is $x < -5$

So that $|x - 3| = \ldots\ldots\ldots\ldots$ if $x - 3 \geq 0$, that is $x\ldots\ldots\ldots\ldots$ and

$|x - 3| = \ldots\ldots\ldots\ldots$ if $x - 3 < 0$, that is $x\ldots\ldots\ldots\ldots$

The answer is in the following frame

4

> $|x - 3| = x - 3$ if $x - 3 \geq 0$, that is $x \geq 3$ and
> $|x - 3| = -(x - 3)$ if $x - 3 < 0$, that is $x < 3$

Now we shall look at the graphical properties of the absolute value function.

Next frame

5 Graphs

Using the spreadsheet to plot the graph of $y = |x|$ we can take advantage of the built-in function **ABS** which finds the absolute value.

Fill cells **A1** to **A21** with numbers in the range -5 to 5 with Step Value 0.5. In cell **B1** type in $= \textbf{ABS(A1)}$.

Now click the cell **B1** to place the highlight there. Move the cursor so that it is over the small black square, bottom right of the highlight, hold down the left-hand mouse button and drag the small square down until you reach cell **B21**. Release the mouse button and all the cells **B1 ... B21** fill with the appropriate copies of the formula. This is a much easier and quicker way to copy into a column of cells.

Highlight cells **A1 ... B21** and construct the graph of $y = |x|$ using the Chart Wizard and selecting the XY (Scatter) option to produce the following graph.

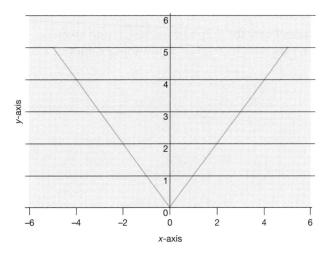

Notice that *Excel* does not do the best job possible with this graph. The point at the origin should be a sharp point whereas the graph produced by *Excel* appears to be rounded there. This demonstrates another danger of using *Excel* as anything other than a guide.

Next frame

Inequalities

6

Notice that a line drawn parallel to the *x*-axis, through the point $y = 2$ intersects the graph at $x = \pm 2$. This graphically demonstrates that if:

$y < 2$, that is $|x| < 2$ then $-2 < x < 2$ and if
$y > 2$, that is $|x| > 2$ then $x < -2$ or $x > 2$

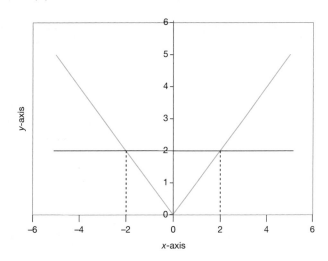

Indeed, if:

$|x| < a$ where $a > 0$ then $-a < x < a$ and if

$|x| > a$ where $a > 0$ then $x < -a$ or $x > a$

You try one. Draw the graph of $y = |x - 1|$ and ascertain from the graph those values of x for which $|x - 1| < 3$ and those values of x for which $|x - 1| > 3$.

Next frame

7

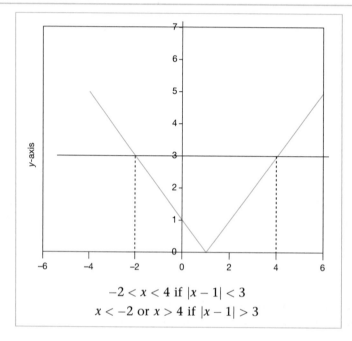

$-2 < x < 4$ if $|x - 1| < 3$

$x < -2$ or $x > 4$ if $|x - 1| > 3$

The graph has the same shape but is shifted to the right 1 unit along the x-axis. The values of x for which $|x - 1| > 3$ are clearly $-2 < x < 4$ and for $|x - 1| > 3$ they are $x < -2$ or $x > 4$.

These graphical considerations are all very well but we do really need to be able to derive these results algebraically and this we now do before returning to an interactive graphical construction.

Less-than inequalities

If $|x| < 3$ then this means that the values of x lie between ± 3, that is $-3 < x < 3$. Now, extending this, the values of x that satisfy the inequality $|x - 1| < 3$ are those where the values of $x - 1$ lie between ± 3. That is:

$-3 < x - 1 < 3$

By adding 1 to both sides of each of the two inequalities we find:

$-3 + 1 < x < 3 + 1$ that is $-2 < x < 4$

In general if $-a < x \pm b < a$ then $-a \mp b < x < a \mp b$.

Now you try one. The values of x that satisfy the inequality $|x + 4| < 1$ are . . . .

The result is in the next frame

8

$$-5 < x < -3$$

Because

$|x + 4| < 1$ means that the values of $x + 4$ lie between ± 1. That is:

$-1 < x + 4 < 1$

By subtracting 4 from both sides of each of the two inequalities we find:

$-1 - 4 < x < 1 - 4$ that is $-5 < x < -3$

Just as we added and subtracted across inequalities in the last two examples, so we can multiply and divide by a positive number. For example:

$|5x - 1| < 2$ means that the values of $5x - 1$ lie between ± 2

That is:

$-2 < 5x - 1 < 2$

By adding 1 to both sides of each of the two inequalities we find:

$-2 + 1 < 5x < 2 + 1$ that is $-1 < 5x < 3$

Now, by dividing each of the two inequalities by 5 we find:

$-\dfrac{1}{5} < x < \dfrac{3}{5}$

Now you try one. The values of x that satisfy the inequality $\left|\dfrac{x}{2} + 6\right| < 9$ are . . . .

The result is in the next frame

9

$$-30 < x < 6$$

Because

$\left|\dfrac{x}{2} + 6\right| < 9$ means that the values of $\dfrac{x}{2} + 6$ lie between ± 9. That is:

$-9 < \dfrac{x}{2} + 6 < 9$

By subtracting 6 from both sides of each of the two inequalities we find:

$-9 - 6 < \dfrac{x}{2} < 9 - 6$ that is $-15 < \dfrac{x}{2} < 3$

Finally, by multiplying by 2 across both sides of each of the two inequalities, we find:

$-30 < x < 6$

Multiplying and dividing an inequality by a negative number raises an issue. For example, if $x < 2$ then the value $x = 1$ will satisfy this inequality because $1 < 2$.

If, however, we multiply both sides of the inequality by -1 we cannot assert that $-x < -2$ and that the value $x = 1$ will still satisfy the inequality because $-1 > -2$.

However, we can solve this problem if *we switch the inequality when we multiply or divide by a negative number*. If $x < 2$ then muliplying by -1 gives $-x > -2$ and $x = 1$ satisfies both inequalities.

So, the values of x that satisfy the inequality $|7 - 2x| < 9$ are . . . .

Next frame

10

$$-1 < x < 8$$

Because

The values of x are such that $-9 < 7 - 2x < 9$. That is $-9 - 7 < -2x < 9 - 7$ so that $-16 < -2x < 2$ and so, by dividing both sides of the inequalities by -2 we see that $8 > x > -1$. *Notice the switch in the inequalities because we have divided both sides of the inequalities by a negative number*. This is then written as $-1 < x < 8$.

One more just to make sure. The values of x that satisfy the inequality $|5 - x/3| < 11$ are . . . .

Next frame

11

$$-18 < x < 48$$

Because

The values of x are such that $-11 < 5 - x/3 < 11$. That is $-11 - 5 < -x/3 < 11 - 5$ so that $-16 < -x/3 < 6$ and so $48 > x > -18$. *Again, notice the switch in the inequalities because we have multiplied both sides of the inequality by a negative number -3*. This is then written as $-18 < x < 48$.

Greater-than inequalities

If $|x| > 4$ then this means that $x > 4$ or $x < -4$. Now, extending this, the values of x that satisfy the inequality $|x + 3| > 7$ are those where:

$x + 3 > 7$ or $x + 3 < -7$

That is:

$x > 7 - 3$ or $x < -7 - 3$

and so:

$x > 4$ or $x < -10$

In general, if $|x \pm b| > a$ then $x > a \mp b$ or $x < -a \mp b$.

Now you try one. The values of x that satisfy the inequality $|x - 1| > 3$ are . . . .

Next frame

12

$$x < -2 \text{ or } x > 4$$

Because

The values of x are such that $x - 1 < -3$ or $x - 1 > 3$. That is $x < -3 + 1$ or $x > 3 + 1$ so that $x < -2$ or $x > 4$.

As we have already seen in Frame 9 of this Unit, multiplying and dividing an inequality by a negative number raises an issue which is resolved by *switching the inequality*. Therefore the values of x that satisfy the inequality $|5 - x| > 2$ can be found as follows.

The values of x that satisfy the inequality $|5 - x| > 2$ are those where:

$5 - x > 2$ or $5 - x < -2$

That is:

$-x > 2 - 5$ or $-x < -2 - 5$

and so:

$-x > -3$ or $-x < -7$

Multiplying by -1 then yields:

$x < 3$ or $x > 7$

Try one yourself. The values of x that satisfy the inequality $|7 - 2x| > 9$ are . . . .

Next frame

13

$$x > 8 \text{ or } x < -1$$

Because

The values of x are such that $7 - 2x < -9$ or $7 - 2x > 9$. That is $-2x < -16$ or $-2x > 2$ so that $x > 8$ or $x < -1$. *Notice again the switch in the inequalities because we have divided by a negative number.*

Now we look at the construction of an interactive graph.

Next frame

14 Interaction

The spreadsheet can be used to demonstrate how changing various features of an equation affect the graph but before we can do this we need to know the difference between the cell addresses **A1** and **A1** when used within a formula. The first address is called a relative address and the spreadsheet interprets it not as an actual address but as an address relative to the cell in which it is written. For example, if **A1** forms part of a formula in cell **B1** then the spreadsheet understands **A1** to refer to the cell on the same row, namely row **1** but one column to the left. So if the formula were then to be copied into cell **B2** the reference **A1** would automatically be copied as **A2** – the cell on the same row as **B2** but one column to the left. The contents of cell **B2** would then use the contents of cell **A2** and not those of **A1**. The address **A1**, on the other hand, refers to the actual cell at that address. We shall see the effect of this now.

> Fill cells **A1** to **A21** with numbers in the range −5 to 5 with Step Value 0·5.
> In cell **B1** type in **= ABS(C1*(A1 + D1)) + E1**.

That is $y = |a(x + b)| + c$. We are going to use cells **C1**, **D1** and **E1** to vary the parameters a, b and c.

> In cell **C1** enter the number 1.
> In cell **D1** enter the number 0.
> In cell **E1** enter the number 0.
> Copy the formula in **B1** into cells **B2 ... B21**.
> Draw the graph.

The result is the graph of $y = |x|$. That is $y = |a(x + b)| + c$ where $a = 1$, $b = 0$ and $c = 0$.

The graph has been created by using the fixed values of 1, 0 and 0 for a, b and c which are located in cells **C1**, **D1** and **E1** respectively. The formula refers to them by their absolute addresses **C1**, **D1** and **E1**. The relative address **A1** has been used in the formula to refer to the value of x, the variable values of which are located in cells **A1...A21**. If on the Command Bar you select **View** and then **Formula Bar** you display the *Formula Bar*. Now move the cursor down column **B** from row **1** to row **21**. You will see in the Formula Bar that the relative address changes but the absolute addresses do not.

Now:

> (a) In cell **D1** enter the number 1 ($b = 1$). What happens?
> (b) In cell **E1** enter the number 3 ($c = 3$). What happens?
> (c) In cell **C1** enter the number 4 ($a = 4$). What happens?

Next frame

> (a) The graph moves 1 unit to the left
> (b) The graph moves 3 units up
> (c) The gradients of the arms of the graph increase

15

This latter is noticeable from the change in the scale on the vertical axis. By changing the numbers in cells **C1**, **D1** and **E1** in this way you will be able to see the immediate effect on the graph.

Let's pause now and summarize the work on absolute values

 # Review summary

Unit 16

16

1 *Absolute values*: When numbers are plotted on a straight line the distance of a given number from zero is called the absolute value or modulus of that number. The absolute value is denoted by placing the number between two vertical lines. For example $|-3| = 3$, which reads that the absolute value of -3 is 3.

2 *Graphs*: Using an Excel spreadsheet to plot a graph of absolute values necessitates the use of the **ABS** function. The graph of $y = |x|$ consists of two straight lines forming a 'V' with the point at the origin. The graph of $y = |x - a|$ consists of the same two straight lines forming a 'V' with the point $x = a$.

3 *Inequalities*: If, for $a > 0$,

  (a) $|x| < a$ then $-a < x < a$

  (b) $|x \pm b| < a$ then $-a \mp b < x < a \mp b$

  (c) $|x| > a$ then $x > a$ or $x < -a$

  (d) $|x \pm b| > a$ then $x > a \mp b$ or $x < -a \mp b$

  (e) $-ax > b$ then $x < -\dfrac{b}{a}$

  (f) $-ax < b$ then $x > -\dfrac{b}{a}$

4 *Interaction*: A spreadsheet has two types of address. The first is the relative address which is not an actual address, but an address relative to the cell in which it is written. For example if **A1** forms part of a formula in cell **B1** then the spreadsheet understands **A1** to refer to the cell in the same row (here row **1**) but one column to the left. If the formula were then copied to cell **B2** the reference **A1** would automatically be copied as **A2** (that is the same row but one column to the left).

    The address **A1**, on the other hand, refers to the actual cell at that address. It is called an absolute address because it refers to an actual cell.

    The spreadsheet can be made interactive by using relative and absolute addresses.

Review exercise   **Unit 16**

17

1 Describe the region of the x–y plane that corresponds to each of the following:

(a) $y > 1 - 4x$ (b) $y \leq \dfrac{3}{x} - 5x$ (c) $2x + 3y > 6$

2 What values of x satisfy each of the following?

(a) $|2 + x| > 5$ (b) $|3 - x| < 4$ (c) $|3x - 2| \geq 6$

Complete the questions. Take your time.
Look back at the Unit if necessary, but don't rush.
The answers and working are in the next frame.

18

1 (a) $y > 1 - 4x$

The equation $y = 1 - 4x$ is the equation of a straight line with gradient -4 that crosses the y-axis at $y = 1$. The inequality $y > 1 - 4x$ defines points for a given x-value whose y-values are greater than the y-value of the point that lies on the line. Therefore the inequality defines the region above the line.

(b) $y \leq \dfrac{3}{x} - 5x$

The equation $y = \dfrac{3}{x} - 5x$ is the equation of a curve. The inequality $y \leq \dfrac{3}{x} - 5x$ defines points for a given x-value whose y-values are less than or equal to the y-value of the point that lies on the line. Therefore the inequality defines the region on and below the curve.

(c) $2x + 3y > 6$

The inequality $2x + 3y > 6$ can be rewritten as follows:

subtracting $2x$ from both sides gives $3y > 6 - 2x$

dividing both sides by 3 gives $y > 2 - (2/3)x$

The equation $2x + 3y = 6$ is, therefore, the equation of a straight line with gradient $-2/3$ that crosses the y-axis at $y = 2$.

This inequality defines points for a given x-value whose y-values are greater than the y-value of the point that lies on the line. Therefore the inequality defines the region above the line.

2 (a) $|2 + x| > 5$ can be written as:

$(2 + x) > 5$ or $(2 + x) < -5$ that is $x > 5 - 2$
or $x < -5 - 2$ so $x > 3$ or $x < -7$

(b) $|3 - x| < 4$ can be written as:

$-4 < (3 - x) < 4$ that is $-4 - 3 < -x < 4 - 3$
so $-7 < -x < 1$, therefore $-7 < -x$ and $-x < 1$
so that $7 > x$ and $x > -1$ therefore $-1 < x < 7$

▷

(c) $|3x - 2| \geq 6$ can be written as:

$$(3x - 2) \geq 6 \quad \text{or} \quad (3x - 2) \leq -6 \quad \text{that is} \quad x \geq \frac{6 + 2}{3} \quad \text{or} \quad x \leq \frac{-6 + 2}{3}$$

$$\text{so} \quad x \geq \frac{8}{3} \quad \text{or} \quad x \leq -\frac{4}{3}$$

Now for the Review test

Review test

Unit 16

1 What values of x satisfy each of the following?

(a) $|x + 2| < 5$ (b) $|x + 2| > 5$ (c) $|2x - 3| < 4$ (d) $|8 - 5x| > 12$

19

You have now come to the end of this Module. A list of **Can You?** questions follows for you to gauge your understanding of the material in the Module. You will notice that these questions match the **Learning outcomes** listed at the beginning of the Module. Now try the **Test exercise**. *Work through the questions at your own pace, there is no need to hurry.* A set of **Further problems** provides valuable additional practice.

20

Can You?

Checklist: Module 4

1

Check this list before and after you try the end of Module test.

On a scale of 1 to 5 how confident are you that you can:

- Construct a collection of ordered pairs of numbers from an equation?
  Yes ☐ ☐ ☐ ☐ ☐ *No*

- Plot points associated with ordered pairs of numbers against Cartesian axes and generate graphs?
  Yes ☐ ☐ ☐ ☐ ☐ *No*

- Appreciate the existence of asymptotes to curves and discontinuities?
  Yes ☐ ☐ ☐ ☐ ☐ *No*

- Use an electronic spreadsheet to draw Cartesian graphs of equations?
  Yes ☐ ☐ ☐ ☐ ☐ *No*

- Describe regions of the x–y plane that are represented by inequalities?
  Yes ☐ ☐ ☐ ☐ ☐ *No*

- Draw graphs of and algebraically manipulate the absolute value or modulus function?
  Yes ☐ ☐ ☐ ☐ ☐ *No*

Test exercise 4

2

1 Given the equation:

$$x^2 + y^2 = 1$$

(a) Transpose the equation to make y the subject of the transposed equation.

(b) Construct ordered pairs of numbers corresponding to the integer values of x where $-1 \leq x \leq 1$ in intervals of 0·2.

(c) Plot the ordered pairs of numbers on a Cartesian graph and join the points plotted with a continuous curve.

2 Plot the graph of:

(a) $y = x - \dfrac{1}{x}$ for $-3 \leq x \leq 3$.

What is happening near to $x = 0$?

(b) $y = \begin{cases} 2 - x & : x \leq 0 \\ x^2 + 2 & : x > 0 \end{cases}$ for $-4 \leq x \leq 4$

3 Use a spreadsheet to draw the Cartesian graphs of:

(a) $y = 3x^2 + 7x - 6$ for $-4 \leq x \leq 5\cdot2$ with step value 0·4.

(b) $y = x^3 + x^2 - x + 1$ for $-2 \leq x \leq 4\cdot9$ with step value 0·3.

4 Describe the region of the x–y plane that corresponds to each of the following:

(a) $y < 2 - 3x$ (b) $y > x - \dfrac{2}{x}$ (c) $x + y \leq 1$

5 What values of x satisfy each of the following?

(a) $|x + 8| < 13$ (b) $|x + 8| > 13$

(c) $|4x - 5| < 8$ (d) $|6 - 3x| > 14$

 Further problems 4

3

1 Given $x^2 - y^2 = 0$ find the equation giving y in terms of x and plot the graph of this equation for $-3 \leq x \leq 3$.

2 Using a spreadsheet plot the graph of:

$$y = x^3 + 10x^2 + 10x - 1$$

for $-10 \leq x \leq 2$ with a step value of 0·5.

3 Using a spreadsheet plot the graph of:

$$y = \dfrac{x^3}{1 - x^2}$$

for $-2 \leq x \leq 2\cdot6$ with a step value of 0·2. Draw a sketch of this graph on a sheet of graph paper indicating discontinuities and asymptotic behaviour more accurately.

4 Given the equation $x^2 + y^2 = 4$ transpose it to find y in terms of x. With the aid of a spreadsheet describe the shape that this equation describes.

5 Given the equation:

$$\left(\frac{x}{2}\right)^2 + \left(\frac{y}{4}\right)^2 = 1$$

transpose it to find y in terms of x. With the aid of a spreadsheet describe the shape that this equation describes.

6 Given the equation $x^2 + y^2 + 2x + 2y + 1 = 0$ transpose it to find y in terms of x. With the aid of a spreadsheet describe the shape that this equation describes.

7 Describe the region of the x–y plane that corresponds to each of the following:

(a) $y \geq 2x + 4$ (b) $y < 3 - x$ (c) $3x - 4y \geq 1$ (d) $x^2 + y^2 > 2$

8 Draw the graph of $f(x) = |3x - 4| + 2$ and describe the differences between that graph and the graph of $g(x) = |4 - 3x| + 2$.

9 What values of x between 0 and 2π radians satisfy each of the following:

(a) $|\sin x| < 0.5$

(b) $|\cos x| > 0.5$

10 Show that $|a - b| \leq c$ is equivalent to $b - c \leq a \leq b + c$.

11 Verify the rule that for two real numbers x and y then
$|x + y| \leq |x| + |y|$.

12 Given that for two real numbers x and y, $|x + y| \leq |x| + |y|$, deduce that $|x - y| \leq |x| - |y|$.

Linear equations and simultaneous linear equations

Linear equations

Unit 17

1

A linear equation in a single variable (unknown) involves powers of the variable no higher than the first. A linear equation is also referred to as a simple equation.

Solution of simple equations

The solution of simple equations consists essentially of simplifying the expressions on each side of the equation to obtain an equation of the form $ax + b = cx + d$ giving $ax - cx = d - b$ and

hence $x = \dfrac{d - b}{a - c}$ provided $a \neq c$.

Example

If $6x + 7 + 5x - 2 + 4x - 1 = 36 + 7x$

then $15x + 4 = 7x + 36$

                    so $8x = 32$ therefore $x = 4$

Similarly:

if $5(x - 1) + 3(2x + 9) - 2 = 4(3x - 1) + 2(4x + 3)$

then $x = \ldots\ldots\ldots\ldots$

$$\boxed{x = 2}$$

2

Because

$5(x - 1) + 3(2x + 9) - 2 = 4(3x - 1) + 2(4x + 3)$

so $5x - 5 + 6x + 27 - 2 = 12x - 4 + 8x + 6$

                $11x + 20 = 20x + 2$

            so $18 = 9x$ therefore $x = 2$

Equations which appear not to be simple equations sometimes develop into simple equations during simplification. For example:

$(4x + 1)(x + 3) - (x + 5)(x - 3) = (3x + 1)(x - 4)$

$(4x^2 + 13x + 3) - (x^2 + 2x - 15) = 3x^2 - 11x - 4$

            so $3x^2 + 11x + 18 = 3x^2 - 11x - 4$

$3x^2$ can now be subtracted from each side, giving:

$11x + 18 = -11x - 4$

    so $22x = -22$ therefore $x = -1$

▷

It is always wise to check the result by substituting this value for x in the original equation:

$$\text{LHS} = (4x+1)(x+3) - (x+5)(x-3)$$
$$= (-3)(2) - (4)(-4) = -6 + 16 = 10$$
$$\text{RHS} = (3x+1)(x-4) = (-2)(-5) = 10 \quad \therefore \ \text{LHS} = \text{RHS}$$

So, in the same way, solving the equation:

$$(4x+3)(3x-1) - (5x-3)(x+2) = (7x+9)(x-3)$$

we have $x = \dots\dots\dots$

3

$$\boxed{x = -3}$$

Simplification gives $7x^2 - 2x + 3 = 7x^2 - 12x - 27$

Hence $10x = -30 \quad \therefore \ x = -3$

Where simple equations involve algebraic fractions, the first step is to eliminate the denominators by multiplying throughout by the LCM (Lowest Common Multiple) of the denominators. For example, to solve:

$$\frac{x+2}{2} - \frac{x+5}{3} = \frac{2x-5}{4} + \frac{x+3}{6}$$

The LCM of 2, 3, 4 and 6 is 12:

$$\frac{12(x+2)}{2} - \frac{12(x+5)}{3} = \frac{12(2x-5)}{4} + \frac{12(x+3)}{6}$$
$$6(x+2) - 4(x+5) = 3(2x-5) + 2(x+3)$$
$$\therefore \ x = \dots\dots\dots$$

4

$$\boxed{x = \frac{1}{6}}$$

That was easy enough. Now let us look at this one.

To solve $\dfrac{4}{x-3} + \dfrac{2}{x} = \dfrac{6}{x-5}$

Here, the LCM of the denominators is $x(x-3)(x-5)$. So, multiplying throughout by the LCM, we have

$$\frac{4x(x-3)(x-5)}{x-3} + \frac{2x(x-3)(x-5)}{x} = \frac{6x(x-3)(x-5)}{x-5}$$

After cancelling where possible:

$$4x(x-5) + 2(x-3)(x-5) = 6x(x-3)$$

so, finishing it off, $x = \dots\dots\dots$

5

$$x = \frac{5}{3}$$

Because
$$4x^2 - 20x + 2(x^2 - 8x + 15) = 6x^2 - 18x$$
$$4x^2 - 20x + 2x^2 - 16x + 30 = 6x^2 - 18x$$
$$\therefore \ 6x^2 - 36x + 30 = 6x^2 - 18x$$

Subtracting the $6x^2$ term from both sides:

$$-36x + 30 = -18x \quad \therefore \ 30 = 18x \quad \therefore \ x = \frac{5}{3}$$

Working in the same way: $\dfrac{3}{x-2} + \dfrac{5}{x-3} - \dfrac{8}{x+3} = 0$

gives the solution $x = \ldots\ldots\ldots\ldots$

6

$$x = \frac{7}{3}$$

The LCM of the denominators is $(x-2)(x-3)(x+3)$, so

$$\frac{3(x-2)(x-3)(x+3)}{x-2} + \frac{5(x-2)(x-3)(x+3)}{x-3} - \frac{8(x-2)(x-3)(x+3)}{x+3} = 0$$

$$\therefore \ 3(x-3)(x+3) + 5(x-2)(x+3) - 8(x-2)(x-3) = 0$$
$$3(x^2 - 9) + 5(x^2 + x - 6) - 8(x^2 - 5x + 6) = 0$$
$$3x^2 - 27 + 5x^2 + 5x - 30 - 8x^2 + 40x - 48 = 0$$
$$\therefore \ 45x - 105 = 0$$
$$\therefore \ x = \frac{105}{45} = \frac{7}{3} \quad \therefore \ x = \frac{7}{3}$$

Now on to Frame 7

Simultaneous linear equations with two unknowns

7

A linear equation in two variables has an infinite number of solutions. For example, the two-variable linear equation $y - x = 3$ can be transposed to read:

$$y = x + 3$$

Any one of an infinite number of x-values can be substituted into this equation and each one has a corresponding y-value. However, for two such equations there may be just one pair of x- and y-values that satisfy both equations *simultaneously*.

Solution by substitution

To solve the pair of equations

$$5x + 2y = 14 \qquad\qquad\qquad\qquad\qquad (1)$$
$$3x - 4y = 24 \qquad\qquad\qquad\qquad\qquad (2)$$

From (1): $5x + 2y = 14$ $\therefore$ $2y = 14 - 5x$ $\therefore$ $y = 7 - \dfrac{5x}{2}$

If we substitute this for y in (2), we get:

$$3x - 4\left(7 - \frac{5x}{2}\right) = 24$$
$$\therefore\ 3x - 28 + 10x = 24$$
$$13x = 52 \quad \therefore\ x = 4$$

If we now substitute this value for x in the other original equation, i.e. (1), we get:

$$5(4) + 2y = 14$$
$$20 + 2y = 14 \quad \therefore\ 2y = -6 \quad \therefore\ y = -3$$
$$\therefore\ \text{We have } x = 4,\ y = -3$$

As a check, we can substitute both these values in (1) and (2):

$$(1) \quad 5x + 2y = 5(4) + 2(-3) = 20 - 6\ = 14 \ \checkmark$$
$$(2) \quad 3x - 4y = 3(4) - 4(-3) = 12 + 12 = 24 \ \checkmark$$
$$\therefore\ x = 4,\ y = -3 \text{ is the required solution.}$$

Another example:

To solve

$$3x + 4y = 9 \qquad\qquad\qquad\qquad\qquad (1)$$
$$2x + 3y = 8 \qquad\qquad\qquad\qquad\qquad (2)$$

Proceeding as before, determine the values of x and y and check the results by substitution in the given equations.

$$x = \dots\dots\dots\dots,\quad y = \dots\dots\dots\dots$$

$$\boxed{x = -5, \ y = 6}$$

Because

$$(1) \qquad\qquad 3x + 4y = 9 \ \therefore \ 4y = 9 - 3x \ \text{so} \ y = \frac{9}{4} - \frac{3x}{4}$$

Substituting this for y in (2) yields:

$$2x + 3\left(\frac{9}{4} - \frac{3x}{4}\right) = 8 \ \text{that is} \ 2x + \frac{27}{4} - \frac{9x}{4} = 8 \ \therefore \ -\frac{x}{4} = \frac{5}{4} \ \therefore \ x = -5$$

Substituting this value for x in (1) yields:

$$-15 + 4y = 9 \ \therefore \ 4y = 24 \ \therefore \ y = 6$$

Solution by equating coefficients

To solve $\quad 3x + 2y = 16 \qquad\qquad\qquad\qquad\qquad\qquad\qquad\qquad (1)$

$\qquad\qquad\quad 4x - 3y = 10 \qquad\qquad\qquad\qquad\qquad\qquad\qquad\qquad (2)$

If we multiply both sides of (1) by 3 (the coefficient of y in (2)) and we multiply both sides of (2) by 2 (the coefficient of y in (1)) then we have

$$9x + 6y = 48$$
$$8x - 6y = 20$$

If we now add these two lines together, the y-term disappears:

$$\therefore \ 17x = 68 \ \therefore \ x = 4$$

Substituting this result, $x = 4$, in either of the original equations, provides the value of y.

In (1) $\quad 3(4) + 2y = 16 \ \therefore \ 12 + 2y = 16 \ \therefore \ y = 2$

$$\therefore \ x = 4, \ y = 2$$

Check in (2): $\quad 4(4) - 3(2) = 16 - 6 = 10 \ \checkmark$

Had the y-terms been of the same sign, we should, of course, have subtracted one line from the other to eliminate one of the variables.

Another example:

To solve $\quad 3x + y = 18 \qquad\qquad\qquad\qquad\qquad\qquad\qquad\qquad\qquad (1)$

$\qquad\qquad\quad 4x + 2y = 21 \qquad\qquad\qquad\qquad\qquad\qquad\qquad\qquad\qquad (2)$

Working as before: $x = \ldots\ldots\ldots\ldots, \quad y = \ldots\ldots\ldots\ldots$

9

$$x = 7{\cdot}5, \ y = -4{\cdot}5$$

(1)   × 2 $6x + 2y = 36$
(2) $4x + 2y = 21$

Subtract: $2x \quad = 15$ ∴ $x = 7{\cdot}5$

Substitute $x = 7{\cdot}5$ in (1):

$$3(7{\cdot}5) + y = 18 \qquad 22{\cdot}5 + y = 18$$
$$∴ \ y = 18 - 22{\cdot}5 = -4{\cdot}5 \qquad ∴ \ y = -4{\cdot}5$$

Check in (2): $4x + 2y = 4(7{\cdot}5) + 2(-4{\cdot}5) = 30 - 9 = 21$ ✓

   ∴ $x = 7{\cdot}5, \ y = -4{\cdot}5$

10 And one more – on your own.

   Solve $7x - 4y = 23$ (1)
              $4x - 3y = 11$ (2)

Working as before: $x = \ldots\ldots\ldots\ldots, \quad y = \ldots\ldots\ldots\ldots$

11

$$x = 5, \ y = 3$$

(1)   × 3 $21x - 12y = 69$
(2)   × 4 $16x - 12y = 44$

Subtract: $5x \quad = 25$ ∴ $x = 5$

Substitute in (2): $20 - 3y = 11$ ∴ $3y = 9$ ∴ $y = 3$

                                             ∴ $x = 5, \ y = 3$

Check in (1): $7x - 4y = 35 - 12 = 23$ ✓

Simultaneous linear equations with three unknowns

With three unknowns and three equations the method of solution is just an extension of the work with two unknowns.

12

Example 1

To solve

$$3x + 2y - z = 19 \qquad (1)$$
$$4x - y + 2z = 4 \qquad (2)$$
$$2x + 4y - 5z = 32 \qquad (3)$$

We take a pair of equations and eliminate one of the variables using the method in Frame 7:

$$3x + 2y - z = 19 \qquad (1)$$
$$4x - y + 2z = 4 \qquad (2)$$

$(1) \times 2$ $6x + 4y - 2z = 38$

(2) $4x - y + 2z = 4$

Add: $10x + 3y = 42 \qquad (4)$

Now take another pair, e.g. (1) and (3):

$(1) \times 5$ $15x + 10y - 5z = 95$

(3) $2x + 4y - 5z = 32$

Subtract: $13x + 6y = 63 \qquad (5)$

We can now solve equations (4) and (5) for values of x and y in the usual way.

$$x = \ldots\ldots\ldots\ldots, \quad y = \ldots\ldots\ldots\ldots$$

13

$$\boxed{x = 3,\ y = 4}$$

Because $10x + 3y = 42$ (4)

                  $13x + 6y = 63$ (5)

$(4) \times 2$ $20x + 6y = 84$

(5) $13x + 6y = 63$

Subtract: $7x \quad\ = 21 \quad \therefore\ x = 3$

Substitute in (4): $30 + 3y = 42 \quad \therefore\ 3y = 12 \quad \therefore\ y = 4$

Then we substitute these values in one of the original equations to obtain the value of z:

    e.g.   (2) $4x - y + 2z = 12 - 4 + 2z = 4$

                 $\therefore \quad 2z = -4 \quad \therefore\ z = -2$

                    $\therefore\ x = 3,\ y = 4,\ z = -2$

Finally, substitute all three values in the other two original equations as a check procedure:

(1) $3x + 2y - \ z = 9 + 8 + 2 = 19$

(3) $2x + 4y - 5z = 6 + 16 + 10 = 32$ – so all is well.

The working is clearly longer than with only two unknowns, but the method is no more difficult.

    Here is another.

Example 2

    Solve $5x - 3y - 2z = 31$ (1)

              $2x + 6y + 3z = \ \ 4$ (2)

              $4x + 2y - \ z = 30$ (3)

Work through it in just the same way and, as usual, check the results.

    $x = \ldots\ldots\ldots\ldots,\quad y = \ldots\ldots\ldots\ldots,\quad z = \ldots\ldots\ldots\ldots$

14

$$x = 5, y = 2, z = -6$$

(1) × 3	$15x - 9y - 6z = 93$	
(2) × 2	$4x + 12y + 6z = 8$	
	$19x + 3y \quad\quad = 101$	(4)
(1)	$5x - 3y - 2z = 31$	
(3) × 2	$8x + 4y - 2z = 60$	
	$3x + 7y \quad\quad = 29$	(5)

Solving (4) and (5) and substituting back gives the results above:

$$x = 5, \quad y = 2, \quad z = -6$$

Sometimes, the given equations need to be simplified before the method of solution can be carried out. **15**

Pre-simplification

Example 1

Solve the pair of equations:

$2(x + 2y) + 3(3x - y) = 38$	(1)
$4(3x + 2y) - 3(x + 5y) = -8$	(2)

$2x + 4y + 9x - 3y = 38$	$\therefore \ 11x + y = 38$	(3)
$12x + 8y - 3x - 15y = -8$	$\therefore \ 9x - 7y = -8$	(4)

The pair of equations can now be solved in the usual manner.

$$x = \ldots\ldots, \quad y = \ldots\ldots\ldots$$

16

$$x = 3,\ y = 5$$

Because (3) × 7 $\quad 77x + 7y = 266$

(4) $\qquad\qquad \dfrac{9x - 7y = -8}{}$

$\qquad\qquad\qquad 86x \qquad = 258 \quad \therefore\ x = 3$

Substitute in (3): $\qquad 33 + y = 38 \quad \therefore\ y = 5$

Check in (4): $\qquad 9(3) - 7(5) = 27 - 35 = -8\ \checkmark$

$\qquad \therefore\ x = 3,\ y = 5$

Example 2

$$\frac{2x-1}{5} + \frac{x-2y}{10} = \frac{x+1}{4} \tag{1}$$

$$\frac{3y+2}{3} + \frac{4x-3y}{2} = \frac{5x+4}{4} \tag{2}$$

For (1) LCM = 20 $\quad \therefore\ \dfrac{20(2x-1)}{5} + \dfrac{20(x-2y)}{10} = \dfrac{20(x+1)}{4}$

$$4(2x-1) + 2(x-2y) = 5(x+1)$$

$$8x - 4 + 2x - 4y = 5x + 5$$

$$\therefore\ 5x - 4y = 9 \tag{3}$$

Similarly for (2), the simplified equation is

. . . .

17

$$9x - 6y = 4$$

Because

$$\frac{3y+2}{3} + \frac{4x-3y}{2} = \frac{5x+4}{4} \qquad \text{LCM} = 12$$

$$\therefore\ \frac{12(3y+2)}{3} + \frac{12(4x-3y)}{2} = \frac{12(5x+4)}{4}$$

$$4(3y+2) + 6(4x-3y) = 3(5x+4)$$

$$12y + 8 + 24x - 18y = 15x + 12$$

$$\therefore\ 24x - 6y + 8 = 15x + 12$$

$$\therefore\ 9x - 6y = 4 \tag{4}$$

So we have $\qquad 5x - 4y = 9 \tag{3}$

$\qquad\qquad\qquad 9x - 6y = 4 \tag{4}$

Finishing off, we get

$$x = \ldots\ldots\ldots\ldots,\quad y = \ldots\ldots\ldots\ldots$$

18

$$x = -19/3, \ y = -61/6$$

Because

$$9 \times (5x - 4y = 9) \text{ gives } 45x - 36y = 81$$
$$5 \times (9x - 6y = 4) \text{ gives } 45x - 30y = 20$$

Subtracting gives $-6y = 61$ so that $y = -\dfrac{61}{6}$. Substitution then gives:

$$5x = 9 + 4y = 9 - \frac{244}{6} = -\frac{190}{6}, \text{ so } x = -\frac{190}{30} = -\frac{19}{3}$$

That was easy enough.

Example 3

$$5(x + 2y) - 4(3x + 4z) - 2(x + 3y - 5z) = 16$$
$$2(3x - y) + 3(x - 2z) + 4(2x - 3y + z) = -16$$
$$4(y - 2z) + 2(2x - 4y - 3) - 3(x + 4y - 2z) = -62$$

Simplifying these three equations, we get

. . . .

19

$$\begin{array}{rcl} -9x + \ 4y - 6z &=& 16 \\ 17x - 14y - 2z &=& -16 \\ x - 16y - 2z &=& -56 \end{array}$$

Solving this set of three by equating coefficients, we obtain

$$x = \ldots\ldots\ldots\ldots, \quad y = \ldots\ldots\ldots\ldots, \quad z = \ldots\ldots\ldots\ldots$$

20

$$\boxed{x = 2, \, y = 4, \, z = -3}$$

Here is the working as a check:

$$-9x + 4y - 6z = 16 \tag{1}$$
$$17x - 14y - 2z = -16 \tag{2}$$
$$x - 16y - 2z = -56 \tag{3}$$

(1) $\qquad -9x + 4y - 6z = 16$
$(2) \times 3$ $\qquad 51x - 42y - 6z = -48$

Subtracting: $\qquad 60x - 46y = -64 \quad \therefore \; 30x - 23y = -32 \tag{4}$

(2) $\qquad 17x - 14y - 2z = -16$
(3) $\qquad x - 16y - 2z = -56$

Subtracting: $\qquad 16x + 2y = 40 \quad \therefore \; 8x + y = 20 \tag{5}$

$\therefore \; (4)$ $\qquad 30x - 23y = -32$
$(5) \times 23$ $\qquad 184x + 23y = 460$

Add: $\qquad 214x = 428 \quad \therefore \; x = 2$

Substitute in (5): $\qquad\qquad 16 + y = 20 \quad \therefore \; y = 4$

Substitute for x and y in (3):

$$2 - 64 - 2z = -56$$
$$-62 - 2z = -56 \quad \therefore \; z + 31 = 28$$
$$\therefore \; z = -3$$

Check by substituting all values in (1):

$$-9(2) + 4(4) - 6(-3) = -18 + 16 + 18 = 16 \quad \checkmark$$
$$\therefore \; x = 2, \, y = 4, \, z = -3$$

 # Review summary
                                                                   Unit 17

21

1 A *linear equation* involves powers of the variables no higher than the first. A linear equation in a single variable is also referred to as a *simple equation*.

2 Simple equations are solved by isolating the variable on one side of the equation.

3 A linear equation in two variables has an infinite number of solutions. For two such equations there may be only one solution that satisfies both of them simultaneously.

4 Simultaneous equations can be solved:

(a) by substitution

(b) by elimination.

Review exercise

1 Solve the following linear equations:

(a) $2(x-1) - 4(x+2) = 3(x+5) + (x-1)$

(b) $\dfrac{x-1}{2} - \dfrac{x+1}{2} = 5 - \dfrac{x+2}{4}$

(c) $\dfrac{3}{x+2} - \dfrac{5}{x} = -\dfrac{2}{x-1}$

2 Solve the following pairs of simultaneous equations:

(a) $x - y = 2$ by elimination

    $2x + 3y = 9$

(b) $4x + 2y = 10$ by substitution

    $3x - 5y = 1$

3 Solve the following set of three equations in three unknowns:

$x + y + z = 6$

$2x - y + 3z = 9$

$x + 2y - 3z = -4$

4 Simplify and solve the following set of simultaneous equations:

$2(x+2y) - 3(x+2z) - 4(y-2z) = -1$

$3(2x-y) + 2(3y-4z) - 5(3x-2z) = -7$

$-(x-y) + (x-z) + (y+z) = -2$

Complete all four questions. Take your time, there is no need to rush.
If necessary, look back at the Unit.
The answers and working are in the next frame.

---

1 (a) $2(x-1) - 4(x+2) = 3(x+5) + (x-1)$

Eliminate the brackets to give $2x - 2 - 4x - 8 = 3x + 15 + x - 1$.

Simplify each side to give $-2x - 10 = 4x + 14$.

Add $2x - 14$ to both sides of the equation to give $-24 = 6x$ so that $x = -4$.

(b) $\dfrac{x-1}{2} - \dfrac{x+1}{2} = 5 - \dfrac{x+2}{4}$

Multiply throughout by 4 to give $2(x-1) - 2(x+1) = 20 - x - 2$.

Simplify to give $-4 = 18 - x$, therefore $x = 22$.

22

23

(c) $\dfrac{3}{x+2} - \dfrac{5}{x} = -\dfrac{2}{x-1}$

Multiply throughout by $x(x+2)(x-1)$ to give $3x(x-1) - 5(x+2)(x-1) = -2x(x+2)$.

Eliminate the brackets to give $3x^2 - 3x - 5x^2 - 5x + 10 = -2x^2 - 4x$.

Simplify to give $-2x^2 - 8x + 10 = -2x^2 - 4x$.

Add $2x^2$ to both sides to give $-8x + 10 = -4x$. Add $8x$ to both sides of the equation to give $10 = 4x$ so that $x = \dfrac{5}{2}$.

2 (a) $x - y = 2$ [1]

     $2x + 3y = 9$ [2]   Multiply [1] by -2

     $-2x + 2y = -4$ Add [2] to give $5y = 5$ so that $y = 1$.
     Substitution in [1] gives $x = 3$

(b) $4x + 2y = 10$ [1]

     $3x - 5y = 1$ [2]   Subtract $4x$ from both sides of [1] and then divide by 2 to give: $y = 5 - 2x$.
     Substitute this expression for y in [2] to give $3x - 5(5 - 2x) = 1$.
     Simplify to give $13x - 25 = 1$ so that $x = 2$ and $y = 5 - 4 = 1$.

3 $x + y + z = 6$ [1]

    $2x - y + 3z = 9$ [2]

    $x + 2y - 3z = -4$ [3]   Add [1] and [3] and subtract [2] to give $(2x + 3y - 2z) - (2x - y + 3z) = (6 - 4) - 9$. That is

    $4y - 5z = -7$ [4]   Subtract [1] from [3] to give

    $y - 4z = -10$ [5]   Multiply this by 4 to give

    $4y - 16z = -40$ [6]   Subtract [6] from [4] to give $11z = 33$ so that $z = 3$.
    Substitute this value in [5] to give $y = -10 + 12 = 2$ so from [1], $x = 6 - 3 - 2 = 1$.

4 $2(x + 2y) - 3(x + 2z) - 4(y - 2z) = -1$ [1]

    $3(2x - y) + 2(3y - 4z) - 5(3x - 2z) = -7$ [2]

    $-(x - y) + (x - z) + (y + z) = -2$ [3]   Eliminate the brackets to give

    $-x + 2z = -1$ [4]

    $-9x + 3y + 2z = -7$ [5]

    $2y = -2$ [6]   Thus $y = -1$. Substituting this value in [5] gives

    $-9x + 2z = -4$ Subtracting [4] gives $-8x = -3$ so that $x = \dfrac{3}{8}$. Substitution in [4] gives $z = -\dfrac{5}{16}$.

Now for the Review test

Review test   **Unit 17**

1 Solve for x: **24**

(a) $3(x - 1) + 2(3 - 2x) = 6(x + 4) - 7$

(b) $\dfrac{x - 2}{3} - \dfrac{3 - x}{2} = 1 + \dfrac{4 - 5x}{4}$

(c) $\dfrac{4}{x - 1} + \dfrac{2}{x + 1} = \dfrac{6}{x}$

2 Solve the following pair of simultaneous equations for x and y:

$$3x - 4y = 7$$
$$5x - 2y = 7$$

3 Solve the following set of three equations in three unknowns:

$$x - 2y + 3z = 10$$
$$3x - 2y + z = 2$$
$$4x + 5y + 2z = 29$$

You have now come to the end of this Module. A list of **Can You?** questions   **25**
follows for you to gauge your understanding of the material in the Module.
You will notice that these questions match the **Learning outcomes** listed at
the beginning. Now try the **Test exercise**. *Work through the questions at your
own pace, there is no need to hurry.* A set of **Further problems** provides
additional valuable practice.

 Can You?

Checklist: Module 5   **1**

Check this list before and after you try the end of Module test.

On a scale of 1 to 5 how confident are you that you can:

● Solve any linear equation?

Yes ☐ ☐ ☐ ☐ ☐ No

● Solve simultaneous linear equations in two unknowns?

Yes ☐ ☐ ☐ ☐ ☐ No

● Solve simultaneous linear equations in three unknowns?

Yes ☐ ☐ ☐ ☐ ☐ No

Test exercise 5

2

1 Solve the following linear equations:

(a) $4(x + 5) - 6(2x + 3) = 3(x + 14) - 2(5 - x) + 9$

(b) $\dfrac{2x + 1}{3} - \dfrac{2x + 5}{5} = 2 + \dfrac{x - 1}{6}$

(c) $\dfrac{2}{x - 2} + \dfrac{3}{x} = \dfrac{5}{x - 4}$

(d) $(4x - 3)(3x - 1) - (7x + 2)(x + 1) = (5x + 1)(x - 2) - 10$

2 Solve the following pairs of simultaneous equations:

(a) $2x + 3y = 7$

$5x - 2y = 8$ by substitution

(b) $4x + 2y = 5$

$3x + y = 9$ by elimination

3 Solve the following set of three equations in three unknowns:

$2x + 3y - z = -5$

$x - 4y + 2z = 21$

$5x + 2y - 3z = -4$

4 Simplify and solve the following set of simultaneous equations:

$4(x + 3y) - 2(4x + 3z) - 3(x - 2y - 4z) = 17$

$2(4x - 3y) + 5(x - 4z) + 4(x - 3y + 2z) = 23$

$3(y + 4z) + 4(2x - y - z) + 2(x + 3y - 2z) = 5$

 Further problems 5

3

Solve the following equations, numbered **1** to **15**:

1 $(4x + 5)(4x - 3) + (5 - 4x)(5 + 4x) = (3x - 2)(3x + 2) - (9x - 5)(x + 1)$

2 $(3x - 1)(x - 3) - (4x - 5)(3x - 4) = 6(x - 7) - (3x - 5)^2$

3 $\dfrac{4x + 7}{3} - \dfrac{x - 7}{5} = \dfrac{x + 6}{9}$

4 $\dfrac{2x - 5}{2} + \dfrac{5x + 2}{6} = \dfrac{2x + 3}{4} - \dfrac{x + 5}{3}$

5 $(3x + 1)(x - 2) - (5x - 2)(2x - 1) = 16 - (7x + 3)(x + 2)$

6 $\dfrac{3x - 4}{3} + \dfrac{2x - 5}{8} = \dfrac{4x + 5}{6} - \dfrac{x + 2}{4}$

7 $8x - 5y = 10$

$6x - 4y = 11$

8 $\dfrac{2}{x + 3} + \dfrac{5}{x} = \dfrac{7}{x - 2}$

9 $(3 - 4x)(3 + 4x) = 3(3x - 2)(x + 1) - (5x - 3)(5x + 3)$

10 $3x - 2y + z = 2$

$x - 3y + 2z = 1$

$2x + y - 3z = -5$

11 $2x - y + 3z = 20$

$x - 6y - z = -41$

$3x + 6y + 2z = 70$

12 $3x + 2y + 5z = 2$

$5x + 3y - 2z = 4$

$2x - 5y - 3z = 14$

13 $5x - 6y + 3z = -9$

$2x - 3y + 2z = -5$

$3x - 7y + 5z = -16$

14 $\dfrac{3x + 2}{4} - \dfrac{x + 2y}{2} = \dfrac{x - 3}{12}$

$\dfrac{2y + 1}{5} + \dfrac{x - 3y}{4} = \dfrac{3x + 1}{10}$

15 $\dfrac{x - 5}{x + 5} + \dfrac{x - 7}{x + 7} = 2$

16   Simplify each side separately and hence determine the solution:

$$\frac{x - 2}{x - 4} - \frac{x - 4}{x - 6} = \frac{x - 1}{x - 3} - \frac{x - 3}{x - 5}$$

17   Solve the simultaneous equations:

$$\frac{3x - 2y}{2} = \frac{2x + y}{7} + \frac{3}{2}$$

$$7 - \frac{2x - y}{6} = x + \frac{y}{4}$$

18   Writing u for $\dfrac{1}{x + 8y}$ and v for $\dfrac{1}{8x - y}$, solve the following equations for u and v, and hence determine the values of x and y:

$$\frac{2}{x + 8y} - \frac{1}{8x - y} = 4; \qquad \frac{1}{x + 8y} + \frac{2}{8x - y} = 7$$

19 Solve the pair of equations:

$$\frac{6}{x-2y} - \frac{15}{x+y} = 0{\cdot}5$$

$$\frac{12}{x-2y} - \frac{9}{x+y} = -0{\cdot}4$$

20 Solve:

$$\frac{4}{x+1} + \frac{7}{4(2-y)} = 9$$

$$\frac{3}{2(x+1)} + \frac{7}{2-y} = 7$$

Polynomial equations

Learning outcomes

When you have completed this Module you will be able to:

- Solve quadratic equations by factors, by completing the square and by formula
- Solve cubic equations with at least one linear factor
- Solve fourth-order equations with at least two linear factors

Units

Polynomial equations   Unit 18

1   In Module 3 we looked at polynomial equations. In particular we selected a value for the variable x in a polynomial expression and found the resulting value of the polynomial expression. In other words, we *evaluated* the polynomial expression. Here we reverse the process by giving the polynomial expression a value of zero and finding those values of x which satisfy the resulting equation. We start with quadratic equations.

On now to the second frame

Quadratic equations, $ax^2 + bx + c = 0$

2   Solution by factors

We dealt at some length in Modules 2 and 3 with the representation of a quadratic expression as a product of two simple linear factors, where such factors exist.

For example, $x^2 + 5x - 14$ can be factorized into $(x + 7)(x - 2)$. The equation $x^2 + 5x - 14 = 0$ can therefore be written in the form:

$$(x + 7)(x - 2) = 0$$

and this equation is satisfied if either factor has a zero value.

$\therefore$ $x + 7 = 0$ or $x - 2 = 0$

i.e. $x = -7$ or $x = 2$

are the solutions of the given equation $x^2 + 5x - 14 = 0$.

By way of revision, then, you can solve the following equations with no trouble:

(a) $x^2 - 9x + 18 = 0$

(b) $x^2 + 11x + 28 = 0$

(c) $x^2 + 5x - 24 = 0$

(d) $x^2 - 4x - 21 = 0$

3

(a) $x = 3$ or $x = 6$
(b) $x = -4$ or $x = -7$
(c) $x = 3$ or $x = -8$
(d) $x = 7$ or $x = -3$

Not all quadratic expressions can be factorized as two simple linear factors. Remember that the test for the availability of factors with $ax^2 + bx + c$ is to calculate whether $(b^2 - 4ac)$ is ....

4

a perfect square

The test should always be applied at the begining to see whether, in fact, simple linear factors exist.

For example, with $x^2 + 8x + 15$, $(b^2 - 4ac) = $ ....

5

4, i.e. 2^2

So $x^2 + 8x + 15$ can be written as the product of two simple linear factors.

But with $x^2 + 8x + 20$, $a = 1$, $b = 8$, $c = 20$
and $(b^2 - 4ac) = 8^2 - 4 \times 1 \times 20 = 64 - 80 = -16$, which is not a perfect square.

So $x^2 + 8x + 20$ cannot be written as the product of two simple linear factors.

On to the next frame

6

If the coefficient of the x^2 term is other than unity, the factorization process is a trifle more involved, but follows the routine already established in Module 2. You will remember the method.

Example 1

To solve $2x^2 - 3x - 5 = 0$

In this case, $ax^2 + bx + c = 0$, $a = 2$, $b = -3$, $c = -5$.

(a) Test for simple factors:

$$(b^2 - 4ac) = (-3)^2 - 4 \times 2 \times (-5)$$
$$= 9 + 40 = 49 = 7^2 \qquad \therefore \text{simple factors exist.}$$

(b) $|ac| = 10$. Possible factors of 10 are $(1, 10)$ and $(2, 5)$.

　　c is negative: $\therefore$ factors differ by $|b|$, i.e. 3
　　　　　　　　　$\therefore$ Required factors are $(2, 5)$

　　c is negative: $\therefore$ factors are of different sign, the numerically larger having the sign of b, i.e. negative.

$$\therefore 2x^2 - 3x - 5 = 2x^2 + 2x - 5x - 5$$
$$= 2x(x+1) - 5(x+1)$$
$$= (x+1)(2x-5)$$

$\therefore$ The equation $2x^2 - 3x - 5 = 0$ becomes $(x+1)(2x-5) = 0$

$\therefore x + 1 = 0$ or $2x - 5 = 0$ 　$\therefore x = -1$ or $x = 2{\cdot}5$

Example 2

In the same way, the equation:

　$3x^2 + 14x + 8 = 0$

has solutions $x = \ldots\ldots\ldots$ or $x = \ldots\ldots\ldots$

7

$$\boxed{x = -4 \text{ or } x = -\frac{2}{3}}$$

Because $3x^2 + 14x + 8 = 0$ 　$a = 3, b = 14, c = 8$

Test for simple factors: $(b^2 - 4ac) = 14^2 - 4 \times 3 \times 8 = 196 - 96 = 100 = 10^2$
　　　　　　　　　　　　　　　$=$ perfect square, $\therefore$ simple factors exist.

$|ac| = 24$. Possible factors of 24 are $(1, 24), (2, 12), (3, 8)$ and $(4, 6)$.
c is positive: $\therefore$ factors add up to $|b|$, i.e. 14 $\therefore (2, 12)$
c is positive: $\therefore$ both factors have same sign as b, i.e. positive.

$\therefore 3x^2 + 14x + 8 = 3x^2 + 2x + 12x + 8 = 0$
　　　　　　　　$= x(3x + 2) + 4(3x + 2) = 0$
　　　　　　　　$= (3x + 2)(x + 4) = 0$

$\therefore x + 4 = 0$ or $3x + 2 = 0$ 　$\therefore x = -4$ or $x = -\dfrac{2}{3}$

They are all done in the same way – provided that simple linear factors exist, so always test to begin with by evaluating $(b^2 - 4ac)$.

Here is another: work right through it on your own. Solve the equation $4x^2 - 16x + 15 = 0$

　　　　　　$x = \ldots\ldots\ldots$ or $x = \ldots\ldots\ldots$

$$x = \frac{3}{2} \text{ or } x = \frac{5}{2}$$

8

Solution by completing the square

We have already seen that some quadratic equations are incapable of being factorized into two simple factors. In such cases, another method of solution must be employed. The following example will show the procedure.

Solve $x^2 - 6x - 4 = 0$ $a = 1$, $b = -6$, $c = -4$.

$(b^2 - 4ac) = 36 - 4 \times 1 \times (-4) = 36 + 16 = 52$. Not a perfect square.
$\therefore$ No simple factors.

So: $x^2 - 6x - 4 = 0$

Add 4 to both sides: $x^2 - 6x = 4$

Add to each side the square of half the coefficient of x:

$x^2 - 6x + (-3)^2 = 4 + (-3)^2$

$x^2 - 6x + 9 = 4 + 9 = 13$

$\therefore (x - 3)^2 = 13$

$\therefore x - 3 = \pm\sqrt{13}$. Remember to include the two signs.

$\therefore x = 3 \pm \sqrt{13} = 3 \pm 3{\cdot}6056 \therefore x = 6{\cdot}606 \text{ or } x = -0{\cdot}606$

Now this one. Solve $x^2 + 8x + 5 = 0$ by the method of completing the square.

9

First we take the constant term to the right-hand side:

$x^2 + 8x + 5 = 0$

$x^2 + 8x \quad = -5$

Then we add to both sides ....

10

the square of half the coefficient of x

$x^2 + 8x + 4^2 = -5 + 4^2$

$\therefore x^2 + 8x + 16 = -5 + 16 = 11$

$\therefore (x + 4)^2 = 11$

And now we can finish it off, finally getting:

$x = \ldots\ldots\ldots\ldots \text{ or } x = \ldots\ldots\ldots\ldots$

11

$$x = -0.683 \text{ or } x = -7.317$$

If the coefficient of the squared term is not unity, the first step is to divide both sides of the equation by the existing coefficient.

For example: $2x^2 + 10x - 7 = 0$

Dividing throughout by 2: $x^2 + 5x - 3.5 = 0$

We then proceed as before, which finally gives:

$$x = \ldots\ldots\ldots \text{ or } x = \ldots\ldots\ldots$$

12

$$x = 0.622 \text{ or } x = -5.622$$

Because

$$x^2 + 5x - 3.5 = 0$$
$$\therefore x^2 + 5x = 3.5$$
$$x^2 + 5x + 2.5^2 = 3.5 + 6.25 = 9.75$$
$$\therefore (x + 2.5)^2 = 9.75$$
$$\therefore x + 2.5 = \pm\sqrt{9.75} = \pm 3.122$$
$$\therefore x = -2.5 \pm 3.122 \quad \therefore x = 0.622 \text{ or } x = -5.622$$

One more. Solve the equation $4x^2 - 16x + 3 = 0$ by completing the square.

$$x = \ldots\ldots\ldots \text{ or } x = \ldots\ldots\ldots$$

13

$$x = 0.197 \text{ or } x = 3.803$$

Because

$$4x^2 - 16x + 3 = 0 \quad \therefore x^2 - 4x + 0.75 = 0$$
$$\therefore x^2 - 4x = -0.75$$
$$x^2 - 4x + (-2)^2 = -0.75 + (-2)^2$$
$$x^2 - 4x + 4 = -0.75 + 4 = 3.25$$
$$\therefore (x - 2)^2 = 3.25$$
$$\therefore x - 2 = \pm\sqrt{3.25} = \pm 1.8028$$
$$\therefore x = 2 \pm 1.803 \quad \therefore x = 0.197 \text{ or } x = 3.803$$

On to the next topic

Solution by formula

We can establish a formula for the solution of the general quadratic equation $ax^2 + bx + c = 0$ which is based on the method of completing the square:

$$ax^2 + bx + c = 0$$

Dividing throughout by the coefficient of x, i.e. a:

$$x^2 + \frac{b}{a}x + \frac{c}{a} = 0$$

Subtracting $\dfrac{c}{a}$ from each side gives $x^2 + \dfrac{b}{a}x = -\dfrac{c}{a}$

We then add to each side the square of half the coefficient of x:

$$x^2 + \frac{b}{a}x + \left(\frac{b}{2a}\right)^2 = -\frac{c}{a} + \left(\frac{b}{2a}\right)^2$$

$$x^2 + \frac{b}{a}x + \frac{b^2}{4a^2} = \frac{b^2}{4a^2} - \frac{c}{a}$$

$$\left(x + \frac{b}{2a}\right)^2 = \frac{b^2 - 4ac}{4a^2}$$

$$\therefore x + \frac{b}{2a} = \pm\sqrt{\frac{b^2 - 4ac}{4a^2}} = \pm\frac{\sqrt{b^2 - 4ac}}{2a} \quad \therefore x = -\frac{b}{2a} \pm \frac{\sqrt{b^2 - 4ac}}{2a}$$

$$\therefore \text{If } ax^2 + bx + c = 0,\ x = \frac{-b \pm \sqrt{b^2 - 4ac}}{2a}$$

Substituting the values of a, b and c for any particular quadratic equation gives the solutions of the equation.

Make a note of the formula: it is important

As an example, we shall solve the equation $2x^2 - 3x - 4 = 0$.

15

Here $a = 2$, $b = -3$, $c = -4$ and $x = \dfrac{-b \pm \sqrt{b^2 - 4ac}}{2a}$

$$x = \frac{3 \pm \sqrt{9 - 4 \times 2 \times (-4)}}{4} = \frac{3 \pm \sqrt{9 + 32}}{4} = \frac{3 \pm \sqrt{41}}{4}$$

$$= \frac{3 \pm 6 \cdot 403}{4} = \frac{-3 \cdot 403}{4} \text{ or } \frac{9 \cdot 403}{4}$$

$$\therefore x = -0 \cdot 851 \text{ or } x = 2 \cdot 351$$

It is just a case of careful substitution. You need, of course, to remember the formula. For

$$ax^2 + bx + c + 0 \quad x = \ldots\ldots\ldots\ldots$$

16

$$x = \frac{-b \pm \sqrt{b^2 - 4ac}}{2a}$$

As an excrcise, use the formula method to solve the following:

(a) $5x^2 + 12x + 3 = 0$ (c) $x^2 + 15x - 7 = 0$

(b) $3x^2 - 10x + 4 = 0$ (d) $6x^2 - 8x - 9 = 0$

17

(a) $x = -2\cdot117$ or $x = -0\cdot283$ (c) $x = 0\cdot453$ or $x = -15\cdot453$

(b) $x = 0\cdot465$ or $x = 2\cdot869$ (d) $x = -0\cdot728$ or $x = 2\cdot061$

At this stage of our work, we can with advantage bring together a number of the items that we have studied earlier.

Solution of cubic equations having at least one linear factor in the algebraic expression

18

In Module 3, we dealt with the factorization of cubic polynomials, with application of the remainder theorem and factor theorem, and the evaluation of the polynomial functions by nesting. These we can now reapply to the solution of cubic equations.

So move on to the next frame for some examples

19

Example 1

To solve the cubic equation:

$$2x^3 - 11x^2 + 18x - 8 = 0$$

The first step is to find a linear factor of the cubic expression:

$$f(x) = 2x^3 - 11x^2 + 18x - 8$$

by application of the remainder theorem. To facilitate the calculation, we first write $f(x)$ in nested form:

$$f(x) = [(2x - 11)x + 18]x - 8$$

Now we seek a value for x ($x = k$) which gives a zero remainder on division by $(x - k)$. We therefore evaluate $f(1)$, $f(-1)$, $f(2)$ ... etc.

$f(1) = 1$ $\therefore (x - 1)$ is not a factor of $f(x)$

$f(-1) = -39$ $\therefore (x + 1)$ is not a factor of $f(x)$

$f(2) = 0$ $\therefore (x - 2)$ *is* a factor of $f(x)$

We therefore divide $f(x)$ by $(x - 2)$ to determine the remaining factor, which is . . . .

20

$$2x^2 - 7x + 4$$

Because

$$
\begin{array}{r}
2x^2 - 7x + 4 \\
x - 2 \overline{\smash{\big)}\ 2x^3 - 11x^2 + 18x - 8} \\
\underline{2x^3 - 4x^2} \\
-7x^2 + 18x \\
\underline{-7x^2 + 14x} \\
4x - 8 \\
\underline{4x - 8} \\
\bullet \quad \bullet
\end{array}
$$

$\therefore f(x) = (x - 2)(2x^2 - 7x + 4)$ and the cubic equation is now written:

$$(x - 2)(2x^2 - 7x + 4) = 0$$

which gives $x - 2 = 0$ or $2x^2 - 7x + 4 = 0$.

$\therefore x = 2$ and the quadratic equation can be solved in the usual way giving
$x = \ldots\ldots\ldots$ or $x = \ldots\ldots\ldots$

$$x = 0{\cdot}719 \text{ or } x = 2{\cdot}781$$

$$2x^2 - 7x + 4 = 0 \quad \therefore x = \frac{7 \pm \sqrt{49 - 32}}{4} = \frac{7 \pm \sqrt{17}}{4} = \frac{7 \pm 4{\cdot}1231}{4}$$

$$= \frac{2{\cdot}8769}{4} \text{ or } \frac{11{\cdot}1231}{4}$$

$$\therefore x = 0{\cdot}719 \text{ or } x = 2{\cdot}781$$

$$\therefore 2x^3 - 11x^2 + 18x - 8 = 0 \text{ has the solutions}$$

$$x = 2, \ x = 0{\cdot}719, \ x = 2{\cdot}781$$

The whole method depends on the given expression in the equation having at least one linear factor.

Here is another.

Example 2

Solve the equation $3x^3 + 12x^2 + 13x + 4 = 0$

First, in nested form $f(x) = \ldots\ldots\ldots$

22

$$f(x) = [(3x + 12)x + 13]x + 4$$

Now evaluate $f(1), f(-1), f(2), \ldots$

$f(1) = 32 \quad \therefore (x - 1)$ is not a factor of $f(x)$

$f(-1) = \ldots\ldots\ldots\ldots$

23

$$f(-1) = 0$$

$\therefore (x + 1)$ *is* a factor of $f(x)$. Then, by long division, the remaining factor of $f(x)$ is $\ldots\ldots\ldots\ldots$

24

$$3x^2 + 9x + 4$$

$\therefore$ The equation $3x^3 + 12x^2 + 13x + 4 = 0$ can be written

$$(x + 1)(3x^2 + 9x + 4) = 0$$

so that $\quad x + 1 = 0$ or $3x^2 + 9x + 4 = 0$

which gives $x = -1$ or $x = \ldots\ldots\ldots\ldots$ or $x = \ldots\ldots\ldots\ldots$

25

$$x = -2\cdot457 \text{ or } x = -0\cdot543$$

$$3x^2 + 9x + 4 = 0 \quad \therefore x = \frac{-9 \pm \sqrt{81 - 48}}{6} = \frac{-9 \pm \sqrt{33}}{6}$$

$$= \frac{-9 \pm 5\cdot7446}{6} = \frac{-14\cdot7446}{6} \text{ or } \frac{-3\cdot2554}{6}$$

$$= -2\cdot4574 \text{ or } -0\cdot5426$$

The complete solutions of $3x^3 + 12x^2 + 13x + 4 = 0$ are

$$x = -1, x = -2\cdot457, x = -0\cdot543$$

And now one more.

Example 3

Solve the equation $5x^3 + 2x^2 - 26x - 20 = 0$

Working through the method step by step, as before:

$$x = \ldots\ldots\ldots\ldots \text{ or } x = \ldots\ldots\ldots\ldots \text{ or } x = \ldots\ldots\ldots\ldots$$

$$x = -2 \text{ or } x = -0{\cdot}825 \text{ or } x = 2{\cdot}425$$

26

Here is the working, just as a check:

$$f(x) = 5x^3 + 2x^2 - 26x - 20 = 0 \quad \therefore f(x) = [(5x+2)x - 26]x - 20$$
$$f(1) = -39 \quad \therefore (x-1) \text{ is not a factor of } f(x)$$
$$f(-1) = 3 \quad \therefore (x+1) \text{ is not a factor of } f(x)$$
$$f(2) = -24 \quad \therefore (x-2) \text{ is not a factor of } f(x)$$
$$f(-2) = 0 \quad \therefore (x+2) \text{ is a factor of } f(x):$$

$$
\begin{array}{r}
5x^2 - 8x - 10 \\
x+2 \overline{\smash{)}\ 5x^3 + 2x^2 - 26x - 20} \\
\underline{5x^3 + 10x^2} \\
-8x^2 - 26x \\
\underline{-8x^2 - 16x} \\
-10x - 20 \\
\underline{-10x - 20} \\
\bullet \qquad \bullet
\end{array}
$$

$$f(x) = (x+2)(5x^2 - 8x - 10) = 0 \quad \therefore x+2 = 0 \text{ or } 5x^2 - 8x - 10 = 0$$

$$\therefore x = -2 \text{ or } x = \frac{8 \pm \sqrt{64 + 200}}{10} = \frac{8 \pm \sqrt{264}}{10}$$

$$= \frac{8 \pm 16{\cdot}248}{10} = -0{\cdot}825 \text{ or } 2{\cdot}425$$

$$\therefore x = -2 \text{ or } x = -0{\cdot}825 \text{ or } x = 2{\cdot}425$$

Next frame

Solution of fourth-order equations having at least two linear factors in the algebraic function

27

The method here is practically the same as that for solving cubic equations which we have just considered. The only difference is that we have to find two simple linear factors of $f(x)$. An example will reveal all.

Example 1

To solve the equation $4x^4 - 19x^3 + 24x^2 + x - 10 = 0$

(a) As before, express the polynomial in nested form:

$$f(x) = \{[(4x - 19)x + 24]x + 1\}x - 10$$

(b) Determine $f(1)$, $f(-1)$, $f(2)$ etc.

$f(1) = 0 \quad \therefore (x - 1)$ *is* a factor of $f(x)$:

$$
\begin{array}{r}
4x^3 - 15x^2 + 9x + 10 \\
x - 1 \overline{\smash{)}\,4x^4 - 19x^3 + 24x^2 + x - 10} \\
4x^4 - 4x^3 \\
\hline
-15x^3 + 24x^2 \\
-15x^3 + 15x^2 \\
\hline
9x^2 + x \\
9x^2 - 9x \\
\hline
10x - 10 \\
10x - 10 \\
\hline
\bullet \qquad \bullet
\end{array}
$$

$\therefore f(x) = (x - 1)(4x^3 - 15x^2 + 9x + 10) = 0$
$\therefore x = 1$ or $4x^3 - 15x^2 + 9x + 10 = 0$, i.e. $F(x) = 0$
$F(x) = [(4x - 15)x + 9]x + 10$
$F(1) = 8 \quad \therefore (x - 1)$ is not a factor of $F(x)$
$F(-1) = -18 \quad \therefore (x + 1)$ is not a factor of $F(x)$
$F(2) = 0 \quad \therefore (x - 2)$ *is* a factor of $F(x)$

Division of $F(x)$ by $(x - 2)$ by long division gives $(4x^2 - 7x - 5)$:
$F(x) = (x - 2)(4x^2 - 7x - 5)$
$\therefore f(x) = (x - 1)(x - 2)(4x^2 - 7x - 5) = 0$
$\therefore x = 1$, $x = 2$, or $4x^2 - 7x - 5 = 0$

Solving the quadratic by formula gives solutions $x = -0.545$ or $x = 2.295$.
$\therefore x = 1, \; x = 2, \; x = -0.545, \; x = 2.295$

Now let us deal with another example

Example 2

28

Solve $2x^4 - 4x^3 - 23x^2 - 11x + 6 = 0$

$f(x) = \{[(2x - 4)x - 23]x - 11\}x + 6$

$f(1) = -30 \quad \therefore (x - 1)$ is not a factor of $f(x)$

$f(-1) = 0 \quad \therefore (x + 1)$ *is* a factor of $f(x)$:

$$
\begin{array}{r}
2x^3 - 6x^2 - 17x + 6 \\
x + 1 \overline{\smash{\big)}\ 2x^4 - 4x^3 - 23x^2 - 11x + 6} \\
\underline{2x^4 + 2x^3} \\
-6x^3 - 23x^2 \\
\underline{-6x^3 - 6x^2} \\
-17x^2 - 11x \\
\underline{-17x^2 - 17x} \\
6x + 6 \\
\underline{6x + 6} \\
\bullet \quad \bullet
\end{array}
$$

$\therefore f(x) = (x + 1)(2x^3 - 6x^2 - 17x + 6) = 0$

$\therefore x = -1$ or $2x^3 - 6x^2 - 17x + 6 = 0$, i.e. $F(x) = 0$

$F(x) = [(2x - 6)x - 17]x + 6$

$F(1) = -15 \quad \therefore (x - 1)$ is not a factor of $F(x)$

Now you can continue and finish it off:

$x = -1, \ x = \dots\dots\dots, \ x = \dots\dots\dots, \ x = \dots\dots\dots$

$$\boxed{x = -2, \ x = 0{\cdot}321, \ x = 4{\cdot}679}$$

29

Because

$F(-1) = 15 \quad \therefore (x + 1)$ is not a factor of $F(x)$

$F(2) \ = -36 \quad \therefore (x - 2)$ is not a factor of $F(x)$

$F(-2) = 0 \quad \therefore (x + 2)$ *is* a factor of $F(x)$

Division of $F(x)$ by $(x + 2)$ by long division gives:

$F(x) = (x + 2)(2x^2 - 10x + 3)$

$\therefore f(x) = (x + 1)(x + 2)(2x^2 - 10x + 3) = 0$

$\therefore x = -1, \ x = -2, \ x = \dfrac{10 \pm \sqrt{100 - 24}}{4} = \dfrac{10 \pm \sqrt{76}}{4}$

$\qquad = \dfrac{10 \pm 8{\cdot}7178}{4} = \dfrac{1{\cdot}2822}{4} \text{ or } \dfrac{18{\cdot}7178}{4}$

$\qquad = 0{\cdot}3206 \text{ or } 4{\cdot}6794$

$\therefore x = -1, \ x = -2, \ x = 0{\cdot}321, \ x = 4{\cdot}679$

30

Example 3

Solve $f(x) = 0$ when $f(x) = 3x^4 + 2x^3 - 15x^2 + 12x - 2$

$$f(x) = 3x^4 + 2x^3 - 15x^2 + 12x - 2$$
$$= \{[(3x + 2)x - 15]x + 12\}x - 2$$

Now you can work through the solution, taking the same steps as before. There are no snags, so you will have no trouble:

$$x = \ldots\ldots\ldots\ldots$$

31

$$\boxed{x = 1,\ x = 1,\ x = -2\cdot897,\ x = 0\cdot230}$$

Because $f(1) = 0$, $\therefore (x - 1)$ *is* a factor of $f(x)$:

$$
\begin{array}{r}
3x^3 + 5x^2 - 10x + 2 \\
x - 1\,\overline{\smash{\big)}\,3x^4 + 2x^3 - 15x^2 + 12x - 2} \\
\underline{3x^4 - 3x^3} \\
5x^3 - 15x^2 \\
\underline{5x^3 - 5x^2} \\
-10x^2 + 12x \\
\underline{-10x^2 + 10x} \\
2x - 2 \\
\underline{2x - 2} \\
\cdot \quad \cdot
\end{array}
$$

$\therefore f(x) = (x - 1)(3x^3 + 5x^2 - 10x + 2) = (x - 1) \times F(x) = 0$
$F(x) = [(3x + 5)x - 10]x + 2$
$F(1) = 0 \quad \therefore (x - 1)$ *is* a factor of $F(x)$:

$$
\begin{array}{r}
3x^2 + 8x - 2 \\
x - 1\,\overline{\smash{\big)}\,3x^3 + 5x^2 - 10x + 2} \\
\underline{3x^3 - 3x^2} \\
8x^2 - 10x \\
\underline{8x^2 - 8x} \\
-2x + 2 \\
\underline{-2x + 2} \\
\cdot \quad \cdot
\end{array}
$$

$\therefore f(x) = (x - 1)(x - 1)(3x^2 + 8x - 2) = 0$

Solving the quadratic by formula gives $x = -2\cdot8968$ or $x = 0\cdot2301$

$\therefore x = 1,\ x = 1,\ x = -2\cdot897,\ x = 0\cdot230$

All correct?

Finally, one more for good measure.

32

Example 4

Solve the equation $2x^4 + 3x^3 - 13x^2 - 6x + 8 = 0$

Work through it using the same method as before.

$$x = \ldots\ldots\ldots$$

$$\boxed{x = -1, \ x = 2, \ x = 0{\cdot}637, \ x = -3{\cdot}137}$$

33

Here is an outline of the solution:

$$f(x) = 2x^4 + 3x^3 - 13x^2 - 6x + 8$$
$$= \{[(2x + 3)x - 13]\,x - 6\}x + 8$$
$$f(-1) = 0 \quad \therefore (x + 1) \text{ is a factor of } f(x).$$

Division of $f(x)$ by $(x + 1)$ gives:

$$f(x) = (x + 1)(2x^3 + x^2 - 14x + 8) = (x + 1) \times F(x)$$
$$F(x) = 2x^3 + x^2 - 14x + 8 = [(2x + 1)x - 14]x + 8$$
$$F(2) = 0 \quad \therefore (x - 2) \text{ is a factor of } F(x).$$

Division of $F(x)$ by $(x - 2)$ gives $F(x) = (x - 2)(2x^2 + 5x - 4)$.

$$\therefore f(x) = (x + 1)(x - 2)(2x^2 + 5x - 4) = 0$$

Solution of the quadratic by formula shows $x = 0{\cdot}637$ or $x = -3{\cdot}137$.

$$\therefore x = -1, \ x = 2, \ x = 0{\cdot}637, \ x = -3{\cdot}137$$

The methods we have used for solving cubic and fourth-order equations have depended on the initial finding of one or more simple factors by application of the remainder theorem.

34

There are, however, many cubic and fourth-order equations which have no simple factors and other methods of solution are necessary. These are more advanced and will not be dealt with here.

At the point let us pause and summarize the main facts on polynomial equations

Review summary

35

1 *Quadratic equations* $ax^2 + bx + c = 0$

(a) *Solution by factors*

Test for availability of factors: evaluate $(b^2 - 4ac)$.

If the value of $(b^2 - 4ac)$ is a perfect square, simple linear factors exist. If $(b^2 - 4ac)$ is not a perfect square, there are no simple linear factors.

(b) *Solution by completing the square*

(i)   Remove the constant term to the RHS.

(ii)   Add to both sides the square of half the coefficient of x. This completes the square on the LHS.

(iii) Take the square root of each side – including both signs.

(iv) Simplify the results to find the values of x.

(c) *Solution by formula*

Evaluate $x = \dfrac{-b \pm \sqrt{b^2 - 4ac}}{2a}$, to obtain two values for x.

2 *Cubic equations* – with at least one linear factor

(a) Rewrite the polynomial function, $f(x)$, in nested form.

(b) Apply the remainder theorem by substituting values for x until a value, $x = k$, gives zero remainder. Then $(x - k)$ is a factor of $f(x)$.

(c) By long division, determine the remaining quadratic factor of $f(x)$, i.e. $(x - k)(ax^2 + bx + c) = 0$. Then $x = k$, and $ax^2 + bx + c = 0$ can be solved by the usual methods. $\therefore x = k$, $x = x_1$, $x = x_2$.

3 *Fourth-order equations* – with at least two linear factors.

(a) Find the first linear factor as in section **2** above.

(b) Divide by $(x - k)$ to obtain the remaining cubic expression. Then $f(x) = (x - k) \times F(x)$ where $F(x)$ is a cubic.

(c) $f(x) = 0$ $\therefore (x - k) \times F(x) = 0$ $\therefore x = k$ or $F(x) = 0$.

(d) The cubic $F(x)$ is now solved as in section **2** above, giving:

$$(x - m)(ax^2 + bx + c) = 0$$

$$\therefore x = k, \ x = m \text{ and } ax^2 + bx + c = 0$$

giving the four solutions $x = k$, $x = m$, $x = x_1$ and $x = x_2$.

Review exercise

36

1 Solve for x by factorizing:
$$12x^2 - 5x - 2 = 0$$

2 Solve for x (accurate to 3 dp) by completing the square:
$$3x^2 - 4x - 2 = 0$$

3 Solve for x (accurate to 3 dp) by using the formula:
$$x^2 - 3x + 1 = 0$$

4 Solve for x:
$$x^3 - 5x^2 + 7x - 2 = 0$$

5 Solve for x:
$$x^4 + x^3 - 2x^2 - x + 1 = 0$$

Complete all five questions. Take your time, there is no need to rush.
If necessary, refer back to the Unit.
The answers and working are in the next frame.

37

1 $12x^2 - 5x - 2 = 0$
Here, $ax^2 + bx + c = 0$ where $a = 12$, $b = -5$, $c = -2$.

(a) Test for factors: $(b^2 - 4ac) = (-5)^2 - [4 \times 12 \times (-2)] = 25 + 96 = 121$ so that factors exist.

(b) $|ac| = 24$. Possible factors of 24 are (1, 24), (2, 12), (3, 8) and (4, 6).
c is negative $\therefore$ factors differ by $|b| = 5$ $\therefore$ Required factors are (3, 8)
c is negative $\therefore$ factors of different sign, the numerically larger having the same sign of b, that is negative, therefore:

$$12x^2 - 5x - 2 = 12x^2 - 8x + 3x - 2$$
$$= 4x(3x - 2) + (3x - 2)$$
$$= (4x + 1)(3x - 2) = 0$$

So that $(4x + 1) = 0$ or $(3x - 2) = 0$. That is $x = -1/4$ or $x = 2/3$.

2 $3x^2 - 4x - 2 = 0$, that is $x^2 - (4/3)x - (2/3) = 0$.

Here $a = 1$, $b = -4/3$, $c = -2/3$. So that $(b^2 - 4ac) = \dfrac{16}{9} + 4 \times 1 \times \dfrac{2}{3} = \dfrac{40}{9}$

– not a perfect square so we proceed by completing the square:

$$x^2 - (4/3)x - 2/3 = 0$$
$$x^2 - (4/3)x = 2/3$$
$$x^2 - (4/3)x + (2/3)^2 = 2/3 + (2/3)^2$$
$$x^2 - (4/3)x + 4/9 = 10/9$$
$$(x - 2/3)^2 = 10/9$$

So that $x - 2/3 = \pm\sqrt{10/9}$ giving $x = 2/3 \pm \sqrt{10/9} = 0.6667 \pm 1.0541$ to 4 dp.
Therefore $x = 1.721$ or $x = -0.387$.

3 $x^2 - 3x + 1 = 0$. Here $a = 1$, $b = -3$, $c = 1$ where:

$$x = \frac{-b \pm \sqrt{b^2 - 4ac}}{2a}$$

so that $x = \dfrac{3 \pm \sqrt{9 - 4}}{2} = \dfrac{3 \pm \sqrt{5}}{2} = 2 \cdot 618$ or $0 \cdot 382$

4 $x^3 - 5x^2 + 7x - 2 = 0$.

Let $f(x) = x^3 - 5x^2 + 7x - 2$
$$= ((x - 5)x + 7)x - 2 \text{ so}$$
$$f(2) = ((2 - 5)2 + 7)2 - 2 = 0$$

Thus $x - 2$ is a factor of $f(x)$ and so $x = 2$ is a solution of the cubic. To find the other two we first need to divide the cubic by this factor:

$$\frac{x^3 - 5x^2 + 7x - 2}{x - 2} = x^2 - 3x + 1$$

That is $x^3 - 5x^2 + 7x - 2 = (x - 2)(x^2 - 3x + 1) = 0$. So, from the previous question, the complete solution is $x = 2$, $x = 2 \cdot 6180$ or $x = 0 \cdot 3820$ to 4 dp.

5 $x^4 + x^3 - 2x^2 - x + 1 = 0$.

Let $f(x) = x^4 + x^3 - 2x^2 - x + 1$
$$= (((x + 1)x - 2)x - 1)x + 1 \text{ so}$$
$$f(1) = (((1 + 1)1 - 2)1 - 1)1 + 1 = 0 \text{ and}$$
$$f(-1) = ((([-1] + 1)[-1] - 2)[-1] - 1)[-1] + 1 = 0$$

Thus $x - 1$ and $x + 1$ are factors of $f(x)$. Now $(x - 1)(x + 1) = x^2 - 1$ and:

$$\frac{x^4 + x^3 - 2x^2 - x + 1}{x^2 - 1} = x^2 + x - 1$$

so that $x^4 + x^3 - 2x^2 - x + 1 = (x^2 - 1)(x^2 + x - 1)$ giving the solution of $x^4 + x^3 - 2x^2 - x + 1 = 0$ as $x = \pm 1$, $x = 0 \cdot 618$ or $x = -1 \cdot 618$.

Now for the Review test

 # Review test   **Unit 18**

---

38   **1** Solve x by factorizing:

$$6x^2 - 13x + 6 = 0$$

2 Solve for x (accurate to 3 dp) by completing the square:

$$3x^2 + 4x - 1 = 0$$

3 Solve x (accurate to 3 dp) by using the formula:

$$x^2 + 3x - 1 = 0$$

4 Solve for x:

$$x^3 - 8x^2 + 18x - 9 = 0$$

5 Solve for x:

$$2x^4 - 9x^3 + 12x^2 - 3x - 2 = 0$$

You have now come to the end of this Module. A list of **Can You?** questions **39** follows for you to gauge your understanding of the material in the Module. You will notice that these questions match the **Learning outcomes** listed at the begining of the Module. Now try the **Test exercise.** *Work through the questions at your own pace, there is no need to hurry.* A set of **Further problems** provides additional valuable practice.

Can You?

Test exercise 6

1 Solve the following equations by the method of factors: **2**

(a) $x^2 + 11x + 18 = 0$ (d) $2x^2 + 13x + 20 = 0$

(b) $x^2 - 13x + 42 = 0$ (e) $3x^2 + 5x - 12 = 0$

(c) $x^2 + 4x - 21 = 0$ (f) $5x^2 - 26x + 24 = 0$

2 Solve the following by completing the square:

(a) $2x^2 - 4x - 3 = 0$ (c) $3x^2 + 12x - 18 = 0$

(b) $5x^2 + 10x + 2 = 0$

3 Solve by means of the formula:

(a) $4x^2 + 11x + 2 = 0$ (c) $5x^2 - 9x - 4 = 0$

(b) $6x^2 - 3x - 5 = 0$

4 Solve the cubic equation:

$2x^3 + 11x^2 + 6x - 3 = 0$

5 Solve the fourth-order equation:

$4x^4 - 18x^3 + 23x^2 - 3x - 6 = 0$

Further problems 6

3

1 Solve the following by the method of factors:
   (a) $x^2 + 3x - 40 = 0$ (c) $x^2 + 10x + 24 = 0$
   (b) $x^2 - 11x + 28 = 0$ (d) $x^2 - 4x - 45 = 0$

2 Solve the following:
   (a) $4x^2 - 5x - 6 = 0$ (d) $7x^2 + 4x - 5 = 0$
   (b) $3x^2 + 7x + 3 = 0$ (e) $6x^2 - 15x + 7 = 0$
   (c) $5x^2 + 8x + 2 = 0$ (f) $8x^2 + 11x - 3 = 0$

3 Solve the following cubic equations:
   (a) $5x^3 + 14x^2 + 7x - 2 = 0$ (d) $2x^3 + 4x^2 - 3x - 3 = 0$
   (b) $4x^3 + 7x^2 - 6x - 5 = 0$ (e) $4x^3 + 2x^2 - 17x - 6 = 0$
   (c) $3x^3 - 2x^2 - 21x - 10 = 0$ (f) $4x^3 - 7x^2 - 17x + 6 = 0$

4 Solve the following fourth-order equations:
   (a) $2x^4 - 4x^3 - 23x^2 - 11x + 6 = 0$
   (b) $5x^4 + 8x^3 - 8x^2 - 8x + 3 = 0$
   (c) $4x^4 - 3x^3 - 30x^2 + 41x - 12 = 0$
   (d) $2x^4 + 14x^3 + 33x^2 + 31x + 10 = 0$
   (e) $3x^4 - 6x^3 - 25x^2 + 44x + 12 = 0$
   (f) $5x^4 - 12x^3 - 6x^2 + 17x + 6 = 0$
   (g) $2x^4 - 16x^3 + 37x^2 - 29x + 6 = 0$

Module 7

Partial fractions

Learning outcomes

When you have completed this Module you will be able to:

- Factorize the denominator of an algebraic fraction into its factors
- Separate an algebraic fraction into its partial fractions
- Recognize the rules of partial fractions

Units

Partial fractions   **Unit 19**

1

To simplify an arithmetical expression consisting of a number of fractions, we first convert the individual fractions to a new form having a common denominator which is the LCM of the individual denominators.

With $\dfrac{2}{5} - \dfrac{3}{4} + \dfrac{1}{2}$ the LCM of the denominators, 5, 4 and 2, is 20.

$$\therefore \; \frac{2}{5} - \frac{3}{4} + \frac{1}{2} = \frac{8 - 15 + 10}{20} = \frac{3}{20}$$

In just the same way, algebraic fractions can be combined by converting them to a new denominator which is the LCM of the individual denominators.

For example, with $\dfrac{2}{x-3} - \dfrac{4}{x-1}$, the LCM of the denominators is $(x-3)(x-1)$. Therefore:

$$\frac{2}{x-3} - \frac{4}{x-1} = \frac{2(x-1) - 4(x-3)}{(x-3)(x-1)}$$

$$= \frac{2x - 2 - 4x + 12}{(x-3)(x-1)}$$

$$= \frac{10 - 2x}{(x-3)(x-1)}$$

In practice, the reverse process is often required. That is, presented with a somewhat cumbersome algebraic fraction there is a need to express this as a number of simpler component fractions.

From the previous example:

$$\frac{2}{x-3} - \frac{4}{x-1} = \frac{10 - 2x}{(x-3)(x-1)}$$

The two simple fractions on the left-hand side are called the *partial fractions* of the expression on the right-hand side. What follows describes how these partial fractions can be obtained from the original fraction.

So, on to the next frame

2

Let us consider a simple case and proceed step by step.
To separate:

$$\frac{8x - 28}{x^2 - 6x + 8}$$

into its partial fractions we must first factorize the denominator into its prime factors. You can do this:

$$x^2 - 6x + 8 = (\ldots\ldots\ldots\ldots)(\ldots\ldots\ldots\ldots)$$

$$(x-2)(x-4)$$

Therefore:

$$\frac{8x-28}{x^2-6x+8} = \frac{8x-28}{(x-2)(x-4)}$$

We now assume that each simple factor in the denominator gives rise to a single partial fraction. That is, we assume that we can write:

$$\frac{8x-28}{(x-2)(x-4)} = \frac{A}{x-2} + \frac{B}{x-4}$$

where A and B are constants. We shall now show that this assumption is valid by finding the values of A and B. First we add the two partial fractions on the RHS to give:

$$\frac{8x-28}{(x-2)(x-4)} = \ldots\ldots\ldots\ldots$$

The answer is in the next frame

$$\frac{8x-28}{x^2-6x+8} = \frac{A(x-4)+B(x-2)}{(x-2)(x-4)}$$

Because

$$\frac{A}{x-2} + \frac{B}{x-4} = \frac{A(x-4)}{(x-2)(x-4)} + \frac{B(x-2)}{(x-2)(x-4)}$$
$$= \frac{A(x-4)+B(x-2)}{(x-2)(x-4)}$$

The equation $\dfrac{8x-28}{x^2-6x+8} = \dfrac{A(x-4)+B(x-2)}{(x-2)(x-4)}$ is an identity because the RHS is just an alternative way of writing the LHS. Also, because the denominator on the RHS is an alternative way of writing the denominator on the LHS, the same must be true of the numerators. Consequently, equating the numerators we find that ....

Next frame

5

$$8x-28 \equiv A(x-4)+B(x-2)$$

Because this is an identity it must be true for all values of x. It is convenient to choose a value of x that makes one of the brackets zero. For example:

Letting $x=4$ gives $B = \ldots\ldots\ldots\ldots$

6

$$B = 2$$

Because

$32 - 28 = A(0) + B(2)$ so that $4 = 2B$ giving $B = 2$.

Similarly if we let $x = 2$ then $A = \ldots\ldots\ldots\ldots$

7

$$A = 6$$

Because

$16 - 28 = A(-2) + B(0)$ so that $-12 = -2A$ giving $A = 6$.

Therefore:

$$\frac{8x - 28}{(x - 2)(x - 4)} = \frac{6}{x - 2} + \frac{2}{x - 4}$$

the required partial fraction breakdown.

This example has demonstrated the basic process whereby the partial fractions of a given rational expression can be obtained. There is, however, one important proviso that has not been mentioned:

To effect the partial fraction breakdown of a rational algebraic expression it is necessary for the degree of the numerator to be less than the degree of the denominator.

If, in the original algebraic rational expression, the degree of the numerator is not less than the degree of the denominator then we divide out by long division. This gives a polynomial with a rational remainder where the remainder has a numerator with degree less than the denominator. The remainder can then be broken down into its partial fractions. In the following frames we consider some examples of this type.

8

Example 1

Express $\dfrac{x^2 + 3x - 10}{x^2 - 2x - 3}$ in partial fractions.

The first consideration is $\ldots\ldots\ldots\ldots$

9

Is the numerator of lower degree than the denominator?

No, it is not, so we have to divide out by long division:

$$
\begin{array}{r}
1 \\
x^2 - 2x - 3\;\big)\;\overline{x^2 + 3x\ -10} \\
x^2 - 2x\ -\ 3 \\
\hline
5x -\ 7
\end{array}
$$

$$
\therefore\ \frac{x^2 + 3x - 10}{x^2 - 2x - 3} = 1 + \frac{5x - 7}{x^2 - 2x - 3}
$$

Now we factorize the denominator into its prime factors, which gives . . . .

$(x + 1)(x - 3)$

$$
\therefore\ \frac{x^2 + 3x - 10}{x^2 - 2x - 3} = 1 + \frac{5x - 7}{(x + 1)(x - 3)}
$$

The remaining fraction will give partial fractions of the form:

$$
\frac{5x - 7}{(x + 1)(x - 3)} = \frac{A}{x + 1} + \frac{B}{x - 3}
$$

Multiplying both sides by the denominator $(x + 1)(x - 3)$:

$$
5x - 7 = \ldots\ldots\ldots\ldots
$$

11

$A(x - 3) + B(x + 1)$

$5x - 7 \equiv A(x - 3) + B(x + 1)$ is an identity, since the RHS is the LHS merely written in a different form. Therefore the statement is true for any value of x we choose to substitute.

As was said previously, it is convenient to select a value for x that makes one of the brackets zero. So if we put $x = 3$ in both sides of the identity, we get . . . .

$15 - 7 = A(0) + B(4)$

i.e. $8 = 4B$ $\therefore B = 2$

Similarly, if we substitute $x = -1$, we get . . . .

13

$$-5 - 7 = A(-4) + B(0)$$

$$\therefore \ -12 = -4A \qquad \therefore A = 3$$

$$\therefore \ \frac{5x - 7}{(x + 1)(x - 3)} = \frac{3}{x + 1} + \frac{2}{x - 3}$$

So, collecting our results together:

$$\frac{x^2 + 3x - 10}{x^2 - 2x - 3} = 1 + \frac{5x - 7}{x^2 - 2x - 3} = 1 + \frac{5x - 7}{(x + 1)(x - 3)}$$

$$= 1 + \frac{3}{x + 1} + \frac{2}{x - 3}$$

Example 2

Express $\dfrac{2x^2 + 18x + 31}{x^2 + 5x + 6}$ in partial fractions.

The first step is . . . .

14

$$\boxed{\text{to divide the numerator by the denominator}}$$

since the numerator is not of lower degree than that of the denominator.

$$\therefore \ \frac{2x^2 + 18x + 31}{x^2 + 5x + 6} = 2 + \frac{8x + 19}{x^2 + 5x + 6}. \text{ Now we attend to } \frac{8x + 19}{x^2 + 5x + 6}.$$

Factorizing the denominator, we have . . . .

15

$$\boxed{\dfrac{8x + 19}{(x + 2)(x + 3)}}$$

so the form of the partial fractions will be . . . .

16

$$\boxed{\dfrac{A}{x + 2} + \dfrac{B}{x + 3}}$$

i.e. $\dfrac{8x + 19}{(x + 2)(x + 3)} = \dfrac{A}{x + 2} + \dfrac{B}{x + 3}$

You can now multiply both sides by the denominator $(x + 2)(x + 3)$ and finish it off:

$$\frac{2x^2 + 18x + 31}{x^2 + 5x + 6} = \ldots\ldots\ldots\ldots$$

$$2 + \frac{3}{x+2} + \frac{5}{x+3}$$

So move on to the next frame

Example 3

Express $\dfrac{3x^3 - x^2 - 13x - 13}{x^2 - x - 6}$ in partial fractions.

Applying the rules, we first divide out:

$$\frac{3x^3 - x^2 - 13x - 13}{x^2 - x - 6} = \cdots\cdots$$

$$3x + 2 + \frac{7x - 1}{x^2 - x - 6}$$

Now we attend to $\dfrac{7x - 1}{x^2 - x - 6}$ in the normal way.

Finish it off

$$3x + 2 + \frac{3}{x+2} + \frac{4}{x-3}$$

Because

$$\frac{7x - 1}{(x+2)(x-3)} = \frac{A}{x+2} + \frac{B}{x-3}$$

$$\therefore 7x - 1 = A(x - 3) + B(x + 2)$$

$$x = 3 \qquad 20 = A(0) + B(5) \qquad \therefore B = 4$$

$$x = -2 \qquad -15 = A(-5) + B(0) \qquad \therefore A = 3$$

Remembering to include the polynomial part:

$$\therefore \frac{3x^3 - x^2 - 13x - 13}{x^2 - x - 6} = 3x + 2 + \frac{3}{x+2} + \frac{4}{x-3}$$

Now one more entirely on your own just like the last one.

Example 4

Express $\dfrac{2x^3 + 3x^2 - 54x + 50}{x^2 + 2x - 24}$ in partial fractions.

Work right through it: then check with the next frame.

$$\frac{2x^3 + 3x^2 - 54x + 50}{x^2 + 2x - 24} = \cdots\cdots$$

21

$$2x - 1 + \frac{1}{x-4} - \frac{5}{x+6}$$

Here it is:

$$\begin{array}{r} 2x - 1 \\ x^2 + 2x - 24 \overline{\smash{\big)}\ 2x^3 + 3x^2 - 54x + 50} \\ 2x^3 + 4x^2 - 48x \\ \hline - x^2 - 6x + 50 \\ - x^2 - 2x + 24 \\ \hline - 4x + 26 \end{array}$$

$$\therefore \frac{2x^3 + 3x^2 - 54x + 50}{x^2 + 2x - 24} = 2x - 1 - \frac{4x - 26}{x^2 + 2x - 24}$$

$$\frac{4x - 26}{(x-4)(x+6)} = \frac{A}{x-4} + \frac{B}{x+6} \qquad \therefore 4x - 26 = A(x+6) + B(x-4)$$

$$x = 4 \qquad -10 = A(10) + B(0) \qquad \therefore A = -1$$

$$x = -6 \qquad -50 = A(0) + B(-10) \qquad \therefore B = 5$$

$$\therefore \frac{2x^3 + 3x^2 - 54x + 50}{x^2 + 2x - 24} = 2x - 1 - \left\{ -\frac{1}{x-4} + \frac{5}{x+6} \right\}$$

$$= 2x - 1 + \frac{1}{x-4} - \frac{5}{x+6}$$

At this point let us pause and summarize the main facts so far on the breaking into partial fractions of rational algebraic expressions with denominators in the form of a product of two simple factors

Review summary

Unit 19

1 To effect the partial fraction breakdown of a rational algebraic expression it is **22** necessary for the degree of the numerator to be less than the degree of the denominator. In such an expression whose denominator can be expressed as a product of simple prime factors, each of the form $ax + b$:

(a) Write the rational expression with the denominator given as a product of its prime factors.

(b) Each factor then gives rise to a partial fraction of the form $\dfrac{A}{ax + b}$

where A is a constant whose value is to be determined.

(c) Add the partial fractions together to form a single algebraic fraction whose numerator contains the unknown constants and whose denominator is identical to that of the original expression.

(d) Equate the numerator so obtained with the numerator of the original algebraic fraction.

(e) By substituting appropriate values of x in this equation determine the values of the unknown constants.

2 If, in the original algebraic rational expression, the degree of the numerator is not less than the degree of the denominator then we divide out by long division. This gives a polynomial with a rational remainder where the remainder has a numerator with degree less than the denominator. The remainder can then be broken down into its partial fractions.

 # Review exercise

Unit 19

Express the following in partial fractions: **23**

1 $\dfrac{x + 7}{x^2 - 7x + 10}$

2 $\dfrac{10x + 37}{x^2 + 3x - 28}$

3 $\dfrac{3x^2 - 8x - 63}{x^2 - 3x - 10}$

4 $\dfrac{2x^2 + 6x - 35}{x^2 - x - 12}$

Complete all four questions. Take your time, there is no need to rush.
If necessary, refer back to the Unit.
The answers and working are in the next frame.

24

1 $\dfrac{x+7}{x^2-7x+10} = \dfrac{x+7}{(x-2)(x-5)} = \dfrac{A}{x-2} + \dfrac{B}{x-5}$

$\therefore x+7 = A(x-5) + B(x-2)$

$x=5 \qquad 12 = A(0) + B(3) \qquad \therefore B = 4$

$x=2 \qquad 9 = A(-3) + B(0) \qquad \therefore A = -3$

$\dfrac{x+7}{x^2-7x+10} = \dfrac{4}{x-5} - \dfrac{3}{x-2}$

2 $\dfrac{10x+37}{x^2+3x-28} = \dfrac{10x+37}{(x-4)(x+7)} = \dfrac{A}{x-4} + \dfrac{B}{x+7}$

$\therefore 10x+37 = A(x+7) + B(x-4)$

$x=-7 \qquad -33 = A(0) + B(-11) \qquad \therefore B = 3$

$x=4 \qquad 77 = A(11) + B(0) \qquad \therefore A = 7$

$\therefore \dfrac{10x+37}{x^2+3x-28} = \dfrac{7}{x-4} + \dfrac{3}{x+7}$

3 $\dfrac{3x^2-8x-63}{x^2-3x-10} = 3 + \dfrac{x-33}{(x+2)(x-5)}$

$\dfrac{x-33}{(x+2)(x-5)} = \dfrac{A}{x+2} + \dfrac{B}{x-5}$

$\therefore x-33 = A(x-5) + B(x+2)$

$x=5 \qquad -28 = A(0) + B(7) \qquad \therefore B = -4$

$x=-2 \qquad -35 = A(-7) + B(0) \qquad \therefore A = 5$

$\therefore \dfrac{3x^2-8x-63}{x^2-3x-10} = 3 + \dfrac{5}{x+2} - \dfrac{4}{x-5}$

4 $\dfrac{2x^2+6x-35}{x^2-x-12} = 2 + \dfrac{8x-11}{(x+3)(x-4)}$

$\dfrac{8x-11}{(x+3)(x-4)} = \dfrac{A}{x+3} + \dfrac{B}{x-4}$

$\therefore 8x-11 = A(x-4) + B(x+3)$

$x=4 \qquad 21 = A(0) + B(7) \qquad \therefore B = 3$

$x=-3 \qquad -35 = A(-7) + B(0) \qquad \therefore A = 5$

$\therefore \dfrac{2x^2+6x-35}{x^2-x-12} = 2 + \dfrac{5}{x+3} + \dfrac{3}{x-4}$

Now for the Review test

Review test

Unit 19

Express each of the following in partial fraction form:

1 $\dfrac{3x + 9}{x^2 + 8x + 12}$

2 $\dfrac{x^2 + x + 1}{x^2 + 3x + 2}$

25

Denominators with repeated and quadratic factors

Unit 20

Now let's look at a rational algebraic fraction where the denominator contains a quadratic factor that will not factorize into two simple factors.

Express $\dfrac{15x^2 - x + 2}{(x - 5)(3x^2 + 4x - 2)}$ in partial fractions.

Here the degree of the numerator is less than the degree of the denominator so no initial division is required. However, the denominator contains a quadratic factor that cannot be factorized further into simple factors. The usual test confirms this because $(b^2 - 4ac) = 16 - 4 \times 3 \times (-2) = 40$ which is not a perfect square. In this situation there is a rule that applies:

An irreducible quadratic factor in the denominator of the original rational expression of the form $(ax^2 + bx + c)$ gives rise to a partial fraction of the form $\dfrac{Ax + B}{ax^2 + bx + c}$

$$\therefore \quad \frac{15x^2 - x + 2}{(x - 5)(3x^2 + 4x - 2)} = \frac{A}{x - 5} + \frac{Bx + C}{3x^2 + 4x - 2}$$

Multiplying throughout by the denominator:

$$15x^2 - x + 2 = A(3x^2 + 4x - 2) + (Bx + C)(x - 5)$$
$$= 3Ax^2 + 4Ax - 2A + Bx^2 + Cx - 5Bx - 5C$$

Collecting up like terms on the RHS gives:

$$15x^2 - x + 2 = \ldots\ldots\ldots\ldots$$

1

2

$$15x^2 - x + 2 = (3A + B)x^2 + (4A - 5B + C)x - 2A - 5C$$

This is an identity, so we can equate coefficients of like terms on each side:

$$[x^2] \quad 15 = 3A + B \tag{1}$$
$$\text{constant term} \quad [CT] \quad 2 = -2A - 5C \tag{2}$$
$$[x] \quad -1 = 4A - 5B + C \tag{3}$$

From (1): $B = 15 - 3A$

From (2): $2 = -2A - 5C \quad \therefore 5C = -2A - 2 \quad \therefore C = \dfrac{-(2A + 2)}{5}$

Substituting for B and C in (3), we have:

$$-1 = 4A - 5(15 - 3A) - \frac{2A + 2}{5}$$
$$\therefore -5 = 20A - 25(15 - 3A) - (2A + 2)$$
$$-5 = 20A - 375 + 75A - 2A - 2 = 93A - 377$$
$$\therefore 93A = 377 - 5 = 372 \quad \therefore A = 4$$

Sub. in (1): $15 = 12 + B \quad \therefore B = 3$

Sub. in (3): $-1 = 16 - 15 + C \quad \therefore C = -2$

$$\therefore \frac{15x^2 - x + 2}{(x - 5)(3x^2 + 4x - 2)} = \frac{4}{x - 5} + \frac{3x - 2}{3x^2 + 4x - 2}$$

Now on to another example

3

Express $\dfrac{7x^2 - 18x - 7}{(x - 4)(2x^2 - 6x + 3)}$ in partial fractions.

Here again, for the factor $(2x^2 - 6x + 3)$, $(b^2 - 4ac) = 36 - 24 = 12$ which is not a perfect square. $\therefore (2x^2 - 6x + 3)$ is irreducible.

$\therefore$ The partial fractions of $\dfrac{7x^2 - 18x - 7}{(x - 4)(2x^2 - 6x + 3)}$ will be of the form ....

4

$$\frac{A}{x - 4} + \frac{Bx + C}{2x^2 - 6x + 3}$$

Multiplying throughout by the complete denominator:

$$7x^2 - 18x - 7 = \ldots\ldots\ldots$$

5

$$A(2x^2 - 6x + 3) + (Bx + C)(x - 4)$$

Then multiply out and collect up like terms, and that gives:

$$7x^2 - 18x - 7 = \ldots\ldots\ldots$$

6

$$7x^2 - 18x - 7 = (2A + B)x^2 - (6A + 4B - C)x + 3A - 4C$$

Now you can equate coefficients of like terms on each side and finish it. The required partial fractions for

$$\frac{7x^2 - 18x - 7}{(x - 4)(2x^2 - 6x + 3)} = \ldots\ldots\ldots$$

7

$$\frac{3}{x - 4} + \frac{x + 4}{2x^2 - 6x + 3}$$

Because

$$[x^2] \qquad 7 = 2A + B \quad \therefore B = 7 - 2A \qquad\qquad\qquad\qquad (1)$$

$$[CT] \qquad -7 = 3A - 4C \quad \therefore C = \frac{3A + 7}{4} \qquad\qquad\qquad (2)$$

$$[x] \quad -18 = -\left(6A + 28 - 8A - \frac{3A + 7}{4}\right)$$

$$\therefore 72 = 24A + 112 - 32A - 3A - 7$$

$$= -11A + 105 \quad \therefore 11A = 33 \quad \therefore A = 3$$

Substitution in (1) and (2) gives $B = 1$ and $C = 4$.

$$\therefore \quad \frac{7x^2 - 18x - 7}{(x - 4)(2x^2 - 6x + 3)} = \frac{3}{x - 4} + \frac{x + 4}{2x^2 - 6x + 3}$$

Next frame

8

Now let's look at a rational algebraic fraction where the denominator contains a repeated simple factor.

Express $\dfrac{35x - 14}{(7x - 2)^2}$ in partial fractions.

Again, there is a rule that applies:

Repeated factors in the denominator of the algebraic expression of the form $(ax + b)^2$ *give partial fractions of the form* $\dfrac{A}{ax + b} + \dfrac{B}{(ax + b)^2}$. *Similarly* $(ax + b)^3$ *gives rise to partial fractions of the form:*

$$\frac{A}{ax + b} + \frac{B}{(ax + b)^2} + \frac{C}{(ax + b)^3}$$

Consequently, we write:

$$\frac{35x - 14}{(7x - 2)^2} = \frac{A}{7x - 2} + \frac{B}{(7x - 2)^2}$$

Then we multiply throughout as usual by the original denominator:

$$35x - 14 = A(7x - 2) + B$$
$$= 7Ax - 2A + B$$

Now we simply equate coefficients and A and B are found:

$$\frac{35x - 14}{(7x - 2)^2} = \ldots\ldots\ldots\ldots$$

9

$$\boxed{\frac{5}{7x - 2} - \frac{4}{(7x - 2)^2}}$$

Similarly:

$$\frac{42x + 44}{(6x + 5)^2} \text{ in partial fractions} = \ldots\ldots\ldots\ldots$$

Complete it

10

$$\boxed{\frac{7}{6x + 5} + \frac{9}{(6x + 5)^2}}$$

$$\frac{42x + 44}{(6x + 5)^2} = \frac{A}{6x + 5} + \frac{B}{(6x + 5)^2}$$
$$\therefore \ 42x + 44 = A(6x + 5) + B = 6Ax + 5A + B$$
$$[x] \quad 42 = 6A \quad \therefore A = 7$$
$$[CT] \quad 44 = 5A + B = 35 + B \quad \therefore B = 9$$
$$\therefore \ \frac{42x + 44}{(6x + 5)^2} = \frac{7}{6x + 5} + \frac{9}{(6x + 5)^2}$$

And now this one:

$$\text{Express } \frac{18x^2 + 3x + 6}{(3x + 1)^3} \text{ in partial fractions.}$$

Complete the work with that one and then check with the next frame

11

$$\frac{2}{3x+1} - \frac{3}{(3x+1)^2} + \frac{7}{(3x+1)^3}$$

Here $\dfrac{18x^2 + 3x + 6}{(3x+1)^3} = \dfrac{A}{3x+1} + \dfrac{B}{(3x+1)^2} + \dfrac{C}{(3x+1)^3}$

$\therefore\ 18x^2 + 3x + 6 = A(3x+1)^2 + B(3x+1) + C$

$$= A(9x^2 + 6x + 1) + B(3x+1) + C$$
$$= 9Ax^2 + 6Ax + A + 3Bx + B + C$$
$$= 9Ax^2 + (6A + 3B)x + (A + B + C)$$

Equating coefficients:

$[x^2]$ $\quad 18 = 9A$ $\qquad\qquad \therefore\ A = 2$

$[x]$ $\quad\ \ 3 = 6A + 3B$ $\qquad 3 = 12 + 3B \quad 3B = -9 \quad \therefore\ B = -3$

$[CT]$ $\quad 6 = A + B + C$ $\qquad 6 = 2 - 3 + C \quad \therefore\ C = 7$

$\therefore\ \dfrac{18x^2 + 3x + 6}{(3x+1)^3} = \dfrac{2}{3x+1} - \dfrac{3}{(3x+1)^2} + \dfrac{7}{(3x+1)^3}$

Now determine the partial fractions of $\dfrac{20x^2 - 54x + 35}{(2x-3)^3}$

The working is just the same as with the previous example:

$$\frac{20x^2 - 54x + 35}{(2x-3)^3} = \ldots\ldots\ldots\ldots$$

12

$$\frac{5}{2x-3} + \frac{3}{(2x-3)^2} - \frac{1}{(2x-3)^3}$$

Because

$$\frac{20x^2 - 54x + 35}{(2x-3)^3} = \frac{A}{2x-3} + \frac{B}{(2x-3)^2} + \frac{C}{(2x-3)^3}$$

$\therefore\ 20x^2 - 54x + 35 = A(2x-3)^2 + B(2x-3) + C$

Multiplying out and collecting up like terms:

$$20x^2 - 54x + 35 = 4Ax^2 - (12A - 2B)x + (9A - 3B + C)$$

Then, equating coefficients in the usual way:

$A = 5;\ B = 3;\ C = -1$

$\therefore\ \dfrac{20x^2 - 54x + 35}{(2x-3)^3} = \dfrac{5}{2x-3} + \dfrac{3}{(2x-3)^2} - \dfrac{1}{(2x-3)^3}$

Next frame

13

Let us now consider the case where the denominator is a cubic expression with different linear factors.

Express $\dfrac{10x^2 + 7x - 42}{(x - 2)(x + 4)(x - 1)}$ in partial fractions.

Here the factors are all different, so the partial fractions will be ....

14

$$\frac{A}{x - 2} + \frac{B}{x + 4} + \frac{C}{x - 1}$$

$$\frac{10x^2 + 7x - 42}{(x - 2)(x + 4)(x - 1)} = \frac{A}{x - 2} + \frac{B}{x + 4} + \frac{C}{x - 1}$$

$$\therefore\ 10x^2 + 7x - 42 = A(x + 4)(x - 1) + B(x - 2)(x - 1) + C(x + 4)(x - 2)$$

$$= A(x^2 + 3x - 4) + B(x^2 - 3x + 2) + C(x^2 + 2x - 8)$$

Multiplying out and collecting up like terms:

$$10x^2 + 7x - 42 = \ldots\ldots\ldots$$

15

$$(A + B + C)x^2 + (3A - 3B + 2C)x - (4A - 2B + 8C)$$

Equating coefficients:

$$[x^2] \qquad 10 = A + B + C \tag{1}$$
$$[x] \qquad 7 = 3A - 3B + 2C \tag{2}$$
$$[CT] \quad -42 = -4A + 2B - 8C \tag{3}$$

Solving these three simultaneous equations provides the values of A, B and C:

$(1) \times 2 \qquad 2A + 2B + 2C = 20$

$(2) \qquad\quad\ \ 3A - 3B + 2C = \ \ 7$

$$\overline{-A + 5B \qquad\ \ = 13} \tag{4}$$

$(2) \times 4 \qquad 12A - 12B + 8C = \ \ 28$

$(3) \qquad\quad\ \ -4A + \ 2B - 8C = -42$

$$\overline{8A - 10B \qquad\ = -14} \tag{5}$$

Now we can solve (4) and (5) to find A and B, and then substitute in (1) to find C. So finally:

$$\frac{10x^2 + 7x - 42}{(x - 2)(x + 4)(x - 1)} = \ldots\ldots\ldots$$

16

$$\frac{2}{x-2} + \frac{3}{x+4} + \frac{5}{x-1}$$

In this latest example, the denominator has conveniently been given as the product of three linear factors. It may well be that this could be given as a cubic expression, in which case factorization would have to be carried out using the remainder theorem before further progress could be made. Here then is an example which brings us to the peak of this Module.

Determine the partial fractions of $\dfrac{8x^2 - 14x - 10}{x^3 - 4x^2 + x + 6}$.

First we see that no initial division is necessary. Then we have to factorize the denominator into its prime factors, as we did in Module 3.

So, putting $f(x) = x^3 - 4x^2 + x + 6$, we determine the three simple factors of $f(x)$, if they exist. These are . . . .

17

$$(x + 1)(x - 2)(x - 3)$$

Because

$f(x) = x^3 - 4x^2 + x + 6 = [(x - 4)x + 1]x + 6$ in nested form.

$f(1) = 4$ $\therefore (x - 1)$ is not a factor

$f(-1) = 0$ $\therefore (x + 1)$ *is* a factor of $f(x)$:

$$
\begin{array}{r}
x^2 \;\; - 5x + 6 \\
x+1 \,\overline{\smash{\big)}\, x^3 \;\; - 4x^2 + \;\; x + 6} \\
\underline{x^3 \;\; + \;\; x^2} \\
- 5x^2 + \;\; x \\
\underline{- 5x^2 - 5x} \\
6x + 6 \\
\underline{6x + 6} \\
\cdot \quad \cdot
\end{array}
$$

$\therefore f(x) = (x + 1)(x^2 - 5x + 6) = (x + 1)(x - 2)(x - 3)$

$\therefore \dfrac{8x^2 - 14x - 10}{x^3 - 4x^2 + x + 6} = \dfrac{8x^2 - 14x - 10}{(x + 1)(x - 2)(x - 3)}$

and now we can proceed as in the previous example. Work right through it and then check the results with the following frame.

$$\frac{8x^2 - 14x - 10}{(x + 1)(x - 2)(x - 3)} = \ldots\ldots\ldots\ldots$$

18

$$\boxed{\dfrac{1}{x+1}+\dfrac{2}{x-2}+\dfrac{5}{x-3}}$$

Because

$$\frac{8x^2-14x-10}{(x+1)(x-2)(x-3)}=\frac{A}{x+1}+\frac{B}{x-2}+\frac{C}{x-3}$$

$$\therefore\ 8x^2-14x-10=A(x-2)(x-3)+B(x+1)(x-3)+C(x+1)(x-2)$$

$$=A(x^2-5x+6)+B(x^2-2x-3)+C(x^2-x-2)$$

$$=(A+B+C)x^2-(5A+2B+C)x+(6A-3B-2C)$$

$[x^2]$	$A+B+C=8$	(1)
$[x]$	$5A+2B+C=14$	(2)
$[CT]$	$6A-3B-2C=-10$	(3)

$(1)\times2 \qquad 2A+2B+2C=\ \ 16$
$(3) \qquad\ \ \ \ 6A-3B-2C=-10$

$$\overline{\ \ 8A-\ \ B\qquad=\ \ 6\ \ }\qquad\qquad(4)$$

$(2) \qquad\ \ \ \ 5A+2B+C=14$
$(1) \qquad\ \ \ \ \ A+\ B+C=\ \ 8$

$$\overline{\ \ 4A+B\qquad\ \ =\ \ 6\ \ }\qquad\qquad(5)$$

$(5) \qquad\ \ \ \ \ 4A+B=\ \ 6$
$(6) \qquad\ \ \ \ \ 8A-B=\ \ 6$

$$\overline{\ \ 12A\qquad\ \ =12\ \ }\qquad\therefore A=1$$

$(5) \qquad\ \ \ 4A+B=6 \qquad 4+B=6 \qquad \therefore B=2$
$(1) \qquad\ \ \ A+B+C=8 \qquad 1+2+C=8 \qquad \therefore C=5$

$$\therefore\frac{8x^2-14x-10}{x^3-4x^2+x+6}=\frac{8x^2-14x-10}{(x+1)(x-2)(x-3)}$$

$$=\frac{1}{x+1}+\frac{2}{x-2}+\frac{5}{x-3}$$

At this point let us pause and summarize the main facts on the breaking into partial fractions of rational algebraic expressions with denominators containing irreducible quadratic factors or repeated simple factors

Review summary

Unit 20

19

To effect the partial fraction breakdown of a rational algebraic expression it is necessary for the degree of the numerator to be less than the degree of the denominator. In such an expression whose denominator contains:

1 A quadratic factor of the form $ax^2 + bx + c$ which cannot be expressed as a product of simple factors, the partial fraction breakdown gives rise to a partial fraction of the form $\dfrac{Ax + B}{ax^2 + bx + c}$

2 Repeated factors of the form $(ax + b)^2$, the partial fraction breakdown gives rise to a partial fraction of the form $\dfrac{A}{ax + b} + \dfrac{B}{(ax + b)^2}$. Similarly $(ax + b)^3$ gives rise to partial fractions of the form

$$\dfrac{A}{ax + b} + \dfrac{B}{(ax + b)^2} + \dfrac{C}{(ax + b)^3}$$

 # Review exercise

Unit 20

Express in partial fractions:

20

1 $\dfrac{32x^2 - 28x - 5}{(4x - 3)^3}$

2 $\dfrac{9x^2 + 48x + 18}{(2x + 1)(x^2 + 8x + 3)}$

3 $\dfrac{12x^2 + 36x + 6}{x^3 + 6x^2 + 3x - 10}$

Complete the questions.
Refer back to the Unit if necessary, but don't rush.
The answers and working are in the next frame.

21

1 $\dfrac{32x^2 - 28x - 5}{(4x - 3)^3} = \dfrac{A}{4x - 3} + \dfrac{B}{(4x - 3)^2} + \dfrac{C}{(4x - 3)^3}$

$= \dfrac{A(4x - 3)^2 + B(4x - 3) + C}{(4x - 3)^3}$

$\therefore 32x^2 - 28x - 5 = A(4x - 3)^2 + B(4x - 3) + C$

$= 16Ax^2 + (4B - 24A)x + (9A + C - 3B)$

Therefore: $A = 2$, $B = 5$, $C = -8$ giving

$\dfrac{32x^2 - 28x - 5}{(4x - 3)^3} = \dfrac{2}{4x - 3} + \dfrac{5}{(4x - 3)^2} - \dfrac{8}{(4x - 3)^3}$

2 $\dfrac{9x^2 + 48x + 18}{(2x+1)(x^2 + 8x + 3)} = \dfrac{A}{(2x+1)} + \dfrac{Bx+C}{x^2 + 8x + 3}$

$\qquad\qquad = \dfrac{A(x^2 + 8x + 3) + (Bx+C)(2x+1)}{(2x+1)(x^2 + 8x + 3)}$

and so $9x^2 + 48x + 18 = A(x^2 + 8x + 3) + (Bx+C)(2x+1)$

$\qquad\qquad\qquad\qquad\quad = (A+2B)x^2 + (8A+B+2C)x + (3A+C)$

Therefore: $A = 5$, $B = 2$, $C = 3$ giving

$\dfrac{9x^2 + 48x + 18}{(2x+1)(x^2 + 8x + 3)} = \dfrac{5}{2x+1} + \dfrac{2x+3}{x^2 + 8x + 3}$

3 $\dfrac{12x^2 + 36x + 6}{x^3 + 6x^2 + 3x - 10} = \dfrac{A}{x-1} + \dfrac{B}{x+2} + \dfrac{C}{x+5}$

$\qquad\qquad = \dfrac{A(x+2)(x+5) + B(x-1)(x+5) + C(x-1)(x+2)}{x^3 + 6x^2 + 3x - 10}$

and so $12x^2 + 36x + 6 = A(x+2)(x+5) + B(x-1)(x+5) + C(x-1)(x+2)$

$\qquad\qquad\qquad\qquad\quad = (A+B+C)x^2 + (7A+4B+C)x + (10A-5B-2C)$

Therefore: $A = 3$, $B = 2$, $C = 7$ giving

$\dfrac{12x^2 + 36x + 6}{x^3 + 6x^2 + 3x - 10} = \dfrac{3}{x-1} + \dfrac{2}{x+2} + \dfrac{7}{x+5}$

Now for the Review test

Review test   **Unit 20**

22 Express each of the following in partial fractions:

1 $\dfrac{7x^2 + 6x + 5}{(x+1)(x^2 + x + 1)}$

2 $\dfrac{5x+6}{(x-1)^2}$

3 $\dfrac{2x^3 - 5x + 13}{(x+4)^2}$

23 You have now come to the end of this Module. A list of **Can You?** questions follows for you to gauge your understanding of the material in the Module. You will notice that these questions match the **Learning outcomes** listed at the beginning of the Module. Now try the **Test exercise**. *Work through the questions at your own pace, there is no need to hurry.* A set of **Further problems** provides valuable additional practice.

Can You?

 # Test exercise 7

Express each of the following in partial fractions: **2**

1 $\dfrac{x - 14}{x^2 - 10x + 24}$

2 $\dfrac{13x - 7}{10x^2 - 11x + 3}$

3 $\dfrac{4x^2 + 9x - 73}{x^2 + x - 20}$

4 $\dfrac{6x^2 + 19x - 11}{(x + 1)(x^2 + 5x - 2)}$

5 $\dfrac{10x - 13}{(2x - 3)^2}$

6 $\dfrac{3x^2 - 34x + 97}{(x - 5)^3}$

7 $\dfrac{9x^2 - 16x + 34}{(x + 4)(2x - 3)(3x + 1)}$

8 $\dfrac{8x^2 + 27x + 13}{x^3 + 4x^2 + x - 6}$

 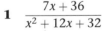 # Further problems 7

Express each of the following in partial fractions: **3**

1 $\dfrac{7x + 36}{x^2 + 12x + 32}$

2 $\dfrac{5x - 2}{x^2 - 3x - 28}$

3 $\dfrac{x + 7}{x^2 - 7x + 10}$

4 $\dfrac{3x - 9}{x^2 - 3x - 18}$

5 $\dfrac{7x - 9}{2x^2 - 7x - 15}$

6 $\dfrac{14x}{6x^2 - x - 2}$

7 $\dfrac{13x - 7}{10x^2 - 11x + 3}$

8 $\dfrac{7x - 7}{6x^2 + 11x + 3}$

9 $\dfrac{18x + 20}{(3x + 4)^2}$

10 $\dfrac{35x + 17}{(5x + 2)^2}$

11 $\dfrac{12x - 16}{(4x - 5)^2}$

12 $\dfrac{5x^2 - 13x + 5}{(x - 2)^3}$

13 $\dfrac{75x^2 + 35x - 4}{(5x + 2)^3}$

14 $\dfrac{64x^2 - 148x + 78}{(4x - 5)^3}$

15 $\dfrac{8x^2 + x - 3}{(x + 2)(x - 1)^2}$

16 $\dfrac{4x^2 - 24x + 11}{(x + 2)(x - 3)^2}$

17 $\dfrac{14x^2 + 31x + 5}{(x - 1)(2x + 3)^2}$

18 $\dfrac{4x^2 - 47x + 141}{x^2 - 13x + 40}$

19 $\dfrac{5x^2 - 77}{x^2 - 2x - 15}$

20 $\dfrac{8x^2 - 19x - 24}{(x - 2)(x^2 - 2x - 5)}$

21 $\dfrac{5x^2 + 9x - 1}{(2x + 3)(x^2 + 5x + 2)}$

22 $\dfrac{7x^2 - 18x - 7}{(3x - 1)(2x^2 - 4x - 5)}$

23 $\dfrac{11x + 23}{(x + 1)(x + 2)(x + 3)}$

24 $\dfrac{5x^2 + 28x + 47}{(x - 1)(x + 3)(x + 4)}$

25 $\dfrac{8x^2 - 60x - 43}{(4x + 1)(2x + 3)(3x - 2)}$

26 $\dfrac{74x^2 - 39x - 6}{(2x - 1)(3x + 2)(4x - 3)}$

27 $\dfrac{16x^2 - x + 3}{6x^3 - 5x^2 - 2x + 1}$

28 $\dfrac{11x^2 + 11x - 32}{6x^3 + 5x^2 - 16x - 15}$

29 $\dfrac{4x^2 - 3x - 4}{6x^3 - 29x^2 + 46x - 24}$

30 $\dfrac{2x^2 + 85x + 36}{20x^3 + 47x^2 + 11x - 6}$

Trigonometry

Learning outcomes

When you have completed this Module you will be able to:

- Convert angles measured in degrees, minutes and seconds into decimal degrees
- Convert degrees into radians and vice versa
- Use a calculator to determine the values of trigonometric ratios for any acute angle
- Verify trigonometric identities

Units

Angles and trigonometric ratios

1 Rotation

When a straight line is rotated about a point it sweeps out an angle that can be measured either in *degrees* or in *radians*. By convention a straight line rotating through a *full angle* and returning to its starting position is said to have rotated through 360 degrees – 360° – where each degree is subdivided into 60 minutes – 60′ – and each minute further subdivided into 60 seconds – 60″. A *straight angle* is half of this, namely 180° and a *right angle* is half of this again, namely 90°. Any angle less than 90° is called an *acute* angle and any angle greater than 90° is called an *obtuse* angle.

An angle that is measured in degrees, minutes and seconds can be converted to a decimal degree as follows:

$$45°36'18'' = 45° + \left(\frac{36}{60}\right)° + \left(\frac{18}{60 \times 60}\right)°$$
$$= (45 + 0.6 + 0.005)°$$
$$= 45.605°$$

That was easy, so the decimal form of 53°29′7″ to 3 dp is ....

The answer is in the next frame

2

$$\boxed{53.485°}$$

Because

$$53°29'7'' = 53° + \left(\frac{29}{60}\right)° + \left(\frac{7}{60 \times 60}\right)°$$
$$= (53 + 0.48\dot{3} + 0.0019\dot{4})°$$
$$= 53.485° \text{ to 3 dp}$$

How about the other way? For example, to convert 18·478° to degrees, minutes and seconds we proceed as follows:

$18.478° = 18° + (0.478 \times 60)'$ Multiply the fractional part of the degree by 60

$= 18° + 28.68'$

$= 18° + 28' + (0.68 \times 60)''$ Multiply the fractional part of the minute by 60

$= 18° + 28' + 40.8''$

$= 18°28'41''$ to the nearest second

So that 236·986° = ... (in degrees, minutes and seconds)

Next frame

3

$$236°59'10''$$

Because

$$236{\cdot}986° = 236° + (0{\cdot}986 \times 60)'$$
$$= 236° + 59{\cdot}16'$$
$$= 236° + 59' + (0{\cdot}16 \times 60)''$$
$$= 236° + 59' + 9{\cdot}6''$$
$$= 236°59'10'' \text{ to the nearest second}$$

Move now to the next frame

Radians

4

The degree is an arbitrary measure of rotation. No-one really knows why the number 360 was chosen for the number of degrees in a full rotation. Indeed, a measure sometimes used by the military is the *grad* and there are 400 grads in a full rotation. A better measure of rotation, that is not arbitrary like the degree or the grad, is the radian which is defined as a ratio. If a straight line of length *r* rotates about one end so that the other end describes an arc of length *r*, the line is said to have rotated through 1 radian – 1 rad.

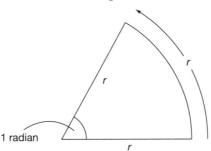

Because the arc described when the line rotates through a full angle is the circumference of a circle which measures $2\pi r$, the number of radians in a full angle is 2π rad. Consequently, relating degrees to radians we see that:

$$360° = 2\pi \text{ rad}$$
$$= 6{\cdot}2831\ldots \text{ rad}$$

So that $1° = \ldots\ldots\ldots\ldots$ rad (to 3 sig fig)

The answer is in the next frame

5

$$\boxed{0 \cdot 0175 \text{ rad}}$$

Because

$$360° = 2\pi \text{ rad, so } 1° = \frac{2\pi}{360} = \frac{\pi}{180} = 0 \cdot 0175 \text{ rad to 3 sig fig}$$

Often, when degrees are transformed to radians they are given as multiples of π. For example:

$$360° = 2\pi \text{ rad, so that } 180° = \pi \text{ rad, } 90° = \pi/2 \text{ rad, } 45° = \pi/4 \text{ rad and so on}$$

So, 30°, 120° and 270° are given in multiples of π as ......, ..., ...

Answers in the next frame

6

$$\boxed{\pi/6 \text{ rad, } 2\pi/3 \text{ rad, } 3\pi/2 \text{ rad}}$$

Because

$$30° = 180°/6 = \pi/6 \text{ rad}$$
$$120° = 2 \times 60° = 2 \times (180°/3) = 2\pi/3 \text{ rad}$$
$$270° = 3 \times 90° = 3 \times (180°/2) = 3\pi/2 \text{ rad}$$

Also, 1 rad = ... degrees (to 3 dp)

Check your answer in the next frame

7

$$\boxed{57 \cdot 296°}$$

Because

$$2\pi \text{ rad } = 360°, \text{ so 1 rad } = \frac{360}{2\pi} = \frac{180}{\pi} = 57 \cdot 296°$$

So, the degree equivalents of 2·34 rad, $\pi/3$ rad, $5\pi/6$ rad and $7\pi/4$ rad are ....

Check with the next frame

8

$$\boxed{134 \cdot 1° \text{ to 1 dp, } 60°, 150° \text{ and } 315°}$$

Because

$$2 \cdot 34 \text{ rad } = \left(2 \cdot 34 \times \frac{180}{\pi}\right)° = 134 \cdot 1° \text{ to 1 dp}$$

$$\pi/3 \text{ rad } = \left(\frac{\pi}{3} \times \frac{180}{\pi}\right)° = 60°$$

$$5\pi/6 \text{ rad } = \left(\frac{5\pi}{6} \times \frac{180}{\pi}\right)° = 150°$$

$$7\pi/4 \text{ rad } = \left(\frac{7\pi}{4} \times \frac{180}{\pi}\right)° = 315°$$

Move to the next frame

Triangles

All triangles possess shape and size. The shape of a triangle is governed by the three angles (which always add up to 180°) and the size by the lengths of the three sides. Two triangles can possess the same shape – possess the same angles – but be of different sizes. We say that two such triangles are *similar*. It is the similarity of figures of different sizes that permits an artist to draw a picture of a scene that looks like the real thing – the lengths of the corresponding lines in the picture and the scene are obviously different but the corresponding angles in the picture and the scene are the same.

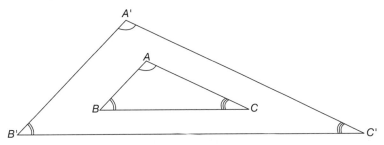

A significant feature of similar figures is that lengths of corresponding sides are all in the same ratio so that, for example, in the similar triangles ABC and $A'B'C'$ in the figure:

$$\frac{AB}{A'B'} = \frac{AC}{A'C'} = \frac{BC}{B'C'}$$

So from a knowledge of the ratios of the sides of a given triangle we can make deductions about any triangle that is similar to it. For example, if in triangle ABC of the above figure:

$AB = 2$ cm, $AC = 5$ cm and $BC = 4$ cm

and in triangle $A'B'C'$, $A'B' = 3$ cm, the length of $A'C'$ can be found as follows:

Since $\dfrac{AB}{A'B'} = \dfrac{AC}{A'C'}$ and $\dfrac{AB}{A'B'} = \dfrac{2}{3}$ then $\dfrac{AC}{A'C'} = \dfrac{5}{A'C'} = \dfrac{2}{3}$ giving

$$A'C' = \frac{5 \times 3}{2} = 7 \cdot 5 \text{ cm}$$

This means that the length of $B'C' = $ ....

Check your answer in the next frame

10

$$\boxed{6 \text{ cm}}$$

Because

$$\frac{AB}{A'B'} = \frac{BC}{B'C'} = \frac{2}{3} \text{ then } \frac{4}{B'C'} = \frac{2}{3} \text{ so that } B'C' = \frac{4 \times 3}{2} = 6 \text{ cm}$$

Ratios between side lengths of one given triangle are also equal to the corresponding ratios between side lengths of a similar triangle. We can prove this using the figure in Frame 9 where:

$$\frac{AB}{A'B'} = \frac{AC}{A'C'}$$

By multiplying both sides of this equation by $\dfrac{A'B'}{AC}$ we find that:

$$\frac{AB}{A'B'} \times \frac{A'B'}{AC} = \frac{AC}{A'C'} \times \frac{A'B'}{AC}, \text{ that is } \frac{AB}{AC} = \frac{A'B'}{A'C'}$$

so that the ratio between sides AB and AC in the smaller triangle is equal to the ratio between the two corresponding sides $A'B'$ and $A'C'$ in the larger, similar triangle.

Show that $\dfrac{AB}{BC} = \dfrac{A'B'}{B'C'}$

The working is in the next frame

11

From Frame 9:

$$\frac{AB}{A'B'} = \frac{BC}{B'C'} \text{ then multiplying both sides of this equation by } \frac{A'B'}{BC}$$

we find that:

$$\frac{AB}{A'B'} \times \frac{A'B'}{BC} = \frac{BC}{B'C'} \times \frac{A'B'}{BC}, \text{ that is } \frac{AB}{BC} = \frac{A'B'}{B'C'}$$

Similarly, $\dfrac{AC}{BC} = \ldots\ldots\ldots$

Next frame

12

$$\boxed{\dfrac{A'C'}{B'C'}}$$

Because

$$\frac{AC}{A'C'} = \frac{BC}{B'C'} \text{ so multiplying both sides of this equation by } \frac{A'C'}{BC} \text{ gives:}$$

$$\frac{AC}{A'C'} \times \frac{A'C'}{BC} = \frac{BC}{B'C'} \times \frac{A'C'}{BC} \text{, that is } \frac{AC}{BC} = \frac{A'C'}{B'C'}$$

All triangles whose corresponding ratios of side lengths are equal have the same shape – they are similar triangles because corresponding angles are equal. Consequently, while the lengths of the sides of a triangle dictate the size of the triangle, the *ratios* of the side lengths dictate the angles of the triangle.

Because we need to know the properties of similar triangles we shall now link these ratios of side lengths to specific angles by using a right-angled triangle; the ratios are then called the *trigonometric ratios*.

On now to the next frame

Trigonometric ratios

13

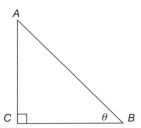

Triangle *ABC* has a right angle at *C*, as denoted by the small square. Because of this the triangle *ABC* is called a *right-angled* triangle. In this triangle the angle at *B* is denoted by θ where side *AC* is *opposite* θ, side *BC* is *adjacent* to θ and side *AB* is called the *hypotenuse*. We define the trigonometric ratios:

$$\textit{sine of angle } \theta \text{ as } \frac{\text{opposite}}{\text{hypotenuse}} = \frac{AC}{AB} - \text{ this ratio is denoted by } \sin\theta$$

$$\textit{cosine of angle } \theta \text{ as } \frac{\text{adjacent}}{\text{hypotenuse}} = \frac{BC}{AB} - \text{ this ratio is denoted by } \cos\theta$$

$$\textit{tangent of angle } \theta \text{ as } \frac{\text{opposite}}{\text{adjacent}} = \frac{AC}{BC} - \text{ this ratio is denoted by } \tan\theta$$

Every angle possesses its respective set of values for the trigonometric ratios and these are most easily found by using a calculator. For example, with the calculator in degree mode, enter 58 and press the *sin* key to display 0·84804 . . . which is the value of sin 58° (that is the ratio of the opposite side over the hypotenuse of all right-angled triangles with an angle of 58°). Now, with your calculator in radian mode enter 2 and press the *sin* key to display 0·90929 . . . which is the value of sin 2 rad – ordinarily we shall omit the rad and just write sin 2. Similar results are obtained using the *cos* key to find the cosine of an angle and the *tan* key to find the tangent of an angle.

Use a calculator in degree mode to find to 4 dp the values of:

(a) sin 27°
(b) cos 84°
(c) tan 43°

The answers are in the next frame

14

(a) 0·4540
(b) 0·1045
(c) 0·9325

That was easy enough. Now use a calculator in radian mode to find to 4 dp the values of the following where the angles are measured in radians:

(a) cos 1·321
(b) tan 0·013
(c) sin $\pi/6$

Check with the next frame

15

(a) 0·2472
(b) 0·0130
(c) 0·5000

We can now use these ratios to find unknowns. For example, a ladder of length 3 m leans against a vertical wall at an angle of 56° to the horizontal.

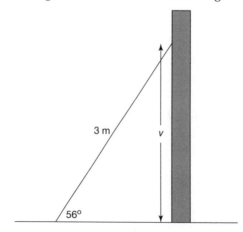

The vertical height of the ladder can now be found as follows. Dividing the vertical height v (the opposite) by the length of the ladder (the hypotenuse) gives the sine of the angle of inclination 56°. That is:

$$\frac{\text{vertical height}}{\text{length of ladder}} = \sin 56°. \text{ That is } \frac{v}{3} = 0·82903\ldots \text{ giving the vertical height}$$

v as

$$3 \times 0·82903\ldots = 2·49 \text{ m (to 3 sig fig)}$$

So if a ladder of length L leans against a vertical wall at an angle of 60° to the horizontal with the top of the ladder 4·5 m above the ground, the length of the ladder is:

$$L = \ldots\ldots$$

The answer is in the next frame

16

$$\boxed{5·20 \text{ m}}$$

Because

$$\frac{\text{vertical height}}{L} = \frac{4·5}{L} = \sin 60° = 0·8660\ldots$$

$$\text{so that } L = \frac{4·5}{0·8660} = 5·20 \text{ m (to 2 dp)}$$

Next frame

Reciprocal ratios

17

In addition to the three trigonometrical ratios there are three *reciprocal ratios*, namely:

$$\operatorname{cosec}\theta = \frac{1}{\sin\theta}, \ \sec\theta = \frac{1}{\cos\theta} \text{ and } \cot\theta = \frac{1}{\tan\theta} \equiv \frac{\cos\theta}{\sin\theta}$$

The values of these for a given angle can also be found using a calculator by finding the appropriate trigonometric ratio and then pressing the *reciprocal* key – the $\frac{1}{x}$ key.

So that, to 4 dp:

    (a) $\cot 12° = \ldots\ldots\ldots\ldots$
    (b) $\sec 37° = \ldots\ldots\ldots\ldots$
    (c) $\operatorname{cosec} 71° = \ldots\ldots\ldots\ldots$

Next frame

18

> (a) 4·7046
> (b) 1·2521
> (c) 1·0576

Because

(a) $\tan 12° = 0·21255\ldots$ and the reciprocal of that is 4·7046 to 4 dp
(b) $\cos 37° = 0·79863\ldots$ and the reciprocal of that is 1·2521 to 4 dp
(c) $\sin 71° = 0·94551\ldots$ and the reciprocal of that is 1·0576 to 4 dp

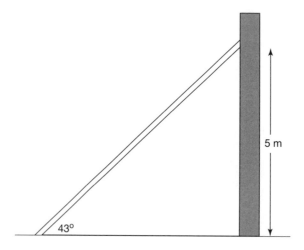

To strengthen a vertical wall a strut has to be placed 5 m up the wall and inclined at an angle of 43° to the ground. To do this the length of the strut must be . . . .

Check the next frame

19

> 7·33 m

Because

$$\frac{\text{length of strut}}{5} = \frac{1}{\sin 43°} = \operatorname{cosec} 43° \text{ that is } \frac{L}{5} = 1·4662\ldots$$
giving $L = 7·33$ to 2 dp

Now go to the next frame

Pythagoras' theorem

All right-angled triangles have a property in common that is expressed in Pythagoras' theorem:

> *The square on the hypotenuse of a right-angled triangle is equal to the sum of the squares on the other two sides*

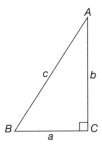

So in the figure:

$$a^2 + b^2 = c^2$$

Notice how the letter for each side length corresponds to the opposite angle (*a* is opposite angle *A* etc.); this is the common convention.

So, if a right-angled triangle has a hypotenuse of length 8 and one other side of length 3, the length of the third side to 3 dp is ....

Check your answer in the next frame

$$7 \cdot 416$$

Because

If *a* represents the length of the third side then:

$$a^2 + 3^2 = 8^2 \text{ so } a^2 = 64 - 9 = 55 \text{ giving } a = 7 \cdot 416 \text{ to 3 dp}$$

Here's another. Is the triangle with sides 7, 24 and 25 a right-angled triangle?

Answer in the next frame

22

$$\boxed{\text{Yes}}$$

Because

Squaring the lengths of the sides gives:

$7^2 = 49$, $24^2 = 576$ and $25^2 = 625$.

Now, $49 + 576 = 625$ so that $7^2 + 24^2 = 25^2$

The sum of the squares of the lengths of the two smaller sides is equal to the square on the longest side. Because the lengths satisfy Pythagoras' theorem, the triangle is right-angled.

How about the triangle with sides 5, 11 and 12? Is this a right-angled triangle?

Check in the next frame

23

$$\boxed{\text{No}}$$

Because

$5^2 = 25$ and $11^2 = 121$ so $5^2 + 11^2 = 146 \neq 12^2$. The squares of the smaller sides do not add up to the square of the longest side so the triangle does not satisfy Pythagoras' theorem and so is not a right-angled triangle.

Next frame

24   Special triangles

Two right-angled triangles are of special interest because the trigonometric ratios of their angles can be given in surd or fractional form. The first is the right-angled *isosceles* triangle (an isosceles triangle is any triangle with two sides of equal length). Since the angles of any triangle add up to 180°, the angles in a right-angled isosceles triangle are 90°, 45° and 45° (or, in radians, $\pi/2$, $\pi/4$ and $\pi/4$) with side lengths, therefore, in the ratio $1 : 1 : \sqrt{2}$ (by Pythagoras' theorem).

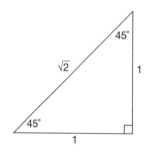

Here we see that:

$$\sin 45° = \frac{\text{opposite}}{\text{hypotenuse}} = \frac{1}{\sqrt{2}}, \quad \cos 45° = \frac{\text{adjacent}}{\text{hypotenuse}} = \frac{1}{\sqrt{2}}$$

$$\text{and} \quad \tan 45° = \frac{\text{opposite}}{\text{adjacent}} = \frac{1}{1} = 1$$

Or, measuring the angles in radians:

$$\sin \pi/4 = \cos \pi/4 = \frac{1}{\sqrt{2}} \quad \text{and} \quad \tan \pi/4 = 1$$

Now, a problem using these ratios:

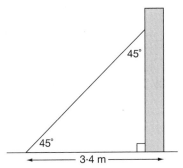

A prop in the form of an isosceles triangle constructed out of timber is placed against a vertical wall. If the length of the side along the horizontal ground is 3·4 m the length of the hypotenuse to 2 dp is obtained as follows:

$$\frac{\text{ground length}}{\text{hypotenuse}} = \frac{3·4}{\text{hypotenuse}} = \cos 45° = \frac{1}{\sqrt{2}}$$

so that:

$$\text{hypotenuse} = \sqrt{2} \times 3·4 = 4·81 \text{ m}$$

Now one for you to try.

A bicycle frame is in the form of an isosceles triangle with the horizontal crossbar forming the hypotenuse. If the crossbar is 53 cm long, the length of each of the other two sides to the nearest mm is . . . .

The answer is in the next frame

25

$$\boxed{37·5 \text{ cm}}$$

Because

$$\frac{\text{side length}}{\text{hypotenuse}} = \frac{\text{side length}}{53} = \cos 45° = \frac{1}{\sqrt{2}} = 0·7071 \ldots$$

so that:

$$\text{side length} = 53 \times 0·7071 = 37·5 \text{ cm}$$

Next frame for some more surd forms

26 Half equilateral triangles

An equilateral triangle is a triangle whose sides are all the same length and whose angles are all equal to 60° (or, in radians, $\pi/3$).

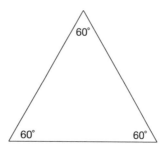

The second right-angled triangle of interest is the *half equilateral* triangle with side lengths (again, by Pythagoras) in the ratio $1 : \sqrt{3} : 2$.

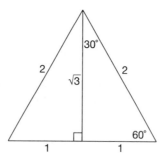

Here we see that:

$$\sin 30° = \cos 60° = \frac{1}{2}, \ \sin 60° = \cos 30° = \frac{\sqrt{3}}{2} \text{ and } \tan 60° = \frac{1}{\tan 30°} = \sqrt{3}$$

Again, if we measure the angles in radians:

$$\sin \pi/6 = \cos \pi/3 = \frac{1}{2}, \ \sin \pi/3 = \cos \pi/6 = \frac{\sqrt{3}}{2} \text{ and } \tan \pi/3 = \frac{1}{\tan \pi/6} = \sqrt{3}$$

Here's an example using these new ratios.

A tree casts a horizontal shadow $8\sqrt{3}$ m long. If a line were to be drawn from the end of the shadow to the top of the tree it would be inclined to the horizontal at 60°. The height of the tree is obtained as follows:

$$\frac{\text{height of tree}}{\text{length of shadow}} = \tan 60° = \sqrt{3}$$

so that

$$\text{height of tree} = \sqrt{3} \times \text{length of shadow} = \sqrt{3} \times 8\sqrt{3} = 8 \times 3 = 24 \text{ m}$$

Now try this one.

When a small tent is erected the front forms an equilateral triangle. If the tent pole is $\sqrt{3}$ m long, the lengths of the sides of the tent are both . . . .

Check your answer in the next frame

27

$$\boxed{2 \text{ m}}$$

Because

$$\frac{\text{length of tent pole}}{\text{length of tent side}} = \frac{\sqrt{3}}{L} = \sin 60° = \frac{\sqrt{3}}{2}$$

so that $L = 2$ m.

Review summary   **Unit 21**

28

1. Angles can be measured in degrees, minutes and seconds or radians where a full angle has a magnitude of 360° or 2π radians.

2. Similar triangles have the same shape – the same angles – but different sizes.

3. Ratios of the sides of one triangle have the same values as ratios of the corresponding sides in a similar triangle.

4. The trigonometric ratios are defined within a right-angled triangle. They are:

$$\sin\theta = \frac{\text{opposite}}{\text{hypotenuse}}$$

$$\cos\theta = \frac{\text{adjacent}}{\text{hypotenuse}}$$

$$\tan\theta = \frac{\sin\theta}{\cos\theta} = \frac{\text{opposite}}{\text{adjacent}}$$

and their reciprocal ratios are:

$$\operatorname{cosec}\theta = 1/\sin\theta$$
$$\sec\theta = 1/\cos\theta$$
$$\cot\theta = 1/\tan\theta$$

5. Pythagoras' theorem states that:

The square on the hypotenuse of a right-angled triangle is equal to the sum of the squares on the other two sides

$$a^2 + b^2 = c^2$$

where a and b are the lengths of the two smaller sides and c is the length of the hypotenuse.

6. The right-angled isosceles triangle has angles 90°, 45° and 45° (or, in radians, $\pi/2$, $\pi/4$ and $\pi/4$), and sides in the ratio $1 : 1 : \sqrt{2}$.

7. The right-angled half equilateral triangle has angles 90°, 60° and 30° (or, in radians, $\pi/2$, $\pi/3$ and $\pi/6$), and sides in the ratio $1 : \sqrt{3} : 2$.

Review exercise

Unit 21

29

1   Convert the angle $164°49'13''$ to decimal degree format.

2   Convert the angle $87·375°$ to degrees, minutes and seconds.

3   Convert the following to radians to 2 dp:
(a) $73°$ (b) $18·34°$ (c) $240°$

4   Convert the following to degrees to 2 dp:
(a) $3·721$ rad   (b) $7\pi/6$ rad   (c) $11\pi/12$ rad

5   Find the value of each of the following to 4 dp:
(a) $\sin 32°$ (b) $\cos \pi/12$ (c) $\tan 2\pi/5$
(d) $\sec 57·8°$ (e) $\operatorname{cosec} 13·33°$ (f) $\cot 0·99$ rad

6   Given one side and the hypotenuse of a right-angled triangle as 5·6 cm and 12·3 cm respectively, find the length of the other side.

7   Show that the triangle with sides 9 m, 40 m and 41 m is a right-angled triangle.

8   A rod of length $7\sqrt{2}$ cm is inclined to the horizontal at an angle of $\pi/4$ radians. A shadow is cast immediately below it from a lamp directly overhead. What is the length of the shadow? What is the new length of the shadow if the rod's inclination is changed to $\pi/3$ to the vertical?

Complete all eight questions. Take your time, there is no need to rush.
If necessary, refer back to the Unit.
The answers and working are in the next frame.

---

1   $164·8203°$ to 4 dp.

30

2   $87°50'15''$

3   (a) $1·27$ rad   (b) $0·32$ (c) $4\pi/3$ rad $= 4·19$ rad

4   (a) $213·20°$ (b) $210°$ (c) $165°$

5   (a) $0·5299$ (b) $0·9659$ (c) $3·0777$
(d) $1·8766$ (e) $4·3373$ (f) $0·6563$

6   If the sides are a, b and c where c is the hypotenuse then $a^2 + b^2 = c^2$. That is, $(5·6)^2 + b^2 = (12·3)^2$ so that $b = \sqrt{(12·3)^2 - (5·6)^2} = 11·0$ to 1 dp.

7   $40^2 + 9^2 = 1681 = 41^2$ thereby satisfying Pythagoras' theorem.

8   If l is the length of the shadow then $\dfrac{l}{7\sqrt{2}} = \cos \pi/4 = \dfrac{1}{\sqrt{2}}$ so that $l = 7$ cm.

If the angle is $\pi/3$ to the vertical then $\dfrac{l}{7\sqrt{2}} = \sin \pi/3 = \dfrac{\sqrt{3}}{2}$ so that

$l = \dfrac{7\sqrt{3}\sqrt{2}}{2} = 7\sqrt{\dfrac{3}{2}} = 8·6$ cm.

Now for the Review test

Review test   Unit 21

31 **1** Convert the angle $253°18'42''$ to decimal degree format.

2 Convert the angle $73·415°$ to degrees, minutes and seconds.

3 Convert the following to radians to 2 dp:
(a) $47°$ (b) $12·61°$ (c) $135°$ (as a multiple of π)

4 Convert the following to degrees to 2 dp:
(a) $4·621$ rad   (b) $9\pi/4$ rad   (c) $13\pi/5$ rad

5 Find the value of each of the following to 4 dp:
(a) $\cos 24°$ (b) $\sin 5\pi/12$ (c) $\cot \pi/3$
(d) $\operatorname{cosec} 17·9°$ (e) $\sec 5·42°$ (f) $\tan 3·24$ rad

6 Given one side and the hypotenuse of a right-angled triangle as $5·6$ cm and $12·3$ cm, find the length of the other side.

7 Show that the triangle with sides 7 m, 24 m and 25 m is a right-angled triangle.

8 A ship sails 12 km due north of a port and then sails 14 km due east. How far is the ship from the port? How much further east will it have sailed when it is 30 km from the port, assuming it keeps on the same course?

Trigonometric identities   Unit 22

1 The fundamental identity

Given the right-angled triangle with vertices A, B and C, sides opposite the vertices of a, b and hypotenuse c and angle θ at B then:

$$a^2 + b^2 = c^2$$

Dividing both sides by c^2 gives:

$$\left(\frac{a}{c}\right)^2 + \left(\frac{b}{c}\right)^2 = 1$$

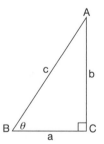

Because $\dfrac{a}{c} = \cos\theta$ and $\dfrac{b}{c} = \sin\theta$ this equation can be written as:

$$\cos^2\theta + \sin^2\theta = 1$$

where the notation $\cos^2\theta \triangleq (\cos\theta)^2$ and $\sin^2\theta \triangleq (\sin\theta)^2$. Since this equation is true for any angle θ the equation is in fact an identity (refer to Frame 5 of Unit 11, Module 3):

$$\cos^2\theta + \sin^2\theta \equiv 1$$

and is called the *fundamental trigonometrical identity*.

▶

For example, to show that the triangle with sides 3 cm, 4 cm and 5 cm is a right-angled triangle it is sufficient to show that it satisfies the fundamental trigonometrical identity. That is, taking the side of length 3 cm to be adjacent to θ (the side with length 5 cm is obviously the hypotenuse as it is the longest side) then:

$$\cos\theta = \frac{3}{5} \text{ and } \sin\theta = \frac{4}{5} \text{ and so}$$

$$\cos^2\theta + \sin^2\theta = \left(\frac{3}{5}\right)^2 + \left(\frac{4}{5}\right)^2 = \frac{9}{25} + \frac{16}{25}$$

$$= \frac{25}{25} = 1$$

Is the triangle with sides of length 8 cm, 12 cm and 10 cm a right-angled triangle?

The answer is in the next frame

| No | **2** |

Because

$$\text{Letting } \cos\theta = \frac{8}{12} \text{ and } \sin\theta = \frac{10}{12}, \text{ then}$$

$$\cos^2\theta + \sin^2\theta = \left(\frac{8}{12}\right)^2 + \left(\frac{10}{12}\right)^2$$

$$= \frac{64}{144} + \frac{100}{144} = \frac{164}{144} \neq 1$$

Since the fundamental trigonometric identity is not satisfied this is not a right-angled triangle.

Move to the next frame

Two more identities

3

Two more identities can be derived directly from the fundamental identity; dividing both sides of the fundamental identity by $\cos^2\theta$ gives the identity . . . .

Check your answer in the next frame

| $1 + \tan^2\theta \equiv \sec^2\theta$ | **4** |

Because

$$\frac{\cos^2\theta}{\cos^2\theta} + \frac{\sin^2\theta}{\cos^2\theta} \equiv \frac{1}{\cos^2\theta} \text{ that is } 1 + \tan^2\theta \equiv \sec^2\theta$$

Dividing the fundamental identity by $\sin^2\theta$ gives a third identity . . . .

Next frame

5

$$\boxed{\cot^2 \theta + 1 \equiv \mathrm{cosec}^2 \theta}$$

Because

$$\frac{\cos^2 \theta}{\sin^2 \theta} + \frac{\sin^2 \theta}{\sin^2 \theta} \equiv \frac{1}{\sin^2 \theta} \text{ that is } \cot^2 \theta + 1 \equiv \mathrm{cosec}^2 \theta$$

Using these three identities and the definitions of the trigonometric ratios it is possible to demonstrate the validity of other identities. For example, to demonstrate the validity of the identity:

$$\frac{1}{1 - \cos \theta} + \frac{1}{1 + \cos \theta} \equiv 2\,\mathrm{cosec}^2 \theta$$

we start with the left-hand side of this identity and demonstrate that it is equivalent to the right-hand side:

$$\text{LHS} = \frac{1}{1 - \cos \theta} + \frac{1}{1 + \cos \theta}$$

$$\equiv \frac{1 + \cos \theta + 1 - \cos \theta}{(1 - \cos \theta)(1 + \cos \theta)} \qquad \text{Adding the two fractions together}$$

$$\equiv \frac{2}{1 - \cos^2 \theta}$$

$$\equiv \frac{2}{\sin^2 \theta} \qquad \text{From the fundamental identity}$$

$$\equiv 2\,\mathrm{cosec}^2 \theta$$

$$= \text{RHS}$$

Try this one. Show that:

$$\tan \theta + \cot \theta \equiv \sec \theta \, \mathrm{cosec}\, \theta$$

Next frame

6

We proceed as follows:

$$\text{LHS} = \tan \theta + \cot \theta$$

$$\equiv \frac{\sin \theta}{\cos \theta} + \frac{\cos \theta}{\sin \theta} \qquad \text{Writing explicitly in terms of sines and cosines}$$

$$\equiv \frac{\sin^2 \theta + \cos^2 \theta}{\cos \theta \sin \theta} \qquad \text{Adding the two fractions together}$$

$$\equiv \frac{1}{\cos \theta \sin \theta} \qquad \text{Since } \sin^2 \theta + \cos^2 \theta = 1 \text{ (the fundamental identity)}$$

$$\equiv \sec \theta \, \mathrm{cosec}\, \theta$$

$$= \text{RHS}$$

So demonstrate the validity of each of the following identities:

(a) $\tan^2 \theta - \sin^2 \theta \equiv \sin^4 \theta \sec^2 \theta$ (b) $\dfrac{1 + \sin \theta}{\cos \theta} \equiv \dfrac{\cos \theta}{1 - \sin \theta}$

Take care with the second one – it is done by performing an operation on both sides first.

The answers are in the next frame

(a) LHS $= \tan^2 \theta - \sin^2 \theta$

$$\equiv \frac{\sin^2 \theta}{\cos^2 \theta} - \sin^2 \theta \qquad \text{Writing explicitly in terms of sines and cosines}$$

$$\equiv \sin^2 \theta \sec^2 \theta - \sin^2 \theta$$

$$\equiv \sin^2 \theta (\sec^2 \theta - 1) \qquad \text{Factorizing out the } \sin^2 \theta$$

$$\equiv \sin^2 \theta \tan^2 \theta \qquad \text{Using the identity } 1 + \tan^2 \theta \equiv \sec^2 \theta$$

$$\equiv \sin^2 \theta \frac{\sin^2 \theta}{\cos^2 \theta}$$

$$\equiv \sin^4 \theta \sec^2 \theta$$

$$= \text{RHS}$$

(b) $\dfrac{1 + \sin \theta}{\cos \theta} \equiv \dfrac{\cos \theta}{1 - \sin \theta}$

Multiplying both sides by $\cos \theta (1 - \sin \theta)$ transforms the identity into:

$(1 - \sin \theta)(1 + \sin \theta) \equiv \cos^2 \theta$. From this we find that:

$$\text{LHS} = (1 - \sin \theta)(1 + \sin \theta)$$

$$\equiv 1 - \sin^2 \theta$$

$$\equiv \cos^2 \theta \qquad \text{since } \cos^2 \theta + \sin^2 \theta \equiv 1$$

$$= \text{RHS}$$

Move on now to the next frame

Identities for compound angles

The trigonometric ratios of the sum or difference of two angles can be given in terms of the ratios of the individual angles. For example, the cosine of a sum of angles is given by:

$$\cos(\theta + \phi) \equiv \cos \theta \cos \phi - \sin \theta \sin \phi$$

To demonstrate the validity of this, consider the following figure:

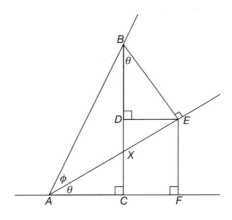

(Notice that in triangles AXC and BXE, $\angle C = \angle E$ as both are right angles, and $\angle AXC = \angle BXE$ as they are equal and opposite. Consequently, the third angles must also be equal so that $\angle EBX = \angle CAX = \theta$.)

Hence we see that:

$$\cos(\theta + \phi) = \frac{AC}{AB} \qquad \text{Adjacent over hypotenuse}$$

$$= \frac{AF - CF}{AB}$$

$$= \frac{AF - DE}{AB} \qquad \text{Because } DE = CF$$

$$= \frac{AF}{AB} - \frac{DE}{AB} \qquad \text{Separating out the fraction}$$

Now, $\cos\theta = \dfrac{AF}{AE}$ so that $AF = AE\cos\theta$. Similarly, $\sin\theta = \dfrac{DE}{BE}$ so that $DE = BE\sin\theta$.

This means that:

$$\cos(\theta + \phi) = \frac{AF}{AB} - \frac{DE}{AB}$$

$$= \frac{AE\cos\theta}{AB} - \frac{BE\sin\theta}{AB}. \text{ Now, } \frac{AE}{AB} = \cos\phi \text{ and } \frac{BE}{AB} = \sin\phi, \text{ therefore}$$

$$\cos(\theta + \phi) \equiv \cos\theta\cos\phi - \sin\theta\sin\phi$$

A similar identity can be demonstrated for the difference of two angles, namely:

$$\cos(\theta - \phi) \equiv \cos\theta\cos\phi + \sin\theta\sin\phi$$

Using these identities it is possible to obtain the cosine of angles other than $30°$, $60°$ and $45°$ in surd form. For example:

$$\cos 75° = \cos(45° + 30°) \qquad \text{Expressing } 75° \text{ in angles where we know the}$$
$$\text{surd form for the trigonometric ratios}$$

$$= \cos 45° \cos 30° - \sin 45° \sin 30° \qquad \text{Using the new formula}$$

$$= \frac{1}{\sqrt{2}} \times \frac{\sqrt{3}}{2} - \frac{1}{\sqrt{2}} \times \frac{1}{2}$$

$$= \frac{\sqrt{3} - 1}{2\sqrt{2}}$$

So the value of $\cos 15°$ in surd form is . . . .

The answer is in the next frame

9

$$\boxed{\dfrac{1+\sqrt{3}}{2\sqrt{2}}}$$

Because

$$\cos 15° = \cos(60° - 45°)$$
$$= \cos 60° \cos 45° + \sin 60° \sin 45°$$
$$= \frac{1}{2} \times \frac{1}{\sqrt{2}} + \frac{\sqrt{3}}{2} \times \frac{1}{\sqrt{2}}$$
$$= \frac{1+\sqrt{3}}{2\sqrt{2}}$$

Just as it is possible to derive the cosine of a sum of angles, it is also possible to derive other trigonometric ratios of sums and differences of angles. In the next frame a list of such identities is given for future reference.

Trigonometric formulas

Sums and differences of angles

10

$$\cos(\theta + \phi) \equiv \cos\theta \cos\phi - \sin\theta \sin\phi \qquad \sin(\theta + \phi) \equiv \sin\theta \cos\phi + \cos\theta \sin\phi$$
$$\cos(\theta - \phi) \equiv \cos\theta \cos\phi + \sin\theta \sin\phi \qquad \sin(\theta - \phi) \equiv \sin\theta \cos\phi - \cos\theta \sin\phi$$

$$\tan(\theta + \phi) \equiv \frac{\sin(\theta + \phi)}{\cos(\theta + \phi)} \equiv \frac{\sin\theta \cos\phi + \cos\theta \sin\phi}{\cos\theta \cos\phi - \sin\theta \sin\phi} \qquad \begin{array}{l}\text{Now divide numerator and} \\ \text{denominator by } \cos\theta \cos\phi\end{array}$$

$$\equiv \frac{\tan\theta + \tan\phi}{1 - \tan\theta \tan\phi}$$

$$\tan(\theta - \phi) \equiv \frac{\tan\theta - \tan\phi}{1 + \tan\theta \tan\phi}$$

Double angles

Double angle formulas come from the above formulas for sums when $\theta = \phi$:

$$\sin 2\theta \equiv 2 \sin\theta \cos\theta$$
$$\cos 2\theta \equiv \cos^2\theta - \sin^2\theta \equiv 2\cos^2\theta - 1 \equiv 1 - 2\sin^2\theta$$
$$\tan 2\theta \equiv \frac{2\tan\theta}{1 - \tan^2\theta}$$

For future reference we now list identities for sums, differences and products of the trigonometric ratios. Each of these can be proved by using the earlier identities and showing that RHS $\equiv$ LHS (rather than showing LHS $\equiv$ RHS as we have done up till now).

Sums and differences of ratios

$$\sin\theta + \sin\phi \equiv 2\sin\frac{\theta+\phi}{2}\cos\frac{\theta-\phi}{2}$$

$$\sin\theta - \sin\phi \equiv 2\cos\frac{\theta+\phi}{2}\sin\frac{\theta-\phi}{2}$$

$$\cos\theta + \cos\phi \equiv 2\cos\frac{\theta+\phi}{2}\cos\frac{\theta-\phi}{2}$$

$$\cos\theta - \cos\phi \equiv -2\sin\frac{\theta+\phi}{2}\sin\frac{\theta-\phi}{2}$$

Products of ratios

$$2\sin\theta\cos\phi \equiv \sin(\theta+\phi) + \sin(\theta-\phi)$$

$$2\cos\theta\cos\phi \equiv \cos(\theta+\phi) + \cos(\theta-\phi)$$

$$2\sin\theta\sin\phi \equiv \cos(\theta-\phi) - \cos(\theta+\phi)$$

 # Review summary   **Unit 22**

11   **1**   The fundamental trigonometric identity is $\cos^2\theta + \sin^2\theta \equiv 1$ and is derived from *Pythagoras'* theorem.

2   Trigonometric identities can be verified using both the fundamental identity and the definitions of the trigonometric ratios.

 # Review exercise   **Unit 22**

12   **1**   Use the fundamental trigonometric identity to show that:

(a) the triangle with sides 5 cm, 12 cm and 13 cm is a right-angled triangle.

(b) the triangle with sides 7 cm, 15 cm and 16 cm is not a right-angled triangle.

2   Verify each of the following identities:

(a) $1 - \dfrac{\sin\theta\tan\theta}{1 + \sec\theta} \equiv \cos\theta$

(b) $\sin\theta + \sin\phi \equiv 2\sin\dfrac{\theta+\phi}{2}\cos\dfrac{\theta-\phi}{2}$

Complete the questions. Take your time.
Refer back to the Unit if necessary but don't rush.
The answers and working are in the next frame.

1 (a) $5^2 + 12^2 = 25 + 144 = 169 = 13^3$

(b) $7^2 + 15^2 = 49 + 225 = 274 \neq 16^2$

2 (a) LHS $= 1 - \dfrac{\sin\theta\tan\theta}{1 + \sec\theta}$

$\equiv \dfrac{1 + \sec\theta - \sin\theta\tan\theta}{1 + \sec\theta}$

$\equiv \dfrac{\cos\theta + 1 - \sin^2\theta}{\cos\theta + 1}$ multiplying top and bottom by $\cos\theta$

$\equiv \dfrac{\cos\theta + \cos^2\theta}{\cos\theta + 1}$

$\equiv \dfrac{\cos\theta(1 + \cos\theta)}{\cos\theta + 1}$

$\equiv \cos\theta$

$= $ RHS

(b) RHS $= 2\sin\dfrac{\theta + \phi}{2}\cos\dfrac{\theta - \phi}{2}$

$\equiv 2\left(\sin\dfrac{\theta}{2}\cos\dfrac{\phi}{2} + \sin\dfrac{\phi}{2}\cos\dfrac{\theta}{2}\right)\left(\cos\dfrac{\theta}{2}\cos\dfrac{\phi}{2} + \sin\dfrac{\phi}{2}\sin\dfrac{\theta}{2}\right)$

$\equiv 2\left(\sin\dfrac{\theta}{2}\cos\dfrac{\theta}{2}\cos^2\dfrac{\phi}{2} + \sin\dfrac{\phi}{2}\cos\dfrac{\phi}{2}\cos^2\dfrac{\theta}{2}\right.$

$\left. + \sin\dfrac{\phi}{2}\cos\dfrac{\phi}{2}\sin^2\dfrac{\theta}{2} + \sin^2\dfrac{\phi}{2}\sin\dfrac{\theta}{2}\cos\dfrac{\theta}{2}\right)$

$\equiv \sin\theta\cos^2\dfrac{\phi}{2} + \sin\phi\cos^2\dfrac{\theta}{2} + \sin\phi\sin^2\dfrac{\theta}{2} + \sin^2\dfrac{\phi}{2}\sin\theta$

$\equiv \sin\theta\cos^2\dfrac{\phi}{2} + \sin^2\dfrac{\phi}{2}\sin\theta + \sin\phi\cos^2\dfrac{\theta}{2} + \sin\phi\sin^2\dfrac{\theta}{2}$

$\equiv \sin\theta\left(\cos^2\dfrac{\phi}{2} + \sin^2\dfrac{\phi}{2}\right) + \sin\phi\left(\cos^2\dfrac{\theta}{2} + \sin^2\dfrac{\theta}{2}\right)$

$\equiv \sin\theta + \sin\phi$

$= $ LHS

Now for the Review test

13

Review test   **Unit 22**

14 **1** Verify each of the following trigonometric identities:

(a) $(\sin\theta - \cos\theta)^2 + (\sin\theta + \cos\theta)^2 \equiv 2$

(b) $(1 - \cos\theta)^{\frac{1}{2}}(1 + \cos\theta)^{\frac{1}{2}} \equiv \sin\theta$

(c) $\tan\theta + \sec\theta \equiv \dfrac{1}{\sec\theta - \tan\theta}$

(d) $\cos\theta - \cos\phi \equiv -2\sin\dfrac{\theta + \phi}{2}\sin\dfrac{\theta - \phi}{2}$

2 Show that:

(a) $\tan 75° = \dfrac{\sqrt{3} + 1}{\sqrt{3} - 1}$

(b) $\sin 15° = \dfrac{\sqrt{3} - 1}{2\sqrt{2}}$

15 You have now come to the end of this Module. A list of **Can You?** questions follows for you to gauge your understanding of the material in the Module. You will notice that these questions match the **Learning outcomes** listed at the beginning of the Module. Then try the **Test exercise**. *Work through the questions at your own pace, there is no need to hurry.* A set of **Further problems** provides additional valuable practice.

 Can You?

1 Checklist: Module 8

Check this list before and after you try the end of Module test.

On a scale of 1 to 5 how confident are you that you can:

● Convert angles measured in degrees, minutes and seconds into decimal degrees?

   Yes ☐ ☐ ☐ ☐ ☐ *No*

● Convert degrees into radians and vice versa?

   Yes ☐ ☐ ☐ ☐ ☐ *No*

● Use a calculator to determine the values of trigonometric ratios for any acute angle?

   Yes ☐ ☐ ☐ ☐ ☐ *No*

● Verify trigonometric identities?

   Yes ☐ ☐ ☐ ☐ ☐ *No*

Test exercise 8

2

1 Convert the angle $39°57'2''$ to decimal degree format.

2 Convert the angle $52·505°$ to degrees, minutes and seconds.

3 Convert the following to radians to 2 dp:
   (a) $84°$ (b) $69·12°$ (c) $240°$ (as a multiple of π)

4 Convert the following to degrees to 2 dp:
   (a) $2·139\,\text{rad}$ (b) $5\pi/3\,\text{rad}$ (c) $9\pi/10\,\text{rad}$

5 Find the value of each of the following to 4 dp:
   (a) $\cos 18°$ (b) $\sin \pi/11$ (c) $\cos 2\pi/7$
   (d) $\cot 48·7°$ (e) $\operatorname{cosec} 1·04\,\text{rad}$ (f) $\sec 0·85\,\text{rad}$

6 Given one side and the hypotenuse of a right-angled triangle as $4·3\,\text{cm}$ and $11·2\,\text{cm}$, find the length of the other side.

7 Show that the triangle with sides $9\,\text{cm}$, $12\,\text{cm}$ and $15\,\text{cm}$ is a right-angled triangle.

8 A triangle has its three sides in the ratio $1 : 0·6 : 0·8$. Is it a right-angled triangle?

9 Verify each of the following trigonometric identities:

   (a) $\dfrac{(\cos\theta - \sin\theta)^2}{\cos\theta} \equiv \sec\theta - 2\sin\theta$

   (b) $\dfrac{\operatorname{cosec}\theta\,\sec\theta}{\cot\theta} \equiv 1 + \tan^2\theta$

   (c) $2\sin\theta\cos\phi \equiv \sin(\theta + \phi) + \sin(\theta - \phi)$

10 Show that:

   (a) $\sin 75° = \dfrac{1 + \sqrt{3}}{2\sqrt{2}}$

   (b) $\tan 15° = \dfrac{\sqrt{3} - 1}{\sqrt{3} + 1}$

Further problems 8

3

1 Convert the angle $81°18'23''$ to decimal degree format.

2 Convert the angle $63·216°$ to degrees, minutes and seconds.

3 Convert the following to radians to 2 dp:
   (a) $31°$ (b) $48·15°$ (c) $225°$ (as a multiple of π)

4 Convert the following to degrees to 2 dp:
   (a) $1·784\,\text{rad}$ (b) $3\pi/4\,\text{rad}$ (c) $4\pi/5\,\text{rad}$

5 Find the value of each of the following to 4 dp:
   (a) $\tan 27°$ (b) $\sin \pi/5$ (c) $\tan 4\pi/9$
   (d) $\sec 89·2°$ (e) $\operatorname{cosec} 0·04°$ (f) $\cot 1·18$ rad

6 Given one side and the hypotenuse of a right-angled triangle as 6·4 cm and 9·1 cm, find the length of the other side.

7 Show that the triangle with sides 5 cm, 11 cm and 12 cm is not a right-angled triangle.

8 What is the length of the diagonal of a square of side length $\sqrt{2}$?

9 Verify each of the following trigonometric identities:

   (a) $\dfrac{\cos \theta - 1}{\sec \theta + \tan \theta} + \dfrac{\cos \theta + 1}{\sec \theta - \tan \theta} \equiv 2(1 + \tan \theta)$

   (b) $\sin^3 \theta - \cos^3 \theta \equiv (\sin \theta - \cos \theta)(1 + \sin \theta \cos \theta)$

   (c) $\operatorname{cosec}^2 \theta - \operatorname{cosec} \theta \equiv \dfrac{\cot^2 \theta}{1 + \sin \theta}$

   (d) $\cot \theta \cos \theta + \tan \theta \sin \theta \equiv (\operatorname{cosec} \theta + \sec \theta)(1 - \sin \theta \cos \theta)$

   (e) $\dfrac{\cos \theta + \sin \theta}{\cos \theta - \sin \theta} \equiv 1 + \dfrac{2 \tan \theta}{1 - \tan \theta}$

   (f) $(\sin \theta - \cos \theta)^2 + (\sin \theta + \cos \theta)^2 \equiv 2$

   (g) $\sqrt{\dfrac{1 + \tan^2 \theta}{1 + \cot^2 \theta}} \equiv \tan \theta$

Functions

Processing numbers   Unit 23

1

The equation that states that y *is equal to some expression in* x, written as:

$$y = f(x)$$

has been described with the words 'y *is a function of* x'. Despite being widely used and commonly accepted, this description is not strictly correct as will be seen in Frame 3. Put simply, for all the functions that you have considered so far, both x and y are *numbers*.

Take out your calculator and enter the number:

5   this is x, the *input* number

Now press the x^2 key and the display changes to:

25   this is y, the *output* number where $y = x^2$

The *function* is a *rule* embodied in a *set of instructions* within the calculator that changed the 5 to 25, activated by you pressing the x^2 key. A diagram can be constructed to represent this:

The box labelled f represents the function. The notation ^2 inside the box means *raising to the power 2* and describes the rule – what the set of instructions will do when activated. The diagram tells you that the input number x is *processed* by the function f to produce the output number $y = f(x)$. So that $y = f(x)$ is the *result* of function f acting on x.

So, use diagrams and describe the functions appropriate to each of the following equations:

(a) $y = \dfrac{1}{x}$ (b) $y = x - 6$

(c) $y = 4x$ (d) $y = \sin x$

Just follow the reasoning above, the answers are in the next frame

2

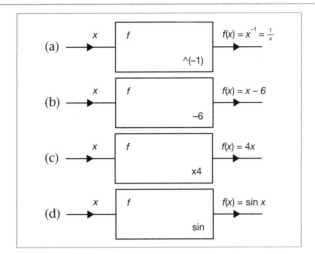

(a) Function f produces the reciprocal of the input

(b) Function f subtracts 6 from the input

(c) Function f multiplies the input by 4

(d) Function f produces the sine of the input

Let's now expand this idea

Functions are rules but not all rules are functions

3

A function of a variable x is a *rule* that describes how a value of the variable x is manipulated to generate a value of the variable y. The rule is often expressed in the form of an equation $y = f(x)$ with the proviso that for any input x there is a unique value for y. Different outputs are associated with different inputs – the function is said to be *single valued* because for a given input there is only one output. For example, the equation:

$$y = 2x + 3$$

expresses the rule '*multiply the value of x by two and add three*' and this rule is the function. On the other hand, the equation:

$y = x^{\frac{1}{2}}$ which is the same as $y = \pm\sqrt{x}$

expresses the rule '*take the positive and negative square roots of the value of x*'. This rule is not a function because to each value of the input $x > 0$ there are two different values of output y.

The graph of $y = \pm\sqrt{x}$ illustrates this quite clearly:

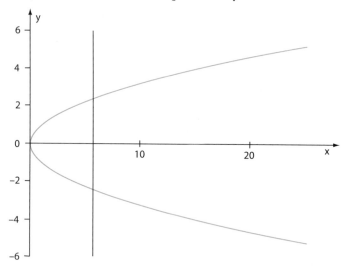

If a vertical line is drawn through the x-axis (for $x > 0$) it intersects the graph at more than one point, that is for a given value of x there is more than one corresponding value of y. The fact that for $x = 0$ the vertical line intersects the graph at only *one* point does not matter – that there are other points where the vertical line intersects the graph in more than one point is sufficient to bar this from being the graph of a function. Notice that $y = x^{\frac{1}{2}}$ has no real values of y for $x < 0$.

Also note that your calculator only gives a single answer to $x^{\frac{1}{2}}$ because it is, in fact, calculating $\sqrt{x}$.

So, which of the following equations express rules that are functions?

(a) $y = 5x^2 + 2x^{-\frac{1}{4}}$

(b) $y = 7x^{\frac{1}{3}} - 3x^{-1}$

Next frame

4

(a) $y = 5x^2 + 2x^{-\frac{1}{4}}$ does not

(b) $y = 7x^{\frac{1}{3}} - 3x^{-1}$ does

(a) $y = 5x^2 + 2x^{-\frac{1}{4}}$ does not express a function because to each value of x ($x > 0$) there are two values of $x^{-\frac{1}{4}}$, positive and negative because $x^{-\frac{1}{4}} \equiv (x^{-\frac{1}{2}})^{\frac{1}{2}} \equiv \pm\sqrt{x^{-\frac{1}{2}}}$. Indeed, *any* even root produces two values.

(b) $y = 7x^{\frac{1}{3}} - 3x^{-1}$ does express a function because to each value of x ($x \neq 0$) there is just one value of y.

All the input numbers x that a function can process are collectively called the function's *domain*. The complete collection of numbers y that correspond to the numbers in the domain is called the *range* (or *co-domain*) of the function. For example, if:

$y = \sqrt{1 - x^2}$ where both x and y are real numbers

the domain is $-1 \le x \le 1$ because these are the only values of x for which y has a real value. The range is $0 \le y \le 1$ because 0 and 1 are the minimum and maximum values of y over the domain. Other functions may, for some purpose or other, be defined on a restricted domain. For example, if we specify:

$y = x^3$, $\quad -2 \le x < 3$ $\qquad$ (the function is defined only for the restricted set of x-values given)

the domain is given as $-2 \le x < 3$ and the range as $-8 \le y < 27$ because -8 and 27 are the minimum and maximum values of y over the domain.

So the domains and ranges of each of the following are:

(a) $y = x^3 \quad -5 \le x < 4$ $\qquad$ (b) $y = x^4$ $\qquad$ (c) $y = \dfrac{1}{(x-1)(x+2)} \quad 0 \le x \le 6$

Take care with the domain of (c)
The answers are in the next frame

5

(a) $y = x^3$, $\quad -5 \le x < 4$
  domain $-5 \le x < 4$, range $-125 \le y < 64$
(b) $y = x^4$
  domain $-\infty < x < \infty$, range $0 \le y < \infty$
(c) $y = \dfrac{1}{(x-1)(x+2)}$, $\quad 0 \le x \le 6$
  domain $0 \le x < 1$ and $1 < x \le 6$,
  range $-\infty < y \le -0.5$, $0.025 \le y < \infty$

Because
(a) The domain is given as $-5 \le x < 4$ and the range as $-125 \le y < 64$ because -125 and 64 are the minimum and maximum values of y over the domain.
(b) The domain is not given and is assumed to consist of all finite values of x, that is, $-\infty < x < \infty$. The range values are all positive because of the even power.
(c) The domain is $0 \le x < 1$ and $1 < x \le 6$ since y is not defined when $x = 1$ where there is a vertical asymptote. To the left of the asymptote $(0 \le x < 1)$ the y-values range from $y = -0.5$ when $x = 0$ and increase negatively towards $-\infty$ as $x \to 1$. To the right of the asymptote $1 < x \le 6$ the y-values range from infinitely large and positive to 0.025 when $x = 6$. If you plot the graph on your spreadsheet this will be evident.

Next frame

6 Combining functions

Functions can be added, subtracted, multiplied and divided provided care is taken over their common domains. For example:

If $f(x) = x^2 - 1, \quad -2 \le x < 4$

and $g(x) = \dfrac{2}{x+3}, \quad 0 < x \le 5$

then, for example

(a) $h(x) = f(x) + g(x) = x^2 - 1 + \dfrac{2}{x+3}$

with the new common domain $0 < x < 4$

because $g(x)$ is not defined for $-2 \le x \le 0$, $f(x)$ is not defined for $4 \le x \le 5$, so $0 < x < 4$ is the common domain between them.

(b) $k(x) = \dfrac{g(x)}{f(x)} = \dfrac{2}{(x+3)(x^2-1)}$ with the domain $0 < x < 4$ and $x \ne 1$

because $g(x)$ is not defined for $-2 \le x \le 0$, $f(x)$ is not defined for $4 \le x \le 5$ and $k(x)$ is not defined when $x = 1$.

So if:

$$f(x) = \dfrac{2x}{x^3 - 1}, \text{ where } -3 < x < 3 \text{ and } x \ne 1 \text{ and}$$

$$g(x) = \dfrac{4x - 8}{x + 5}, \quad 0 < x \le 6 \text{ then } h(x) = \dfrac{f(x)}{g(x)} \text{ is } \ldots\ldots\ldots$$

The answer is in the next frame

7

$$h(x) = \dfrac{f(x)}{g(x)} = \dfrac{2x(x+5)}{(x^3-1)(4x-8)} \text{ where } 0 < x < 3, \ x \ne 1 \text{ and } x \ne 2$$

Because when $x = 1$ or 2, $h(x)$ is not defined; when $-3 < x \le 0$, $g(x)$ is not defined; and when $3 \le x \le 6$, $f(x)$ is not defined.

Inverses of functions

8

The process of generating the output of a function is assumed to be reversible so that what has been constructed can be de-constructed. The effect can be described by reversing the flow of information through the diagram so that, for example, if:

$$y = f(x) = x + 5$$

the flow is then reversed by making the output the input and *retrieving the original input as the new output*:

The reverse process is different because instead of adding 5 to the input, 5 is now subtracted from the input. The rule that describes the reversed process is called the *inverse of the function* which is labelled as either f^{-1} or arc$f$. That is: $f^{-1}(x) = x - 5$:

The notation f^{-1} is very commonly used but care must be taken to remember that the -1 does not mean that it is in any way related to the reciprocal of f.

Try some. Find $f^{-1}(x)$ in each of the following cases:

(a) $f(x) = 6x$ (b) $f(x) = x^3$ (c) $f(x) = \dfrac{x}{2}$

Draw the diagram, reverse the flow and find the inverse of the function in each case

9

$$\text{(a) } f^{-1}(x) = \frac{x}{6}$$
$$\text{(b) } f^{-1}(x) = x^{\frac{1}{3}}$$
$$\text{(c) } f^{-1}(x) = 2x$$

Because

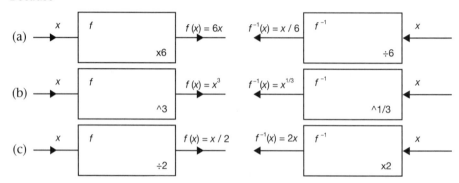

(a)

$f(x) = 6x$ $f^{-1}(x) = x/6$ f^{-1}

×6   ÷6

(b)

$f(x) = x^3$ $f^{-1}(x) = x^{1/3}$ f^{-1}

^3   ^1/3

(c)

$f(x) = x/2$ $f^{-1}(x) = 2x$ f^{-1}

÷2   ×2

The inverses of the arithmetic operations are just as you would expect:

addition and subtraction are inverses of each other
multiplication and division are inverses of each other
raising to the power k and raising to a power 1/k are inverses of each other

Now, can you think of two functions that are each identical to their inverse?

Think carefully

10

$$f(x) = x \text{ and } f(x) = \frac{1}{x}$$

Because the function with output $f(x) = x$ does not alter the input at all so the inverse does not either, and the function with output $f(x) = \frac{1}{x}$ is its own inverse because the reciprocal of the reciprocal of a number is the number:

$$\frac{1}{1/x} = x$$

x f $f(x) = 1/x$ $f^{-1}(x) = 1/x$ f^{-1} x

^(−1)   ^(−1)

Let's progress

Graphs of inverses   **11**

The diagram of the inverse of a function can be drawn by reversing the flow of information and this is the same as interchanging the contents of each ordered pair generated by the function. As a result, when the ordered pairs generated by the inverse of a function are plotted, the graph takes up the shape of the original function but reflected in the line $y = x$. Let's try it. Use your spreadsheet to plot $y = x^3$ and the inverse $y = x^{\frac{1}{3}}$. If you are unfamiliar with the use of a spreadsheet, read Module 4 first where the spreadsheet is introduced.

What you are about to do is a little involved, so follow the
instructions to the letter and take it slowly and carefully

The graph of $y = x^3$ **12**

Open up your spreadsheet

Enter $-1\cdot1$ in cell **A1**
Highlight **A1** to **A24**
Click **Edit-Fill-Series** and enter the **step value** as $0\cdot1$

The cells **A1** to **A24** then fill with the numbers $-1\cdot1$ to $1\cdot2$.

In cell **B1** enter the formula **=A1^3** and press **Enter**

Cell **B1** now contains the cube of the contents of cell **A1**

Make **B1** the active cell
Click **Edit-Copy**   This copies the contents of B1 to the Clipboard
Highlight **B2** to **B24**
Click **Edit-Paste**   This pastes the contents of the Clipboard to B2
                                 to B24

Each of the cells **B1** to **B24** contains the cube of the contents of the adjacent cell in the **A** column.

Highlight the block of cells **A1** to **B24**
Click the *Chart Wizard* button to create an **XY (Scatter)** graph with joined-up points

▷

The graph you obtain will look like that depicted below:

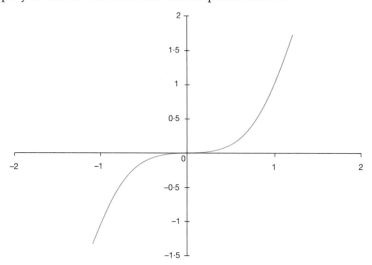

Now for the graph of $y = x^{\frac{1}{3}}$

13 The graph of $y = x^{\frac{1}{3}}$

Keep the data you already have on the spreadsheet, you are going to use it:

Highlight cells **A1** to **A24**
Click **Edit-Copy**   This copies the contents of A1 to A24 to the
                             Clipboard
Place the cursor in cell **B26**
Click **Edit-Paste**   This pastes the contents of the Clipboard from
                             B26 to B49

The cells **B26** to **B49** then fill with the same values as those in cells **A1** to **A24**

Highlight cells **B1** to **B24**
Click **Edit-Copy**   This copies the contents of B1 to B24 to the
                             Clipboard
Place the cursor in cell **A26**
Click **Edit-Paste Special**
In the *Paste Special* window select **Values** and click **OK**

The cells **A26** to **A49** then fill with the same values as those in cells **B1** to **B24**. Because the cells **B1** to **B24** contain formulas, using **Paste Special** rather than simply **Paste** ensures that you copy the values rather than the formulas.

What you now have are the original ordered pairs for the first function reversed in readiness to draw the graph of the inverse of the function.

Notice that row 25 is empty. This is essential because later on you are going to obtain a plot of two curves on the same graph.

▶

For now you must first clear away the old graph:

Click the boundary of the graph to display the handles
Click **Edit-Clear-All**

and the graph disappears. Now, to draw the new graph:

Highlight the block of cells **A26** to **B49**
Click the *Chart Wizard* button to create an **XY (Scatter)** graph with joined-up points

The graph you obtain will look like that depicted to the right:

Same shape as the previous one but a different orientation.

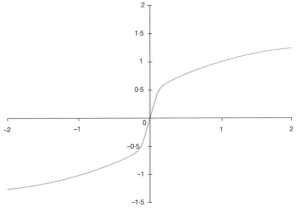

Now for both the graph of $y = x^3$ and $y = x^{\frac{1}{3}}$ together

The graphs of $y = x^3$ and $y = x^{\frac{1}{3}}$ plotted together

14

Clear away the graph you have just drawn. Then:

Highlight the block of cells **A1** to **B49**
Click the *Chart Wizard* button to create an **XY (Scatter)** graph with joined-up points

The graph you obtain will look like that depicted to the right:

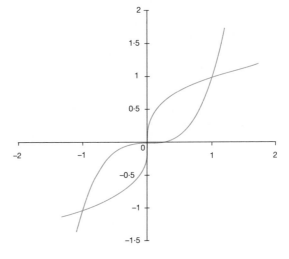

Now you can see that the two graphs are each a reflection of the other in the line $y = x$. To firmly convince yourself of this:

Place the cursor in cell **A51** and enter the number $-1 \cdot 1$
Enter the number $-1 \cdot 1$ in cell **B51**
Enter the number $1 \cdot 2$ in cell **A52**
Enter the number $1 \cdot 2$ in cell **B52**

You now have two points with which to plot the straight line $y = x$. *Notice again, row 50 this time is empty.*

Clear away the last graph. Then:

Highlight the block of cells **A1** to **B52**
Click the *Chart Wizard* button to create an **XY (Scatter)** graph with joined-up points

The graph you obtain will look like that depicted below:

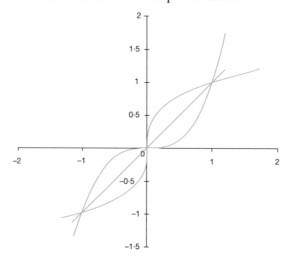

As you can see, the graphs are symmetric about the sloping line $y = x$. We say they are **reflection symmetric about** $y = x$ because each one could be considered as a reflection of the other in a double-sided mirror lying along this line.

Now you try one. Use the spreadsheet to plot the graphs of $y = x^2$ and its inverse $y = x^{\frac{1}{2}}$. You do not need to start from scratch, just use the sheet you have already used and change the contents of cell **B1** to the formula **=A1^2**, copy this down the **B** column to **B24** and then **Paste Special** these values into cells **A26** to **A49**.

15

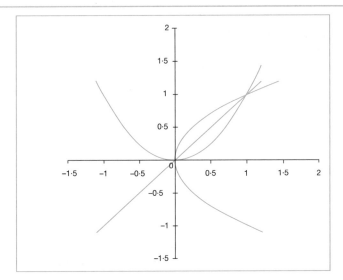

The graph of the inverse of the square function is a parabola on its side. However, as you have seen earlier, this is not a graph of a function. If, however, the bottom branch of this graph is removed, what is left is the graph of the function expressed by $y = \sqrt{x}$ which is called the *inverse function* because it is single valued.

Plot the graph of $y = x^4 - x^2 + 1$ by simply changing the formula in **B1** and copying it into cells **B2** to **B24**. So:

   (a) What does the inverse of the function look like?

   (b) Is the inverse of the function the inverse function?

16

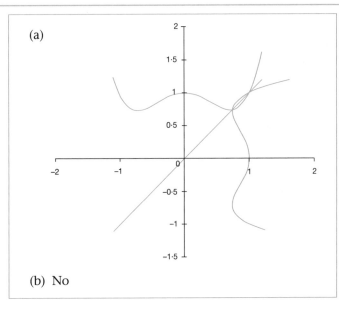

(a)

(b) No

The answer to (b) is 'no' because

> The inverse of the function is not single valued so it cannot be a function. The inverse function would have to be obtained by removing parts of the inverse of the function to obtain a function that was single valued.

At this point let us pause and summarize the main facts so far on functions and their inverses

 # Review summary

Unit 23

17

1 A function is a rule expressed in the form $y = f(x)$ with the proviso that for each value of x there is a unique value of y.

2 The collection of permitted input values to a function is called the *domain* of the function and the collection of corresponding output values is called the *range*.

3 The inverse of a function is a rule that associates range values to domain values of the original function.

 # Review exercise

Unit 23

18

1 Which of the following equations expresses a rule that is a function:
   (a) $y = 6x - 2$
   (b) $y = \sqrt{x^3}$
   (c) $y = \left(\dfrac{3x}{x^2 + 3} \right)^{\frac{5}{2}}$

2 Given the two functions f and g expressed by:
   $$f(x) = 2x - 1 \text{ for } -2 < x < 4$$
   and
   $$g(x) = \frac{4}{x - 2} \text{ for } 3 < x < 5,$$
   find the domain and range of:
   (a) $h(x) = f(x) - g(x)$
   (b) $k(x) = -\dfrac{2f(x)}{g(x)}$

▶

3 Use your spreadsheet to draw each of the following and their inverses. Is the inverse a function?

(a) $y = x^6$ Use the data from the text and just change the formula

(b) $y = -3x$ Use the data from the text and just change the formula

(c) $y = \sqrt{x^3}$ Enter 0 in cell **A1** and **Edit-Fill-Series** with *step value* 0·1

Complete all three questions. Take your time, there is no need to rush.
If necessary, refer back to the Unit.
The answers and working are in the next frame.

---

1 (a) $y = 6x - 2$ expresses a rule that is a function because to each value of x there is only one value of y.   **19**

(b) $y = \sqrt{x^3}$ expresses a rule that is a function because to each value of x there is only one value of y. The surd sign $\sqrt{}$ stands for the positive square root.

(c) $y = \left(\dfrac{3x}{x^2 + 3}\right)^{\frac{5}{2}}$ expresses a rule that is not a function because to each positive value of the bracket there are two values of y. The power 5/2 represents raising to the power 5 and taking the square root, and there are always two square roots to each positive number.

2 (a) $h(x) = f(x) - g(x) = 2x - 1 - \dfrac{4}{x - 2}$ for $3 < x < 4$ because $g(x)$ is not defined for $-2 < x \leq 3$ and $f(x)$ is not defined for $4 \leq x < 5$. Range $1 < h(x) < 5$.

(b) $k(x) = -\dfrac{2f(x)}{g(x)} = -\dfrac{(2x - 1)(x - 2)}{2}$ for $3 < x < 4$.

Range $-7 < k(x) < -5/2$.

3 (a) $y = x^6$ has an inverse $y = x^{\frac{1}{6}}$. This does not express a function because there are always two values to an even root (see Frame 4 of this Unit).

(b) $y = -3x$ has an inverse $y = -\dfrac{x}{3}$. This does express a function because there is only one value of y to each value of x.

(c) $y = \sqrt{x^3}$ has an inverse $y = x^{\frac{2}{3}}$ because $\sqrt{x^3}$ represents the positive value of $y = x^{\frac{3}{2}}$. The inverse does express a function.

Now for the Review test

Review test   **Unit 23**

20

1 Which of the following equations expresses a rule that is a function?

(a) $y = 1 - x^2$ (b) $y = -\sqrt{x^4}$ (c) $y = x^{\frac{1}{6}}$

2 Given the two functions f and g expressed by:

$$f(x) = \frac{1}{4-x} \text{ for } 0 \le x < 4 \text{ and } g(x) = x - 3 \text{ for } 0 < x \le 5$$

find the domain and range of functions h and k where

(a) $h(x) = f(x) + 3g(x)$ (b) $k(x) = \dfrac{f(x)}{2g(x)}$

3 Use your spreadsheet to draw each of the following and their inverses. Is the inverse a function?

(a) $y = x^5$ (b) $y = -3x^2$ (c) $y = \sqrt{1-x^2}$

Composition – 'function of a function'   **Unit 24**

1

Chains of functions can be built up where the output from one function forms the input to the next function in the chain. Take out your calculator again and this time enter the number:

4

Now press the key – the reciprocal key – and the display changes to:

0·25   the reciprocal of 4

Now press the key and the display changes to:

0·0625   the square of 0·25

Here, the number 4 was the input to the reciprocal function and the number 0·25 was the output. This same number 0·25 was then the input to the squaring function with output 0·0625. This can be represented by the following diagram:

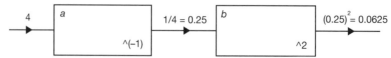

Notice that the two functions have been named a and b, but any letter can be used to label a function.

At the same time the *total* processing by f could be said to be that the number 4 was input and the number 0·0625 was output:

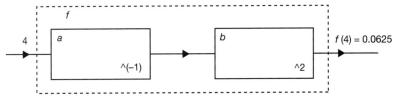

So the function f is *composed* of the two functions a and b where $a(x) = \dfrac{1}{x}$, $b(x) = x^2$ and $f(x) = \left(\dfrac{1}{x}\right)^2$. It is said that f is the *composition* of a and b, written as:

$$f = b \circ a$$

and read as *b of a*. Notice that the functions a and b are written down algebraically in the reverse order from the order in which they are given in the diagram. This is because in the diagram the input to the composition enters on the left, whereas algebraically the input is placed to the right:

$$f(x) = b \circ a(x)$$

So that $f(x) = b \circ a(x)$, which is read as f *of x equals b of a of x*. An alternative notation, more commonly used, is:

$$f(x) = b[a(x)]$$

and f is described as being a *function of a function*.

Now you try. Given that $a(x) = x + 3$, $b(x) = 4x$ find the functions f and g where:

(a) $f(x) = b[a(x)]$
(b) $g(x) = a[b(x)]$

Stick with what you know, draw the boxes and see what you find

2

(a) $f(x) = 4x + 12$
(b) $g(x) = 4x + 3$

Because

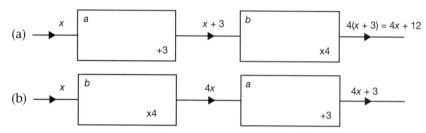

Notice how these two examples show that $b[a(x)]$ is different from $a[b(x)]$. That is, the order of composition matters.

Now, how about something a little more complicated? Given the three functions a, b and c where $a(x) = x^3$, $b(x) = 2x$ and $c(x) = x + 4$, find each of the following as expressions in x:

(a) $f(x) = a(b[c(x)])$ (b) $g(x) = c(a[b(x)])$ (c) $h(x) = a(a[c(x)])$

Remember, draw the boxes and follow the logic

3

(a) $f(x) = 8(x + 4)^3$
(b) $g(x) = 8x^3 + 4$
(c) $h(x) = (x + 4)^9$

Because

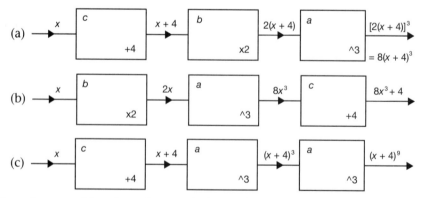

How about working the other way? Given the expression $f(x)$ for the output from a composition of functions, how do you decompose it into its component functions? This is particularly easy because you already know how to do it even though you may not yet realize it.

Let's look at a specific example first ▶

Given the output from a composition of functions as $f(x) = 6x - 4$ ask yourself how, given a calculator, would you find the value of $f(2)$? You would:

enter the number 2	the input	x
multiply by 6 to give 12	the first function	$a(x) = 6x$ input times 6
subtract 4 to give 8	the second function	$b(x) = 6x - 4$ input minus 4

so that $f(x) = b[a(x)]$. The very act of using a calculator to enumerate the output from a composition requires you to decompose the composition automatically as you go.

Try it yourself. Decompose the composition with output $f(x) = (x + 5)^4$.

Get your calculator out and find the output for a specific input

4

$$f(x) = b[a(x)] \text{ where } a(x) = x + 5 \text{ and } b(x) = x^4$$

Because

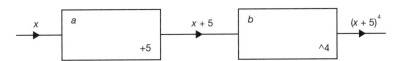

Notice that this decomposition is not unique. You could have defined $b(x) = x^2$ in which case the composition would have been $f(x) = b(b[a(x)])$.

Just to make sure you are clear about this, decompose the composition with output $f(x) = 3(2x + 7)^2$.

Use your calculator and take it steadily. There are four functions here

5

$$f(x) = d[c(b[a(x)])] \text{ where } a(x) = 2x$$
$$b(x) = x + 7, \; c(x) = x^2 \text{ and } d(x) = 3x$$

Because

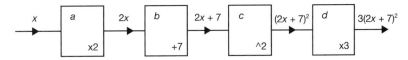

Let's keep going

6 Inverses of compositions

As has been stated before, the diagram of the inverse of a function can be drawn as the function with the information flowing through it in the reverse direction, and the same applies to a composition.

For example, consider the function f with output $f(x) = (3x - 5)^{\frac{1}{3}}$. By decomposing f you find that:

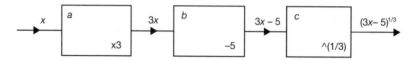

where $a(x) = 3x$, $b(x) = x - 5$ and $c(x) = x^{\frac{1}{3}}$, so $f(x) = c(b[a(x)])$. Each of the three functions in the composition has an inverse, namely:

$$a^{-1}(x) = \frac{x}{3}$$
$$b^{-1}(x) = x + 5$$
$$c^{-1}(x) = x^3$$

By reversing the flow of information through the diagram we find that:

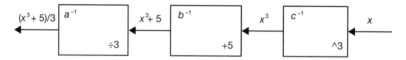

so that

$$f^{-1}(x) = a^{-1}(b^{-1}[c^{-1}(x)]) = (x^3 + 5)/3$$

Notice the *reversal* of the order of the components:

$$f(x) = c(b[a(x)]), \qquad f^{-1}(x) = a^{-1}(b^{-1}[c^{-1}(x)])$$

Now you try this one. Find the inverse of the function f with output

$$f(x) = \left(\frac{x + 2}{4}\right)^5.$$

Answer in the next frame

7

$$\boxed{f^{-1}(x) = 4x^{\frac{1}{5}} - 2}$$

Because

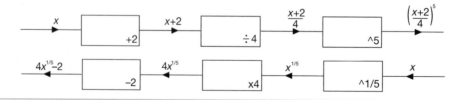

Review summary Unit 24

1 *Composition:* Chains of functions can be built up where the output from one function forms the input to the next function in the chain. **8**

2 *Inverses of compositions:* Just as with a single function, the inverse of a composition of functions can be obtained by reversing the flow of information through the chain of functions. The inverses of the functions in the chain then act in reverse order. That is, if:

$$f(x) = a(b[c(d[x])]) \text{ then } f^{-1}(x) = d^{-1}(c^{-1}[b^{-1}(a^{-1}[x])])$$

 Review exercise Unit 24

1 Given that $a(x) = 4x$, $b(x) = x^2$, $c(x) = x - 5$ and $d(x) = \sqrt{x}$ find: **9**
   (a) $f(x) = a[b(c[d(x)])]$
   (b) $f(x) = a(a[d(x)])$
   (c) $f(x) = b[c(b[c(x)])]$

2 Given that $f(x) = (2x - 3)^3 - 3$, decompose f into its component functions and find its inverse. Is the inverse a function?

Complete both of these questions, referring to the Unit if you need to.
You can check your answers and working in the next frame.

1 (a) $f(x) = a[b(c[d(x)])] = 4(\sqrt{x} - 5)^2$ **10**
   (b) $f(x) = a(a[d(x)]) = 16\sqrt{x}$
   (c) $f(x) = b[c(b[c(x)])] = ((x - 5)^2 - 5)^2 = x^4 - 20x^3 + 140x^2 - 400x + 400$

2 $f = b \circ c \circ b \circ a$ so that $f(x) = b[c(b[a(x)])]$ where $a(x) = 2x, b(x) = x - 3$ and $c(x) = x^3$. The inverse is $f^{-1}(x) = a^{-1}[b^{-1}(c^{-1}[b^{-1}(x)])]$ so that:

$$f^{-1}(x) = a^{-1}[b^{-1}(c^{-1}[b^{-1}(x)])] = \frac{(x + 3)^{\frac{1}{3}} + 3}{2} \text{ where } a^{-1}(x) = x/2,$$

$b^{-1}(x) = x + 3$ and $c^{-1}(x) = x^{\frac{1}{3}}$. The inverse is a function.

So far our work on functions has centred around *algebraic functions*. This is just one category of function. We shall move on and consider other types of function and their specific properties.

But first, the Review test

Review test   Unit 24

11 **1** Given that $a(x) = -2x$, $b(x) = x^3$, $c(x) = x - 1$ and $d(x) = \sqrt{x}$ find:
   (a) $f(x) = a[b(c[d(x)])]$
   (b) $f(x) = a(a[d(x)])$
   (c) $f(x) = b[c(b[c(x)])]$

2 Given that $f(x) = (3x + 4)^2 + 4$, decompose f into its component functions and find its inverse. Is the inverse a function?

Trigonometric functions   Unit 25

1 Rotation

In Module 8 the trigonometric ratios were defined for the two acute angles in a right-angled triangle. These definitions can be extended to form *trigonometric functions* that are valid for *any* angle and yet retain all the properties of the original ratios. Start with the circle generated by the end point A of a straight line OA of unit length rotating anticlockwise about the end O as shown in the diagram below:

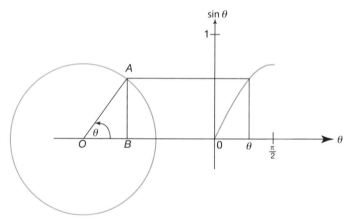

For angles θ where $0 < \theta < \pi/2$ radians you already know that:

$$\sin\theta = \frac{AB}{OA} = AB \text{ since } OA = 1$$

That is, the value of the trigonometric ratio $\sin\theta$ is equal to the height of A above B. The *sine* function with output $\sin\theta$ is now defined as the height of A above B *for any angle* θ $(0 \le \theta < \infty)$.

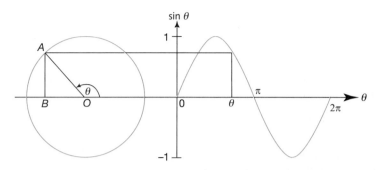

Notice that when *A* is *below B* the height is *negative*. The definition of the sine function can be further extended by taking into account negative angles, which represent a clockwise rotation of the line *OA* giving the complete graph of the sine function as in the diagram below:

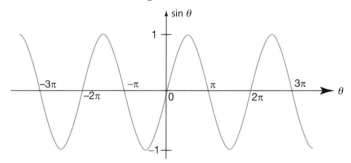

As you can see from this diagram, the value of $\sin \theta$ ranges from $+1$ to -1 depending upon the value of θ. You can reproduce this graph using a spreadsheet – in cells **A1** to **A21** enter the numbers -10 to 10 in steps of 1 and in cell **B1** enter the formula **=sin(A1)** and copy this into cells **B2** to **B21**. Use the *Chart Wizard* to draw the graph.

Just as before, you can use a calculator to find the values of the sine of an angle. So the sine of $153°$ is ....

Remember to put your calculator in degree mode

0·4540 to 4 dp

2

and the sine of $-\pi/4$ radians is ....

Remember to put your calculator in radian mode

3

$$-0\!\cdot\!7071 \text{ to } 4 \text{ dp}$$

By the same reasoning, referring back to the first diagram in Frame 1 of this Unit, for angles θ where $0 < \theta < \pi/2$ radians you already know that:

$$\cos\theta = \frac{OB}{OA} = OB \text{ since } OA = 1$$

This time, the value of the trigonometric ratio $\cos\theta$ is equal to the distance from O to B. The *cosine* function with output $\cos\theta$ is now defined as the distance from O to B *for any angle* θ $(-\infty < \theta < \infty)$.

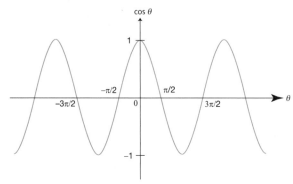

Notice that when B is to the left of O the distance from O to B is negative.

Again, you can reproduce this graph using a spreadsheet – in cells **A1** to **A21** enter the numbers -10 to 10 in steps of 1 and in cell **B1** enter the formula **=cos (A1)** and copy this into cells **B2** to **B21**. Use the *Chart Wizard* to draw the graph.

A calculator is used to find the values of the cosine of an angle. So the cosine of $-272°$ is ....

Remember to put your calculator in degree mode

4

$$0\!\cdot\!0349 \text{ to } 4 \text{ dp}$$

and the cosine of $2\pi/3$ radians is ....

Remember to put your calculator in radian mode

5

$$-0\!\cdot\!5$$

Now to put these two functions together

The tangent

The third basic trigonometric function, the tangent, is defined as the ratio of the sine to the cosine:

$$\tan \theta = \frac{\sin \theta}{\cos \theta}$$

Because $\cos \theta = 0$ whenever θ is an odd multiple of $\pi/2$, the tangent is not defined at these points. Instead the graph has vertical asymptotes as seen below:

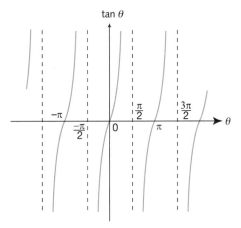

You can plot a single branch of the tangent function using your spreadsheet. Enter -1.5 in cell **A1** and **Edit-Fill-Series** down to **A21** with *step value* 0.15, then use the function **=tan (A1)** in cell **B1** and copy down to **B21**. Use the *Chart Wizard* to create the graph.

Next frame

Also, just as before, you can use a calculator to find the values of the tangent of an angle. So the tangent of $333°$ is .... to 4 dp.

7

Remember to put your calculator in degree mode

-0.5095 to 4 dp

and the tangent of $-6\pi/5$ radians is .... to 4 dp.

Remember to put your calculator in radian mode

9

$$-0{\cdot}7265 \text{ to } 4 \text{ dp}$$

Make a note of the diagrams in Frames 1, 3 and 6. It is essential that you are able to draw sketch graphs of these functions.

For the sine and cosine functions the repeated sinusoidal wave pattern is easily remembered, all you then have to remember is that each rises and falls between $+1$ and -1 and crosses the horizontal axis:

(a) every whole multiple of π for the sine function

(b) every odd multiple of $\pi/2$ for the cosine function

The repeated *branch* pattern of the tangent function is also easily remembered, all you then have to remember is that it rises from $-\infty$ to $+\infty$, crosses the horizontal axis every even multiple of $\pi/2$ and has a vertical asymptote every odd multiple of $\pi/2$.

Now to look at some common properties of these trigonometric functions

10 Period

As can be seen from the graph in Frame 6, the tangent function $f(\theta) = \tan\theta$ repeats its output values every $180°$ (π radians). Use your calculator in degree mode to verify that:

$$
\begin{aligned}
f(45) &= \tan 45° &= 1 \\
f(45 + 180) &= \tan 225° &= 1 \\
f(45 + 360) &= \tan 405° &= 1 \\
f(45 + 540) &= \tan 585° &= 1
\end{aligned}
$$

We can write this fact in the form of an equation:

$\tan(45 + 180n) = \tan 45°$ where $n = 1, 2, 3, \ldots$

or, using radians:

$\tan(\pi/4 + n\pi) = \tan \pi/4$ where $n = 1, 2, 3, \ldots$

Indeed, as can be seen from the graph in Frame 6, the tangent function continually repeats the output value corresponding to each input value of θ in the interval $-\pi/2 < \theta < \pi/2$ which is of width π. That is:

$\tan(\theta + n\pi) = \tan\theta$ where $n = 1, 2, 3, \ldots$

The sine function $f(\theta) = \sin\theta$ is periodic with period . . . .

The answer is in the next frame

11

$$2\pi$$

Because

As is evident from the graph of $f(\theta) = \sin\theta$ in Frame 1 the sinusoidal wave pattern consists of a repeated wave of width 2π. The sine function therefore continually repeats the output value corresponding to each input value of θ in the inverval $0 \le \theta < 2\pi$

Similarly, the cosine function $f(\theta) = \cos\theta$ is periodic with period 2π as can be seen in the graph in Frame 3.

Finding the periods of trigonometric functions with more involved outputs requires some manipulation. For example, consider $f(\theta) = \cos 3\theta$. Here we see that, for example:

$$f(\theta) = \cos 3\theta = 1 \text{ when } 3\theta = 0,\ 2\pi,\ 4\pi,\ 6\pi, \dots,\ 2n\pi$$

that is when:

$$\theta = 0,\ 2\pi/3,\ 4\pi/3,\ 6\pi/3\ (2\pi),\ \dots,\ 2n\pi/3$$

So that $f(\theta + 2\pi/3) = \cos 3(\theta + 2\pi/3) = \cos(3\theta + 2\pi) = \cos 3\theta = f(\theta)$. That is, the period of $f(\theta) = \cos 3\theta$ is $2\pi/3$. The output of $f(\theta) = \cos 3\theta$ certainly repeats itself over 2π radians but within 2π the basic sinusoidal shape is repeated three times.

So the period of $\cos 4\theta$ is ....

Answer in the next frame

$$\frac{\pi}{2}$$

Because

$$\cos 4\theta = \cos(4\theta + 2\pi) = \cos 4\left(\theta + \frac{2\pi}{4}\right) = \cos 4\left(\theta + \frac{\pi}{2}\right)$$

And the period of $\tan 5\theta$ = ...

Answer in the next frame

13

$$\boxed{\dfrac{\pi}{5}}$$

Because

$$\tan 5\theta = \tan(5\theta + \pi) = \tan 5(\theta + \pi/5)$$

Now, try another one. The period of $\sin(\theta/3) = \ldots\ldots\ldots\ldots$

Just follow the same procedure. The answer may surprise you

14

$$\boxed{6\pi}$$

Because

$$\sin(\theta/3) = \sin(\theta/3 + 2\pi) = \sin\frac{1}{3}(\theta + 6\pi)$$

The answer is not 2π because the basic sinusoidal shape is only completed over the interval of 6π radians. If you are still not convinced of all this, use the spreadsheet to plot their graphs. Just one more before moving on.

The period of $\cos(\theta/2 + \pi/3) = \ldots\ldots\ldots\ldots$

Just follow the procedure

15

$$\boxed{4\pi}$$

Because

$$\cos(\theta/2 + \pi/3) = \cos(\theta/2 + \pi/3 + 2\pi) = \cos\left(\frac{1}{2}[\theta + 4\pi] + \pi/3\right)$$

The $\pi/3$ has no effect on the period, it just shifts the basic sinusoidal shape $\pi/3$ radians to the left.

Move on

16 **Amplitude**

Every periodic function possesses an *amplitude* that is given as *the difference between the maximum value and the average value of the output taken over a single period*. For example, the average value of the output from the cosine function is zero (it ranges between $+1$ and -1) and the maximum value of the output is 1, so the amplitude is $1 - 0 = 1$.

So the amplitude of $4\cos(2\theta - 3)$ is $\ldots\ldots\ldots\ldots$

Next frame

17

4

Because

> The maximum and minimum values of the cosine function are $+1$ and -1
> respectively, so the output here ranges from $+4$ to -4 with an average of
> zero. The maximum value is 4 so that the amplitude is $4 - 0 = 4$.

Periodic functions are not always trigonometric functions. For example, the
function with the graph shown in the diagram below is also periodic:

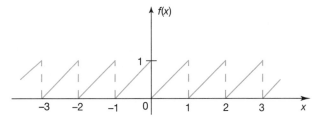

The straight line branch between $x = 0$ and $x = 1$ repeats itself indefinitely.
For $0 \le x < 1$ the output from f is given as $f(x) = x$. The output from f for
$1 \le x < 2$ matches the output for $0 \le x < 1$. That is:

$f(x + 1) = f(x)$ for $0 \le x < 1$

So for example, $f(1{\cdot}5) = f(0{\cdot}5 + 1) = f(0{\cdot}5) = 0{\cdot}5$
   The output from f for $2 \le x < 3$ also matches the output for $0 \le x < 1$.
That is:

$f(x + 2) = f(x)$ for $0 \le x < 1$

So that, for example, $f(2{\cdot}5) = f(0{\cdot}5 + 2) = f(0{\cdot}5) = 0{\cdot}5$
   This means that we can give the prescription for the function as:

$f(x) = x$ for $0 \le x < 1$
$f(x + n) = f(x)$ for any integer n

For a periodic function of this type with period P where the first branch of the
function is given for $a \le x < a + P$ we can say that:

$f(x) = $ some expression in x for $a \le x < a + P$
$f(x + nP) = f(x)$

Because of its shape, the specific function we have considered is called a
sawtooth wave.

The amplitude of this sawtooth wave is . . . .

                                                          Remember the definition of amplitude

18

$$\boxed{\dfrac{1}{2}}$$

Because

The amplitude is given as the *difference between the maximum value and the average value of the output taken over a single period.* Here the maximum value of the output is 1 and the average output is $\frac{1}{2}$, so the amplitude is $1 - 1/2 = 1/2$.

Next frame

19   **Phase difference**

The phase difference of a periodic function is the interval of the input by which the output leads or lags behind the *reference function.* For example, the plots of $y = \sin x$ and $y = \sin(x + \pi/4)$ on the same graph are shown below:

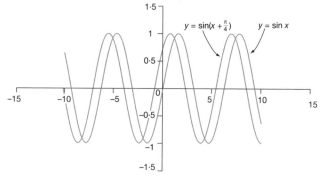

The diagram shows that $y = \sin(x + \pi/4)$ has the identical shape to $y = \sin x$ but is *leading* it by $\pi/4$ radians. It might appear to lag behind when you look at the diagram but it is, in fact, leading because when $x = 0$ then $\sin(x + \pi/4)$ already has the value $\sin \pi/4$, whereas $\sin x$ only has the value $\sin 0$. It is said that $y = \sin(x + \pi/4)$ leads with a *phase difference* of $\pi/4$ radians relative to the reference function $y = \sin x$. A function with a negative phase difference is said to *lag behind* the reference function. So that $y = \sin(x - \pi/4)$ lags behind $y = \sin x$ with a phase difference of $-\pi/4$.

So the phase difference of $y = \sin(x - \pi/6)$ relative to $y = \sin x$ is ....

Next frame

20

$$-\pi/6 \text{ radians}$$

Because

The graph of $y = \sin(x - \pi/6)$ *lags behind* $y = \sin x$ by $\pi/6$ radians.

The phase difference of $y = \cos x$ relative to the reference function $y = \sin x$ is . . . .

Think how $\cos x$ relates to $\sin x$

21

$$\pi/2 \text{ radians}$$

Because

$\cos x = \sin(x + \pi/2)$ and $\sin(x + \pi/2)$ *leads* $\sin x$ by $\pi/2$ radians.

Finally, try this. The phase difference of $y = \sin(3x + \pi/8)$ relative to $y = \sin 3x$ is . . . .

Take care to compare like with like – plot the graph if necessary

22

$$\pi/24 \text{ radians}$$

Because

$\sin(3x + \pi/8) = \sin(3[x + \pi/24])$ and $\sin(3[x + \pi/24])$ *leads* $\sin 3x$ by $\pi/24$ radians.

At this point let us pause and summarize the main facts so far on trigonometric functions

 # Review summary
Unit 25

23

1 The definitions of the trigonometric ratios, valid for angles greater than $0°$ and less than $90°$, can be extended to the trigonometric functions valid for any angle.

2 The trigonometric functions possess periods, amplitudes and phases.

Review exercise   Unit 25

24

1 Use a calculator to find the value of each of the following (take care to ensure that your calculator is in the correct mode):
   (a) $\sin(3\pi/4)$ (b) $\operatorname{cosec}(-\pi/13)$ (c) $\tan(125°)$
   (d) $\cot(-30°)$ (e) $\cos(-5\pi/7)$ (f) $\sec(18\pi/11)$

2 Find the period, amplitude and phase (in radians) of each of the following:
   (a) $f(\theta) = 3\sin 9\theta$ (b) $f(\theta) = -7\cos(5\theta - 3)$
   (c) $f(\theta) = \tan(2 - \theta)$ (d) $f(\theta) = -\cot(3\theta - 4)$

3 A function is defined by the following prescription:
   $$f(x) = -x + 4, \qquad 0 \le x < 3, \qquad f(x + 3) = f(x)$$
   Plot a graph of this function for $-9 \le x < 9$ and find:
   (a) the period
   (b) the amplitude
   (c) the phase of $f(x) + 2$ with respect to $f(x)$

Complete all three questions. Take your time, there is no need to rush.
If necessary, refer back to the Unit.
The answers and working are in the next frame.

25

1 Using your calculator you will find:
   (a) 0.7071 (b) -4.1786 (c) -1.4281
   (d) -1.7321 (e) -0.6235 (f) 2.4072

2 (a) $3\sin 9\theta = 3\sin(9\theta + 2\pi) = 3\sin 9(\theta + 2\pi/9)$ so the period of $f(\theta)$ is $2\pi/9$ and the phase is 0. The maximum value of $f(\theta)$ is 3 and the average value is 0, so the amplitude of $f(\theta)$ is 3.

   (b) $-7\cos(5\theta - 3) = -7\cos(5\theta - 3 + 2\pi) = -7\sin 5(\theta + 2\pi/5 - 3/5)$ so the period of $f(\theta)$ is $2\pi/5$, the phase is $-3/5$ and the amplitude is 7.

   (c) $\tan(2 - \theta) = \tan(2 - \theta + \pi) = \tan(-\theta + 2 + \pi)$ so the period of $f(\theta)$ is π and $f(\theta)$ leads $\tan(-\theta)$ by the phase 2 with an infinite amplitude.

   (d) $-\cot(3\theta - 4) = \cot(-3\theta + 4 + \pi) = \cot 3(-\theta + 4/3 + \pi/3)$ so the period of $f(\theta)$ is $\pi/3$ and $f(\theta)$ leads $\cot(-3\theta)$ by the phase 4/3 with an infinite amplitude.

3 (a) 3   (b) 1.5   (c) 0

Now for the Review test

Review test

1 Use a calculator to find the value of each of the following:

26

(a) $\cos(-5\pi/3)$ (b) $\sec(115°)$ (c) $\tan(-13°)$

2 Find the period, amplitude and phase of each of the following:

(a) $f(\theta) = 2\cos 6\theta$ (b) $f(\theta) = -2\tan(2\theta - 2)$ (c) $f(\theta) = \cos(\pi - \theta)$

3 A function is defined by the following prescription:

$$f(x) = 4 - x^2,\ 0 \le x < 2,\ f(x \pm 2) = f(x)$$

Plot a graph of this function for $-6 \le x \le 6$ and find:

(a) the period

(b) the amplitude

(c) the phase of $f(x) + 2$ with respect to $f(x)$

Inverse trigonometric functions   Unit 26

If the graph of $y = \sin x$ is reflected in the line $y = x$, the graph of the inverse of the sine function is what results:

1

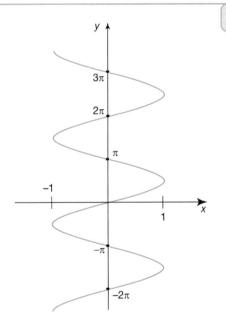

However, as you can see, this is not a function because there is more than one value of y corresponding to a given value of x. If you cut off the upper and lower parts of the graph you obtain a single-valued function and it is this that is the *inverse sine function*:

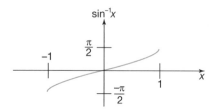

In a similar manner you can obtain the *inverse cosine function* and the *inverse tangent function*:

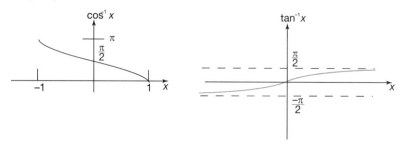

As in the case of the trigonometric functions, the values of these inverse functions are found using a calculator.

So that:

(a) $\sin^{-1}(0.5) = \ldots\ldots\ldots$

(b) $\tan^{-1}(-3.5) = \ldots\ldots\ldots$

(c) $\sec^{-1}(10) = \ldots\ldots\ldots$ (refer to Module 8, Frame 17 of Unit 21 for the definition of $\sec\theta$)

Next frame for the answer

2

$$(a)\ \ 30° \qquad (b)\ \ -74.05° \qquad (c)\ \ 84.26°$$

Because

(c) If $\sec^{-1}(10) = \theta$ then $\sec\theta = 10 = \dfrac{1}{\cos\theta}$ so that $\cos\theta = 0.1$ and

$\theta = \cos^{-1}(0.1) = 84.26°$

So remember $\sec^{-1}\theta = \cos^{-1}\dfrac{1}{\theta}$.

Similar results are obtained for $\operatorname{cosec}^{-1}\theta$ and $\cot^{-1}\theta$:

$$\operatorname{cosec}^{-1}\theta = \sin^{-1}\frac{1}{\theta} \quad \text{and} \quad \cot^{-1}\theta = \tan^{-1}\frac{1}{\theta}$$

Now to use these functions and their inverses to solve equations

Trigonometric equations   **3**

A simple trigonometric equation is one that involves just a single trigono-metric expression. For example, the equation:

$\sin x = 0$ is a simple trigonometric equation.

To solve this equation we look at the sine graph in Frame 1 of Unit 25 of this Module, where we can see that $\sin x = 0$ when:

$x = \ldots, -3\pi, -2\pi, -\pi, 0, \pi, 2\pi, 3\pi, \ldots$, that is when $x = \pm n\pi$

So the values of x that satisfy the trignometric equation $\cos x = -1$ are

. . . .

> *Recall that if n is a natural number then*
> *$2n$ is an even number and so $2n + 1$ is an odd number*

$$\pm(2n + 1)\pi$$

4

Because

From the cosine graph in Frame 3 of Unit 25 of this Module, we can see that $\cos x = -1$ when $x = \pm\pi, \pm 3\pi, \pm 5\pi, \ldots$, that is when $x = \pm(2n + 1)\pi$ – an odd multiple of π.

Let us look at a less simple problem. To find the solution to:

$$\cos x = \frac{\sqrt{3}}{2}$$

we recall that $\sqrt{3}$ appeared when we were dealing with the half equilateral triangle in Frame 26 of Unit 21 of Module 8. There we saw that:

$$\cos\frac{\pi}{6} = \frac{\sqrt{3}}{2}$$

and so $x = \frac{\pi}{6}$ is a solution to the equation. By drawing a horizontal line through $\sqrt{3}/2$ on the vertical axis, we can see that the values $x = \frac{\pi}{6} \pm 2n\pi$ are also solutions to the equation. However, these are not the only solutions, as can be seen from the graph below:

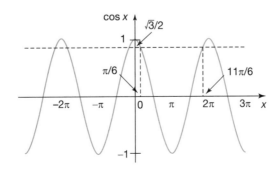

By the symmetry of the cosine curve $x = 2\pi - \dfrac{\pi}{6} = \dfrac{11\pi}{6}$ is also a solution, as are the values $x = \dfrac{11\pi}{6} \pm 2n\pi$.

Try one yourself. The solutions to the equation $\sin x = \dfrac{1}{\sqrt{2}}$ are ....

The answer is in the next frame

5

$$\dfrac{\pi}{4} \pm 2n\pi \text{ and } \dfrac{3\pi}{4} \pm 2n\pi$$

Because

Recalling that $\sqrt{2}$ appeared when we were dealing with the right-angled isosceles triangle in Frame 24 of Unit 21 of Module 8, we saw that:

$$\sin\dfrac{\pi}{4} = \dfrac{1}{\sqrt{2}}$$

So $x = \dfrac{\pi}{4}$ is a solution to the equation, as are the values $x = \dfrac{\pi}{4} \pm 2n\pi$.

However, these are not the only solutions, as can be seen from the graph below:

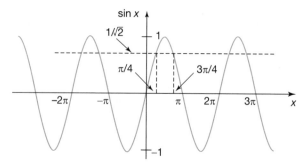

By the symmetry of the sine curve $x = \pi - \dfrac{\pi}{4} = \dfrac{3\pi}{4}$ is also a solution, as are the values $x = \dfrac{3\pi}{4} \pm 2n\pi$.

Just to make sure try this one. The solution to the equation $\tan x = \sqrt{3}$ is ....

The answer is in the next frame

6

$$\boxed{\dfrac{\pi}{3} \pm n\pi}$$

Because

From Frame 26 of Unit 21 of Module 8, we saw that $\tan\dfrac{\pi}{3} = \sqrt{3}$ and so $x = \dfrac{\pi}{3}$ is a solution.

Since the period of the tangent function is π the values $x = \dfrac{\pi}{3} \pm n\pi$ are also solutions.

Move now to the next frame

Using inverse functions

7

Solutions to these trigonometric equations can also be found by using a calculator provided care is taken in interpreting the results. For example, to solve:

$$\sin x = 0.1336$$

we can write an equivalent equation using the inverse sine as:

$$x = \sin^{-1} 0.1336$$

If now you use your calculator you will find that:

$$x = 0.1340 \text{ radians or } 7.6777° \text{ (both to 4 dp)}$$

This is all you can expect from the inverse sine because, as we saw in Frame 1 of this Unit, the range of the inverse sine function is restricted to be between $-\pi/2$ and $\pi/2$ radians. However, if we look again at the graph of the sine function in Frame 1 of Unit 25, we can say that as the sine function is periodic with period 2π then:

$$x = 0.1340 \pm 2n\pi \text{ radians or } 7.6777° \pm n \times 360°$$

Also, from the symmetry displayed on the graph of the sine function:

$$x = (\pi - 0.1340) \pm 2n\pi = 3.0076 \pm 2n\pi \text{ radians or } x = 172.3223° \pm n \times 360°$$

Try one yourself. The solution to the equation:

$$\cos x = 0.4257 \text{ is } \ldots\ldots\ldots\ldots \text{ radians to 4 dp.}$$

The answer is in the next frame

8

$$1 \cdot 1311 \pm 2n\pi \text{ and } 5 \cdot 1521 \pm 2n\pi$$

Because

An equivalent equation using the inverse cosine is:

$$x = \cos^{-1} 0 \cdot 4257$$

Using a calculator gives:

$$x = 1 \cdot 1311 \text{ radians or } 64 \cdot 8050° \text{ (both to 4 dp)}$$

Therefore, due to the periodicity of the cosine function:

$$x = 1 \cdot 1311 \pm 2n\pi \text{ radians}$$

And, due to the symmetry displayed on the graph of the cosine function:

$$x = (2\pi - 1 \cdot 1311) \pm 2n\pi = 5 \cdot 1521 \pm 2n\pi \text{ radians}$$

And another one. The solutions to the equation:

$$4 \sin 3x = 2$$

can be found first by dividing through by 4 to give $\sin 3x = \dfrac{2}{4} = \dfrac{1}{2}$

From the half equilateral triangle in Frame 26 of Unit 21 of Module 8 we know that if

$$\sin 3x = \frac{1}{2} \text{ then } 3x = \frac{\pi}{6} \pm 2n\pi \text{ radians or } 30° \pm n \times 360°$$

and so, dividing through by 3:

$$x = \frac{\pi}{18} \pm \frac{2n\pi}{3} \text{ radians or } 10° \pm n \times 120°$$

In addition, because of the symmetry shown by the graph of the sine function:

$$3x = \left(\pi - \frac{\pi}{6}\right) \pm 2n\pi \text{ radians or } (180 - 30)° \pm n \times 360°.$$

That is:

$$3x = \frac{5\pi}{6} \pm 2n\pi \text{ radians or } 150° \pm n \times 360°$$

and so

$$x = \frac{5\pi}{18} \pm 2n\pi \text{ radians or } 50° \pm n \times 120°$$

Try one for yourself. The solution to the equation:

$$9 \cos 5x = 3 \text{ is } \ldots\ldots\ldots \text{ radians to 4 dp}$$

You will have to use the inverse cosine for this one.
The answer is in the next frame.

$$\boxed{0{\cdot}2462 \pm 2n\pi/5 \text{ and } 1{\cdot}0104 \pm 2n\pi/5}$$ **9**

Because

To solve $9\cos 5x = 3$ first divide by 9 to give $\cos 5x = 0{\cdot}3333$. that is:

$5x = \cos^{-1} 0{\cdot}3333 = 1{\cdot}2310$ radians to 4 dp.

That is:

$5x = 1{\cdot}2310 \pm 2n\pi$ due to the periodicity

and

$5x = (2\pi - 1{\cdot}2310) \pm 2n\pi = 5{\cdot}0522 \pm 2n\pi$ due to the symmetries of the cosine graph

Therefore

$$x = \frac{1{\cdot}2310}{5} \pm \frac{2n\pi}{5} \quad \text{and} \quad x = \frac{5{\cdot}0522}{5} \pm \frac{2n\pi}{5}$$
$$= 0{\cdot}2462 \pm \frac{2n\pi}{5} \qquad\qquad = 1{\cdot}0104 \pm \frac{2n\pi}{5}$$

At this point let us pause and summarize the main facts so far on inverse trigonometric functions and trigonometric equations

 # Review summary **Unit 26**

1 The inverse trigonometric functions have restricted ranges. Remember this **10** when using inverse trigonometric functions.

2 A simple trigonometric equation is one that involves just one trigonometric expression.

3 Simple trigonometric equations are solved using a combination of known properties of special triangles, the graphs of the trigonometric functions and the inverse trigonometric functions.

 # Review exercise **Unit 26**

1 Solve the following simple trigonometric equations: **11**

(a) $\sin x = 0{\cdot}5432$

(b) $\cos 3x = 1$

(c) $\sqrt{3}\tan 4x = 1$

Complete the questions. Take one step at a time, there is no need to rush. If you need to, refer back to the Unit. The answers and working are in the next frame.

12   **1** (a) $\sin x = 0.5432$

An equivalent equation is $x = \sin^{-1} 0.5432 = 0.5742$ radians or $32.9017°$.
From the periodicity of the sine function there is an infinity of solutions:

   $x = 0.5742 \pm 2n\pi$ radians or $32.9017° \pm n \times 360°$

From the symmetries of the graph of the sine function:

   $x = (\pi - 0.5742) \pm 2n\pi$ radians or $(180° - 32.9017°) \pm n \times 360°$.

That is:

   $x = 2.5673 \pm 2n\pi$ radians or $147.0983° \pm n \times 360°$ are also solutions.

(b) $\cos 3x = 1$

From the graph of the cosine function $3x = 0 \pm 2n\pi = \pm 2n\pi$ so that $x = \pm\dfrac{2n\pi}{3}$.

(c) $\sqrt{3}\tan 4x = 1$ that is $\tan 4x = \dfrac{1}{\sqrt{3}}$

From the half equilateral triangle, since $\tan 4x = \dfrac{1}{\sqrt{3}}$ then:

   $4x = \dfrac{\pi}{6} \pm n\pi$ so that $x = \dfrac{\pi}{24} \pm \dfrac{n\pi}{4}$

Now for the Review test

Review test   **Unit 26**

13   **1** Solve the following simple trigonometric equations:

(a) $\cos x = 0.1133$

(b) $\sin 2x = 1/\sqrt{2}$

(c) $4\tan 3x = 0.5$

Exponential and logarithmic functions

Exponential functions

1

An *exponent* is a power. For example:

$8 = 2^3$

The number 2 is called the *base* and the number 3 is called the *exponent*. An exponential function is one where the independent variable is a power. For example:

$y = 2^x$

is an exponential function of base 2 with the following graph:

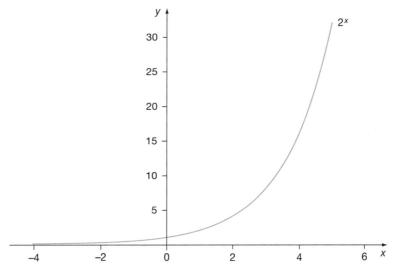

The features of this graph are:

- no negative values of y
- very small positive values of y for negative values of x
- the graph crosses the vertical axis at $y = 1$
- rapidly increasing positive values of y for positive and increasing values of x

These features are common to all exponential functions provided the base is a positive number. The general exponential function is written as:

$y = a^x$ where $a > 0$

The two exponential functions available on a calculator are:

The base 10 exponential function $y = 10^x$
The exponential function $y = e^x$ [sometimes written $y = \exp(x)$] where $e = 2 \cdot 7182818\ldots$ is called the exponential number.

Let's try a few.

To find the value of $y = 10^{1 \cdot 2}$ use the 10^x key and enter $x = 1 \cdot 2$. The result is 157·8489 to 4 dp.

To find the value of $y = e^{-2 \cdot 3}$ use the e^x (or exp) key and enter $x = -2 \cdot 3$. The result is 0·1003 to 4 dp.

You try some. The values of $y = 10^x$ when:

(a) $x = 0 \cdot 321$ is .... to 4 dp
(b) $x = -4 \cdot 26$ is .... to 4 sig fig.

The answers are in the next frame

2

> (a) 2·0941
> (b) 0·00005495

Because

Using the 10^x key and entering $x = 0 \cdot 321$ gives the result 2·0941 to 4 dp.

Using the 10^x key and entering $x = -4 \cdot 26$ gives the result 0·00005459 to 4 sig fig.

And two more. The values of $y = e^x$ when:

(a) $x = 5 \cdot 332$ is .... to 4 dp
(b) $x = -0 \cdot 01$ is .... to 4 sig fig.

The answers are in the next frame

3

> (a) 206·8513
> (b) 0·9900

Because

Using the e^x key and entering $x = 5 \cdot 332$ gives the result 206·8513 to 4 dp.

Using the e^x key and entering $x = -0 \cdot 01$ gives the result 0·9900 to 4 sig fig.

Move to the next frame

4   **Logarithmic functions**

In Frame 7 of Unit 8 of Module 2, it was stated that if a, b or c are real numbers where:

$a = b^c$ and $b > 1$

the power c is called the logarithm of the number a to the base b and is written:

$c = \log_b a$ spoken as c is the log of a to the base b

In the example:

$8 = 2^3$

the number 2 is the base and the power 3 is the logarithm of 8 to the base 2.

▶

We write:

$$3 = \log_2 8$$

A logarithmic function is a function of the form:

$$y = \log_a x \text{ where } a > 0$$

The graph of $y = \log_2 x$ is given below:

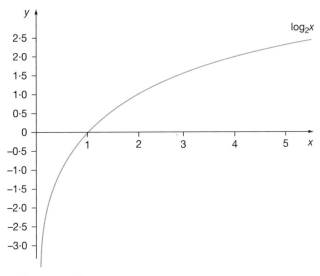

The features of this graph are:

- the logarithm is not defined for negative values of x
- for values of x less than 1 the logarithm is negative
- the graph crosses the horizontal axis at $x = 1$
- increasing positive values of y for positive and increasing values of x

These features are common to all logarithmic functions.

The function $y = \log_2 x$ is the inverse function of the exponential function $y = 2^x$. The same applies in reverse: the exponential function $y = 2^x$ is the inverse function of the logarithmic function $y = \log_2 x$. Indeed,

functions $y = \log_a x$ and $y = a^x$ are each the inverse function of the other.

Two particular functions have their own special notation. These are the exponential function whose base is the exponential number e where:

$$y = e^x \text{ which can also be written as } y = \exp(x)$$

and its inverse

$$y = \log_e x \text{ which can also be written as } y = \ln x, \text{ as found on a calculator.}$$

A third special notation arises by tradition – logarithms to the base 10 have the base omitted. That is:

$$y = \log_{10} x \text{ can also be written as } y = \log x, \text{ also as found on a calculator.}$$

Just try some:

(a) $e^{1\cdot 34} = \ldots\ldots\ldots$ to 3 dp
(b) $\ln 3\cdot 11 = \ldots\ldots\ldots$ to 3 dp
(c) $\ln(\exp[2\cdot 871]) = \ldots\ldots\ldots$ to 3 dp
(d) $\log 4\cdot 75 = \ldots\ldots\ldots$ to 3 dp

The answers are in the next frame

5

| (a) 3·819 | (b) 1·135 | (c) 2·871 | (d) 0·677 |

Because

(a) Using the exp key (or e key), $e^{1\cdot 34} = 3\cdot 819$ to 3 dp.

(b) Using the ln key, $\ln 3\cdot 11 = 1\cdot 135$ to 3 dp.

(c) Using the ln key and the exp key (or e key), $\ln(\exp[2\cdot 871]) = 2\cdot 871$ to 3 dp. You would expect this result because the ln and exp functions are inverses of each other.

(d) Using the log key, $\log 4\cdot 75 = 0\cdot 677$ to 3 dp.

Move to the next frame

6   Indicial equations

An *indicial equation* is an equation where the variable appears as an index and the solution of such an equation requires the application of logarithms.

Example 1

Here is a simple case. We have to find the value of x, given that $12^{2x} = 35\cdot 4$.

Taking logs of both sides – and using $\log(A^n) = n\log A$ we have

$$(2x)\log 12 = \log 35\cdot 4$$

i.e. $(2x)1\cdot 0792 = 1\cdot 5490$

$$2\cdot 1584x = 1\cdot 5490$$

$$\therefore\ x = \frac{1\cdot 5490}{2\cdot 1584} = 0\cdot 71766$$

$$\therefore\ x = 0\cdot 7177 \text{ to 4 sig fig}$$

Example 2

Solve the equation $4^{3x-2} = 26^{x+1}$

The first line in this solution is $\ldots\ldots\ldots$

7

| $(3x-2)\log 4 = (x+1)\log 26$ |

$$\therefore\ (3x-2)0\cdot 6021 = (x+1)1\cdot 4150$$

Multiplying out and collecting up, we eventually get

$x = \ldots\ldots\ldots$ to 4 sig fig

8

$$\boxed{6\cdot694}$$

Because we have

$$(3x - 2)0\cdot6021 = (x + 1)1\cdot4150$$
$$1\cdot8063x - 1\cdot2042 = 1\cdot4150x + 1\cdot4150$$
$$(1\cdot8063 - 1\cdot4150)x = (1\cdot4150 + 1\cdot2042)$$
$$0\cdot3913x = 2\cdot6192$$

$$\therefore \; x = \frac{2\cdot6192}{0\cdot3913} = 6\cdot6936$$

$$\therefore \; x = 6\cdot694 \text{ to 4 sig fig}$$

Care must be taken to apply the rules of logarithms rigidly.

Now we will deal with another example

Example 3

9

Solve the equation $5\cdot4^{x+3} \times 8\cdot2^{2x-1} = 4\cdot8^{3x}$

We recall that $\log(A \times B) = \log A + \log B$

Therefore, we have $\log\{5\cdot4^{x+3}\} + \log\{8\cdot2^{2x-1}\} = \log\{4\cdot8^{3x}\}$

i.e. $\quad (x + 3)\log 5\cdot4 + (2x - 1)\log 8\cdot2 = 3x\log 4\cdot8$

You can finish it off, finally getting

$$x = \ldots\ldots\ldots \text{ to 4 sig fig}$$

10

$$\boxed{2\cdot485}$$

Here is the working as a check:

$$(x + 3)0\cdot7324 + (2x - 1)0\cdot9138 = (3x)0\cdot6812$$
$$0\cdot7324x + 2\cdot1972 + 1\cdot8276x - 0\cdot9138 = 2\cdot0436x$$
$$(0\cdot7324 + 1\cdot8276)x + (2\cdot1972 - 0\cdot9138) = 2\cdot0436x$$
$$\therefore \; 2\cdot5600x + 1\cdot2834 = 2\cdot0436x$$
$$(2\cdot5600 - 2\cdot0436)x = -1\cdot2834$$
$$0\cdot5164x = -1\cdot2834$$

$$x = -\frac{1\cdot2834}{0\cdot5164} = -2\cdot4853 \quad \therefore \; x = -2\cdot485 \text{ to 4 sig fig}$$

Finally, here is one to do all on your own.

Example 4

Solve the equation $7(14\cdot3^{x+5}) \times 6\cdot4^{2x} = 294$

Work right through it, giving the result to 4 sig fig.

$$x = \ldots\ldots\ldots$$

11

$$\boxed{-1{\cdot}501}$$

Check the working:

$$7(14{\cdot}3^{x+5}) \times 6{\cdot}4^{2x} = 294$$

$$\therefore \ \log 7 + (x+5)\log 14{\cdot}3 + (2x)\log 6{\cdot}4 = \log 294$$

$$0{\cdot}8451 + (x+5)1{\cdot}1553 + (2x)0{\cdot}8062 = 2{\cdot}4683$$

$$0{\cdot}8451 + 1{\cdot}1553x + 5{\cdot}7765 + 1{\cdot}6124x = 2{\cdot}4683$$

$$(1{\cdot}1553 + 1{\cdot}6124)x + (0{\cdot}8451 + 5{\cdot}7765) = 2{\cdot}4683$$

$$2{\cdot}7677x + 6{\cdot}6216 = 2{\cdot}4683$$

$$2{\cdot}7677x = 2{\cdot}4683 - 6{\cdot}6216 = -4{\cdot}1533$$

$$\therefore \ x = -\frac{4{\cdot}1533}{2{\cdot}7677} = -1{\cdot}5006$$

$$x = -1{\cdot}501 \text{ to 4 sig fig}$$

12

Some indicial equations may need a little manipulation.

Example 5

Solve the equation $2^{2x} - 6 \times 2^x + 8 = 0$

Because $2^{2x} = (2^x)^2$ this is an equation that is quadratic in 2^x. We can, therefore, write $y = 2^x$ and substitute y for 2^x in the equation to give:

$$y^2 - 6y + 8 = 0$$

which factorizes to give:

$$(y-2)(y-4) = 0$$

so that $y = 2$ or $y = 4$. That is $2^x = 2$ or $2^x = 4$ so that $x = 1$ or $x = 2$.

Try this one. Solve $2 \times 3^{2x} - 6 \times 3^x + 4 = 0$.

The answer is in Frame 13

13

$$x = 0.631 \text{ to 3 dp or } x = 0$$

Because $3^{2x} = (3^x)^2$ this is an equation that is quadratic in 3^x. We can, therefore, write $y = 3^x$ and substitute y for 3^x in the equation to give:

$$2y^2 - 6y + 4 = 0$$

which factorizes to give:

$$(2y - 4)(y - 1) = 0$$

so that $y = 2$ or $y = 1$. That is $3^x = 2$ or $3^x = 1$ so that $x \log 3 = \log 2$ or $x = 0$.

That is $x = \dfrac{\log 2}{\log 3} = 0.631$ to 3 dp or $x = 0$.

At this point let us pause and summarize the main facts so far on exponential and logarithmic functions

Review summary

Unit 27

14

1 Any function of the form $f(x) = a^x$ is called an *exponential* function.

2 The exponential function $f(x) = a^x$ and the logarithmic function $g(x) = \log_a x$ are *mutual inverses*: $f^{-1}(x) = g(x)$ and $g^{-1}(x) = f(x)$.

Review exercise

Unit 27

15

1 Solve the following indicial equations giving the results to 4 dp:

    (a) $13^{3x} = 8.4$ (b) $2.8^{2x+1} \times 9.4^{2x-1} = 6.3^{4x}$

    (c) $7^{2x} - 9 \times 7^x + 14 = 0$

Complete the question. Take your time, there is no need to rush.
If necessary, refer back to the Unit.
The answers and working are in the next frame.

16

1 (a) $13^{3x} = 8\cdot 4$

Taking logs of both sides gives: $3x\log 13 = \log 8\cdot 4$ so that:

$$x = \frac{\log 8\cdot 4}{3\log 13} = \frac{0\cdot 9242\ldots}{3\cdot 3418\ldots} = 0\cdot 277 \text{ to 3 dp}$$

(b) $2\cdot 8^{2x+1} \times 9\cdot 4^{2x-1} = 6\cdot 3^{4x}$

Taking logs:

$(2x+1)\log 2\cdot 8 + (2x-1)\log 9\cdot 4 = 4x\log 6\cdot 3$

Factorizing x gives:

$x(2\log 2\cdot 8 + 2\log 9\cdot 4 - 4\log 6\cdot 3) = \log 9\cdot 4 - \log 2\cdot 8$

So that:

$$x = \frac{\log 9\cdot 4 - \log 2\cdot 8}{2\log 2\cdot 8 + 2\log 9\cdot 4 - 4\log 6\cdot 3}$$

$$= \frac{0\cdot 9731\ldots - 0\cdot 4471\ldots}{2(0\cdot 4471\ldots) + 2(0\cdot 9731\ldots) - 4(0\cdot 7993\ldots)} = -1\cdot 474 \text{ to 3 dp}$$

(c) $7^{2x} - 9\times 7^x + 14 = 0$

This equation is quadratic in 7^x so let $y = 7^x$ and rewrite the equation as:

$y^2 - 9y + 14 = 0$ which factorizes to $(y-2)(y-7) = 0$

with solution: $y = 2$ or $y = 7$ that is $7^x = 2$ or $7^x = 7$

so that $x = \dfrac{\log 2}{\log 7} = 0\cdot 356$ to 3 dp or $x = 1$

Now for the Review test

 # Review test

17

1 Find the value of x corresponding to each of the following:

(a) $2^{-2x} = 1$ (b) $\exp(-4x) = e^3$ (c) $e^{\frac{1}{x}} = 2\cdot 34$

(d) $\log_2 x = 0\cdot 4$ (e) $\ln x = 2$

2 Solve for x:

(a) $2^{x-5}5^{x+1} = 62\cdot 5$

(b) $e^{2x} - e^{x+3} - e^{x+1} + e^4 = 0$

(c) $\log(x^2) = \log\left(\dfrac{1}{x}\right) - \log 5$

(d) $\log_4(3 - x^2) = -6\cdot 2$

Odd and even functions   **Unit 28**

If, by replacing x by $-x$ in $f(x)$ the expression does not change its value, f is called an *even* function. For example, if:

$f(x) = x^2$ then $f(-x) = (-x)^2 = x^2 = f(x)$ so that f is an even function.

On the other hand, if $f(-x) = -f(x)$ then f is called an *odd* function. For example, if:

$f(x) = x^3$ then $f(-x) = (-x)^3 = -x^3 = -f(x)$ so that f is an odd function

Because $\sin(-\theta) = -\sin\theta$ the sine function is an odd function and because $\cos(-\theta) = \cos\theta$ the cosine function is an even function. Notice how the graph of the cosine function is reflection symmetric about the vertical axis through $\theta = 0$ in the diagram in Frame 3 of Unit 25 of this Module. All even functions possess this type of symmetry. The graph of the sine function is rotation symmetric about the origin as it goes into itself under a rotation of 180° about this point, as can be seen from the third diagram in Frame 1 of Unit 25 of this Module. All odd functions possess this type of antisymmetry. Notice also that $\tan(-\theta) = -\tan\theta$ so that the tangent function, like the sine function, is odd and has an antisymmetric graph (see the diagram in Frame 6 of Unit 25 of this Module).

Odd and even parts

Not every function is either even or odd but many can be written as the sum of an even part and an odd part. If, given $f(x)$ where $f(-x)$ is also defined then:

$f_e(x) = \dfrac{f(x) + f(-x)}{2}$ is even and $f_o(x) = \dfrac{f(x) - f(-x)}{2}$ is odd.

Furthermore $f_e(x)$ is called the *even part* of $f(x)$ and

$f_o(x)$ is called the *odd part* of $f(x)$.

For example, if $f(x) = 3x^2 - 2x + 1$ then $f(-x) = 3(-x)^2 - 2(-x) + 1$ $= 3x^2 + 2x + 1$ so that the even and odd parts of $f(x)$ are:

$f_e(x) = \dfrac{(3x^2 - 2x + 1) + (3x^2 + 2x + 1)}{2} = 3x^2 + 1$ and

$f_o(x) = \dfrac{(3x^2 - 2x + 1) - (3x^2 + 2x + 1)}{2} = -2x$

So, the even and odd parts of $f(x) = x^3 - 2x^2 - 3x + 4$ are . . . .

Apply the two formulas; the answer is in the next frame

3

$$f_e(x) = -2x^2 + 4$$
$$f_o(x) = x^3 - 3x$$

Because

$$f_e(x) = \frac{f(x) + f(-x)}{2}$$

$$= \frac{(x^3 - 2x^2 - 3x + 4) + ((-x)^3 - 2(-x)^2 - 3(-x) + 4)}{2}$$

$$= \frac{x^3 - 2x^2 - 3x + 4 - x^3 - 2x^2 + 3x + 4}{2} = \frac{-2x^2 + 4 - 2x^2 + 4}{2}$$

$$= -2x^2 + 4$$

and

$$f_o(x) = \frac{f(x) - f(-x)}{2}$$

$$= \frac{(x^3 - 2x^2 - 3x + 4) - ((-x)^3 - 2(-x)^2 - 3(-x) + 4)}{2}$$

$$= \frac{x^3 - 2x^2 - 3x + 4 + x^3 + 2x^2 - 3x - 4}{2} = \frac{x^3 - 3x + x^3 - 3x}{2}$$

$$= x^3 - 3x$$

Make a note here that *even polynomial functions consist of only even powers* and *odd polynomial functions consist of only odd powers.*

So the odd and even parts of $f(x) = x^3(x^2 - 3x + 5)$ are ....

Answer in next frame

4

$$f_e(x) = -3x^4 \qquad f_o(x) = x^5 + 5x^3$$

Because

Even polynomial functions consist of only even powers and odd polynomial functions consist of only odd powers. Consequently, the even part of $f(x)$ consists of even powers only and the odd part of $f(x)$ consists of odd powers only.

Now try this. The even and odd parts of $f(x) = \dfrac{1}{x-1}$ are ....

Next frame

5

$$f_e(x) = \frac{1}{(x-1)(x+1)}$$
$$f_o(x) = \frac{x}{(x-1)(x+1)}$$

Because

$$f_e(x) = \frac{f(x) + f(-x)}{2}$$

$$= \frac{1}{2}\left(\frac{1}{(x-1)} + \frac{1}{(-x-1)}\right) = \frac{1}{2}\left(\frac{1}{(x-1)} - \frac{1}{(x+1)}\right)$$

$$= \frac{1}{2}\frac{x+1-(x-1)}{(x-1)(x+1)} = \frac{1}{2}\frac{2}{(x-1)(x+1)}$$

$$= \frac{1}{(x-1)(x+1)}$$

and

$$f_o(x) = \frac{f(x) - f(-x)}{2}$$

$$= \frac{1}{2}\left(\frac{1}{(x-1)} - \frac{1}{(-x-1)}\right) = \frac{1}{2}\left(\frac{1}{(x-1)} + \frac{1}{(x+1)}\right)$$

$$= \frac{1}{2}\frac{x+1+x-1}{(x-1)(x+1)} = \frac{1}{2}\frac{2x}{(x-1)(x+1)}$$

$$= \frac{x}{(x-1)(x+1)}$$

Next frame

Odd and even parts of the exponential function

6

The exponential function is neither odd nor even but it can be written as a sum of an odd part and an even part.

That is, $\exp_e(x) = \dfrac{\exp(x) + \exp(-x)}{2}$ and $\exp_o(x) = \dfrac{\exp(x) - \exp(-x)}{2}$. These two functions are known as the *hyperbolic cosine* and the *hyperbolic sine* respectively:

$$\cosh x = \frac{e^x + e^{-x}}{2} \quad \text{and} \quad \sinh x = \frac{e^x - e^{-x}}{2}$$

Using these two functions the hyperbolic tangent can also be defined:

$$\tanh x = \frac{e^x - e^{-x}}{e^x + e^{-x}}$$

The logarithmic function $y = \log_a x$ is neither odd nor even and indeed does not possess even and odd parts because $\log_a(-x)$ is not defined.

At this point let us pause and summarize the main facts so far on odd and even functions

Review summary   Unit 28

7

1 If $f(-x) = f(x)$ then f is called an *even* function and if $f(-x) = -f(x)$ then f is called an *odd* function.

2 Even polynomial functions consist of only even powers and odd polynomial functions consist of only odd powers.

3 If $f_e(x) = \dfrac{f(x) + f(-x)}{2}$ can be defined it is called the *even* part of $f(x)$ and if

$f_o(x) = \dfrac{f(x) - f(-x)}{2}$ can be defined it is called the *odd* part of $f(x)$. Notice that $f(x) = f_e(x) + f_o(x)$.

 Review exercise   Unit 28

8

1 Show that $f(x) = |x|$ is an even function.

2 Find the odd and even parts of $f(x) = x^2 - x^3 + 2$

Complete the questions.
Refer back to the Unit if necessary, but don't rush.
The answers and working are in the next frame.

9

1 $f(x) = |x|$ and $f(-x) = |-x| = |x|$. Therefore $f(-x) = f(x)$ and so $f(x) = |x|$ is an even function.

2 $f(x) = x^2 - x^3 + 2$ and $f(-x) = (-x)^2 - (-x)^3 + 2$. Therefore:

$$f_e(x) = \frac{f(x) + f(-x)}{2} = \frac{(x^2 - x^3 + 2) + (x^2 + x^3 + 2)}{2} = x^2 + 2$$

and $\quad f_o(x) = \dfrac{f(x) - f(-x)}{2} = \dfrac{(x^2 - x^3 + 2) - (x^2 + x^3 + 2)}{2} = -x^3$

Now for the Review test

 Review test   Unit 28

10

1 Show that $f(x) = |x^3|$ is an even function.

2 Find the odd and even parts of $f(x) = x(x^2 + x + 1)$

Limits

Unit 29

Limits of functions

1

There are times when a function is not defined for a particular value of x, say $x = x_0$ but it is defined for values of x that are arbitrarily close to x_0. For example, the expression:

$$f(x) = \frac{x^2 - 1}{x - 1}$$

is not defined when $x = 1$ because at that point the denominator is zero and division by zero is not defined. However, we note that:

$$f(x) = \frac{x^2 - 1}{x - 1} = \frac{(x - 1)(x + 1)}{x - 1} = x + 1 \text{ provided that } x \neq 1$$

We still cannot permit the value $x = 1$ because to do so would mean that the cancellation of the $x - 1$ factor would be division by zero. But we can say that as the value of x approaches 1, the value of $f(x)$ approaches 2. Clearly the value of $f(x)$ never actually attains the value of 2 but it does get as close to it as you wish it to be by selecting a value of x sufficiently close to 1. We say that the limit of $\frac{x^2 - 1}{x - 1}$ as x approaches 1 is 2 and we write:

$$\underset{x \to 1}{Lim} \left(\frac{x^2 - 1}{x - 1} \right) = 2$$

You try one, $\underset{x \to -3}{Lim} \left(\frac{x^2 - 9}{x + 3} \right) = \dots\dots\dots$

The answer is in the next frame

$$\boxed{-6}$$

2

Because

$$f(x) = \frac{x^2 - 9}{x + 3}$$
$$= \frac{(x + 3)(x - 3)}{x + 3}$$
$$= x - 3 \text{ provided } x \neq -3$$

So that:

$$\underset{x \to -3}{Lim} \left(\frac{x^2 - 9}{x + 3} \right) = -6$$

The rules of limits

Listed here is a collection of simple rules used for evaluating limits. They are what you would expect with no surprises.

If x_0 is some fixed value of x and if $Lim_{x \to x_0} f(x) = A$ and $Lim_{x \to x_0} g(x) = B$ then:

The limit of the sum (or difference) is equal to the sum (or difference) of the limits

$$Lim_{x \to x_0} [f(x) \pm g(x)] = Lim_{x \to x_0} f(x) \pm Lim_{x \to x_0} g(x) = A \pm B$$

The limit of the product is equal to the product of the limits

$$Lim_{x \to x_0} [f(x)g(x)] = Lim_{x \to x_0} f(x) \, Lim_{x \to x_0} g(x) = AB$$

The limit of the quotient is equal to the quotient of the limits

$$Lim_{x \to x_0} \left[\frac{f(x)}{g(x)} \right] = \frac{Lim_{x \to x_0} f(x)}{Lim_{x \to x_0} g(x)} = \frac{A}{B} \text{ provided } B \neq 0$$

The limit of the composition is equal to the composition of the limit

$$Lim_{x \to x_0} f[g(x)] = f\left[Lim_{x \to x_0} g(x) \right] = f(B) \text{ provided } f(g(x)) \text{ is continuous at } x = x_0$$

So you try these for yourself:

(a) $Lim_{x \to \pi} (x^2 - \sin x)$

(b) $Lim_{x \to \pi} (x^2 \sin x)$

(c) $Lim_{x \to \pi/4} \dfrac{(\tan x)}{(\sin x)}$

(d) $Lim_{x \to 1} \cos(x^2 - 1)$

The answers are in the next frame

3

$$\boxed{\text{(a) } \pi^2 \qquad \text{(b) } 0 \qquad \text{(c) } \sqrt{2} \qquad \text{(d) } 1}$$

Because

(a) $\displaystyle \lim_{x \to \pi} \left(x^2 - \sin x\right) = \lim_{x \to \pi} x^2 - \lim_{x \to \pi} (\sin x) = \pi^2 - 0 = \pi^2$

(b) $\displaystyle \lim_{x \to \pi} \left(x^2 \sin x\right) = \lim_{x \to \pi} (x^2) \lim_{x \to \pi} (\sin x) = \pi^2 \times 0 = 0$

(c) $\displaystyle \lim_{x \to \pi/4} \frac{(\tan x)}{(\sin x)} = \frac{\displaystyle \lim_{x \to \pi/4} \tan x}{\displaystyle \lim_{x \to \pi/4} \sin x} = \frac{1}{1/\sqrt{2}} = \sqrt{2}$

(d) $\displaystyle \lim_{x \to 1} \cos\left(x^2 - 1\right) = \cos\left(\lim_{x \to 1} \left(x^2 - 1\right)\right) = \cos 0 = 1$

All fairly straightforward for these simple problems. Difficulties do occur for the limits of quotients when both the numerator and the denominator are simultaneously zero at the limit point but those problems we shall not consider in this book.

Let's pause and summarize the work on limits

 # Review summary   Unit 29

4

1   The limit of a sum (difference) is equal to the sum (difference) of the limits.
The limit of a product is equal to the product of the limits.
The limit of a quotient is equal to the quotient of the limits provided the denominator limit is not zero.
The limit of a composition is the composition of the limit.

 # Review exercise   Unit 29

5

1   Evaluate:

(a) $\displaystyle \lim_{x \to -1} \left(\frac{x^2 + 2x + 1}{x^2 + 3x + 2}\right)$

(b) $\displaystyle \lim_{x \to \pi/4} (3x - \tan x)$

(c) $\displaystyle \lim_{x \to \pi/2} \left(2x^2 \cos[3x - \pi/2]\right)$

Complete the question, working back over the previous frames if you need to.
Don't rush, take your time.
The answers and working are in the next frame.

6

1 (a) $\underset{x\to-1}{Lim}\left(\dfrac{x^2+2x+1}{x^2+3x+2}\right)=\underset{x\to-1}{Lim}\left(\dfrac{(x+1)^2}{(x+1)(x+2)}\right)$

$\qquad\qquad\qquad =\underset{x\to-1}{Lim}\left(\dfrac{x+1}{x+2}\right)\qquad$ The cancellation is permitted
$\qquad\qquad\qquad\qquad\qquad\qquad\qquad$ since $x\neq-1$

$\qquad\qquad\qquad =\dfrac{0}{1}=0$

(b) $\underset{x\to\pi/4}{Lim}\,(3x-\tan x)=\underset{x\to\pi/4}{Lim}\,(3x)-\underset{x\to\pi/4}{Lim}\,(\tan x)$

$\qquad\qquad\qquad =\dfrac{3\pi}{4}-\tan\dfrac{\pi}{4}=\dfrac{3\pi}{4}-1$

(c) $\underset{x\to\pi/2}{Lim}\,(2x^2\cos[3x-\pi/2])=\underset{x\to\pi/2}{Lim}\,(2x^2)\,\underset{x\to\pi/2}{Lim}\,(\cos[3x-\pi/2])$

$\qquad\qquad\qquad\qquad =\dfrac{2\pi^2}{4}\times\cos\underset{x\to\pi/2}{Lim}\,[3x-\pi/2]$

$\qquad\qquad\qquad\qquad =\dfrac{2\pi^2}{4}\times\cos[3\pi/2-\pi/2]$

$\qquad\qquad\qquad\qquad =\dfrac{2\pi^2}{4}\times\cos\pi=-\dfrac{2\pi^2}{4}$

Now for the Review test

 # Review test   **Unit 29**

7

1 Evaluate each of the following limits:

(a) $\underset{x\to3}{Lim}\left(\dfrac{x^2-9}{x-3}\right)$

(b) $\underset{x\to1/3}{Lim}\,(3x^2-2x+1)$

(c) $\underset{x\to-1}{Lim}\left(\dfrac{(x+1)e^{3x+2}}{x^2-1}\right)$

8

You have now come to the end of this Module. A list of **Can You?** questions follows for you to gauge your understanding of the material in the Module. These questions match the **Learning outcomes** listed at the beginning of the Module. Now try the **Test exercise**. *Work through the questions at your own pace, there is no need to hurry.* A set of **Further problems** provides additional valuable practice.

Can You?

Check this list before and after you try the end of Module test.

On a scale of 1 to 5 how confident are you that you can:

- Identify a function as a rule and recognize rules that are not functions?
  Yes ☐ ☐ ☐ ☐ ☐ *No*

- Determine the domain and range of a function?
  Yes ☐ ☐ ☐ ☐ ☐ *No*

- Construct the inverse of a function and draw its graph?
  Yes ☐ ☐ ☐ ☐ ☐ *No*

- Construct compositions of functions and de-construct them into their component functions?
  Yes ☐ ☐ ☐ ☐ ☐ *No*

- Develop the trigonometric functions from the trigonometric ratios?
  Yes ☐ ☐ ☐ ☐ ☐ *No*

- Find the period, amplitude and phase of a periodic function?
  Yes ☐ ☐ ☐ ☐ ☐ *No*

- Distinguish between the inverse of a trigonometric function and the inverse trigonometric function?
  Yes ☐ ☐ ☐ ☐ ☐ *No*

- Solve trigonometric equations using the inverse trigonometric functions?
  Yes ☐ ☐ ☐ ☐ ☐ *No*

- Recognize that the exponential function and the natural logarithmic function are mutual inverses and solve indicial and logarithmic equations?
  Yes ☐ ☐ ☐ ☐ ☐ *No*

- Find the even and odd parts of a function when they exist?
  Yes ☐ ☐ ☐ ☐ ☐ *No*

- Construct the hyperbolic functions from the odd and even parts of the exponential function?
  Yes ☐ ☐ ☐ ☐ ☐ *No*

- Evaluate limits of simple functions?
  Yes ☐ ☐ ☐ ☐ ☐ *No*

Test exercise 9

2

1 Which of the following equations expresses a rule that is a function?

(a) $y = -x^{\frac{3}{2}}$ (b) $y = x^2 + x + 1$ (c) $y = (\sqrt{x})^3$

2 Given the two functions f and g expressed by:

$$f(x) = \frac{1}{x-2} \text{ for } 2 < x \le 4 \text{ and } g(x) = x - 1 \text{ for } 0 \le x < 3$$

find the domain and range of functions h and k where:

(a) $h(x) = 2f(x) - 3g(x)$ (b) $k(x) = -\dfrac{3f(x)}{5g(x)}$

3 Use your spreadsheet to draw each of the following and their inverses. Is the inverse a function?

(a) $y = 3x^4$ (b) $y = -x^3$ (c) $y = 3 - x^2$

4 Given that $a(x) = 5x$, $b(x) = x^4$, $c(x) = x + 3$ and $d(x) = \sqrt{x}$ find:

(a) $f(x) = a[b(c[d(x)])]$ (b) $f(x) = a(a[d(x)])$

(c) $f(x) = b[c(b[c(x)])]$

5 Given that $f(x) = (5x - 4)^3 - 4$ decompose f into its component functions and find its inverse. Is the inverse a function?

6 Use a calculator to find the value of each of the following:

(a) $\sin(-320°)$ (b) $\operatorname{cosec}(\pi/11)$ (c) $\cot(-\pi/2)$

7 Find the period, amplitude and phase of each of the following:

(a) $f(\theta) = 4\cos 7\theta$ (b) $f(\theta) = -2\sin(2\theta - \pi/2)$

(c) $f(\theta) = \sec(3\theta + 4)$

8 A function is defined by the following prescription:

$$f(x) = 9 - x^2, \quad 0 \le x < 3$$
$$f(x + 3) = f(x)$$

Plot a graph of this function for $-6 \le x \le 6$ and find:

(a) the period

(b) the amplitude

(c) the phase of $f(x) + 2$ with respect to $f(x)$

9 Solve the following trigonometric equations:

(a) $\sin x = -0.1123$

(b) $\tan 7x = 1$

(c) $\sqrt{2}\cos 5x = 1$

10 Find the value of x corresponding to each of the following:

(a) $4^{-x} = 1$ (b) $\exp(3x) = e$

(c) $e^x = 54.32$ (d) $\log_5 x = 4$

(e) $\log_{10}(x - 3) = 0.101$ (f) $\ln 100 = x$

11 Solve for x:

(a) $4^{x+2}5^{x+1} = 32\,000$ (b) $e^{2x} - 5e^x + 6 = 0$

(c) $\log_x 36 = 2$ (d) $\frac{1}{2}\log(x^2) = 5\log 2 - 4\log 3$

(e) $\ln(x^{1/2}) = \ln x + \ln 3$

12 Find the even and odd parts of $f(x) = a^x$.

13 Evaluate:

(a) $\underset{x \to 4}{Lim}\left(\dfrac{x^2 - 8x + 16}{x^2 - 5x + 4}\right)$ (b) $\underset{x \to 1}{Lim}\left(6x^{-1} + \ln x\right)$

(c) $\underset{x \to 1/2}{Lim}\left[(3x^3 - 1)/(4 - 2x^2)\right]$ (d) $\underset{x \to -3}{Lim}\,\dfrac{(1 + 9x^{-2})}{(3x - x^2)}$

(e) $\underset{x \to 1}{Lim}\left(\tan^{-1}\left[\dfrac{x^2 - 1}{2(x - 1)}\right]\right)$

 # Further problems 9

1 Do the graphs of $f(x) = 3\log x$ and $g(x) = \log(3x)$ intersect?   **3**

2 Let $f(x) = \ln\left(\dfrac{1 + x}{1 - x}\right)$.

(a) Find the domain and range of f.

(b) Show that the new function g formed by replacing x in $f(x)$ by $\dfrac{2x}{1 + x^2}$

is given by $g(x) = 2f(x)$.

3 Describe the graph of $x^2 - 9y^2 = 0$.

4 Two functions C and S are defined as the even and odd parts of f where $f(x) = a^x$. Show that:

(a) $[C(x)]^2 - [S(x)]^2 = 1$

(b) $S(2x) = 2S(x)C(x)$

5 Show by using a diagram that, for functions f and g, $(f \circ g)^{-1} = g^{-1} \circ f^{-1}$.

6 Is it possible to find a value of x such that $\log_a(x) = a^x$ for $a > 1$?

7 Given the three functions a, b and c where $a(x) = 6x$, $b(x) = x - 2$ and $c(x) = x^3$ find the inverse of:

(a) $f(x) = a(b[c(x)])$ (b) $f = c \circ b \circ c$ (c) $f = b \circ c \circ a \circ b \circ c$

8 Use your spreadsheet to plot $\sin\theta$ and $\sin 2\theta$ on the same graph. Plot from $\theta = -5$ to $\theta = +5$ with a step value of 0.5.

9 The square sine wave with period 2 is given by the prescription:

$$f(x) = \begin{cases} 1 & 0 \le x < 1 \\ -1 & 1 \le x < 2 \end{cases}$$

Plot this wave for $-4 \le x \le 4$ on a sheet of graph paper.

10 The absolute value of x is given as:
$$|x| = \begin{cases} x & \text{if } x \geq 0 \\ -x & \text{if } x < 0 \end{cases}$$

(a) Plot the graph of $y = |x|$ for $-2 \leq x \leq 2$.

(b) Find the derivative of y.

(c) Does the derivative exist at $x = 0$?

11 Use the spreadsheet to plot the rectified sine wave $f(x) = |\sin x|$ for $-10 \leq x \leq 10$ with step value 1.

12 Use your spreadsheet to plot $f(x) = \dfrac{\sin x}{x}$ for $-40 \leq x \leq 40$ with step value 4.
(You will have to enter the value $f(0) = 1$ specifically in cell **B11**.)

13 Solve the following giving the results to 4 sig fig:

(a) $6\{8^{3x+2}\} = 5^{2x-7}$

(b) $4 \cdot 5^{1-2x} \times 6 \cdot 2^{3x+4} = 12 \cdot 7^{5x}$

(c) $5\{17 \cdot 2^{x+4}\} \times 3\{8 \cdot 6^{2x}\} = 4 \cdot 7^{x-1}$

14 Evaluate each of the following limits:

(a) $\displaystyle \lim_{x \to \pi/2} \left(\frac{(x^2 - \pi/4)\sin(\cos x)}{x - \pi/2} \right)$

(b) $\displaystyle \lim_{x \to -1} \ln\left(\exp\left[\frac{3x^2 + 2x - 1}{x + 1} \right] \right)$

(c) $\displaystyle \lim_{x \to 2+\sqrt{3}} \cos\left(\sin^{-1}\left(\frac{x - 2}{x - \sqrt{3}} \right) \right)$

Matrices

When you have completed this Module you will be able to:

- Use matrices as stores of numbers and manipulate such matrices by adding and subtracting, by multiplying by a scalar and by multiplying two matrices together
- Compute the inverse of a 2×2 matrix and use inverse matrices to solve pairs of simultaneous linear equations
- Expand 2×2 and 3×3 determinants and use Cramer's rule to solve simultaneous linear equations

Matrices

1   Matrix operations

A matrix is a rectangular array of numbers arranged in regular rows and columns. For example:

$$\mathbf{A} = \begin{pmatrix} 1 & 2 \\ 3 & 6 \end{pmatrix} \quad \text{and} \quad \mathbf{B} = \begin{pmatrix} 4 & -3 & 2 \\ 5 & 0 & 1 \end{pmatrix}$$

are both matrices. The first matrix $\mathbf{A}$, having 2 rows and 2 columns is referred to as a 2×2 ('two-by-two') matrix and the second matrix $\mathbf{B}$, having 2 rows and 3 columns is referred to as a 2×3 matrix.

A matrix having i rows and j columns is referred to as an ....

Next frame for the answer

2

$i \times j$ matrix

Because

The matrix has i rows and j columns

Matrices are used for storing numbers. For example, Alice, Ben and Charlie wish to record the quantities of two particular items they have sold during week 1 of January:

January	Week 1	
	Item 1	Item 2
Alice	2	3
Ben	4	5
Charlie	6	7

Provided we remember what each row and each column refers to, the quantities can then be stored in matrix form as:

$$\begin{pmatrix} 2 & 3 \\ 4 & 5 \\ 6 & 7 \end{pmatrix}$$

▶

Similarly for week 2:

January	Week 2	
	Item 1	Item 2
Alice	3	4
Ben	1	8
Charlie	0	2

The quantities are then stored in matrix form as:

$$\begin{pmatrix} \cdots & \cdots \\ \cdots & \cdots \\ \cdots & \cdots \end{pmatrix}$$

Next frame

3

$$\begin{pmatrix} 3 & 4 \\ 1 & 8 \\ 0 & 2 \end{pmatrix}$$

The table containing the sum of these two quantity records is:

January	Week 1		Week 2		Week 1 + Week 2	
	Item 1	Item 2	Item 1	Item 2	Item 1	Item 2
Alice	2	3	3	4	$\cdots$	$\cdots$
Ben	4	5	1	8	$\cdots$	$\cdots$
Charlie	6	7	0	2	$\cdots$	$\cdots$

Next frame

4

January	Week 1		Week 2		Week 1 + Week 2	
	Item 1	Item 2	Item 1	Item 2	Item 1	Item 2
Alice	2	3	3	4	5	7
Ben	4	5	1	8	5	13
Charlie	6	7	0	2	6	9

Because corresponding numbers are added to form the sum.

January	Week 1		Week 2		Week 1 + Week 2	
	Item 1	Item 2	Item 1	Item 2	Item 1	Item 2
Alice	2	3	3	4	$2+3=5$	$3+4=7$
Ben	4	5	1	8	$4+1=5$	$5+8=13$
Charlie	6	7	0	2	$6+0=6$	$7+2=9$

This can be represented in matrix form as:

$$\begin{pmatrix} 2 & 3 \\ 4 & 5 \\ 6 & 7 \end{pmatrix} + \begin{pmatrix} 3 & 4 \\ 1 & 8 \\ 0 & 2 \end{pmatrix} = \begin{pmatrix} \cdots & \cdots \\ \cdots & \cdots \\ \cdots & \cdots \end{pmatrix}$$

There is no trick here, just write down what you would expect the answer to be.
The answer is in the next frame

5

$$\begin{pmatrix} 2 & 3 \\ 4 & 5 \\ 6 & 7 \end{pmatrix} + \begin{pmatrix} 3 & 4 \\ 1 & 8 \\ 0 & 2 \end{pmatrix} = \begin{pmatrix} 5 & 7 \\ 5 & 13 \\ 6 & 9 \end{pmatrix}$$

Because

The numbers in the Week 1 + Week 2 column are:

$$\begin{pmatrix} 5 & 7 \\ 5 & 13 \\ 6 & 9 \end{pmatrix}$$

Move to the next frame

Matrix addition

Two matrices are added by adding their corresponding elements. For example, as we have just seen:

$$\begin{pmatrix} 2 & 3 \\ 4 & 5 \\ 6 & 7 \end{pmatrix} + \begin{pmatrix} 3 & 4 \\ 1 & 8 \\ 0 & 2 \end{pmatrix} = \begin{pmatrix} 2+3 & 3+4 \\ 4+1 & 5+8 \\ 6+0 & 7+2 \end{pmatrix} = \begin{pmatrix} 5 & 7 \\ 5 & 13 \\ 6 & 9 \end{pmatrix}$$

Therefore:

$$\begin{pmatrix} 3 & -4 & 0 \\ 2 & 5 & -7 \end{pmatrix} + \begin{pmatrix} -1 & -7 & 3 \\ 4 & 5 & 2 \end{pmatrix} = \begin{pmatrix} \ldots & \ldots & \ldots \\ \ldots & \ldots & \ldots \end{pmatrix}$$

The answer is in the next frame

$$\begin{pmatrix} 2 & -11 & 3 \\ 6 & 10 & -5 \end{pmatrix}$$

Because

$$\begin{pmatrix} 3 & -4 & 0 \\ 2 & 5 & -7 \end{pmatrix} + \begin{pmatrix} -1 & -7 & 3 \\ 4 & 5 & 2 \end{pmatrix} = \begin{pmatrix} 3+(-1) & -4+(-7) & 0+3 \\ 2+4 & 5+5 & -7+2 \end{pmatrix}$$
$$= \begin{pmatrix} 2 & -11 & 3 \\ 6 & 10 & -5 \end{pmatrix}$$

Notice that the addition of two matrices requires both matrices to have the same number of rows and the same number of columns. If the two matrices have different numbers of rows or columns then the two matrices *cannot be added together*.

Move to the next frame

Matrix subtraction

As you would suspect from the definition of matrix addition, two matrices are subtracted by subtracting their corresponding elements. For example:

$$\begin{pmatrix} 2 & 3 \\ 4 & 5 \\ 6 & 7 \end{pmatrix} - \begin{pmatrix} 3 & 4 \\ 1 & 8 \\ 0 & 2 \end{pmatrix} = \begin{pmatrix} 2-3 & 3-4 \\ 4-1 & 5-8 \\ 6-0 & 7-2 \end{pmatrix} = \begin{pmatrix} -1 & -1 \\ 3 & -3 \\ 6 & 5 \end{pmatrix}$$

Therefore:

$$\begin{pmatrix} 3 & -4 & 0 \\ 2 & 5 & -7 \end{pmatrix} - \begin{pmatrix} -1 & -7 & 3 \\ 4 & 5 & 2 \end{pmatrix} = \begin{pmatrix} \ldots & \ldots & \ldots \\ \ldots & \ldots & \ldots \end{pmatrix}$$

The answer is in the next frame

9

$$\begin{pmatrix} 4 & 3 & -3 \\ -2 & 0 & -9 \end{pmatrix}$$

Because

$$\begin{pmatrix} 3 & -4 & 0 \\ 2 & 5 & -7 \end{pmatrix} - \begin{pmatrix} -1 & -7 & 3 \\ 4 & 5 & 2 \end{pmatrix} = \begin{pmatrix} 3-(-1) & -4-(-7) & 0-3 \\ 2-4 & 5-5 & -7-2 \end{pmatrix}$$

$$= \begin{pmatrix} 4 & 3 & -3 \\ -2 & 0 & -9 \end{pmatrix}$$

Again, notice that the subtraction of two matrices requires both matrices to have the same number of rows and the same number of columns. If the two matrices have different numbers of rows or columns then the two matrices *cannot be subtracted.*

Move to the next frame

10   **Multiplication by a scalar**

Multiplying a matrix by 2 is the same as adding the matrix to itself. For example:

$$2\begin{pmatrix} 1 & 2 \\ 3 & 4 \end{pmatrix} = \begin{pmatrix} 1 & 2 \\ 3 & 4 \end{pmatrix} + \begin{pmatrix} 1 & 2 \\ 3 & 4 \end{pmatrix} = \begin{pmatrix} 2 & 4 \\ 6 & 8 \end{pmatrix} = \begin{pmatrix} 2\times1 & 2\times2 \\ 2\times3 & 2\times4 \end{pmatrix}$$

From this example we can see that multiplying a matrix by any scalar gives the same result as multiplying each element of the matrix by that scalar. Consequently:

$$3\begin{pmatrix} 1 & 0 & -2 & 5 \\ 4 & -6 & 2 & 0 \end{pmatrix} = \begin{pmatrix} \dots & \dots & \dots & \dots \\ \dots & \dots & \dots & \dots \end{pmatrix}$$

Next frame

11

$$\begin{pmatrix} 3 & 0 & -6 & 15 \\ 12 & -18 & 6 & 0 \end{pmatrix}$$

Because

$$3\begin{pmatrix} 1 & 0 & -2 & 5 \\ 4 & -6 & 2 & 0 \end{pmatrix} = \begin{pmatrix} 3\times1 & 3\times0 & 3\times(-2) & 3\times5 \\ 3\times4 & 3\times(-6) & 3\times2 & 3\times0 \end{pmatrix}$$

$$= \begin{pmatrix} 3 & 0 & -6 & 15 \\ 12 & -18 & 6 & 0 \end{pmatrix}$$

Let's move on

Multiplication of matrices

12

Two matrices **A** and **B** can be multiplied together to form the product **AB** *provided matrix* **A** *has as many columns as* **B** *has rows*. For example to multiply a 3×2 matrix by a 2×1 matrix we proceed as follows:

Each number in a given row in **A** is multiplied by the corresponding number in the column of **B**. The products are then added to give the appropriate number in the product matrix. So that, for example, if

$$\mathbf{A} = \begin{pmatrix} 2 & 3 \\ 4 & 5 \\ 6 & 7 \end{pmatrix} \text{ and } \mathbf{B} = \begin{pmatrix} 1 \\ 8 \end{pmatrix} \text{ then the product:}$$

$$\mathbf{AB} = \begin{pmatrix} 2 & 3 \\ 4 & 5 \\ 6 & 7 \end{pmatrix} \times \begin{pmatrix} 1 \\ 8 \end{pmatrix} = \begin{pmatrix} 2 \times 1 + 3 \times 8 \\ 4 \times 1 + 5 \times 8 \\ 6 \times 1 + 7 \times 8 \end{pmatrix}$$

$$= \begin{pmatrix} 26 \\ 44 \\ 62 \end{pmatrix}$$

Try one yourself. Let $\mathbf{A} = \begin{pmatrix} 3 & 6 \\ 1 & 4 \\ 7 & 8 \end{pmatrix}$ and $\mathbf{B} = \begin{pmatrix} 5 \\ 4 \end{pmatrix}$ then the product:

$$\mathbf{AB} = \begin{pmatrix} 3 & 6 \\ 1 & 4 \\ 7 & 8 \end{pmatrix} \times \begin{pmatrix} 5 \\ 4 \end{pmatrix} = \begin{pmatrix} \cdots \\ \cdots \\ \cdots \end{pmatrix}$$

The answer is in the next frame

13

$$\begin{pmatrix} 39 \\ 21 \\ 67 \end{pmatrix}$$

Because

$$\mathbf{AB} = \begin{pmatrix} 3 & 6 \\ 1 & 4 \\ 7 & 8 \end{pmatrix} \times \begin{pmatrix} 5 \\ 4 \end{pmatrix} = \begin{pmatrix} 3 \times 5 + 6 \times 4 \\ 1 \times 5 + 4 \times 4 \\ 7 \times 5 + 8 \times 4 \end{pmatrix}$$

$$= \begin{pmatrix} 39 \\ 21 \\ 67 \end{pmatrix}$$

Notice that:

$$\mathbf{BA} = \begin{pmatrix} 5 \\ 4 \end{pmatrix} \times \begin{pmatrix} 3 & 6 \\ 1 & 4 \\ 7 & 8 \end{pmatrix}$$

cannot be worked out because the number of columns in **B** does not match the number of rows in **A**. In this case we say that **BA** is not defined.

Next frame

14

This procedure can be extended as follows:

$$\begin{pmatrix} 2 & 3 \\ 4 & 5 \\ 6 & 7 \end{pmatrix} \times \begin{pmatrix} 1 & -4 \\ 8 & 9 \end{pmatrix} = \begin{pmatrix} 2 \times 1 + 3 \times 8 & 2 \times (-4) + 3 \times 9 \\ 4 \times 1 + 5 \times 8 & 4 \times (-4) + 5 \times 9 \\ 6 \times 1 + 7 \times 8 & 6 \times (-4) + 7 \times 9 \end{pmatrix}$$

$$= \begin{pmatrix} 26 & 19 \\ 44 & 29 \\ 62 & 39 \end{pmatrix}$$

And another:

$$\begin{pmatrix} 3 & 6 \\ 1 & 4 \\ 7 & 8 \end{pmatrix} \times \begin{pmatrix} 5 & 0 \\ 4 & -2 \end{pmatrix} = \begin{pmatrix} \cdots & \cdots \\ \cdots & \cdots \\ \cdots & \cdots \end{pmatrix}$$

The answer is in the next frame

15

$$\begin{pmatrix} 39 & -12 \\ 21 & -8 \\ 67 & -16 \end{pmatrix}$$

Because

$$\begin{pmatrix} 3 & 6 \\ 1 & 4 \\ 7 & 8 \end{pmatrix} \times \begin{pmatrix} 5 & 0 \\ 4 & -2 \end{pmatrix} = \begin{pmatrix} 3 \times 5 + 6 \times 4 & 3 \times 0 + 6 \times (-2) \\ 1 \times 5 + 4 \times 4 & 1 \times 0 + 4 \times (-2) \\ 7 \times 5 + 8 \times 4 & 7 \times 0 + 8 \times (-2) \end{pmatrix}$$

$$= \begin{pmatrix} 39 & -12 \\ 21 & -8 \\ 67 & -16 \end{pmatrix}$$

Notice that a 3×2 matrix multiplied by a 2×1 matrix produces a 3×1 matrix and a 3×2 matrix multiplied by a 2×2 matrix produces a 3×2 matrix. In general if an $n \times m$ matrix is multiplied by an $m \times p$ matrix it will produce an $n \times p$ matrix:

$$(n \times m) \times (m \times p) \rightarrow (n \times p)$$

noting that the number of columns (m) in the left hand matrix equals the number of rows in the right-hand matrix.

Just to make sure that you are happy with this procedure consider the following product:

$$\begin{pmatrix} 2 & -1 & 6 \\ 4 & 7 & 0 \end{pmatrix} \times \begin{pmatrix} 6 & 3 \\ -2 & 5 \\ 1 & -4 \end{pmatrix}$$

This product will produce an $r \times s$ matrix where $r = \dots\dots\dots$ and $s = \dots\dots\dots$

Next frame for the answer

$$\boxed{r = 2, \ s = 2}$$

16

Because

An $n \times m$ matrix multiplied by an $m \times p$ matrix will produce an $n \times p$ matrix so that a 2×3 matrix multiplied by a 3×2 matrix will produce a 2×2 matrix.

Furthermore:

$$\begin{pmatrix} 2 & -1 & 6 \\ 4 & 7 & 0 \end{pmatrix} \times \begin{pmatrix} 6 & 3 \\ -2 & 5 \\ 1 & -4 \end{pmatrix} = \begin{pmatrix} \dots & \dots \\ \dots & \dots \end{pmatrix}$$

Next frame

$$\boxed{\begin{pmatrix} 20 & -23 \\ 10 & 47 \end{pmatrix}}$$

17

Because

$$\begin{pmatrix} 2 & -1 & 6 \\ 4 & 7 & 0 \end{pmatrix} \times \begin{pmatrix} 6 & 3 \\ -2 & 5 \\ 1 & -4 \end{pmatrix}$$

$$= \begin{pmatrix} 2 \times 6 + (-1) \times (-2) + 6 \times 1 & 2 \times 3 + (-1) \times 5 + 6 \times (-4) \\ 4 \times 6 + 7 \times (-2) + 0 \times 1 & 4 \times 3 + 7 \times 5 + 0 \times (-4) \end{pmatrix}$$

$$= \begin{pmatrix} 20 & -23 \\ 10 & 47 \end{pmatrix}$$

Let us now pause and summarize the main facts so far on matrices

Review summary

18

1 *Matrix*: A matrix is a rectangular array of numbers arranged in regular rows and columns. A matrix with *i* rows and *j* columns is referred to as an $i \times j$ ('*i* by *j*') matrix.

2 *Matrix addition*: Two matrices are added by adding their corresponding elements. This requires both matrices to have the same number of rows and the same number of columns.

3 *Matrix subtraction*: Two matrices are subtracted by subtracting their corresponding elements. This requires both matrices to have the same number of rows and the same number of columns.

4 *Multiplication by a scalar*: A matrix can be multiplied by a scalar by multiplying each number in the matrix by the scalar.

5 *Matrix multiplication*: Two matrices **A** and **B** can be multiplied together to form the product **AB** *provided matrix* **A** *has as many columns as matrix* **B** *has rows*.

Each number in a given row in **A** is multiplied by the corresponding number in the column of **B**. The products are then added to give the appropriate number in the product matrix. In general if an $n \times m$ matrix is multiplied by an $m \times p$ matrix it will produce an $n \times p$ matrix:

$$(n \times m) \times (m \times p) \to (n \times p)$$

noting that the number of columns (*m*) in the left hand matrix equals the number of rows in the right-hand matrix.

 # Review exercise

19

1 Given matrices

$$\mathbf{P} = \begin{pmatrix} 12 & -9 \\ -7 & 11 \end{pmatrix}, \mathbf{Q} = \begin{pmatrix} -3 & 10 \\ 8 & 15 \end{pmatrix}, \mathbf{R} = (-6 \quad 2) \text{ and } \mathbf{S} = \begin{pmatrix} -4 \\ 7 \end{pmatrix}$$

calculate each of the following:

(a) **P + Q**   (b) **Q − P**   (c) **− 3P**   (d) **PQ**

(e) **QP**   (f) **PR**   (g) **RS**

Complete the question. Take your time, there is no need to rush.
If necessary, look back at the Unit.
The answers and working are in the next frame.

1 (a) $\mathbf{P} + \mathbf{Q} = \begin{pmatrix} 12 & -9 \\ -7 & 11 \end{pmatrix} + \begin{pmatrix} -3 & 10 \\ 8 & 15 \end{pmatrix} = \begin{pmatrix} 12 + (-3) & -9 + 10 \\ -7 + 8 & 11 + 15 \end{pmatrix}$

$$= \begin{pmatrix} 9 & 1 \\ 1 & 26 \end{pmatrix}$$

20

(b) $\mathbf{Q} - \mathbf{P} = \begin{pmatrix} -3 & 10 \\ 8 & 15 \end{pmatrix} - \begin{pmatrix} 12 & -9 \\ -7 & 11 \end{pmatrix} = \begin{pmatrix} (-3) - 12 & 10 - (-9) \\ 8 - (-7) & 15 - 11 \end{pmatrix}$

$$= \begin{pmatrix} -15 & 19 \\ 15 & 4 \end{pmatrix}$$

(c) $-3\mathbf{P} = -3 \begin{pmatrix} 12 & -9 \\ -7 & 11 \end{pmatrix} = \begin{pmatrix} (-3) \times 12 & (-3) \times (-9) \\ (-3) \times (-7) & (-3) \times 11 \end{pmatrix} = \begin{pmatrix} -36 & 27 \\ 21 & -33 \end{pmatrix}$

(d) $\mathbf{PQ} = \begin{pmatrix} 12 & -9 \\ -7 & 11 \end{pmatrix} \begin{pmatrix} -3 & 10 \\ 8 & 15 \end{pmatrix}$

$$= \begin{pmatrix} 12 \times (-3) + (-9) \times 8 & 12 \times 10 + (-9) \times 15 \\ (-7) \times (-3) + 11 \times 8 & (-7) \times 10 + 11 \times 15 \end{pmatrix} = \begin{pmatrix} -108 & -15 \\ 109 & 95 \end{pmatrix}$$

(e) $\mathbf{QP} = \begin{pmatrix} -3 & 10 \\ 8 & 15 \end{pmatrix} \begin{pmatrix} 12 & -9 \\ -7 & 11 \end{pmatrix}$

$$= \begin{pmatrix} (-3) \times 12 + 10 \times (-7) & (-3) \times (-9) + 10 \times 11 \\ 8 \times 12 + 15 \times (-7) & 8 \times (-9) + 15 \times 11 \end{pmatrix} = \begin{pmatrix} -106 & 137 \\ -9 & 93 \end{pmatrix}$$

(f) $\mathbf{PR} = \begin{pmatrix} 12 & -9 \\ -7 & 11 \end{pmatrix} (-6 \quad 2)$: it is not possible to multiply a 2×2 by a 1×2 matrix

(g) $\mathbf{RS} = (-6 \quad 2) \begin{pmatrix} -4 \\ 7 \end{pmatrix} = ((-6) \times (-4) + 2 \times 7) = 38$

Now for the Review test

Review test

Unit 30

1 Given matrices

21

$$\mathbf{A} = \begin{pmatrix} -7 & 23 \\ 0 & -5 \end{pmatrix}, \mathbf{B} = \begin{pmatrix} -13 & 1 \\ 9 & -2 \end{pmatrix}, \mathbf{C} = (-6 \quad -11) \text{ and } \mathbf{D} = \begin{pmatrix} 21 \\ -17 \end{pmatrix}$$

calculate each of the following:

(a) $\mathbf{A} + \mathbf{B}$ (b) $\mathbf{B} - \mathbf{A}$ (c) $2\mathbf{A}$ (d) $\mathbf{AB}$

(e) $\mathbf{BA}$ (f) $\mathbf{AD}$ (g) $\mathbf{CB}$

Inverse matrices   **Unit 31**

1 Unit matrices

Any matrix with the same number of rows as columns is called a **square matrix** and any square matrix whose only non-zero elements are unity (1) down the main left to right diagonal (also called the leading diagonal) is called a **unit matrix**. So:

$$\begin{pmatrix} 1 & 0 \\ 0 & 1 \end{pmatrix} \text{ is the } 2 \times 2 \text{ unit matrix, } \begin{pmatrix} 1 & 0 & 0 \\ 0 & 1 & 0 \\ 0 & 0 & 1 \end{pmatrix} \text{ is the } 3 \times 3 \text{ unit matrix}$$

$$\text{and } \begin{pmatrix} 1 & 0 & 0 & 0 \\ 0 & 1 & 0 & 0 \\ 0 & 0 & 1 & 0 \\ 0 & 0 & 0 & 1 \end{pmatrix} \text{ is the } 4 \times 4 \text{ unit matrix.}$$

They are called *unit matrices* because when they multiply another matrix they leave it unchanged. For example:

$$\text{given } \mathbf{A} = \begin{pmatrix} -7 & 3 \\ 4 & 6 \end{pmatrix} \text{ and } \mathbf{I} = \begin{pmatrix} 1 & 0 \\ 0 & 1 \end{pmatrix} \text{ then } \mathbf{AI} = \dots\dots\dots$$

Next frame

2

$$\boxed{\mathbf{A}}$$

Because

$$\mathbf{AI} = \begin{pmatrix} -7 & 3 \\ 4 & 6 \end{pmatrix}\begin{pmatrix} 1 & 0 \\ 0 & 1 \end{pmatrix}$$

$$= \begin{pmatrix} (-7) \times 1 + 3 \times 0 & (-7) \times 0 + 3 \times 1 \\ 4 \times 1 + 6 \times 0 & 4 \times 0 + 6 \times 1 \end{pmatrix} = \begin{pmatrix} -7 & 3 \\ 4 & 6 \end{pmatrix}$$

$$= \mathbf{A}$$

Try another:

$$\text{given } \mathbf{A} = \begin{pmatrix} 1 & 0 & 2 \\ 6 & -3 & 2 \\ 4 & 5 & 0 \end{pmatrix} \text{ and } \mathbf{I} = \begin{pmatrix} 1 & 0 & 0 \\ 0 & 1 & 0 \\ 0 & 0 & 1 \end{pmatrix} \text{ then } \mathbf{IA} = \dots\dots\dots$$

Next frame

A

3

Because

$$\mathbf{IA} = \begin{pmatrix} 1 & 0 & 0 \\ 0 & 1 & 0 \\ 0 & 0 & 1 \end{pmatrix} \begin{pmatrix} 1 & 0 & 2 \\ 6 & -3 & 2 \\ 4 & 5 & 0 \end{pmatrix}$$

$$= \begin{pmatrix} 1\times1+0\times6+0\times4 & 1\times0+0\times(-3)+0\times5 & 1\times2+0\times2+0\times0 \\ 0\times1+1\times6+0\times4 & 0\times0+1\times(-3)+0\times5 & 0\times2+1\times2+0\times0 \\ 0\times1+0\times6+1\times4 & 0\times0+0\times(-3)+1\times5 & 0\times2+0\times2+1\times0 \end{pmatrix}$$

$$= \begin{pmatrix} 1 & 0 & 2 \\ 6 & -3 & 2 \\ 4 & 5 & 0 \end{pmatrix}$$

$$= \mathbf{A}$$

Move on to the next frame

So far we have added, subtracted and multiplied matrices by a scalar. We have even multiplied matrices together. So what of division? We shall answer that question in Frame 7 where we discuss inverse matrices. However, before we can talk of inverse matrices we need to know about determinants.

4

The determinant of a matrix

Every square matrix **A** has a number associated with it called its **determinant**, denoted by:

det **A** or, more simply, $|\mathbf{A}|$

The determinant of the 2×2 matrix $\mathbf{A} = \begin{pmatrix} 1 & 2 \\ 3 & 4 \end{pmatrix}$ is given as:

$$\det \mathbf{A} = \begin{vmatrix} 1 & 2 \\ 3 & 4 \end{vmatrix} = 1 \times 4 - 2 \times 3 = -2$$

This rather mysterious operation is performed by the *multiplication of diagonally opposite numbers* and then the *subtraction of the products*:

$$\begin{matrix} 1 & & 2 \\ & \times & \\ 3 & & 4 \end{matrix} \qquad 1 \times 4 - 2 \times 3 = -2$$

So:

$$\begin{vmatrix} 6 & -3 \\ 5 & 2 \end{vmatrix} = \ldots\ldots\ldots\ldots$$

Next frame

5

27

Because

$$\begin{vmatrix} 6 & -3 \\ 5 & 2 \end{vmatrix} = 6 \times 2 - (-3) \times 5$$

$$= 27$$

Indeed, the general 2×2 matrix $\begin{pmatrix} a & b \\ c & d \end{pmatrix}$ has the determinant:

$$\begin{vmatrix} a & b \\ c & d \end{vmatrix} = \ldots\ldots\ldots\ldots$$

Next frame

6

$ad - bc$

Move on to the next frame

7 ## The inverse of a matrix

Division of matrices cannot be defined but an operation similar in effect to division is defined. Namely, that of the multiplication of a matrix by its inverse. If **A** and **B** are two square matrices such that their product is a unit matrix:

$$\mathbf{AB} = \mathbf{I}$$

where **I** is a unit matrix. Then matrix **B** is called the inverse of matrix **A** and is written as:

$$\mathbf{B} = \mathbf{A}^{-1} \text{ so that } \mathbf{AA}^{-1} = \mathbf{I}$$

It could also be claimed that **A** was the inverse of **B**, written as:

$$\mathbf{A} = \mathbf{B}^{-1} \text{ so that } \mathbf{B}^{-1}\mathbf{B} = \mathbf{I}$$

More correctly **B** is the right-hand inverse of **A** and **A** is the left-hand inverse of **B**. We shall only be concerned with the inverses of 2×2 matrices where the left-hand and right-hand inverses are equal so that:

$$\mathbf{AA}^{-1} = \mathbf{A}^{-1}\mathbf{A} = \mathbf{I}$$

The inverse of a 2 x 2 matrix

The inverse of the 2 × 2 matrix $\mathbf{A} = \begin{pmatrix} a & b \\ c & d \end{pmatrix}$ is given as:

$$\mathbf{A}^{-1} = \frac{1}{\det \mathbf{A}} \begin{pmatrix} d & -b \\ -c & a \end{pmatrix}$$

The two numbers on the leading diagonal of **A** are interchanged, the other two numbers are multiplied by −1 and the whole matrix divided by the determinant of **A**. Notice that if $\det \mathbf{A} = 0$ then *the inverse matrix does not exist.*

Let's try one. The inverse of the matrix $\mathbf{A} = \begin{pmatrix} 4 & 6 \\ 5 & 7 \end{pmatrix}$ is $\mathbf{A}^{-1} = \ldots\ldots\ldots\ldots$

Just follow the procedure stated here.
The answer is in the next frame

8

$$\begin{pmatrix} -3.5 & 3 \\ 2.5 & -2 \end{pmatrix}$$

Because

Given $\mathbf{A} = \begin{pmatrix} 4 & 6 \\ 5 & 7 \end{pmatrix}$ then $\det \mathbf{A} = 4 \times 7 - 6 \times 5 = -2$ and so:

$$\mathbf{A}^{-1} = \frac{1}{\det \mathbf{A}} \begin{pmatrix} 7 & -6 \\ -5 & 4 \end{pmatrix}$$
$$= \frac{1}{-2} \begin{pmatrix} 7 & -6 \\ -5 & 4 \end{pmatrix}$$
$$= \begin{pmatrix} -3.5 & 3 \\ 2.5 & -2 \end{pmatrix}$$

As a check to show that this is indeed the inverse matrix the product:

$$\mathbf{A}^{-1}\mathbf{A} = \ldots\ldots\ldots\ldots$$

Next frame

9

$$\boxed{\text{I}}$$

Because

$$
\begin{aligned}
\mathbf{A}^{-1}\mathbf{A} &= \begin{pmatrix} -3\cdot5 & 3 \\ 2\cdot5 & -2 \end{pmatrix}\begin{pmatrix} 4 & 6 \\ 5 & 7 \end{pmatrix} \\
&= \begin{pmatrix} (-3\cdot5)\times 4 + 3\times 5 & (-3\cdot5)\times 6 + 3\times 7 \\ 2\cdot5\times 4 + (-2)\times 5 & 2\cdot5\times 6 + (-2)\times 7 \end{pmatrix} \\
&= \begin{pmatrix} 1 & 0 \\ 0 & 1 \end{pmatrix} \\
&= \mathbf{I}
\end{aligned}
$$

Now try another. The inverse of the matrix $\mathbf{A} = \begin{pmatrix} 2 & 3 \\ 5 & 1 \end{pmatrix}$ is $\mathbf{A}^{-1} = \dots\dots\dots$

Next frame

10

$$\boxed{\begin{pmatrix} -1/13 & 3/13 \\ 5/13 & -2/13 \end{pmatrix}}$$

Because

Given $\mathbf{A} = \begin{pmatrix} 2 & 3 \\ 5 & 1 \end{pmatrix}$ then $\det \mathbf{A} = 2\times 1 - 3\times 5 = -13$ and so

$$
\begin{aligned}
\mathbf{A}^{-1} &= \frac{1}{\det \mathbf{A}}\begin{pmatrix} 1 & -3 \\ -5 & 2 \end{pmatrix} \\
&= \frac{1}{-13}\begin{pmatrix} 1 & -3 \\ -5 & 2 \end{pmatrix} \\
&= \begin{pmatrix} -1/13 & 3/13 \\ 5/13 & -2/13 \end{pmatrix}
\end{aligned}
$$

Again, as a check to show that this is indeed the inverse matrix the product:

$$\mathbf{A}^{-1}\mathbf{A} = \dots\dots\dots$$

Next frame

$$\boxed{\mathbf{I}}$$

Because

$$\mathbf{A}^{-1}\mathbf{A} = \begin{pmatrix} -1/13 & 3/13 \\ 5/13 & -2/13 \end{pmatrix} \begin{pmatrix} 2 & 3 \\ 5 & 1 \end{pmatrix}$$

$$= \begin{pmatrix} (-1/13) \times 2 + (3/13) \times 5 & (-1/13) \times 3 + (3/13) \times 1 \\ (5/13) \times 2 + (-2/13) \times 5 & (5/13) \times 3 + (-2/13) \times 1 \end{pmatrix}$$

$$= \begin{pmatrix} 1 & 0 \\ 0 & 1 \end{pmatrix}$$

$$= \mathbf{I}$$

Make a note of this particular inverse matrix because we shall use it in the next Unit.

For now, let us pause and summarize the main facts so far on inverse matrices

Review summary

Unit 31

1   *Square matrix*: A matrix with the same number of rows as columns is called a square matrix.

2   *Unit matrix*: A square matrix whose only non-zero elements are unity down the leading diagonal is called a unit matrix. So that:

$$\begin{pmatrix} 1 & 0 \\ 0 & 1 \end{pmatrix} \text{ is the } 2 \times 2 \text{ unit matrix, } \begin{pmatrix} 1 & 0 & 0 \\ 0 & 1 & 0 \\ 0 & 0 & 1 \end{pmatrix} \text{ is the } 3 \times 3 \text{ unit matrix}$$

3   *The determinant of a matrix*: Every square matrix **A** has a number associated with it called its determinant, denoted by:

    det **A** or, more simply, |**A**|

The determinant of the 2×2 matrix:

$$\mathbf{A} = \begin{pmatrix} a & b \\ c & d \end{pmatrix} \text{ is given as } \det \mathbf{A} = \begin{vmatrix} a & b \\ c & d \end{vmatrix} = ad - bc$$

This operation is performed by the *multiplication of diagonally opposite numbers* and then the *subtraction of the products*:

$$\begin{matrix} a & & b \\ & \times & \\ c & & d \end{matrix} \qquad ad - bc$$

4 *The inverse of a matrix*: If **A** and **B** are two square matrices such that their product is a unit matrix:

$$\mathbf{AB} = \mathbf{I}$$

where **I** is a unit matrix, then matrix **B** is called the inverse of matrix **A** and is written as

$$\mathbf{B} = \mathbf{A}^{-1} \text{ so that } \mathbf{AA}^{-1} = \mathbf{I}$$

5 *The inverse of a 2 × 2 matrix*: The inverse of the 2 × 2 matrix $\mathbf{A} = \begin{pmatrix} a & b \\ c & d \end{pmatrix}$ is given as:

$$\mathbf{A}^{-1} = \frac{1}{\det \mathbf{A}} \begin{pmatrix} d & -b \\ -c & a \end{pmatrix}$$

The two numbers on the leading diagonal of **A** are interchanged, the other two numbers are multiplied by −1 and the whole matrix divided by the determinant of **A**. Notice that if $\det \mathbf{A} = 0$ then *the inverse matrix does not exist*.

 # Review exercise   **Unit 31**

13

1 Find the determinant of each of the following matrices:

(a) $\mathbf{A} = \begin{pmatrix} 2 & 5 \\ 3 & 7 \end{pmatrix}$ (b) $\mathbf{B} = \begin{pmatrix} 6 & -4 \\ -1 & -3 \end{pmatrix}$ (c) $\mathbf{C} = \begin{pmatrix} -4 & 8 \\ -1 & 2 \end{pmatrix}$ (d) $\mathbf{D} = \begin{pmatrix} 3 & -5 \end{pmatrix}$

2 Find the inverse of each of the following matrices:

(a) $\mathbf{P} = \begin{pmatrix} 3 & 6 \\ 2 & 8 \end{pmatrix}$ (b) $\mathbf{Q} = \begin{pmatrix} -7 & -5 \\ -3 & 2 \end{pmatrix}$ (c) $\mathbf{C} = \begin{pmatrix} 9 & -3 \\ -6 & 2 \end{pmatrix}$

Complete both questions. Again, take your time, there is no need to rush. If necessary, look back at the Unit. The answers and working are in the next frame.

14

1 (a) $|\mathbf{A}| = \begin{vmatrix} 2 & 5 \\ 3 & 7 \end{vmatrix} = 2 \times 7 - 5 \times 3 = -1$

(b) $|\mathbf{B}| = \begin{vmatrix} 6 & -4 \\ -1 & -3 \end{vmatrix} = 6 \times (-3) - (-4) \times (-1) = -22$

(c) $|\mathbf{C}| = \begin{vmatrix} -4 & 8 \\ -1 & 2 \end{vmatrix} = (-4) \times 2 - 8 \times (-1) = 0$

(d) $|\mathbf{D}|$ does not exist because **D** is not a square matrix.

2 (a) $|\mathbf{P}| = \begin{vmatrix} 3 & 6 \\ 2 & 8 \end{vmatrix} = 3 \times 8 - 6 \times 2 = 12$ and so:

$$\mathbf{P}^{-1} = \begin{pmatrix} 3 & 6 \\ 2 & 8 \end{pmatrix}^{-1}$$

$$= \frac{1}{|\mathbf{P}|} \begin{pmatrix} 8 & -6 \\ -2 & 3 \end{pmatrix}$$

$$= \frac{1}{12} \begin{pmatrix} 8 & -6 \\ -2 & 3 \end{pmatrix} = \begin{pmatrix} 8/12 & -6/12 \\ -2/12 & 3/12 \end{pmatrix} = \begin{pmatrix} 2/3 & -1/2 \\ -1/6 & 1/4 \end{pmatrix}$$

(b) $|\mathbf{Q}| = \begin{vmatrix} -7 & -5 \\ -3 & 2 \end{vmatrix} = (-7) \times 2 - (-5) \times (-3) = -29$ and so:

$$\mathbf{Q}^{-1} = \begin{pmatrix} -7 & -5 \\ -3 & 2 \end{pmatrix}^{-1}$$

$$= \frac{1}{|\mathbf{Q}|} \begin{pmatrix} 2 & 5 \\ 3 & -7 \end{pmatrix}$$

$$= \frac{1}{-29} \begin{pmatrix} 2 & 5 \\ 3 & -7 \end{pmatrix} = \begin{pmatrix} -2/29 & -5/29 \\ -3/29 & 7/29 \end{pmatrix}$$

(c) $|\mathbf{C}| = \begin{pmatrix} 9 & -3 \\ -6 & -2 \end{pmatrix}$

$$\mathbf{C}^{-1} = \begin{pmatrix} 9 & -3 \\ -6 & 2 \end{pmatrix}^{-1}$$

$$= \frac{1}{|\mathbf{C}|} \begin{pmatrix} 2 & 3 \\ 6 & 9 \end{pmatrix}$$

$$= \frac{1}{(9 \times 2 - (-3) \times (-6))} \begin{pmatrix} 2 & 3 \\ 6 & 9 \end{pmatrix} = \frac{1}{0} \begin{pmatrix} 2 & 3 \\ 6 & 9 \end{pmatrix} \text{ which is not defined}$$

Now for the Review test

 # Review test

Unit 31

1 Find the determinant of each of the following matrices: **15**

(a) $\mathbf{A} = \begin{pmatrix} 4 & 6 \\ 3 & 8 \end{pmatrix}$ (b) $\mathbf{B} = \begin{pmatrix} 7 & -5 \\ 3 & -9 \end{pmatrix}$ (c) $\mathbf{C} = \begin{pmatrix} -12 & 3 \\ 4 & -1 \end{pmatrix}$ (d) $\mathbf{D} = \begin{pmatrix} 3 \\ -7 \end{pmatrix}$

2 Find the inverse of each of the following matrices:

(a) $\mathbf{P} = \begin{pmatrix} 1 & 2 \\ 2 & 5 \end{pmatrix}$ (b) $\mathbf{Q} = \begin{pmatrix} 3 & 14 \\ -2 & -9 \end{pmatrix}$ (c) $\mathbf{R} = \begin{pmatrix} 4 & -12 \\ 2 & -6 \end{pmatrix}$

Solving simultaneous linear equations

1   Matrix solution to simultaneous linear equations

A pair of simultaneous linear equations can be written as a matrix equation. Using an inverse matrix the solution to the equations can be straightforwardly found. For example the equations:

$$2x + 3y = 1$$
$$5x + y = -4$$

can be written in the form of a matrix equation as:

$$\begin{pmatrix} 2x + 3y \\ 5x + y \end{pmatrix} = \begin{pmatrix} 1 \\ -4 \end{pmatrix} \text{ that is } \begin{pmatrix} 2 & 3 \\ 5 & 1 \end{pmatrix}\begin{pmatrix} x \\ y \end{pmatrix} = \begin{pmatrix} 1 \\ -4 \end{pmatrix}$$

using the definition of the product of two matrices.

Just to ensure that you can do this let's try another. The matrix form of the two equations:

$$7x - 5y = -3$$
$$2x + 8y = 9$$

is . . . .

Just follow the procedure given.
The answer is in the next frame

2

$$\begin{pmatrix} 7 & -5 \\ 2 & 8 \end{pmatrix}\begin{pmatrix} x \\ y \end{pmatrix} = \begin{pmatrix} -3 \\ 9 \end{pmatrix}$$

Because

$$7x - 5y = -3$$
$$2x + 8y = 9$$

can be written in the form of a matrix equation as

$$\begin{pmatrix} 7x - 5y \\ 2x + 8y \end{pmatrix} = \begin{pmatrix} -3 \\ 9 \end{pmatrix}$$

that is

$$\begin{pmatrix} 7 & -5 \\ 2 & 8 \end{pmatrix}\begin{pmatrix} x \\ y \end{pmatrix} = \begin{pmatrix} -3 \\ 9 \end{pmatrix}$$

And another just to make sure. The matrix form of the two equations:

$$-a + 2b = 0$$
$$5a + 6b = 13$$

is . . . .

Again, just follow the procedure given.
The answer is in the next frame

3

$$\begin{pmatrix} -1 & 2 \\ 5 & 6 \end{pmatrix}\begin{pmatrix} a \\ b \end{pmatrix} = \begin{pmatrix} 0 \\ 13 \end{pmatrix}$$

Because

$\begin{matrix} -a + 2b = 0 \\ 5a + 6b = 13 \end{matrix}$ can be written in the form of a matrix equation as

$$\begin{pmatrix} -a + 2b \\ 5a + 6b \end{pmatrix} = \begin{pmatrix} 0 \\ 13 \end{pmatrix} \quad \text{that is} \quad \begin{pmatrix} -1 & 2 \\ 5 & 6 \end{pmatrix}\begin{pmatrix} a \\ b \end{pmatrix} = \begin{pmatrix} 0 \\ 13 \end{pmatrix}$$

Returning now to the original pair of equations in Frame 1, namely

$\begin{matrix} 2x + 3y = 1 \\ 5x + y = -4 \end{matrix}$ and their matrix form $\begin{pmatrix} 2 & 3 \\ 5 & 1 \end{pmatrix}\begin{pmatrix} x \\ y \end{pmatrix} = \begin{pmatrix} 1 \\ -4 \end{pmatrix}$

Multiplying both sides of this equation from the left by the inverse of the 2×2 matrix it is seen that (recalling the inverse from Frame 10 of Unit 31):

$$\begin{pmatrix} -1/13 & 3/13 \\ 5/13 & -2/13 \end{pmatrix}\begin{pmatrix} 2 & 3 \\ 5 & 1 \end{pmatrix}\begin{pmatrix} x \\ y \end{pmatrix} = \begin{pmatrix} -1/13 & 3/13 \\ 5/13 & -2/13 \end{pmatrix}\begin{pmatrix} 1 \\ -4 \end{pmatrix}$$

that is:

$$\begin{pmatrix} \cdots & \cdots \\ \cdots & \cdots \end{pmatrix}\begin{pmatrix} x \\ y \end{pmatrix} = \begin{pmatrix} \cdots \\ \cdots \end{pmatrix}$$

The answer is in the next frame

4

$$\begin{pmatrix} 1 & 0 \\ 0 & 1 \end{pmatrix}\begin{pmatrix} x \\ y \end{pmatrix} = \begin{pmatrix} -1 \\ 1 \end{pmatrix}$$

Because

On the left-hand side a matrix multiplied by its inverse is a unit matrix and so:

$$\begin{pmatrix} 1 & 0 \\ 0 & 1 \end{pmatrix}\begin{pmatrix} x \\ y \end{pmatrix} = \begin{pmatrix} (-1/13) \times 1 + (3/13) \times (-4) \\ (5/13) \times 1 + (-2/13) \times (-4) \end{pmatrix} = \begin{pmatrix} -1 \\ 1 \end{pmatrix}$$

with the result that $x = \ldots\ldots\ldots$ and $y = \ldots\ldots\ldots$

Next frame

5

$$x = -1 \text{ and } y = 1$$

Because

$\begin{pmatrix} 1 & 0 \\ 0 & 1 \end{pmatrix}\begin{pmatrix} x \\ y \end{pmatrix} = \begin{pmatrix} -1 \\ 1 \end{pmatrix}$ that is $\begin{pmatrix} x \\ y \end{pmatrix} = \begin{pmatrix} -1 \\ 1 \end{pmatrix}$ therefore $x = -1$ and $y = 1$

▷

The principle of this method of solving two simultaneous equations can be described in general terms as follows.

1 The pair of simultaneous equations in variables x and y are represented in matrix form as:

$$\mathbf{AX} = \mathbf{b}$$

where $\mathbf{A}$ is the matrix formed from the coefficients of the two equations, $\mathbf{X} = \begin{pmatrix} x \\ y \end{pmatrix}$ and $\mathbf{b}$ is the 2×1 matrix formed from the constants on the right-hand sides of the equations.

2 Find the inverse $\mathbf{A}^{-1}$ of the coefficient matrix $\mathbf{A}$.

3 Multiply from the left both sides of $\mathbf{AX} = \mathbf{b}$ by the inverse $\mathbf{A}^{-1}$ to give:

$$\mathbf{A}^{-1}\mathbf{AX} = \mathbf{A}^{-1}\mathbf{b}$$

Now, $\mathbf{A}^{-1}\mathbf{A} = \mathbf{I}$, the 2×2 unit matrix, so $\mathbf{A}^{-1}\mathbf{AX} = \mathbf{IX} = \mathbf{X}$. Hence

$$\mathbf{X} = \mathbf{A}^{-1}\mathbf{b}$$

and so the solution is found.

Let's try one from beginning to end. To find the solution to the simultaneous equations:

$$x - y = 8$$
$$2x + 5y = 9$$

1 First put the equations in matrix form as:

....

The answer is in the next frame

6

$$\begin{pmatrix} 1 & -1 \\ 2 & 5 \end{pmatrix}\begin{pmatrix} x \\ y \end{pmatrix} = \begin{pmatrix} 8 \\ 9 \end{pmatrix}$$

Because

$$\begin{matrix} x - y = 8 \\ 2x + 5y = 9 \end{matrix} \text{ in matrix form is } \begin{pmatrix} x - y \\ 2x + 5y \end{pmatrix} = \begin{pmatrix} 8 \\ 9 \end{pmatrix}$$

that is $\begin{pmatrix} 1 & -1 \\ 2 & 5 \end{pmatrix}\begin{pmatrix} x \\ y \end{pmatrix} = \begin{pmatrix} 8 \\ 9 \end{pmatrix}$

This is $\mathbf{AX} = \mathbf{b}$.

2 The inverse of the coefficient matrix $\mathbf{A} = \begin{pmatrix} 1 & -1 \\ 2 & 5 \end{pmatrix}$ is:

$$\mathbf{A}^{-1} = \ldots\ldots\ldots$$

The answer is in the next frame

$$\mathbf{A}^{-1} = \begin{pmatrix} 5/7 & 1/7 \\ -2/7 & 1/7 \end{pmatrix}$$

Because

The determinant of $\mathbf{A} = \begin{pmatrix} 1 & -1 \\ 2 & 5 \end{pmatrix}$ is $\det \mathbf{A} = \begin{vmatrix} 1 & -1 \\ 2 & 5 \end{vmatrix} = 5 - (-2) = 7$

and the inverse of $\mathbf{A} = \begin{pmatrix} 1 & -1 \\ 2 & 5 \end{pmatrix}$ is:

$$\mathbf{A}^{-1} = \frac{1}{\det \mathbf{A}} \begin{pmatrix} 5 & 1 \\ -2 & 1 \end{pmatrix}$$

$$= \frac{1}{7} \begin{pmatrix} 5 & 1 \\ -2 & 1 \end{pmatrix}$$

$$= \begin{pmatrix} 5/7 & 1/7 \\ -2/7 & 1/7 \end{pmatrix}$$

3 Multiplying from the left both sides of the matrix equation $\mathbf{AX} = \mathbf{b}$ by $\mathbf{A}^{-1}$ gives:

$$\mathbf{A}^{-1}\mathbf{AX} = \mathbf{A}^{-1}\mathbf{b}$$

which gives $x = \ldots\ldots\ldots\ldots$ and $y = \ldots\ldots\ldots\ldots$

The answer is in the next frame

$$x = 7 \text{ and } y = -1$$

Because

$$\begin{pmatrix} 5/7 & 1/7 \\ -2/7 & 1/7 \end{pmatrix} \begin{pmatrix} 1 & -1 \\ 2 & 5 \end{pmatrix} = \begin{pmatrix} 5/7 & 1/7 \\ -2/7 & 1/7 \end{pmatrix} \begin{pmatrix} 8 \\ 9 \end{pmatrix}$$

that is $\begin{pmatrix} 1 & 0 \\ 0 & 1 \end{pmatrix} \begin{pmatrix} x \\ y \end{pmatrix} = \begin{pmatrix} 7 \\ -1 \end{pmatrix}$ and so:

$\begin{pmatrix} x \\ y \end{pmatrix} = \begin{pmatrix} 7 \\ -1 \end{pmatrix}$ therefore $x = 7$ and $y = -1$

Just to make sure, try one yourself from beginning to end. The solution to the simultaneous equations:

$$-3x + y = -3$$
$$4x + 7y = 29$$

is:

$$x = \ldots\ldots\ldots\ldots \text{ and } y = \ldots\ldots\ldots\ldots$$

The answer is in the next frame

9

$$\boxed{x = 2 \text{ and } y = 3}$$

Because

1 The equations $\begin{matrix} -3x + y = -3 \\ 4x + 7y = 29 \end{matrix}$ in matrix form are:

$$\begin{pmatrix} -3x + y \\ 4x + 7y \end{pmatrix} = \begin{pmatrix} -3 \\ 29 \end{pmatrix}. \text{ That is } \begin{pmatrix} -3 & 1 \\ 4 & 7 \end{pmatrix}\begin{pmatrix} x \\ y \end{pmatrix} = \begin{pmatrix} -3 \\ 29 \end{pmatrix}$$

2 The determinant of the coefficient matrix $\mathbf{A} = \begin{pmatrix} -3 & 1 \\ 4 & 7 \end{pmatrix}$ is:

$$\det \mathbf{A} = \begin{vmatrix} -3 & 1 \\ 4 & 7 \end{vmatrix} = -21 - 4 = -25$$

and so the inverse of $\mathbf{A} = \begin{pmatrix} -3 & 1 \\ 4 & 7 \end{pmatrix}$ is:

$$\mathbf{A}^{-1} = \frac{1}{\det \mathbf{A}}\begin{pmatrix} 7 & -1 \\ -4 & -3 \end{pmatrix}$$
$$= -\frac{1}{25}\begin{pmatrix} 7 & -1 \\ -4 & -3 \end{pmatrix}$$

3 Then multiplying the matrix equation $\begin{pmatrix} -3 & 1 \\ 4 & 7 \end{pmatrix}\begin{pmatrix} x \\ y \end{pmatrix} = \begin{pmatrix} -3 \\ 29 \end{pmatrix}$ on the left by the inverse gives:

$$-\frac{1}{25}\begin{pmatrix} 7 & -1 \\ -4 & -3 \end{pmatrix}\begin{pmatrix} -3 & 1 \\ 4 & 7 \end{pmatrix}\begin{pmatrix} x \\ y \end{pmatrix} = -\frac{1}{25}\begin{pmatrix} 7 & -1 \\ -4 & -3 \end{pmatrix}\begin{pmatrix} -3 \\ 29 \end{pmatrix}$$

that is:

$$\begin{pmatrix} 1 & 0 \\ 0 & 1 \end{pmatrix}\begin{pmatrix} x \\ y \end{pmatrix} = \begin{pmatrix} x \\ y \end{pmatrix} = -\frac{1}{25}\begin{pmatrix} 7 & -1 \\ -4 & -3 \end{pmatrix}\begin{pmatrix} -3 \\ 29 \end{pmatrix} = -\frac{1}{25}\begin{pmatrix} -50 \\ -75 \end{pmatrix} = \begin{pmatrix} 2 \\ 3 \end{pmatrix}$$

and so:

$x = 2 \text{ and } y = 3$

Move on to the next frame

Determinants and Cramer's rule **10**

Simultaneous linear equations can also be solved using determinants in what is called Cramer's rule but before we come to that rule we need to know three properties of determinants:

1 If two rows or two columns of a determinant are identical then the determinant has a value zero. For example:

$$\begin{vmatrix} 1 & -3 \\ 1 & -3 \end{vmatrix} = 1 \times (-3) - (-3) \times 1 = 0 \text{ and } \begin{vmatrix} 2 & 2 \\ 5 & 5 \end{vmatrix} = 2 \times 5 - 2 \times 5 = 0$$

2 If the numbers in any row or column of a determinant contain a common factor then this can be factored out of the determinant. For example:

$$\begin{vmatrix} 6 & 1 \\ 4 & 5 \end{vmatrix} = 6 \times 5 - 1 \times 4 = 26$$

$$\text{and } \begin{vmatrix} 6 & 1 \\ 4 & 5 \end{vmatrix} = \begin{vmatrix} 2 \times 3 & 1 \\ 2 \times 2 & 5 \end{vmatrix} = 2 \begin{vmatrix} 3 & 1 \\ 2 & 5 \end{vmatrix} = 2(15 - 2) = 26$$

3 If the numbers in any row or column are written as sums of two numbers then the determinant is equal to the sum of the two determinants formed from the components of the sums. For example:

$$\begin{vmatrix} 6 & 1 \\ 4 & 5 \end{vmatrix} = \begin{vmatrix} 1+5 & 1 \\ 2+2 & 5 \end{vmatrix} = \begin{vmatrix} 1 & 1 \\ 2 & 5 \end{vmatrix} + \begin{vmatrix} 5 & 1 \\ 2 & 5 \end{vmatrix} = (5-2) + (25-2) = 26$$

All that has been said here has been exemplified with the determinant of a 2×2 matrix but it is equally true for the determinant of any square matrix. This brings us to Cramer's rule.

Move to the next frame

Cramer's rule **11**

To solve the following pair of simultaneous linear equations:

$$x + y = 3$$
$$2x + 4y = 10$$

we first write them in matrix form as:

. . . .

The answer is in the next frame

12

$$\begin{pmatrix} 1 & 1 \\ 2 & 4 \end{pmatrix}\begin{pmatrix} x \\ y \end{pmatrix} = \begin{pmatrix} 3 \\ 10 \end{pmatrix}$$

Now consider two determinants; the first being the square matrix of coefficients given here and the second being the same matrix except that the first column is replaced by the elements of the column matrix on the right of this equation. That is:

$$\begin{vmatrix} 1 & 1 \\ 2 & 4 \end{vmatrix} \text{ and } \begin{vmatrix} 3 & 1 \\ 10 & 4 \end{vmatrix}$$

Looking at the second of these two determinants:

$$\begin{vmatrix} 3 & 1 \\ 10 & 4 \end{vmatrix} = \begin{vmatrix} x+y & 1 \\ 2x+4y & 4 \end{vmatrix} \text{ because } \begin{array}{l} x+y=3 \\ 2x+4y=10 \end{array}$$

$$= \begin{vmatrix} \cdots & 1 \\ \cdots & 4 \end{vmatrix} + \begin{vmatrix} \cdots & 1 \\ \cdots & 4 \end{vmatrix} \text{ by Property 3 in Frame 10}$$

Next frame

13

$$\begin{vmatrix} x & 1 \\ 2x & 4 \end{vmatrix} + \begin{vmatrix} y & 1 \\ 4y & 4 \end{vmatrix}$$

Because

Property **3**. If the numbers in any row or column are written as sums of two numbers then the determinant is equal to the sum of the two determinants formed from the components of the sums. So that:

$$\begin{vmatrix} x+y & 1 \\ 2x+4y & 4 \end{vmatrix} = \begin{vmatrix} x & 1 \\ 2x & 4 \end{vmatrix} + \begin{vmatrix} y & 1 \\ 4y & 4 \end{vmatrix}$$

Therefore:

$$\begin{vmatrix} 3 & 1 \\ 10 & 4 \end{vmatrix} = \begin{vmatrix} x & 1 \\ 2x & 4 \end{vmatrix} + \begin{vmatrix} y & 1 \\ 4y & 4 \end{vmatrix}$$

and from a consideration of Property **2** in Frame 10:

$$\begin{vmatrix} x & 1 \\ 2x & 4 \end{vmatrix} + \begin{vmatrix} y & 1 \\ 4y & 4 \end{vmatrix} = \cdots \begin{vmatrix} \cdots & 1 \\ \cdots & 4 \end{vmatrix} + \cdots \begin{vmatrix} \cdots & 1 \\ \cdots & 4 \end{vmatrix}$$

Next frame

$$x\begin{vmatrix} 1 & 1 \\ 2 & 4 \end{vmatrix} + y\begin{vmatrix} 1 & 1 \\ 4 & 4 \end{vmatrix}$$

Because

Property **2**.   If the numbers in any row or column of a determinant contain a common factor then this can be factored out of the determinant. So that:

$$\begin{vmatrix} x & 1 \\ 2x & 4 \end{vmatrix} = x\begin{vmatrix} 1 & 1 \\ 2 & 4 \end{vmatrix} \text{ and } \begin{vmatrix} y & 1 \\ 4y & 4 \end{vmatrix} = y\begin{vmatrix} 1 & 1 \\ 4 & 4 \end{vmatrix}$$

Therefore:

$$\begin{vmatrix} 3 & 1 \\ 10 & 4 \end{vmatrix} = x\begin{vmatrix} 1 & 1 \\ 2 & 4 \end{vmatrix} + y\begin{vmatrix} 1 & 1 \\ 4 & 4 \end{vmatrix}$$

By considering Property **1** in Frame 10:

$$x\begin{vmatrix} 1 & 1 \\ 2 & 4 \end{vmatrix} + y\begin{vmatrix} 1 & 1 \\ 4 & 4 \end{vmatrix} = x\begin{vmatrix} 1 & 1 \\ 2 & 4 \end{vmatrix} + \dots\dots\dots\dots$$

Next frame

$$x\begin{vmatrix} 1 & 1 \\ 2 & 4 \end{vmatrix} + 0$$

Because

Property **1**.   If two rows or two columns of a determinant are identical then the determinant has a value zero. So that:

$$x\begin{vmatrix} 1 & 1 \\ 2 & 4 \end{vmatrix} + y\begin{vmatrix} 1 & 1 \\ 4 & 4 \end{vmatrix} = x\begin{vmatrix} 1 & 1 \\ 2 & 4 \end{vmatrix} + 0$$

Therefore:

$$\begin{vmatrix} 3 & 1 \\ 10 & 4 \end{vmatrix} = x\begin{vmatrix} 1 & 1 \\ 2 & 4 \end{vmatrix}$$

which means that:

$$x = \frac{\begin{vmatrix} 3 & 1 \\ 10 & 4 \end{vmatrix}}{\begin{vmatrix} 1 & 1 \\ 2 & 4 \end{vmatrix}} = \frac{12 - 10}{4 - 2} = \frac{2}{2} = 1$$

Similarly, starting again from the square matrix of coefficients in Frame 12 only this time with the *second* column replaced by the elements of the column matrix on the right of this equation, then:

$$\begin{vmatrix} 1 & 3 \\ 2 & 10 \end{vmatrix} = \begin{vmatrix} 1 & x+y \\ 2 & 2x+4y \end{vmatrix} = y\begin{vmatrix} \dots & \dots \\ \dots & \dots \end{vmatrix}$$

Complete the determinant.
The answer is in the next frame

16

$$\begin{vmatrix} 1 & 3 \\ 2 & 10 \end{vmatrix} = y \begin{vmatrix} 1 & 1 \\ 2 & 4 \end{vmatrix}$$

Because

$$\begin{vmatrix} 1 & 3 \\ 2 & 10 \end{vmatrix} = \begin{vmatrix} 1 & x+y \\ 2 & 2x+4y \end{vmatrix}$$

$$= \begin{vmatrix} 1 & x \\ 2 & 2x \end{vmatrix} + \begin{vmatrix} 1 & y \\ 2 & 4y \end{vmatrix} \quad \text{by Property } \mathbf{3}$$

$$= x \begin{vmatrix} 1 & 1 \\ 2 & 2 \end{vmatrix} + y \begin{vmatrix} 1 & 1 \\ 2 & 4 \end{vmatrix} \quad \text{by Property } \mathbf{2}$$

$$= 0 + y \begin{vmatrix} 1 & 1 \\ 2 & 4 \end{vmatrix} \quad \text{by Property } \mathbf{1}$$

and so:

$$y = \frac{\begin{vmatrix} \cdots & \cdots \\ \cdots & \cdots \end{vmatrix}}{\begin{vmatrix} \cdots & \cdots \\ \cdots & \cdots \end{vmatrix}} = \cdots$$

The answer is in the next frame

17

$$y = \frac{\begin{vmatrix} 1 & 3 \\ 2 & 10 \end{vmatrix}}{\begin{vmatrix} 1 & 1 \\ 2 & 4 \end{vmatrix}} = 2$$

Because

$$\begin{vmatrix} 1 & 3 \\ 2 & 10 \end{vmatrix} = y \begin{vmatrix} 1 & 1 \\ 2 & 4 \end{vmatrix} \text{ so } y = \frac{\begin{vmatrix} 1 & 3 \\ 2 & 10 \end{vmatrix}}{\begin{vmatrix} 1 & 1 \\ 2 & 4 \end{vmatrix}} = \frac{10-6}{4-2} = 2$$

To summarize what we have just completed, given the matrix equation:

$$\begin{pmatrix} 1 & 1 \\ 2 & 4 \end{pmatrix} \begin{pmatrix} x \\ y \end{pmatrix} = \begin{pmatrix} 3 \\ 10 \end{pmatrix}$$

then Cramer's rule states that:

x is the determinant of the coefficient matrix with the elements of the first column replaced by the elements of the column matrix divided by the determinant of the coefficient matrix:

$$x = \frac{\begin{vmatrix} 3 & 1 \\ 10 & 4 \end{vmatrix}}{\begin{vmatrix} 1 & 1 \\ 2 & 4 \end{vmatrix}}$$

and y is the determinant of the coefficient matrix with the elements of the second column replaced by the elements of the column matrix divided by the determinant of the coefficient matrix:

$$y = \frac{\begin{vmatrix} 1 & 3 \\ 2 & 10 \end{vmatrix}}{\begin{vmatrix} 1 & 1 \\ 2 & 4 \end{vmatrix}}$$

Let's apply this to another one. The solution to the two simultaneous linear equations:

$$\begin{aligned} 3p - 2q &= 19 \\ 4p + 5q &= 10 \end{aligned}$$ is, by Cramer's rule, $p = \ldots\ldots\ldots$ and $q = \ldots\ldots\ldots$

Just write the two equations in matrix form and then apply Cramer's rule.
The answer is in the next frame

$$\boxed{p = 5 \text{ and } q = -2}$$

18

Because

$$\begin{aligned} 3p - 2q &= 19 \\ 4p + 5q &= 10 \end{aligned}$$ in matrix form is $\begin{pmatrix} 3 & -2 \\ 4 & 5 \end{pmatrix} \begin{pmatrix} p \\ q \end{pmatrix} = \begin{pmatrix} 19 \\ 10 \end{pmatrix}$

so, by Cramer's rule: $p = \dfrac{\begin{vmatrix} 19 & -2 \\ 10 & 5 \end{vmatrix}}{\begin{vmatrix} 3 & -2 \\ 4 & 5 \end{vmatrix}} = 5$ and $q = \dfrac{\begin{vmatrix} 3 & 19 \\ 4 & 10 \end{vmatrix}}{\begin{vmatrix} 3 & -2 \\ 4 & 5 \end{vmatrix}} = -2$

That was straightforward enough – another one just to make sure. The solution to the two simultaneous linear equations:

$$\begin{aligned} 2s - 3t &= -8 \\ s - 2 - t &= 0 \end{aligned}$$ is, by Cramer's rule, $s = \ldots\ldots\ldots$ and $t = \ldots\ldots\ldots$

The answer is in the next frame

19

$$s = 14 \text{ and } t = 12$$

Because

$$2s - 3t = -8$$
$$s - 2 - t = 0$$

rewritten as

$$2s - 3t = -8$$
$$s - t = 2$$

is, in matrix form, $\begin{pmatrix} 2 & -3 \\ 1 & -1 \end{pmatrix} \begin{pmatrix} s \\ t \end{pmatrix} = \begin{pmatrix} -8 \\ 2 \end{pmatrix}$ so by Cramer's rule:

$$s = \frac{\begin{vmatrix} -8 & -3 \\ 2 & -1 \end{vmatrix}}{\begin{vmatrix} 2 & -3 \\ 1 & -1 \end{vmatrix}} = 14 \quad \text{and} \quad t = \frac{\begin{vmatrix} 2 & -8 \\ 1 & 2 \end{vmatrix}}{\begin{vmatrix} 2 & -3 \\ 1 & -1 \end{vmatrix}} = 12$$

Move on to the next frame

20 **Higher order determinants**

Cramer's rule can be extended to solve three or more simultaneous linear equations in three or more variables but we shall restrict ourselves here to three simultaneous linear equations in three variables. However, before this rule can be applied we must learn how to expand 3×3 determinants. Consider the determinant:

$$\begin{vmatrix} 1 & 2 & 3 \\ 6 & 4 & 2 \\ 3 & 1 & 2 \end{vmatrix}$$

We first construct a new array of numbers by adding fourth and fifth columns that are identical to the first and second columns thus:

$$\begin{matrix} 1 & 2 & 3 & 1 & 2 \\ 6 & 4 & 2 & 6 & 4 \\ 3 & 1 & 2 & 3 & 1 \end{matrix}$$

We then multiply together all the numbers down the three left-to-right diagonals:

$$\begin{matrix} 1 & 2 & 3 & 1 & 2 \\ 6 & 4 & 2 & 6 & 4 \\ 3 & 1 & 2 & 3 & 1 \end{matrix} \qquad \begin{aligned} 1 \times 4 \times 2 &= 8 \\ 2 \times 2 \times 3 &= 12 \\ 3 \times 6 \times 1 &= 18 \end{aligned}$$

and then we add these products together:

$$8 + 12 + 18 = 38$$

▶

We next multiply together all the numbers down the three right-to-left diagonals:

$$
\begin{array}{ccccc}
1 & 2 & 3 & 1 & 2 \\
6 & 4 & 2 & 6 & 4 \\
3 & 1 & 2 & 3 & 1
\end{array}
\qquad
\begin{array}{l}
3 \times 4 \times 3 = 36 \\
1 \times 2 \times 1 = 2 \\
2 \times 6 \times 2 = 24
\end{array}
$$

and then we add these products together:

$$36 + 2 + 24 = 62$$

Finally we subtract the second sum from the first to give:

$$38 - 62 = -24$$

And this is the value of the determinant.

So now you try one.

$$
\begin{vmatrix}
1 & -1 & 1 \\
2 & 0 & -1 \\
-1 & 4 & 2
\end{vmatrix} = \dots\dots\dots
$$

The answer is in the next frame

15

21

Because

The array becomes, after duplicating the first two columns as the fourth and fifth columns:

$$
\begin{array}{ccccc}
1 & -1 & 1 & 1 & -1 \\
2 & 0 & -1 & 2 & 0 \\
-1 & 4 & 2 & -1 & 4
\end{array}
$$

From which we obtain:

$$[1 \times 0 \times 2] + [(-1) \times (-1) \times (-1)] + [1 \times 2 \times 4] = 0 + (-1) + 8 = 7$$

Multiplying from right to left then gives:

$$
\begin{array}{ccccc}
1 & -1 & 1 & 1 & -1 \\
2 & 0 & -1 & 2 & 0 \\
-1 & 4 & 2 & -1 & 4
\end{array}
$$

From which we obtain:

$$[1 \times 0 \times (-1)] + [1 \times (-1) \times 4] + [(-1) \times 2 \times 2] = 0 - 4 - 4 = -8$$

The value of the determinant is then the difference of these two numbers, namely:

$$7 - (-8) = 15$$

Make sure that you are happy with this procedure because we shall now use it for solving three simultaneous linear equations in three unknowns by an extended Cramer's rule.

Move to the next frame

22

To solve the three simultaneous linear equations:

$$x - y + z = 6$$
$$2x - z = 1$$
$$-x + 4y + 2z = 0$$

we first write them in matrix form.

Armed with the information you already have you can do this. The equations in matrix form are:

. . . .

The answer is in the next frame

23

$$\begin{pmatrix} 1 & -1 & 1 \\ 2 & 0 & -1 \\ -1 & 4 & 2 \end{pmatrix} \begin{pmatrix} x \\ y \\ z \end{pmatrix} = \begin{pmatrix} 6 \\ 1 \\ 0 \end{pmatrix}$$

Because

$$\begin{array}{lll} x - y + z = 6 & & 1x - 1y + 1z = 6 \\ 2x - z = 1 & \text{can be written as} & 2x + 0y - 1z = 1 \\ -x + 4y + 2z = 0 & & -1x + 4y + 2z = 0 \end{array}$$

so that $\quad \begin{pmatrix} 1 & -1 & 1 \\ 2 & 0 & -1 \\ -1 & 4 & 2 \end{pmatrix} \begin{pmatrix} x \\ y \\ z \end{pmatrix} = \begin{pmatrix} 6 \\ 1 \\ 0 \end{pmatrix}$

From the reasoning earlier for 2×2 matrices it is possible to demonstrate that the solution to this matrix equation using Cramer's rule for 3×3 matrices is:

$$x = \frac{\begin{vmatrix} 6 & -1 & 1 \\ 1 & 0 & -1 \\ 0 & 4 & 2 \end{vmatrix}}{\begin{vmatrix} 1 & -1 & 1 \\ 2 & 0 & -1 \\ -1 & 4 & 2 \end{vmatrix}}$$

replacing the first column in the numerator with the numbers in

$$\begin{pmatrix} 6 \\ 1 \\ 0 \end{pmatrix}$$

So $x = \ldots \ldots \ldots$

Next frame

$$x = 2$$

24

Because

The determinant in the numerator gives rise to the array:

$$
\begin{array}{ccccc}
6 & -1 & 1 & 6 & -1 \\
1 & 0 & -1 & 1 & 0 \\
0 & 4 & 2 & 0 & 4
\end{array}
$$

which yields:

$$[6 \times 0 \times 2] + [(-1) \times (-1) \times 0] + [1 \times 1 \times 4]$$
$$- [1 \times 0 \times 0] - [6 \times (-1) \times 4] - [(-1) \times 1 \times 2]$$
$$= 0 + 0 + 4 - 0 + 24 + 2$$
$$= 30$$

and the determinant in the denominator gives rise to the array:

$$
\begin{array}{ccccc}
1 & -1 & 1 & 1 & -1 \\
2 & 0 & -1 & 2 & 0 \\
-1 & 4 & 2 & -1 & 4
\end{array}
$$

which yields:

$$[1 \times 0 \times 2] + [(-1) \times (-1) \times (-1)] + [1 \times 2 \times 4]$$
$$- [1 \times 0 \times (-1)] - [1 \times (-1) \times 4] - [(-1) \times 2 \times 2]$$
$$= 0 - 1 + 8 - 0 + 4 + 4$$
$$= 15$$

That is $x = \dfrac{30}{15} = 2$

Similarly

$$
y = \frac{\begin{vmatrix} 1 & 6 & 1 \\ 2 & 1 & -1 \\ -1 & 0 & 2 \end{vmatrix}}{\begin{vmatrix} 1 & -1 & 1 \\ 2 & 0 & -1 \\ -1 & 4 & 2 \end{vmatrix}}
$$

replacing the second column in the numerator with the numbers in

$$
\begin{pmatrix} 6 \\ 1 \\ 0 \end{pmatrix}
$$

So $y = \ldots\ldots\ldots$

Next frame

25

$$\boxed{y = -1}$$

Because

The determinant in the numerator gives rise to the array:

$$\begin{array}{ccccc} 1 & 6 & 1 & 1 & 6 \\ 2 & 1 & -1 & 2 & 1 \\ -1 & 0 & 2 & -1 & 0 \end{array}$$

which yields:

$$[1 \times 1 \times 2] + [6 \times (-1) \times (-1)] + [1 \times 2 \times 0]$$
$$- [1 \times 1 \times (-1)] - [1 \times (-1) \times 0] - [6 \times 2 \times 2]$$
$$= 2 + 6 + 0 + 1 - 0 - 24$$
$$= -15$$

So that

$$y = \frac{\begin{vmatrix} 1 & 6 & 1 \\ 2 & 1 & -1 \\ -1 & 0 & 2 \end{vmatrix}}{\begin{vmatrix} 1 & -1 & 1 \\ 2 & 0 & -1 \\ -1 & 4 & 2 \end{vmatrix}} = \frac{-15}{15} = -1$$

To solve for z we could use Cramer's rule again:

$$z = \frac{\begin{vmatrix} 1 & -1 & 6 \\ 2 & 0 & 1 \\ -1 & 4 & 0 \end{vmatrix}}{\begin{vmatrix} 1 & -1 & 1 \\ 2 & 0 & -1 \\ -1 & 4 & 2 \end{vmatrix}}$$

replacing the third column in the numerator with the numbers in

$$\begin{pmatrix} 6 \\ 1 \\ 0 \end{pmatrix}$$

However, it is simpler just to use substitution. Substituting into the first equation $x - y + z = 6$ we find that:

$$z = \ldots\ldots\ldots\ldots$$

Next frame

$$z = 3$$

Because

$x - y + z = 6$, that is $2 - (-1) + z = 6$ and so $z = 3$ and, just as a final check, substituting these values into the third equation:

$-x + 4y + 2z = 0$ yields $-(2) + 4(-1) + 2(3) = -2 - 4 + 6 = 0$

Now you try one from the beginning to the end. The solution to the three simultaneous equations:

$$7x + 9y - 2z = 9$$
$$x - 2y + 3z = -2 \quad \text{is} \quad x = \ldots\ldots, \quad y = \ldots\ldots\ldots \quad \text{and} \quad z = \ldots\ldots\ldots$$
$$-2x - y + z = -3$$

Take your time, there is no rush.
The answer is in the next frame

$$x = 1, \, y = 0 \text{ and } z = -1$$

Because

To solve the three simultaneous linear equations:

$$7x + 9y - 2z = 9$$
$$x - 2y + 3z = -2$$
$$-2x - y + z = -3$$

we first write them in matrix form. That is:

$$7x + 9y - 2z = 9 \qquad\qquad\qquad 7x + 9y - 2z = 9$$
$$x - 2y + 3z = -2 \quad \text{can be written as} \quad 1x - 2y + 3z = -2$$
$$-2x - y + z = -3 \qquad\qquad\qquad -2x - 1y + 1z = -3$$

so that
$$\begin{pmatrix} 7 & 9 & -2 \\ 1 & -2 & 3 \\ -2 & -1 & 1 \end{pmatrix} \begin{pmatrix} x \\ y \\ z \end{pmatrix} = \begin{pmatrix} 9 \\ -2 \\ -3 \end{pmatrix}$$

The solution to this matrix equation using Cramer's rule for 3×3 matrices is then:

$$x = \frac{\begin{vmatrix} 9 & 9 & -2 \\ -2 & -2 & 3 \\ -3 & -1 & 1 \end{vmatrix}}{\begin{vmatrix} 7 & 9 & -2 \\ 1 & -2 & 3 \\ -2 & -1 & 1 \end{vmatrix}}$$

replacing the first column in the numerator with the numbers in

$$\begin{pmatrix} 9 \\ -2 \\ -3 \end{pmatrix}$$

The determinant in the numerator gives rise to the array:

$$\begin{array}{ccccc} 9 & 9 & -2 & 9 & 9 \\ -2 & -2 & 3 & -2 & -2 \\ -3 & -1 & 1 & -3 & -1 \end{array}$$

which yields:

$$[9 \times (-2) \times 1] + [9 \times 3 \times (-3)] + [(-2) \times (-2) \times (-1)]$$
$$= -18 - 81 - 4 = -103$$

left to right and

$$\begin{array}{ccccc} 9 & 9 & -2 & 9 & 9 \\ -2 & -2 & 3 & -2 & -2 \\ -3 & -1 & 1 & -3 & -1 \end{array}$$

which yields:

$$[(-2) \times (-2) \times (-3)] + [9 \times 3 \times (-1)] + [9 \times (-2) \times 1]$$
$$= -12 - 27 - 18 = -57$$

right to left. This gives the value of the numerator determinant as

$$-103 - (-57) = -46$$

The determinant in the denominator gives rise to the array:

$$\begin{array}{ccccc} 7 & 9 & -2 & 7 & 9 \\ 1 & -2 & 3 & 1 & -2 \\ -2 & -1 & 1 & -2 & 1 \end{array}$$

which yields:

$$[7 \times (-2) \times 1] + [9 \times 3 \times (-2)] + [(-2) \times 1 \times (-1)] = -14 - 54 + 2 = -66$$

left to right, and

$$\begin{array}{ccccc} 7 & 9 & -2 & 7 & 9 \\ 1 & -2 & 3 & 1 & -2 \\ -2 & -1 & 1 & -2 & -1 \end{array}$$

which yields:

$$[(-2) \times (-2) \times (-2)] + [7 \times 3 \times (-1)] + [9 \times 1 \times 1] = -8 - 21 + 9 = -20$$

right to left. This gives the value of the numerator determinant as

$$-66 - (-20) = -46$$

So that

$$x = \frac{-46}{-46} = 1$$

Similarly,

$$y = \dfrac{\begin{vmatrix} 7 & 9 & -2 \\ 1 & -2 & 3 \\ -2 & -3 & 1 \end{vmatrix}}{\begin{vmatrix} 7 & 9 & -2 \\ 1 & -2 & 3 \\ -2 & -1 & 1 \end{vmatrix}}$$

replacing the second column in the numerator with the numbers in

$$\begin{pmatrix} 9 \\ -2 \\ -3 \end{pmatrix}$$

The determinant in the numerator gives rise to the array:

$$\begin{matrix} 7 & 9 & -2 & 7 & 9 \\ 1 & -2 & 3 & 1 & -2 \\ -2 & -3 & 1 & -2 & 3 \end{matrix}$$

which yields:

$$[7 \times (-2) \times 1] + [9 \times 3 \times (-2)] + [(-2) \times 1 \times (-3)] = -14 - 54 + 6 = -62$$

left to right and

$$\begin{matrix} 7 & 9 & -2 & 7 & 9 \\ 1 & -2 & 3 & 1 & -2 \\ -2 & -3 & 1 & -2 & -3 \end{matrix}$$

which yields:

$$[(-2) \times (-2) \times (-2)] + [7 \times 3 \times (-3)] + [9 \times 1 \times 1] = -8 - 63 + 9 = -62$$

right to left. This gives the value of the numerator determinant as

$$-62 - (-62) = 0$$

So that

$$y = \dfrac{0}{-46} = 0$$

To solve for z we substitute into the first equation $7x + 9y - 2z = 9$. We find that $7 + 0 - 2z = 9$ and so $z = -1$ and, just as a final check, substituting these values into the third equation:

$$-2x - y + z = -3 \quad \text{yields} \quad -2(1) + 0 + (-1) = -3 \quad \text{as required.}$$

At this point let's pause and summarize the work on solving simultaneous linear equations

Review summary

28

1 *Matrix solution to simultaneous linear equations*: A pair of simultaneous equations can be written as a matrix equation:

$$\begin{array}{l} ax + by = m \\ cx + dy = n \end{array} \quad \text{can be written as} \quad \begin{pmatrix} a & b \\ c & d \end{pmatrix}\begin{pmatrix} x \\ y \end{pmatrix} = \begin{pmatrix} m \\ n \end{pmatrix}. \text{ That is } \mathbf{AX} = \mathbf{b}$$

By multiplying on the left by the inverse matrix $\mathbf{A}^{-1}$ a solution can be obtained:

$$\mathbf{A}^{-1}\mathbf{AX} = \mathbf{IX} = \mathbf{X} = \mathbf{A}^{-1}\mathbf{b}$$

2 *Determinants and Cramer's rule*: Given the pair of simultaneous linear equations:

$$\begin{array}{l} ax + by = m \\ cx + dy = n \end{array} \quad \text{written in matrix form as} \quad \begin{pmatrix} a & b \\ c & d \end{pmatrix}\begin{pmatrix} x \\ y \end{pmatrix} = \begin{pmatrix} m \\ n \end{pmatrix}$$

then Cramer's rule states that

$$x = \frac{\begin{vmatrix} m & b \\ n & d \end{vmatrix}}{\begin{vmatrix} a & b \\ c & d \end{vmatrix}} \quad \text{and} \quad y = \frac{\begin{vmatrix} a & m \\ c & n \end{vmatrix}}{\begin{vmatrix} a & b \\ c & d \end{vmatrix}}$$

3 *Third order determinants*: A third order determinant is evaluated by adding the first two columns in the fourth and fifth column locations and multiplying elements down the diagonals. The products obtained from multiplying down the diagonals from left to right are added together as are the products obtained from multiplying down the diagonals from right to left. These two numbers are then subtracted to give the value of the determinant.

4 *Cramer's rule extended*: Given the three simultaneous linear equations:

$$\begin{array}{l} ax + by + cz = m \\ dx + ey + fz = n \\ gx + hy + jz = p \end{array} \quad \text{written in matrix form as} \quad \begin{pmatrix} a & b & c \\ d & e & f \\ g & h & j \end{pmatrix}\begin{pmatrix} x \\ y \\ z \end{pmatrix} = \begin{pmatrix} m \\ n \\ p \end{pmatrix}$$

then Cramer's rule states that

$$x = \frac{\begin{vmatrix} m & b & c \\ n & e & f \\ p & h & j \end{vmatrix}}{\begin{vmatrix} a & b & c \\ d & e & f \\ g & h & j \end{vmatrix}}, \quad y = \frac{\begin{vmatrix} a & m & c \\ d & n & f \\ g & p & j \end{vmatrix}}{\begin{vmatrix} a & b & c \\ d & e & f \\ g & h & j \end{vmatrix}}, \quad z = \frac{\begin{vmatrix} a & b & m \\ d & e & n \\ g & h & p \end{vmatrix}}{\begin{vmatrix} a & b & c \\ d & e & f \\ g & h & j \end{vmatrix}}$$

Review exercise   **Unit 32**

1 Use matrix inversion to solve the pair of simultaneous linear equations: **29**

$4x - 5y = 2$

$x + 3y = 9$

2 Use Cramer's rule to solve the set of three simultaneous linear equations:

$2x - 3y + z = -3$

$3x + y - 2z = -13$

$x + 3y + 2z = 9$

Complete both questions, working carefully through each one.
If you need to look back at the Unit.
The answers and working are in the next frame

---

1 $\begin{array}{l} 4x - 5y = 2 \\ x + 3y = 9 \end{array}$ can be written as $\begin{pmatrix} 4 & -5 \\ 1 & 3 \end{pmatrix}\begin{pmatrix} x \\ y \end{pmatrix} = \begin{pmatrix} 2 \\ 9 \end{pmatrix}$ **30**

that is $\mathbf{AX} = \mathbf{b}$ where $\mathbf{A} = \begin{pmatrix} 4 & -5 \\ 1 & 3 \end{pmatrix}$.

Therefore $\mathbf{A}^{-1} = \dfrac{1}{4 \times 3 - (-5) \times 1}\begin{pmatrix} 3 & 5 \\ -1 & 4 \end{pmatrix} = \dfrac{1}{17}\begin{pmatrix} 3 & 5 \\ -1 & 4 \end{pmatrix}$ and so:

$\dfrac{1}{17}\begin{pmatrix} 3 & 5 \\ -1 & 4 \end{pmatrix}\begin{pmatrix} 4 & -5 \\ 1 & 3 \end{pmatrix}\begin{pmatrix} x \\ y \end{pmatrix} = \dfrac{1}{17}\begin{pmatrix} 3 & 5 \\ -1 & 4 \end{pmatrix}\begin{pmatrix} 2 \\ 9 \end{pmatrix}$ that is

$\begin{pmatrix} 1 & 0 \\ 0 & 1 \end{pmatrix}\begin{pmatrix} x \\ y \end{pmatrix} = \dfrac{1}{17}\begin{pmatrix} 51 \\ 34 \end{pmatrix}$ so

$\begin{pmatrix} x \\ y \end{pmatrix} = \begin{pmatrix} 3 \\ 2 \end{pmatrix}$

giving $x = 3$ and $y = 2$ as the solution.

2 The three equations:

$\begin{array}{l} 2x - 3y + z = -3 \\ 3x + y - 2z = -13 \\ x + 3y + 2z = 9 \end{array}$ can be written as $\begin{pmatrix} 2 & -3 & 1 \\ 3 & 1 & -2 \\ 1 & 3 & 2 \end{pmatrix}\begin{pmatrix} x \\ y \\ z \end{pmatrix} = \begin{pmatrix} -3 \\ -13 \\ 9 \end{pmatrix}$

so that: $x = \dfrac{\begin{vmatrix} -3 & -3 & 1 \\ -13 & 1 & -2 \\ 9 & 3 & 2 \end{vmatrix}}{\begin{vmatrix} 2 & -3 & 1 \\ 3 & 1 & -2 \\ 1 & 3 & 2 \end{vmatrix}}$

The numerator array is extended to

$$\begin{array}{ccccc} -3 & -3 & 1 & -3 & -3 \\ -13 & 1 & -2 & -13 & 1 \\ 9 & 3 & 2 & 9 & 3 \end{array}$$

and this yields

$$[(-3) \times 1 \times 2] + [(-3) \times (-2) \times 9] + [1 \times (-13) \times 3]$$

$$- [1 \times 1 \times 9] - [(-3) \times (-2) \times 3] - [(-3) \times (-13) \times 2]$$

$$= -6 + 54 - 39 - 9 - 18 - 78$$

$$= -96$$

The denominator array is extended to

$$\begin{array}{ccccc} 2 & -3 & 1 & 2 & -3 \\ 3 & 1 & -2 & 3 & 1 \\ 1 & 3 & 2 & 1 & 3 \end{array}$$

and this yields

$$[2 \times 1 \times 2] + [(-3) \times (-2) \times 1] + [1 \times 3 \times 3]$$

$$- [1 \times 1 \times 1] - [2 \times (-2) \times 3] - [(-3) \times 3 \times 2]$$

$$= 4 + 6 + 9 - 1 + 12 + 18$$

$$= 48$$

Therefore $x = \dfrac{-96}{48} = -2$.

$$y = \frac{\begin{vmatrix} 2 & -3 & 1 \\ 3 & -13 & -2 \\ 1 & 9 & 2 \end{vmatrix}}{48}$$

The numerator array is extended to

$$\begin{array}{ccccc} 2 & -3 & 1 & 2 & -3 \\ 3 & -13 & -2 & 3 & -13 \\ 1 & 9 & 2 & 1 & 9 \end{array}$$

and this yields

$$[2 \times (-13) \times 2] + [(-3) \times (-2) \times 1] + [1 \times 3 \times 9]$$

$$- [1 \times (-13) \times 1] - [2 \times (-2) \times 9] - [(-3) \times 3 \times 2]$$

$$= -52 + 6 + 27 + 13 + 36 + 18$$

$$= 48$$

Therefore $y = \dfrac{48}{48} = 1$.

Substituting in the 1st equation gives $2(-2) - 3(1) + z = -3$ so $z = -3 + 3 + 4$ $= 4$. Check this is correct by substituting these three values into the 3rd equation:

$$x + 3y + 2z = 9 \text{ that is } (-2) + 3(1) + 2(4) = 9$$

Now for the Review test

Review test

1 Use matrix inversion to solve the pair of simultaneous linear equations:

$$9x - 7y = -39$$
$$4x + 3y = 1$$

31

2 Use Cramer's rule to solve the set of three simultaneous linear equations:

$$2x - 3y + 4z = -16$$
$$x + 2y - z = 12$$
$$-2x + y + 3z = -6$$

You have now come to the end of this Module. A list of **Can You?** questions follows for you to gauge your understanding of the material in the Module. You will notice that these questions match the **Learning outcomes** listed at the beginning of the Module. Now try the **Test exercise.** *Work through the questions at your own pace, there is no need to hurry.* A set of **Further problems** provides valuable additional practice.

32

 # Can You?

Checklist: Module 10

1

Check this list before and after you try the end of Module test

On a scale of 1 to 5 how confident are you that you can:

- Use matrices as stores of numbers and manipulate such matrices by adding and subtracting, by multiplying by a scalar and by multiplying two matrices together?

  Yes ☐ ☐ ☐ ☐ ☐ *No*

- Compute the inverse of a 2 × 2 matrix and use inverse matrices to solve pairs of simultaneous linear equations?

  Yes ☐ ☐ ☐ ☐ ☐ *No*

- Expand 2 × 2 and 3 × 3 determinants and use Cramer's rule to solve simultaneous linear equations?

  Yes ☐ ☐ ☐ ☐ ☐ *No*

Test exercise 10

2

1 Given matrices

$$\mathbf{A} = \begin{pmatrix} 1 & 2 \\ -3 & 8 \end{pmatrix}, \mathbf{B} = \begin{pmatrix} 5 & 0 \\ -2 & 4 \end{pmatrix}, \mathbf{C} = (5 \quad 3) \text{ and } \mathbf{D} = \begin{pmatrix} 1 \\ 6 \end{pmatrix}$$

   calculate each of the following:

   (a) $\mathbf{A} + \mathbf{B}$ (b) $\mathbf{B} - \mathbf{A}$ (c) $2\mathbf{A}$ (d) $\mathbf{AB}$ (e) $\mathbf{BA}$ (f) $\mathbf{AD}$ (g) $\mathbf{CB}$

2 Find the determinant of each of the following matrices:

   (a) $\mathbf{A} = \begin{pmatrix} -3 & 1 \\ 4 & 2 \end{pmatrix}$ (b) $\mathbf{B} = \begin{pmatrix} 8 & -6 \\ 2 & -7 \end{pmatrix}$

   (c) $\mathbf{C} = \begin{pmatrix} 15 & 30 \\ -1 & -2 \end{pmatrix}$ (d) $\mathbf{D} = \begin{pmatrix} 8 \\ -2 \end{pmatrix}$

3 Find the inverse of each of the following matrices:

   (a) $\mathbf{P} = \begin{pmatrix} 2 & 6 \\ 2 & 7 \end{pmatrix}$ (b) $\mathbf{Q} = \begin{pmatrix} 5 & 12 \\ -7 & -17 \end{pmatrix}$ (c) $\mathbf{C} = \begin{pmatrix} 3 & -6 \\ 4 & -8 \end{pmatrix}$

4 Use matrix inversion to solve each pair of simultaneous linear equations:

   (a) $\begin{array}{l} x - y = 8 \\ 2x + 5y = 9 \end{array}$ (b) $\begin{array}{l} 2x - 4y = 12 \\ -5x + y = 6 \end{array}$

5 Use Cramer's rule to solve each set of three simultaneous linear equations:

   (a) $\begin{array}{l} 4x - 2y - z = 9 \\ 3x - 2y - 2z = 4 \\ x + 2y + 3z = 7 \end{array}$ (b) $\begin{array}{l} a + b - c = -1 \\ 3a + 4c = 10 \\ 2b + 3a = 2 \end{array}$ (c) $\begin{array}{l} 3s - 2t = 1 \\ 5t + u = 44 \\ 2u - 3s = 3 \end{array}$

Further problems 10

3

1 Given matrices

$$\mathbf{A} = \begin{pmatrix} 3 & 7 \\ 4 & -2 \end{pmatrix}, \mathbf{B} = \begin{pmatrix} -1 & -2 \\ 5 & 0 \end{pmatrix}, \mathbf{C} = (3 \quad -4) \text{ and } \mathbf{D} = \begin{pmatrix} -5 \\ 3 \end{pmatrix}$$

   calculate each of the following:

   (a) $2\mathbf{A} + 3\mathbf{B}$ (b) $4\mathbf{B} - 5\mathbf{A}$ (c) $\mathbf{AB}$ (d) $\mathbf{BA}$ (e) $\mathbf{AD}$
   (f) $\mathbf{CB}$ (g) $\mathbf{CD}$ (h) $\mathbf{DC}$

2 Given matrix

$$\mathbf{A} = \begin{pmatrix} 2 & 4 \\ 1 & 0 \end{pmatrix}$$

   then $\mathbf{A}^2 = \mathbf{A} \times \mathbf{A}$ and $\mathbf{A}^3 = \mathbf{A} \times \mathbf{A} \times \mathbf{A}$. Find $\mathbf{A}^3$ and $\mathbf{A}^2$ and show that
   $\mathbf{A}^3\mathbf{A}^2 = \mathbf{A}^2\mathbf{A}^3$

3 Show that if $\mathbf{A}$ is a 2×2 matrix then $|k\mathbf{A}| = k^2|\mathbf{A}|$, where k is a scalar.

▷

4 Given matrices
$$\mathbf{A} = \begin{pmatrix} 2 & -4 \\ 3 & 5 \end{pmatrix} \text{ and } \mathbf{B} = \begin{pmatrix} 7 & 1 \\ -6 & 0 \end{pmatrix}$$
show that $(\mathbf{AB})^{-1} = \mathbf{B}^{-1}\mathbf{A}^{-1}$

5 Using the matrices of the previous question, show that $|\mathbf{AB}| = |\mathbf{A}| \times |\mathbf{B}|$

6 Use matrix inversion to solve:

(a) $\begin{aligned} 2x + 2y &= 4 \\ 7x + 5y &= 8 \end{aligned}$ (b) $\begin{aligned} 9a - b &= 8 \\ 4a &= 2 \end{aligned}$ (c) $\begin{aligned} 6p - 5q &= 15 \\ -2p + 3q &= -3 \end{aligned}$ (d) $\begin{aligned} -m - 2n &= 4 \\ m + 6n &= -8 \end{aligned}$

7 Solve using Cramer's rule:

(a) $\begin{aligned} 3x - 2y &= 5 \\ 5x + 3y &= 2 \end{aligned}$ (b) $\begin{aligned} 4m + n &= 11 \\ 3n - 7m &= -5 \end{aligned}$

(c) $\begin{aligned} a + b - 2 &= 0 \\ 2a - 3b - 19 &= 0 \end{aligned}$ (d) $\begin{aligned} 2s - 3t &= -8 \\ s - t + 2 &= 0 \end{aligned}$

8 Expand each of the following determinants:

(a) $\begin{vmatrix} 1 & 0 & -1 \\ 0 & -1 & 1 \\ 1 & 1 & 0 \end{vmatrix}$ (b) $\begin{vmatrix} 3 & -2 & 5 \\ 8 & 4 & -6 \\ -7 & 1 & 3 \end{vmatrix}$ (c) $\begin{vmatrix} -2 & 7 & -3 \\ 9 & -4 & 2 \\ 0 & -5 & 0 \end{vmatrix}$ (d) $\begin{vmatrix} 2 & 4 & 6 \\ 2 & 3 & 5 \\ 2 & 7 & 9 \end{vmatrix}$

9 Use Cramer's rule to solve:

(a) $\begin{aligned} x - y + z &= 21 \\ x + y - z &= 7 \\ -x + y + z &= -5 \end{aligned}$ (b) $\begin{aligned} p + 2q - 3r &= 36 \\ -8p - r &= 49 \\ 6q + 5p &= 17 \end{aligned}$

(c) $\begin{aligned} 4n - 5l &= -1 \\ 8m + n &= 9 \\ 2l - 3m &= -1 \end{aligned}$ (d) $\begin{aligned} a + b &= -1 \\ b - c &= 3 \\ c - a &= 2 \end{aligned}$

Vectors

Scalar and vector quantities

Unit 33

Physical quantities can be divided into two main groups, scalar quantities and vector quantities.

1

(a) A *scalar quantity* is one that is defined completely by a single number with appropriate units, e.g. length, area, volume, mass, time, etc. Once the units are stated, the quantity is denoted entirely by its size or *magnitude*.

(b) A *vector quantity* is defined completely when we know not only its magnitude (with units) but also the direction in which it operates, e.g. force, velocity, acceleration. A vector quantity necessarily involves *direction* as well as magnitude.

So (a) a speed of 10 km/h is a scalar quantity, but
    (b) a velocity of '10 km/h due north' is a .... quantity.

---

vector

2

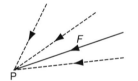

A force F acting at a point P is a vector quantity, since to define it completely we must give:

(a) its magnitude, and also

(b) its ....

---

direction

3

So that:

(a) A temperature of 100°C is a .... quantity.

(b) An acceleration of 9·8 m/s^2 vertically downwards is a .... quantity.

(c) The weight of a 7 kg mass is a .... quantity.

(d) The sum of £500 is a .... quantity.

(e) A north-easterly wind of 20 knots is a .... quantity.

(a) scalar   (b) vector   (c) vector   (d) scalar   (e) vector

4

Since, in (b), (c) and (e) the complete description of the quantity includes not only its magnitude, but also its ....

---

direction

5

Move on to Frame 6

Vector representation

6

A vector quantity can be represented graphically by a line, drawn so that:

(a) the *length* of the line denotes the magnitude of the quantity, according to some stated vector scale

(b) the *direction* of the line denotes the direction in which the vector quantity acts. The sense of the direction is indicated by an arrowhead.

e.g. A horizontal force of 35 N acting to the right, would be indicated by a line ——▸—— and if the chosen vector scale were 1 cm ≡ 10 N, the line would be .... cm long.

7

$$\boxed{3\cdot 5}$$

The vector quantity AB is referred to as

$\overline{AB}$ or **a**

The magnitude of the vector quantity is written $|\overline{AB}|$, or $|\mathbf{a}|$, or simply AB or a.

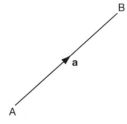

Note that $\overline{BA}$ would represent a vector quantity of the same magnitude but with opposite sense.

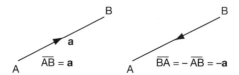

On to Frame 8

8 **Two equal vectors**

If two vectors, **a** and **b**, are said to be equal, they have the same magnitude and the same direction.
If **a** = **b**, then

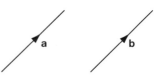

(a) $a = b$ (magnitudes equal)
(b) the direction of **a** = direction of **b**, i.e. the two vectors are parallel and in the same sense.

Similarly, if two vectors **a** and **b** are such that **b** = −**a**, what can we say about:

(a) their magnitudes,

(b) their directions?

(a) Magnitudes are equal

(b) The vectors are parallel but opposite in sense

i.e. if $\mathbf{b} = -\mathbf{a}$, then

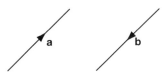

Types of vectors

(a) A *position vector* $\overline{AB}$ occurs when the point A is fixed.

(b) A *line vector* is such that it can slide along its line of action, e.g. a mechanical force acting on a body.

(c) A *free vector* is not restricted in any way. It is completely defined by its magnitude and direction and can be drawn as any one of a set of equal-length parallel lines.

Most of the vectors we shall consider will be free vectors

So on now to Frame 11

Addition of vectors

11

The sum of two vectors, $\overline{AB}$ and $\overline{BC}$, is defined as the single or equivalent or resultant vector $\overline{AC}$

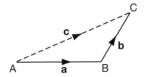

i.e. $\overline{AB} + \overline{BC} = \overline{AC}$

or $\mathbf{a} + \mathbf{b} = \mathbf{c}$

To find the sum of two vectors $\mathbf{a}$ and $\mathbf{b}$ then, we draw them as a chain, starting the second where the first ends: the sum $\mathbf{c}$ is given by the single vector joining the start of the first to the end of the second.

e.g. if $\mathbf{p} \equiv$ a force of 40 N, acting in the direction due east

$\mathbf{q} \equiv$ a force of 30 N, acting in the direction due north

then the magnitude of the vector sum r of these two forces will be . . . .

$$r = 50 \text{ N}$$

Because

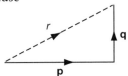

$$r^2 = p^2 + q^2$$
$$= 1600 + 900 = 2500$$
$$r = \sqrt{2500} = 50 \text{ N}$$

The sum of a number of vectors a + b + c + d + ...

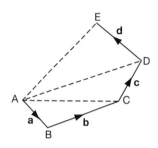

(a) Draw the vectors as a chain.

(b) Then:

$$\mathbf{a} + \mathbf{b} = \overline{AC}$$
$$\overline{AC} + \mathbf{c} = \overline{AD}$$
$$\therefore \quad \mathbf{a} + \mathbf{b} + \mathbf{c} = \overline{AD}$$
$$\overline{AD} + \mathbf{d} = \overline{AE}$$
$$\therefore \quad \mathbf{a} + \mathbf{b} + \mathbf{c} + \mathbf{d} = \overline{AE}$$

i.e. the sum of all vectors, $\mathbf{a}, \mathbf{b}, \mathbf{c}, \mathbf{d}$, is given by the single vector joining the start of the first to the end of the last – in this case, $\overline{AE}$. This follows directly from our previous definition of the sum of two vectors.

Similarly:

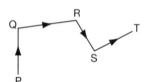

$$\overline{PQ} + \overline{QR} + \overline{RS} + \overline{ST} = \ldots\ldots\ldots\ldots$$

13

$$\boxed{\overline{PT}}$$

Now suppose that in another case, we draw the vector diagram to find the sum of $\mathbf{a}, \mathbf{b}, \mathbf{c}, \mathbf{d}, \mathbf{e}$, and discover that the resulting diagram is, in fact, a closed figure.

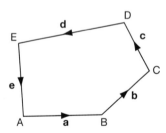

What is the sum of the vectors $\mathbf{a}, \mathbf{b}, \mathbf{c}, \mathbf{d}, \mathbf{e}$ in this case?

Think carefully and when you have decided, move on to Frame 14

14

$$\boxed{\text{Sum of the vectors} = \mathbf{0}}$$

Because we said in the previous case, that the vector sum was given by the single equivalent vector joining the beginning of the first vector to the end of the last.

But, if the vector diagram is a closed figure, the end of the last vector coincides with the beginning of the first, so that the resultant sum is a vector with no magnitude.

Now for some examples:

Find the vector sum $\overline{AB} + \overline{BC} + \overline{CD} + \overline{DE} + \overline{EF}$.

Without drawing a diagram, we can see that the vectors are arranged in a chain, each beginning where the previous one left off. The sum is therefore given by the vector joining the beginning of the first vector to the end of the last.

$$\therefore \text{ Sum} = \overline{AF}$$

In the same way:

$$\overline{AK} + \overline{KL} + \overline{LP} + \overline{PQ} = \ldots\ldots\ldots\ldots$$

$$\boxed{\overline{AQ}}$$

15

Right. Now what about this one?

Find the sum of $\quad \overline{AB} - \overline{CB} + \overline{CD} - \overline{ED}$

We must beware of the negative vectors. Remember that $-\overline{CB} = \overline{BC}$, i.e. the same magnitude and direction but in the opposite sense.
Also $-\overline{ED} = \overline{DE}$

$$\therefore \ \overline{AB} - \overline{CB} + \overline{CD} - \overline{ED} = \overline{AB} + \overline{BC} + \overline{CD} + \overline{DE}$$
$$= \overline{AE}$$

Now you do this one:

Find the vector sum $\quad \overline{AB} + \overline{BC} - \overline{DC} - \overline{AD}$

When you have the result, move on to Frame 16

$$\boxed{0}$$

16

Because

$$\overline{AB} + \overline{BC} - \overline{DC} - \overline{AD} = \overline{AB} + \overline{BC} + \overline{CD} + \overline{DA}$$

and the lettering indicates that the end of the last vector coincides with the beginning of the first. The vector diagram is thus a closed figure and therefore the sum of the vectors is **0**.
Now here are some for you to do:

(a) $\overline{PQ} + \overline{QR} + \overline{RS} + \overline{ST} = \ldots\ldots\ldots\ldots$
(b) $\overline{AC} + \overline{CL} - \overline{ML} = \ldots\ldots\ldots\ldots$
(c) $\overline{GH} + \overline{HJ} + \overline{JK} + \overline{KL} + \overline{LG} = \ldots\ldots\ldots\ldots$
(d) $\overline{AB} + \overline{BC} + \overline{CD} + \overline{DB} = \ldots\ldots\ldots\ldots$

When you have finished all four, check with the results in the next frame

17

Here are the results:

(a) $\overline{PQ} + \overline{QR} + \overline{RS} + \overline{ST} = \overline{PT}$

(b) $\overline{AC} + \overline{CL} - \overline{ML} = \overline{AC} + \overline{CL} + \overline{LM} = \overline{AM}$

(c) $\overline{GH} + \overline{HJ} + \overline{JK} + \overline{KL} + \overline{LG} = \mathbf{0}$

   [Since the end of the last vector coincides
   with the beginning of the first]

(d) $\overline{AB} + \overline{BC} + \overline{CD} + \overline{DB} = \overline{AB}$

The last three vectors form a closed figure and therefore the sum of these three vectors is zero, leaving only $\overline{AB}$ to be considered.

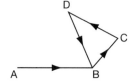

Now on to Frame 18

Components of a given vector

18

Just as $\overline{AB} + \overline{BC} + \overline{CD} + \overline{DE}$ can be replaced by $\overline{AE}$, so any single vector $\overline{PT}$ can be replaced by any number of component vectors so long as they form a chain in the vector diagram, beginning at P and ending at T.

e.g. $\overline{PT} = \mathbf{a} + \mathbf{b} + \mathbf{c} + \mathbf{d}$

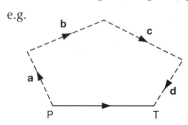

Example 1

ABCD is a quadrilateral, with G and H the mid-points of DA and BC respectively. Show that $\overline{AB} + \overline{DC} = 2\overline{GH}$.

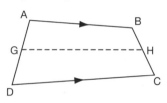

We can replace vector $\overline{AB}$ by any chain of vectors so long as they start at A and end at B e.g. we could say

$$\overline{AB} = \overline{AG} + \overline{GH} + \overline{HB}$$

Similarly, we could say $\overline{DC} = \dots\dots\dots$

$$\boxed{\overline{DC} = \overline{DG} + \overline{GH} + \overline{HC}}$$

So we have:

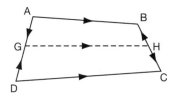

$$\overline{AB} = \overline{AG} + \overline{GH} + \overline{HB}$$
$$\overline{DC} = \overline{DG} + \overline{GH} + \overline{HC}$$

$$\therefore \ \overline{AB} + \overline{DC} = \overline{AG} + \overline{GH} + \overline{HB} + \overline{DG} + \overline{GH} + \overline{HC}$$
$$= 2\overline{GH} + (\overline{AG} + \overline{DG}) + (\overline{HB} + \overline{HC})$$

Now, G is the mid-point of AD. Therefore, vectors $\overline{AG}$ and $\overline{DG}$ are equal in length but opposite in sense.

$$\therefore \ \overline{DG} = -\overline{AG}$$

Similarly $\overline{HC} = -\overline{HB}$

$$\therefore \ \overline{AB} + \overline{DC} = 2\overline{GH} + (\overline{AG} - \overline{AG}) + (\overline{HB} - \overline{HB})$$
$$= 2\overline{GH}$$

Next frame

Example 2

20

Points L, M, N are mid-points of the sides AB, BC, CA of the triangle ABC. Show that:

(a) $\overline{AB} + \overline{BC} + \overline{CA} = \mathbf{0}$

(b) $2\overline{AB} + 3\overline{BC} + \overline{CA} = 2\overline{LC}$

(c) $\overline{AM} + \overline{BN} + \overline{CL} = \mathbf{0}$

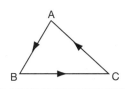

(a)   We can dispose of the first part straight away without any trouble. We can see from the vector diagram that $\overline{AB} + \overline{BC} + \overline{CA} = \mathbf{0}$ since these three vectors form a ....

$$\boxed{\text{closed figure}}$$

Now for part (b):

To show that $2\overline{AB} + 3\overline{BC} + \overline{CA} = 2\overline{LC}$

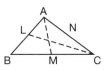 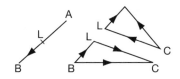

From the figure:

$$\overline{AB} = 2\overline{AL}; \quad \overline{BC} = \overline{BL} + \overline{LC}; \quad \overline{CA} = \overline{CL} + \overline{LA}$$

$$\therefore \ 2\overline{AB} + 3\overline{BC} + \overline{CA} = 4\overline{AL} + 3\overline{BL} + 3\overline{LC} + \overline{CL} + \overline{LA}$$

$$\text{Now } \overline{BL} = -\overline{AL}; \quad \overline{CL} = -\overline{LC}; \quad \overline{LA} = -\overline{AL}$$

Substituting these in the previous line, gives

$$2\overline{AB} + 3\overline{BC} + \overline{CA} = \dots\dots\dots$$

22

$$\boxed{2\overline{LC}}$$

Because

$$2\overline{AB} + 3\overline{BC} + \overline{CA} = 4\overline{AL} + 3\overline{BL} + 3\overline{LC} + \overline{CL} + \overline{LA}$$
$$= 4\overline{AL} - 3\overline{AL} + 3\overline{LC} - \overline{LC} - \overline{AL}$$
$$= 4\overline{AL} - 4\overline{AL} + 3\overline{LC} - \overline{LC}$$
$$= 2\overline{LC}$$

Now part (c):

To prove that $\overline{AM} + \overline{BN} + \overline{CL} = \mathbf{0}$

From the figure in Frame 21, we can say:

$$\overline{AM} = \overline{AB} + \overline{BM}$$
$$\overline{BN} = \overline{BC} + \overline{CN}$$

Similarly $\overline{CL} = \dots\dots\dots$

23

$$\boxed{\overline{CL} = \overline{CA} + \overline{AL}}$$

So $\overline{AM} + \overline{BN} + \overline{CL} = \overline{AB} + \overline{BM} + \overline{BC} + \overline{CN} + \overline{CA} + \overline{AL}$
$$= (\overline{AB} + \overline{BC} + \overline{CA}) + (\overline{BM} + \overline{CN} + \overline{AL})$$
$$= (\overline{AB} + \overline{BC} + \overline{CA}) + \frac{1}{2}(\overline{BC} + \overline{CA} + \overline{AB})$$
$$= \dots\dots\dots$$

Finish it off

24

$$\boxed{\overline{AM} + \overline{BN} + \overline{CL} = \mathbf{0}}$$

Because $\overline{AM} + \overline{BN} + \overline{CL} = (\overline{AB} + \overline{BC} + \overline{CA}) + \frac{1}{2}(\overline{BC} + \overline{CA} + \overline{AB})$

Now $\overline{AB} + \overline{BC} + \overline{CA}$ is a closed figure $\quad \therefore$ Vector sum $= \mathbf{0}$

and $\overline{BC} + \overline{CA} + \overline{AB}$ is a closed figure $\quad \therefore$ Vector sum $= \mathbf{0}$

$\therefore \ \overline{AM} + \overline{BN} + \overline{CL} = \mathbf{0}$

Here is another.

▶

Example 3

ABCD is a quadrilateral in which P and Q are the mid-points of the diagonals AC and BD respectively.

Show that $\quad \overline{AB} + \overline{AD} + \overline{CB} + \overline{CD} = 4\overline{PQ}$

First, just draw the figure.

Then move on to Frame 25

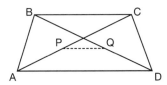

25

To prove that $\overline{AB} + \overline{AD} + \overline{CB} + \overline{CD} = 4\overline{PQ}$

Taking the vectors on the left-hand side, one at a time, we can write:

$\overline{AB} = \overline{AP} + \overline{PQ} + \overline{QB}$

$\overline{AD} = \overline{AP} + \overline{PQ} + \overline{QD}$

$\overline{CB} = \ldots\ldots\ldots\ldots$

$\overline{CD} = \ldots\ldots\ldots\ldots$

$$\boxed{\overline{CB} = \overline{CP} + \overline{PQ} + \overline{QB}; \quad \overline{CD} = \overline{CP} + \overline{PQ} + \overline{QD}}$$

26

Adding all four lines together, we have:

$$\overline{AB} + \overline{AD} + \overline{CB} + \overline{CD} = 4\overline{PQ} + 2\overline{AP} + 2\overline{CP} + 2\overline{QB} + 2\overline{QD}$$
$$= 4\overline{PQ} + 2(\overline{AP} + \overline{CP}) + 2(\overline{QB} + \overline{QD})$$

Now what can we say about $(\overline{AP} + \overline{CP})$?

$$\boxed{\overline{AP} + \overline{CP} = \mathbf{0}}$$

27

Because P is the mid-point of AC $\therefore$ AP = PC

$\therefore \overline{CP} = -\overline{PC} = -\overline{AP}$

$\therefore \overline{AP} + \overline{CP} = \overline{AP} - \overline{AP} = \mathbf{0}$.

In the same way, $(\overline{QB} + \overline{QD}) = \ldots\ldots\ldots\ldots$

$$\boxed{\overline{QB} + \overline{QD} = \mathbf{0}}$$

28

Since Q is the mid-point of BD $\therefore$ $\overline{QD} = -\overline{QB}$

$\therefore \overline{QB} + \overline{QD} = \overline{QB} - \overline{QB} = \mathbf{0}$

$\therefore \overline{AB} + \overline{AD} + \overline{CB} + \overline{CD} = 4\overline{PQ} + \mathbf{0} + \mathbf{0} = 4\overline{PQ}$

29

Here is one more.

Example 4

Prove by vectors that the line joining the mid-points of two sides of a triangle is parallel to the third side and half its length.

Let D and E be the mid-points of AB and AC respectively.

We have $\overline{DE} = \overline{DA} + \overline{AE}$

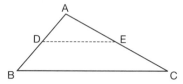

Now express $\overline{DA}$ and $\overline{AE}$ in terms of $\overline{BA}$ and $\overline{AC}$ respectively and see if you can get the required results.

Then on to Frame 30

30

Here is the working. Check through it.

$$\overline{DE} = \overline{DA} + \overline{AE}$$
$$= \frac{1}{2}\overline{BA} + \frac{1}{2}\overline{AC} = \frac{1}{2}(\overline{BA} + \overline{AC})$$
$$\overline{DE} = \frac{1}{2}\overline{BC}$$

∴ $\overline{DE}$ is half the magnitude (length) of $\overline{BC}$ and acts in the same direction.

i.e. DE and BC are parallel.

Now for the next section of the work: move on to Frame 31

31 **Components of a vector in terms of unit vectors**

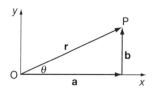

The vector $\overline{OP}$ is defined by its magnitude (r) and its direction (θ). It could also be defined by its two components in the Ox and Oy directions.

i.e. $\overline{OP}$ is equivalent to a vector **a** in the Ox direction + a vector **b** in the Oy direction.

i.e. $\overline{OP} = $ **a** (along Ox) + **b** (along Oy)

If we now define **i** to be a *unit vector* in the Ox direction,

then   **a** = a**i**

Similarly, if we define **j** to be a *unit vector* in the Oy direction,

then   **b** = b**j**

So that the vector OP can be written as:

r = a**i** + b**j**

where **i** and **j** are unit vectors in the Ox and Oy directions.

Let $z_1 = 2\mathbf{i} + 4\mathbf{j}$ and $z_2 = 5\mathbf{i} + 2\mathbf{j}$

32

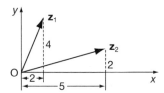

To find $z_1 + z_2$, draw the two vectors in a chain.

$$z_1 + z_2 = \overline{OB} = (2+5)\mathbf{i} + (4+2)\mathbf{j} = 7\mathbf{i} + 6\mathbf{j}$$

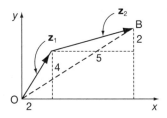

i.e. total up the vector components along O*x*,

and total up the vector components along O*y*

Of course, we can do this without a diagram:

If $z_1 = 3\mathbf{i} + 2\mathbf{j}$ and $z_2 = 4\mathbf{i} + 3\mathbf{j}$

$$z_1 + z_2 = 3\mathbf{i} + 2\mathbf{j} + 4\mathbf{i} + 3\mathbf{j}$$
$$= 7\mathbf{i} + 5\mathbf{j}$$

And in much the same way, $z_2 - z_1 = \ldots\ldots\ldots$

$$\boxed{z_2 - z_1 = \mathbf{i} + \mathbf{j}}$$

33

Because

$$z_2 - z_1 = (4\mathbf{i} + 3\mathbf{j}) - (3\mathbf{i} + 2\mathbf{j})$$
$$= 4\mathbf{i} + 3\mathbf{j} - 3\mathbf{i} - 2\mathbf{j}$$
$$= 1\mathbf{i} + 1\mathbf{j}$$
$$= \mathbf{i} + \mathbf{j}$$

Similarly, if $z_1 = 5\mathbf{i} - 2\mathbf{j}$; $z_2 = 3\mathbf{i} + 3\mathbf{j}$; $z_3 = 4\mathbf{i} - 1\mathbf{j}$

then (a) $z_1 + z_2 + z_3 = \ldots\ldots\ldots$
and (b) $z_1 - z_2 - z_3 = \ldots\ldots\ldots$

When you have the results, move on to Frame 34

34

$$\boxed{\text{(a) } 12\mathbf{i} \qquad \text{(b) } -2\mathbf{i} - 4\mathbf{j}}$$

Here is the working:

(a) $\mathbf{z}_1 + \mathbf{z}_2 + \mathbf{z}_3 = 5\mathbf{i} - 2\mathbf{j} + 3\mathbf{i} + 3\mathbf{j} + 4\mathbf{i} - 1\mathbf{j}$

$$= (5 + 3 + 4)\mathbf{i} + (3 - 2 - 1)\mathbf{j} = 12\mathbf{i}$$

(b) $\mathbf{z}_1 - \mathbf{z}_2 - \mathbf{z}_3 = (5\mathbf{i} - 2\mathbf{j}) - (3\mathbf{i} + 3\mathbf{j}) - (4\mathbf{i} - 1\mathbf{j})$

$$= (5 - 3 - 4)\mathbf{i} + (-2 - 3 + 1)\mathbf{j} = -2\mathbf{i} - 4\mathbf{j}$$

Now this one.

If $\overline{OA} = 3\mathbf{i} + 5\mathbf{j}$ and $\overline{OB} = 5\mathbf{i} - 2\mathbf{j}$, find $\overline{AB}$.

As usual, a diagram will help. Here it is:

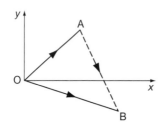

First of all, from the diagram, write down a relationship between the vectors. Then express them in terms of the unit vectors.

$$\overline{AB} = \dots\dots\dots$$

35

$$\boxed{\overline{AB} = 2\mathbf{i} - 7\mathbf{j}}$$

Because we have

$$\overline{OA} + \overline{AB} = \overline{OB} \text{ (from diagram)}$$
$$\therefore \ \overline{AB} = \overline{OB} - \overline{OA}$$
$$= (5\mathbf{i} - 2\mathbf{j}) - (3\mathbf{i} + 5\mathbf{j}) = 2\mathbf{i} - 7\mathbf{j}$$

Let's just pause and summarize the work so far on scalar and vector quantities

---

 # Review summary   Unit 33

36   1   A *scalar* quantity has magnitude only; a *vector* quantity has both magnitude and direction.

   2   The symbols $\mathbf{i}$ and $\mathbf{j}$ denote *unit vectors* that lie along the x- and y-axes respectively.

   If $\overline{OP} = a\mathbf{i} + b\mathbf{j}$, then $|\overline{OP}| = r = \sqrt{a^2 + b^2}$

Review exercise Unit 33

1 If $\overline{OP} = \mathbf{i} - 3\mathbf{j}$, $\overline{OQ} = -2\mathbf{i} + 5\mathbf{j}$, $\overline{OR} = 5\mathbf{i} + 2\mathbf{j}$, find $\overline{PQ}$, $\overline{PR}$ and $\overline{QR}$ and deduce the lengths of the sides of the triangle PQR. **37**

Complete the question. Take your time, there is no need to rush.
If necessary, look back at the Unit.
The answers and working are in the next frame.

---

1 $\overline{OP} = \mathbf{i} - 3\mathbf{j}$, $\overline{OQ} = -2\mathbf{i} + 5\mathbf{j}$, $\overline{OR} = 5\mathbf{i} + 2\mathbf{j}$ so that: **38**

$\overline{OP} + \overline{PQ} = \overline{OQ}$ that is $\overline{PQ} = \overline{OQ} - \overline{OP}$

$$= (-2\mathbf{i} + 5\mathbf{j}) - (\mathbf{i} - 3\mathbf{j})$$

$$= -3\mathbf{i} + 8\mathbf{j}$$

and so $|\overline{PQ}| = \sqrt{(-3)^2 + 8^2} = \sqrt{73}$

$\overline{OP} + \overline{PR} = \overline{OR}$ that is $\overline{PR} = \overline{OR} - \overline{OP}$

$$= (5\mathbf{i} + 2\mathbf{j}) - (\mathbf{i} - 3\mathbf{j})$$

$$= 4\mathbf{i} + 5\mathbf{j}$$

and so $|\overline{PR}| = \sqrt{4^2 + 5^2} = \sqrt{41}$

$\overline{OQ} + \overline{QR} = \overline{OR}$ that is $\overline{QR} = \overline{OR} - \overline{OQ}$

$$= (5\mathbf{i} + 2\mathbf{j}) - (-2\mathbf{i} + 5\mathbf{j})$$

$$= 7\mathbf{i} - 3\mathbf{j}$$

and so $|\overline{QR}| = \sqrt{7^2 + (-3)^2} = \sqrt{58}$

Now for the Review test

 # Review test Unit 33

1 If $\overline{OA} = -2\mathbf{i} - 5\mathbf{j}$, $\overline{OB} = \mathbf{i} - \mathbf{j}$, $\overline{OC} = 3\mathbf{i} - 4\mathbf{j}$, $\overline{OD} = 6\mathbf{i} + \mathbf{j}$, find $\overline{AB}$, $\overline{BC}$, $\overline{CD}$ and $\overline{DA}$ and deduce the lengths of the sides of the quadrilateral ABCD. **39**

Vectors in space   Unit 34

1 So far we have only considered vectors that lie in the *x–y* plane. These vectors are described using the two perpendicular unit vectors **i** and **j** that lie in the *x*- and *y*-directions respectively. To consider vectors in space we need a third direction with an axis that is perpendicular to both the *x*- and *y*-axes. This is provided by the *z*-axis. The *x*-, *y*- and *z*-axes indicate three mutually perpendicular directions for the axes of reference which are defined by the 'right-hand' rule.

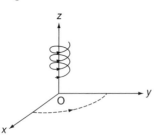

O*x*, O*y*, O*z* form a right-handed set if rotation from O*x* to O*y* takes a right-handed corkscrew action along the positive direction of O*z*.

Similarly, rotation from O*y* to O*z* gives right-hand corkscrew action along the positive direction of ....

2

$$\boxed{Ox}$$

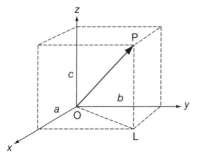

Vector $\overline{OP}$ is defined by its components

a along O*x*

b along O*y*

c along O*z*

Let **i** = unit vector in O*x* direction

j = unit vector in O*y* direction

k = unit vector in O*z* direction

Then $\overline{OP} = a\mathbf{i} + b\mathbf{j} + c\mathbf{k}$

Also $OL^2 = a^2 + b^2$ and $OP^2 = OL^2 + c^2$

$$OP^2 = a^2 + b^2 + c^2$$

So, if $\mathbf{r} = a\mathbf{i} + b\mathbf{j} + c\mathbf{k}$, then $r = \sqrt{a^2 + b^2 + c^2}$

This gives us an easy way of finding the magnitude of a vector expressed in terms of the unit vectors.

Now you can do this one:

If $\overline{PQ} = 4\mathbf{i} + 3\mathbf{j} + 2\mathbf{k}$, then $|\overline{PQ}| = $ ....

$$\boxed{|\overline{PQ}| = \sqrt{29} = 5 \cdot 385}$$

3

Because

$$\overline{PQ} = 4\mathbf{i} + 3\mathbf{j} + 2\mathbf{k}$$
$$|\overline{PQ}| = \sqrt{4^2 + 3^2 + 2^2}$$
$$= \sqrt{16 + 9 + 4}$$
$$= \sqrt{29} = 5 \cdot 385$$

Now move on to Frame 4

Direction cosines

The direction of a vector in three dimensions is determined by the angles which the vector makes with the three axes of reference.

4

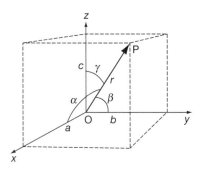

Let $\overline{OP} = \mathbf{r} = a\mathbf{i} + b\mathbf{j} + c\mathbf{k}$

Then

$$\frac{a}{r} = \cos \alpha \qquad \therefore \ a = r \cos \alpha$$

$$\frac{b}{r} = \cos \beta \qquad b = r \cos \beta$$

$$\frac{c}{r} = \cos \gamma \qquad c = r \cos \gamma$$

Also $a^2 + b^2 + c^2 = r^2$

$\therefore \ r^2 \cos^2 \alpha + r^2 \cos^2 \beta + r^2 \cos^2 \gamma = r^2$

$\therefore \ \cos^2 \alpha + \cos^2 \beta + \cos^2 \gamma = 1$

If $l = \cos \alpha$

$m = \cos \beta$

$n = \cos \gamma$ then $l^2 + m^2 + n^2 = 1$

Note: $[l, m, n]$ written in square brackets are called the *direction cosines* of the vector $\overline{OP}$ and are the values of the cosines of the angles which the vector makes with the three axes of reference.

So for the vector $\mathbf{r} = a\mathbf{i} + b\mathbf{j} + c\mathbf{k}$

$$l = \frac{a}{r}; \ m = \frac{b}{r}; \ n = \frac{c}{r}; \ \text{and, of course } r = \sqrt{a^2 + b^2 + c^2}$$

So, with that in mind, find the direction cosines $[l, m, n]$ of the vector

$\mathbf{r} = 3\mathbf{i} - 2\mathbf{j} + 6\mathbf{k}$

Then to Frame 5

$$\mathbf{r} = 3\mathbf{i} - 2\mathbf{j} + 6\mathbf{k}$$

$$\therefore a = 3, \quad b = -2, \quad c = 6, \quad r = \sqrt{9 + 4 + 36}$$

$$\therefore r = \sqrt{49} = 7$$

$$\therefore l = \frac{3}{7}; \quad m = -\frac{2}{7}; \quad n = \frac{6}{7}$$

Just as easy as that!

On to the next frame

Angle between two vectors

Let **a** be one vector with direction cosines $[l, m, n]$
Let **b** be the other vector with direction cosines $[l', m', n']$

We have to find the angle between these two vectors.

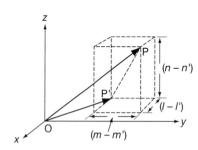

Let $\overline{OP}$ and $\overline{OP'}$ be *unit* vectors parallel to **a** and **b** respectively. Then P has coordinates (l, m, n) and P' has coordinates (l', m', n').

Then

$$(PP')^2 = (l - l')^2 + (m - m')^2 + (n - n')^2$$
$$= l^2 - 2.l.l' + l'^2 + m^2 - 2.m.m' + m'^2 + n^2 - 2n.n' + n'^2$$
$$= (l^2 + m^2 + n^2) + (l'^2 + m'^2 + n'^2) - 2(ll' + mm' + nn')$$

But $(l^2 + m^2 + n^2) = 1$ and $(l'^2 + m'^2 + n'^2) = 1$ as was proved earlier.

$$\therefore \ (PP')^2 = 2 - 2(ll' + mm' + nn') \qquad (a)$$

Also, by the cosine rule:

$$(PP')^2 = OP^2 + OP'^2 - 2.OP.OP'.\cos\theta$$
$$= 1 + 1 - 2.1.1.\cos\theta \qquad \left\{ \begin{array}{l} \overline{OP} \text{ and } \overline{OP'} \text{ are} \\ \text{unit vectors} \end{array} \right\}$$
$$= 2 - 2\cos\theta \qquad (b)$$

So from (a) and (b), we have:

$$(PP')^2 = 2 - 2(ll' + mm' + nn')$$
and $(PP')^2 = 2 - 2\cos\theta$
$$\therefore \qquad \cos\theta = \dots\dots\dots$$

$$\boxed{\cos\theta = ll' + mm' + nn'}$$

7

That is, just sum the products of the corresponding direction cosines of the two given vectors.

So, if $[l, m, n] = [0{\cdot}54, 0{\cdot}83, -0{\cdot}14]$
and $[l', m', n'] = [0{\cdot}25, 0{\cdot}60, 0{\cdot}76]$

the angle between the vectors is $\theta = \ldots\ldots\ldots\ldots$

$$\boxed{\theta = 58°13'}$$

8

Because, we have:

$$
\begin{aligned}
\cos\theta &= ll' &&+ \; mm' &&+ \; nn' \\
&= (0{\cdot}54)(0{\cdot}25) &&+ \; (0{\cdot}83)(0{\cdot}60) &&+ \; (-0{\cdot}14)(0{\cdot}76) \\
&= 0{\cdot}1350 &&+ \; 0{\cdot}4980 &&- \; 0{\cdot}1064 \\
&= 0{\cdot}6330 &&- \; 0{\cdot}1064 && \qquad\qquad = 0{\cdot}5266
\end{aligned}
$$

$\theta = 58°13'$

Note: For *parallel vectors*, $\theta = 0°$ $\therefore$ $ll' + mm' + nn' = 1$

For *perpendicular vectors*, $\theta = 90°$, $\therefore$ $ll' + mm' + nn' = 0$

Now an example for you to work:

Find the angle between the vectors

$$\mathbf{p} = 2\mathbf{i} + 3\mathbf{j} + 4\mathbf{k} \text{ and } \mathbf{q} = 4\mathbf{i} - 3\mathbf{j} + 2\mathbf{k}$$

First of all, find the direction cosines of $\mathbf{p}$. So do that.

$$\boxed{l = \frac{2}{\sqrt{29}} \qquad m = \frac{3}{\sqrt{29}} \qquad n = \frac{4}{\sqrt{29}}}$$

9

Because

$$p = |\mathbf{p}| = \sqrt{2^2 + 3^2 + 4^2} = \sqrt{4 + 9 + 16} = \sqrt{29}$$

$$\therefore l = \frac{a}{p} = \frac{2}{\sqrt{29}}$$

$$m = \frac{b}{p} = \frac{3}{\sqrt{29}}$$

$$n = \frac{c}{p} = \frac{4}{\sqrt{29}}$$

$$\therefore [l, m, n] = \left[\frac{2}{\sqrt{29}}, \frac{3}{\sqrt{29}}, \frac{4}{\sqrt{29}}\right]$$

Now find the direction cosines $[l', m', n']$ of $\mathbf{q}$ in just the same way.

When you have done that move on to the next frame

10

$$l' = \frac{4}{\sqrt{29}} \qquad m' = \frac{-3}{\sqrt{29}} \qquad n' = \frac{2}{\sqrt{29}}$$

Because

$$q = |\mathbf{q}| = \sqrt{4^2 + 3^2 + 2^2} = \sqrt{16 + 9 + 4} = \sqrt{29}$$

$$\therefore \ [l', m', n'] = \left[\frac{4}{\sqrt{29}}, \frac{-3}{\sqrt{29}}, \frac{2}{\sqrt{29}} \right]$$

We already know that, for **p**:

$$[l, m, n] = \left[\frac{2}{\sqrt{29}}, \frac{3}{\sqrt{29}}, \frac{4}{\sqrt{29}} \right]$$

So, using $\cos \theta = ll' + mm' + nn'$, you can finish it off and find the angle θ. Off you go.

11

$$\theta = 76°2'$$

Because

$$\cos \theta = \frac{2}{\sqrt{29}} \cdot \frac{4}{\sqrt{29}} + \frac{3}{\sqrt{29}} \cdot \frac{(-3)}{\sqrt{29}} + \frac{4}{\sqrt{29}} \cdot \frac{2}{\sqrt{29}}$$

$$= \frac{8}{29} - \frac{9}{29} + \frac{8}{29}$$

$$= \frac{7}{29} = 0 \cdot 2414 \ \therefore \ \theta = 76°2'$$

Let's just pause and summarize the work so far on vectors in space

 # Review summary   Unit 34

12

1 The axes of reference, Ox, Oy and Oz, are chosen so that they form a right-handed set. The symbols **i**, **j** and **k** denote *unit vectors* that lie along the x-, y- and z-axes respectively.

    If $\overline{OP} = a\mathbf{i} + b\mathbf{j} + c\mathbf{k}$ then $|\overline{OP}| = r = \sqrt{a^2 + b^2 + c^2}$

2 The *direction cosines* $[l, m, n]$ are the cosines of the angles between the vector and the axes Ox, Oy, Oz respectively.

    For any vector: $l = \dfrac{a}{r}, \ m = \dfrac{b}{r}, \ n = \dfrac{c}{r}$; and $l^2 + m^2 + n^2 = 1$

3 *Angle between two vectors*

    $\cos \theta = ll' + mm' + nn'$

    For perpendicular vectors, $ll' + mm' + nn' = 0$

Review exercise **Unit 34**

1 Find the direction cosines of the vector joining the two points $(3, -1, 5)$ and $(2, 7, -1)$. **13**

2 Find the angle between the vectors joining the points $(3, -1, 5)$ and $(2, 7, -1)$.

Complete both questions.
Look back at the Unit if necessary, but don't rush.
The answers and working are in the next frame.

1 Let $\overline{OA} = (3, -1, 5)$ and $\overline{OB} = (2, 7, -1)$ then $\overline{OA} + \overline{AB} = \overline{OB}$ and so: **14**
$$\overline{AB} = \overline{OB} - \overline{OA}$$
$$= (2, 7, -1) - (3, -1, 5)$$
$$= (-1, 8, -6)$$
Therefore
$$|\overline{AB}| = \sqrt{(-1)^2 + 8^2 + (-6)^2} = \sqrt{101}$$
and the direction cosines are then: $l = \dfrac{-1}{\sqrt{101}}; \; m = \dfrac{8}{\sqrt{101}}; \; n = \dfrac{-6}{\sqrt{101}}$

2 Let $\overline{OA} = (3, -1, 5)$ so that $|\overline{OA}| = \sqrt{3^2 + (-1)^2 + 5^2} = \sqrt{35}$ and so:
$$[l, m, n] = \left[\frac{3}{\sqrt{35}}, \frac{-1}{\sqrt{35}}, \frac{5}{\sqrt{35}}\right]$$
Similarly, let $\overline{OB} = (2, 7, -1)$ so that $|\overline{OB}| = \sqrt{2^2 + 7^2 + (-1)^2} = \sqrt{54}$ and so:
$$[l', m', n'] = \left[\frac{2}{\sqrt{54}}, \frac{7}{\sqrt{54}}, \frac{-1}{\sqrt{54}}\right]$$
This means that:
$$\cos\theta = ll' + mm' + nn'$$
$$= \frac{3}{\sqrt{35}} \times \frac{2}{\sqrt{54}} + \frac{(-1)}{\sqrt{35}} \times \frac{7}{\sqrt{54}} + \frac{5}{\sqrt{35}} \times \frac{(-1)}{\sqrt{54}}$$
$$= \frac{-6}{\sqrt{35 \times 54}}$$
$$= -0.1380 \text{ to 4 dp.}$$
Therefore $\theta = 97.9°$

Now for the Review test

Review test   **Unit 34**

15   **1** Find the direction cosines of the vector joining the two points $(0, 0, -1)$ and $(-1, 1, 0)$.

   2 Find the angle between the vectors joining the points $(0, 0, -1)$ and $(-1, 1, 0)$.

Products of vectors   **Unit 35**

1   Scalar product of two vectors

If **a** and **b** are two vectors, the *scalar product* of **a** and **b** is defined as the *scalar* (number) $ab \cos \theta$ where a and b are the magnitudes of the vectors **a** and **b** and θ is the angle between them.

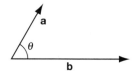

The scalar product is denoted by **a.b** (often called the 'dot product' for obvious reasons).

$$\therefore \ \mathbf{a.b} = ab \cos \theta$$
$$= a \times \text{projection of } \mathbf{b} \text{ on } \mathbf{a}$$
$$= b \times \text{projection of } \mathbf{a} \text{ on } \mathbf{b}$$

In both cases the result is a *scalar* quantity.

For example:

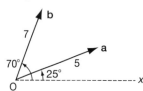

$$\mathbf{a.b} = \dots\dots\dots$$

2

$$\boxed{\mathbf{a.b} = \frac{35\sqrt{2}}{2}}$$

Because we have:

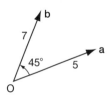

$$\mathbf{a.b} = a.b. \cos \theta$$
$$= 5.7. \cos 45°$$
$$= 35. \frac{1}{\sqrt{2}} = \frac{35\sqrt{2}}{2}$$

Now what about this case:

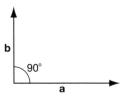

The scalar product of **a** and **b** = **a.b** = ...

$$0$$

3

Because in this case **a.b** = $ab\cos 90° = ab\,0 = 0$. So the scalar product of any two vectors at right-angles to each other is always zero.

And in this case now, with two vectors in the same direction, $\theta = 0°$

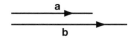

so **a.b** = ...

$$ab$$

4

Because **a.b** = $ab\cos 0° = ab.1 = ab$

Now suppose our two vectors are expressed in terms of the unit vectors **i**, **j** and **k**.

Let   **a** = $a_1\mathbf{i} + a_2\mathbf{j} + a_3\mathbf{k}$
and   **b** = $b_1\mathbf{i} + b_2\mathbf{j} + b_3\mathbf{k}$
Then **a.b** = $(a_1\mathbf{i} + a_2\mathbf{j} + a_3\mathbf{k}).(b_1\mathbf{i} + b_2\mathbf{j} + b_3\mathbf{k})$
= $a_1b_1\mathbf{i.i} + a_1b_2\mathbf{i.j} + a_1b_3\mathbf{i.k} + a_2b_1\mathbf{j.i} + a_2b_2\mathbf{j.j} + a_2b_3\mathbf{j.k}$
$+ a_3b_1\mathbf{k.i} + a_3b_2\mathbf{k.j} + a_3b_3\mathbf{k.k}$

This can now be simplified.

Because **i.i** = $(1)(1)(\cos 0°) = 1$

∴ **i.i** = 1; **j.j** = 1; **k.k** = 1   (a)

Also **i.j** = $(1)(1)(\cos 90°) = 0$

∴ **i.j** = 0; **j.k** = 0; **k.i** = 0   (b)

So, using the results (a) and (b), we get:

$$\mathbf{a.b} = \ldots\ldots\ldots\ldots$$

5

$$\mathbf{a.b} = a_1b_1 + a_2b_2 + a_3b_3$$

Because

$$\mathbf{a.b} = a_1b_1.1 + a_1b_2.0 + a_1b_3.0 + a_2b_1.0 + a_2b_2.1 + a_2b_3.0$$
$$+ a_3b_1.0 + a_3b_2.0 + a_3b_3.1$$
$$\therefore \ \mathbf{a.b} = a_1b_1 + a_2b_2 + a_3b_3$$

i.e. we just sum the products of the coefficients of the unit vectors along the corresponding axes.

For example:

If $\mathbf{a} = 2\mathbf{i} + 3\mathbf{j} + 5\mathbf{k}$ and $\mathbf{b} = 4\mathbf{i} + 1\mathbf{j} + 6\mathbf{k}$

then $\mathbf{a.b} = 2 \times 4 + 3 \times 1 + 5 \times 6$

$$= 8 + 3 + 30$$
$$= 41 \quad \therefore \ \mathbf{a.b} = 41$$

One for you: If $\mathbf{p} = 3\mathbf{i} - 2\mathbf{j} + 1\mathbf{k}; \ \ \mathbf{q} = 2\mathbf{i} + 3\mathbf{j} - 4\mathbf{k}$

then $\mathbf{p.q} = \dots\dots\dots$

6

$$-4$$

Because

$$\mathbf{p.q} = 3 \times 2 + (-2) \times 3 + 1 \times (-4)$$
$$= 6 - 6 - 4$$
$$= -4 \qquad\qquad \therefore \ \mathbf{p.q} = -4$$

Now on to Frame 7

7 **Vector product of two vectors**

The vector product of $\mathbf{a}$ and $\mathbf{b}$ is written $\mathbf{a} \times \mathbf{b}$ (often called the 'cross product') and is defined as a *vector* having magnitude $ab\sin\theta$ where θ is the angle between the two given vectors. The product vector acts in a direction perpendicular to both $\mathbf{a}$ and $\mathbf{b}$ in such a sense that $\mathbf{a}$, $\mathbf{b}$ and $\mathbf{a} \times \mathbf{b}$ form a right-handed set – in that order.

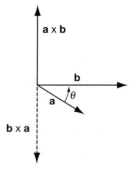

$$|\mathbf{a} \times \mathbf{b}| = ab\sin\theta$$

Note that $\mathbf{b} \times \mathbf{a}$ reverses the direction of rotation and the product vector would now act downwards, i.e.

$$\mathbf{b} \times \mathbf{a} = -(\mathbf{a} \times \mathbf{b})$$

If $\theta = 0°$, then $|\mathbf{a} \times \mathbf{b}| = \dots\dots\dots$

and if $\theta = 90°$, then $|\mathbf{a} \times \mathbf{b}| = \dots\dots\dots$

8

$$\theta = 0°, \quad |\mathbf{a} \times \mathbf{b}| = 0$$
$$\theta = 90°, \quad |\mathbf{a} \times \mathbf{b}| = ab$$

If $\mathbf{a}$ and $\mathbf{b}$ are given in terms of the unit vectors $\mathbf{i}$, $\mathbf{j}$ and $\mathbf{k}$:

$\mathbf{a} = a_1\mathbf{i} + a_2\mathbf{j} + a_3\mathbf{k}$ and $\mathbf{b} = b_1\mathbf{i} + b_2\mathbf{j} + b_3\mathbf{k}$

Then:

$$\mathbf{a} \times \mathbf{b} = a_1b_1\mathbf{i} \times \mathbf{i} + a_1b_2\mathbf{i} \times \mathbf{j} + a_1b_3\mathbf{i} \times \mathbf{k} + a_2b_1\mathbf{j} \times \mathbf{i} + a_2b_2\mathbf{j} \times \mathbf{j}$$
$$+ a_2b_3\mathbf{j} \times \mathbf{k} + a_3b_1\mathbf{k} \times \mathbf{i} + a_3b_2\mathbf{k} \times \mathbf{j} + a_3b_3\mathbf{k} \times \mathbf{k}$$

But $|\mathbf{i} \times \mathbf{i}| = (1)(1)(\sin 0°) = 0 \quad \therefore \quad \mathbf{i} \times \mathbf{i} = \mathbf{j} \times \mathbf{j} = \mathbf{k} \times \mathbf{k} = 0$ (a)

Also $|\mathbf{i} \times \mathbf{j}| = (1)(1)(\sin 90°) = 1$ and $\mathbf{i} \times \mathbf{j}$ is in the direction of $\mathbf{k}$, i.e. $\mathbf{i} \times \mathbf{j} = \mathbf{k}$ (same magnitude and same direction). Therefore:

$\mathbf{i} \times \mathbf{j} = \mathbf{k}$
$\mathbf{j} \times \mathbf{k} = \mathbf{i}$
$\mathbf{k} \times \mathbf{i} = \mathbf{j}$ (b)

And remember too that therefore:

$\mathbf{i} \times \mathbf{j} = -(\mathbf{j} \times \mathbf{i})$
$\mathbf{j} \times \mathbf{k} = -(\mathbf{k} \times \mathbf{j})$
$\mathbf{k} \times \mathbf{i} = -(\mathbf{i} \times \mathbf{k})$ since the sense of rotation is reversed

Now with the results of (a) and (b), and this last reminder, you can simplify the expression for $\mathbf{a} \times \mathbf{b}$.

Remove the zero terms and tidy up what is left.

. . . .

Then on to Frame 9

9

$$\mathbf{a} \times \mathbf{b} = (a_2b_3 - a_3b_2)\mathbf{i} - (a_1b_3 - a_3b_1)\mathbf{j} + (a_1b_2 - a_2b_1)\mathbf{k}$$

Because

$$\mathbf{a} \times \mathbf{b} = a_1b_1 0 + a_1b_2\mathbf{k} + a_1b_3(-\mathbf{j}) + a_2b_1(-\mathbf{k}) + a_2b_2 0 + a_2b_3\mathbf{i}$$
$$+ a_3b_1\mathbf{j} + a_3b_2(-\mathbf{i}) + a_3b_3 0$$
$$\mathbf{a} \times \mathbf{b} = (a_2b_3 - a_3b_2)\mathbf{i} - (a_1b_3 - a_3b_1)\mathbf{j} + (a_1b_2 - a_2b_1)\mathbf{k}$$

Now this is the pattern of a determinant where the first row is made up of the vectors $\mathbf{i}$, $\mathbf{j}$ and $\mathbf{k}$.

So now we have that:

If $\mathbf{a} = a_1\mathbf{i} + a_2\mathbf{j} + a_3\mathbf{k}$ and $\mathbf{b} = b_1\mathbf{i} + b_2\mathbf{j} + b_3\mathbf{k}$ then:

$$\mathbf{a} \times \mathbf{b} = \begin{vmatrix} \mathbf{i} & \mathbf{j} & \mathbf{k} \\ a_1 & a_2 & a_3 \\ b_1 & b_2 & b_3 \end{vmatrix} \rightarrow \begin{matrix} \mathbf{i} & \mathbf{j} & \mathbf{k} & \mathbf{i} & \mathbf{j} \\ a_1 & a_2 & a_3 & a_1 & a_2 \\ b_1 & b_2 & b_3 & b_1 & b_2 \end{matrix}$$

$$= (\mathbf{i}a_2b_3 + \mathbf{j}a_3b_1 + \mathbf{k}a_1b_2) - (\mathbf{k}a_2b_1 + \mathbf{i}a_3b_2 + \mathbf{j}a_1b_3)$$

so that:

$$\mathbf{a} \times \mathbf{b} = \begin{vmatrix} \mathbf{i} & \mathbf{j} & \mathbf{k} \\ a_1 & a_2 & a_3 \\ b_1 & b_2 & b_3 \end{vmatrix} = (a_2b_3 - a_3b_2)\mathbf{i} - (a_1b_3 - a_3b_1)\mathbf{j} + (a_1b_2 - a_2b_1)\mathbf{k}$$

and that is the easiest way to write out the vector product of two vectors.

Notes: (a) The top row consists of the unit vectors in order $\mathbf{i}$, $\mathbf{j}$, $\mathbf{k}$.
(b) The second row consists of the coefficients of $\mathbf{a}$.
(c) The third row consists of the coefficients of $\mathbf{b}$.

For example, if $\mathbf{p} = 2\mathbf{i} + 4\mathbf{j} + 3\mathbf{k}$ and $\mathbf{q} = \mathbf{i} + 5\mathbf{j} - 2\mathbf{k}$, first write down the determinant that represents the vector product $\mathbf{p} \times \mathbf{q}$.

10

$$\mathbf{p} \times \mathbf{q} = \begin{vmatrix} \mathbf{i} & \mathbf{j} & \mathbf{k} \\ 2 & 4 & 3 \\ 1 & 5 & -2 \end{vmatrix} \quad \begin{matrix} \text{Unit vectors} \\ \text{Coefficients of } \mathbf{p} \\ \text{Coefficients of } \mathbf{q} \end{matrix}$$

And now, expanding the determinant, we get:

$$\mathbf{p} \times \mathbf{q} = \dots\dots\dots$$

11

$$\mathbf{p} \times \mathbf{q} = -23\mathbf{i} + 7\mathbf{j} + 6\mathbf{k}$$

Because

$$\mathbf{p} \times \mathbf{q} = \begin{vmatrix} \mathbf{i} & \mathbf{j} & \mathbf{k} \\ 2 & 4 & 3 \\ 1 & 5 & -2 \end{vmatrix}$$

$$= [\mathbf{i} \times 4 \times (-2) + \mathbf{j} \times 3 \times 1 + \mathbf{k} \times 2 \times 5]$$
$$\quad - [\mathbf{i} \times 3 \times 5 + \mathbf{j} \times 2 \times (-2) + \mathbf{k} \times 4 \times 1]$$
$$= \mathbf{i}(-8 - 15) - \mathbf{j}(-4 - 3) + \mathbf{k}(10 - 4)$$
$$= -23\mathbf{i} + 7\mathbf{j} + 6\mathbf{k}$$

So, by way of revision:

(a) *Scalar product* ('dot product')

$\mathbf{a}.\mathbf{b} = ab\cos\theta$ a scalar quantity

(b) *Vector product* ('cross product')

$\mathbf{a} \times \mathbf{b}$ = vector of magnitude $ab\sin\theta$, acting in a direction to make $\mathbf{a}$, $\mathbf{b}$ and $\mathbf{a} \times \mathbf{b}$ a right-handed set. Also:

$$\mathbf{a} \times \mathbf{b} = \begin{vmatrix} \mathbf{i} & \mathbf{j} & \mathbf{k} \\ a_1 & a_2 & a_3 \\ b_1 & b_2 & b_3 \end{vmatrix} = (a_2b_3 - a_3b_2)\mathbf{i} - (a_1b_3 - a_3b_1)\mathbf{j} + (a_1b_2 - a_2b_1)\mathbf{k}$$

And here is one final example on this point.

Find the vector product of $\mathbf{p}$ and $\mathbf{q}$ where:

$\mathbf{p} = 3\mathbf{i} - 4\mathbf{j} + 2\mathbf{k}$ and $\mathbf{q} = 2\mathbf{i} + 5\mathbf{j} - \mathbf{k}$

$$\boxed{\mathbf{p} \times \mathbf{q} = -6\mathbf{i} + 7\mathbf{j} + 23\mathbf{k}}$$

12

Because

$$\mathbf{p} \times \mathbf{q} = \begin{vmatrix} \mathbf{i} & \mathbf{j} & \mathbf{k} \\ 3 & -4 & 2 \\ 2 & 5 & -1 \end{vmatrix}$$

$$= [\mathbf{i} \times (-4) \times (-1) + \mathbf{j} \times 2 \times 2 + \mathbf{k} \times 3 \times 5]$$
$$\quad - [\mathbf{i} \times 2 \times 5 + \mathbf{j} \times 3 \times (-1) + \mathbf{k} \times (-4) \times 2]$$
$$= \mathbf{i}(4 - 10) - \mathbf{j}(-3 - 4) + \mathbf{k}(15 + 8)$$
$$= -6\mathbf{i} + 7\mathbf{j} + 23\mathbf{k}$$

Remember that the order in which the vectors appear in the vector product is important. It is a simple matter to verify that:

$\mathbf{q} \times \mathbf{p} = 6\mathbf{i} - 7\mathbf{j} - 23\mathbf{k} = -(\mathbf{p} \times \mathbf{q})$

At this point let's just pause and summarize the work so far on products of vectors

Review summary   Unit 35

13   1   *Scalar product* ('dot product')
$\mathbf{a}.\mathbf{b} = ab\cos\theta$ where θ is the angle between $\mathbf{a}$ and $\mathbf{b}$.
If $\mathbf{a} = a_1\mathbf{i} + a_2\mathbf{j} + a_3\mathbf{k}$ and $\mathbf{b} = b_1\mathbf{i} + b_2\mathbf{j} + b_3\mathbf{k}$
then $\mathbf{a}.\mathbf{b} = a_1b_1 + a_2b_2 + a_3b_3$

2   *Vector product* ('cross product')
$\mathbf{a} \times \mathbf{b} = (ab\sin\theta)$ in direction perpendicular to $\mathbf{a}$ and $\mathbf{b}$, so that $\mathbf{a}$, $\mathbf{b}$ and $(\mathbf{a} \times \mathbf{b})$ form a right-handed set.

$$\text{Also } \mathbf{a} \times \mathbf{b} = \begin{vmatrix} \mathbf{i} & \mathbf{j} & \mathbf{k} \\ a_1 & a_2 & a_3 \\ b_1 & b_2 & b_3 \end{vmatrix}$$

 Review exercise   Unit 35

14   1   If $\mathbf{a} = \mathbf{i} - \mathbf{j} + \mathbf{k}$ and $\mathbf{b} = 4\mathbf{i} + 2\mathbf{j} + 5\mathbf{k}$ find:
(a) $\mathbf{a}.\mathbf{b}$ (b) $\mathbf{a} \times \mathbf{b}$

2   If $\mathbf{a} = -3\mathbf{i} + 2\mathbf{j} - 6\mathbf{k}$, $\mathbf{b} = \mathbf{i} - 4\mathbf{j} + \mathbf{k}$ and $\mathbf{c} = 7\mathbf{i} + 11\mathbf{j} - 9\mathbf{k}$ determine:

(a) the value of $\mathbf{a}.\mathbf{b}$ and the angle between the vectors $\mathbf{a}$ and $\mathbf{b}$

(b) the magnitude and direction cosines of the vector product $\mathbf{a} \times \mathbf{b}$ and also the angle which this vector product makes with $\mathbf{c}$.

Complete both questions, looking back at the Unit if you need to.
Don't rush, take your time.
The answers and working are in the next frame.

15   1 $\mathbf{a} = \mathbf{i} - \mathbf{j} + \mathbf{k}$ and $\mathbf{b} = 4\mathbf{i} + 2\mathbf{j} + 5\mathbf{k}$ and so:

(a) $\mathbf{a}.\mathbf{b} = (\mathbf{i} - \mathbf{j} + \mathbf{k}).(4\mathbf{i} + 2\mathbf{j} + 5\mathbf{k})$
$= 1 \times 4 + (-1) \times 2 + 1 \times 5$
$= 7$

(b) $\mathbf{a} \times \mathbf{b} = \begin{vmatrix} \mathbf{i} & \mathbf{j} & \mathbf{k} \\ 1 & -1 & 1 \\ 4 & 2 & 5 \end{vmatrix} = (-5 - 2)\mathbf{i} - (5 - 4)\mathbf{j} + (2 - (-4))\mathbf{k}$
$= -7\mathbf{i} - \mathbf{j} + 6\mathbf{k}$

▶

2 $\mathbf{a} = -3\mathbf{i} + 2\mathbf{j} - 6\mathbf{k}$ and $\mathbf{b} = \mathbf{i} - 4\mathbf{j} + \mathbf{k}$ therefore:

(a) $\mathbf{a.b} = (-3\mathbf{i} + 2\mathbf{j} - 6\mathbf{k}).(\mathbf{i} - 4\mathbf{j} + \mathbf{k})$

$$= (-3) \times 1 + 2 \times (-4) + (-6) \times 1$$

$$= -17$$

$$= ab \cos \theta$$

where

$$a = \sqrt{(-3)^2 + 2^2 + (-6)^2} = \sqrt{49} = 7$$

and $b = \sqrt{1^2 + (-4)^2 + 1^2} = \sqrt{18} = 4 \cdot 2$

so that $-17 = 7 \times 4 \cdot 2 \times \cos \theta$ giving $\cos \theta = -\dfrac{17}{29 \cdot 4} = -0 \cdot 58$

therefore $\theta = 125 \cdot 5°$

(b) $\mathbf{a} \times \mathbf{b} = \begin{vmatrix} \mathbf{i} & \mathbf{j} & \mathbf{k} \\ -3 & 2 & -6 \\ 1 & -4 & 1 \end{vmatrix}$

$$= -22\mathbf{i} - 3\mathbf{j} + 10\mathbf{k}$$

and so:

$$|\mathbf{a} \times \mathbf{b}| = |-22\mathbf{i} - 3\mathbf{j} + 10\mathbf{k}| = \sqrt{(-22)^2 + (-3)^2 + 10^2}$$

$$= \sqrt{593} = 24 \cdot 4 \text{ to 1 dp.}$$

The direction cosines are:

$$l = \frac{-22}{24 \cdot 4} = -0 \cdot 90, \; m = \frac{-3}{24 \cdot 4} = -0 \cdot 12, \; n = \frac{10}{24 \cdot 4} = 0 \cdot 41$$

To find the angle that $\mathbf{a} \times \mathbf{b} = -22\mathbf{i} - 3\mathbf{j} + 10\mathbf{k}$ makes with $\mathbf{c} = 7\mathbf{i} + 11\mathbf{j} - 9\mathbf{k}$ we note that:

$$|\mathbf{c}| = |7\mathbf{i} + 11\mathbf{j} - 9\mathbf{k}| = \sqrt{7^2 + 11^2 + (-9)^2} = \sqrt{251} = 15 \cdot 8 \text{ to 1 dp.}$$

Hence the direction cosines of $\mathbf{c}$ are:

$$l' = \frac{7}{15 \cdot 8} = 0 \cdot 44, \; m' = \frac{11}{15 \cdot 8} = 0 \cdot 70, \; n' = \frac{-9}{15 \cdot 8} = -0 \cdot 57$$

From this we find that the cosine of the angle between these two vectors is given by:

$$\cos \theta = ll' + mm' + nn'$$

$$= (-0 \cdot 90)(0 \cdot 44) + (-0 \cdot 12)(0 \cdot 70) + (0 \cdot 41)(-0 \cdot 57)$$

$$= -0 \cdot 71$$

therefore $\theta = 135 \cdot 2°$.

Now for the Review test

Review test   **Unit 35**

16   **1**   If $\mathbf{a} = 2\mathbf{i} + 3\mathbf{j} - 4\mathbf{k}$ and $\mathbf{b} = -\mathbf{i} + \mathbf{j} - \mathbf{k}$ find:
          (a) $\mathbf{a}.\mathbf{b}$ (b) $\mathbf{a} \times \mathbf{b}$
       2   If $\mathbf{a} = 8\mathbf{i} - 5\mathbf{j} + 9\mathbf{k}$, $\mathbf{b} = 10\mathbf{i} + \mathbf{j} - 3\mathbf{k}$ and $\mathbf{c} = -5\mathbf{i} + 13\mathbf{j} - 7\mathbf{k}$ determine:

          (a) the value of $\mathbf{a}.\mathbf{b}$ and the angle between the vectors $\mathbf{a}$ and $\mathbf{b}$

          (b) the magnitude and direction cosines of the vector product $\mathbf{a} \times \mathbf{b}$ and also
              the angle which this vector product makes with $\mathbf{c}$.

17   You have now come to the end of this Module. A list of **Can You?** questions
          follows for you to gauge your understanding of the material in the Module.
          You will notice that these questions match the **Learning outcomes** listed at
          the beginning of the Module. Now try the **Test exercise.** *Work through the
          questions at your own pace, there is no need to hurry.* A set of **Further problems**
          provides valuable additional practice.

Can You?

1   Checklist: Module 11

Check this list before and after you try the end of Module test.

On a scale of 1 to 5 how confident are you that you can:

- Define a vector?
  Yes ☐ ☐ ☐ ☐ ☐   *No*

- Represent a vector by a directed straight line?
  Yes ☐ ☐ ☐ ☐ ☐   *No*

- Add vectors?
  Yes ☐ ☐ ☐ ☐ ☐   *No*

- Write a vector in terms of component vectors?
  Yes ☐ ☐ ☐ ☐ ☐   *No*

- Write a vector in terms of component unit vectors?
  Yes ☐ ☐ ☐ ☐ ☐   *No*

- Set up a coordinate system for representing vectors?
  Yes ☐ ☐ ☐ ☐ ☐   *No*

- Obtain the direction cosines of a vector?
  Yes ☐ ☐ ☐ ☐ ☐   *No*

- Calculate the scalar product of two vectors?

  Yes ☐ ☐ ☐ ☐ ☐ *No*

- Calculate the vector product of two vectors?

  Yes ☐ ☐ ☐ ☐ ☐ *No*

- Determine the angle between two vectors?

  Yes ☐ ☐ ☐ ☐ ☐ *No*

 # Test exercise 11

Take your time: the problems are all straightforward so avoid careless slips. **2** Diagrams often help where appropriate.

1 If $\overline{OA} = 4\mathbf{i} + 3\mathbf{j}$, $\overline{OB} = 6\mathbf{i} - 2\mathbf{j}$, $\overline{OC} = 2\mathbf{i} - \mathbf{j}$, find $\overline{AB}$, $\overline{BC}$ and $\overline{CA}$, and deduce the lengths of the sides of the triangle ABC.

2 Find the direction cosines of the vector joining the two points (4, 2, 2) and (7, 6, 14).

3 If $\mathbf{a} = 2\mathbf{i} + 2\mathbf{j} - \mathbf{k}$ and $\mathbf{b} = 3\mathbf{i} - 6\mathbf{j} + 2\mathbf{k}$, find (a) $\mathbf{a}.\mathbf{b}$ and (b) $\mathbf{a} \times \mathbf{b}$.

4 If $\mathbf{a} = 5\mathbf{i} + 4\mathbf{j} + 2\mathbf{k}$, $\mathbf{b} = 4\mathbf{i} - 5\mathbf{j} + 3\mathbf{k}$ and $\mathbf{c} = 2\mathbf{i} - \mathbf{j} - 2\mathbf{k}$, where $\mathbf{i}$, $\mathbf{j}$, $\mathbf{k}$ are the unit vectors, determine:

   (a) the value of $\mathbf{a}.\mathbf{b}$ and the angle between the vectors $\mathbf{a}$ and $\mathbf{b}$

   (b) the magnitude and the direction cosines of the product vector $\mathbf{a} \times \mathbf{b}$ and also the angle which this product vector makes with the vector $\mathbf{c}$.

 # Further problems 11

1 The centroid of the triangle OAB is denoted by G. If O is the origin and **3** $\overline{OA} = 4\mathbf{i} + 3\mathbf{j}$, $\overline{OB} = 6\mathbf{i} - \mathbf{j}$, find $\overline{OG}$ in terms of the unit vectors, $\mathbf{i}$ and $\mathbf{j}$.

2 Find the direction cosines of the vectors whose direction ratios are (3, 4, 5) and (1, 2, −3). Hence find the angle between the two vectors.

3 Find the modulus and the direction cosines of each of the vectors $3\mathbf{i} + 7\mathbf{j} - 4\mathbf{k}$, $\mathbf{i} - 5\mathbf{j} - 8\mathbf{k}$ and $6\mathbf{i} - 2\mathbf{j} + 12\mathbf{k}$. Find also the modulus and the direction cosines of their sum.

4 If $\mathbf{a} = 2\mathbf{i} + 4\mathbf{j} - 3\mathbf{k}$ and $\mathbf{b} = \mathbf{i} + 3\mathbf{j} + 2\mathbf{k}$, determine the scalar and vector products, and the angle between the two given vectors.

5 If $\overline{OA} = 2\mathbf{i} + 3\mathbf{j} - \mathbf{k}$ and $\overline{OB} = \mathbf{i} - 2\mathbf{j} + 3\mathbf{k}$, determine:

   (a) the value of $\overline{OA}.\overline{OB}$

   (b) the product $\overline{OA} \times \overline{OB}$ in terms of the unit vectors

   (c) the cosine of the angle between $\overline{OA}$ and $\overline{OB}$.

6   Find the cosine of the angle between the vectors $2\mathbf{i} + 3\mathbf{j} - \mathbf{k}$ and $3\mathbf{i} - 5\mathbf{j} + 2\mathbf{k}$.

7   Find the scalar product $\mathbf{a}.\mathbf{b}$ and the vector product $\mathbf{a} \times \mathbf{b}$, when
    (a) $\mathbf{a} = \mathbf{i} + 2\mathbf{j} - \mathbf{k}$, $\mathbf{b} = 2\mathbf{i} + 3\mathbf{j} + \mathbf{k}$
    (b) $\mathbf{a} = 2\mathbf{i} + 3\mathbf{j} + 4\mathbf{k}$, $\mathbf{b} = 5\mathbf{i} - 2\mathbf{j} + \mathbf{k}$.

8   Find the unit vector perpendicular to each of the vectors $2\mathbf{i} - \mathbf{j} + \mathbf{k}$ and $3\mathbf{i} + 4\mathbf{j} - \mathbf{k}$, where $\mathbf{i}$, $\mathbf{j}$, $\mathbf{k}$ are the mutually perpendicular unit vectors. Calculate the sine of the angle between the two vectors.

9   If A is the point $(1, -1, 2)$, B is the point $(-1, 2, 2)$ and C is the point $(4, 3, 0)$, find the direction cosines of $\overline{BA}$ and $\overline{BC}$, and hence show that the angle $ABC = 69°14'$.

10   If $\mathbf{a} = 3\mathbf{i} - \mathbf{j} + 2\mathbf{k}$, $\mathbf{b} = \mathbf{i} + 3\mathbf{j} - 2\mathbf{k}$, determine the magnitude and direction cosines of the product vector $(\mathbf{a} \times \mathbf{b})$ and show that it is perpendicular to a vector $\mathbf{c} = 9\mathbf{i} + 2\mathbf{j} + 2\mathbf{k}$.

11   $\mathbf{a}$ and $\mathbf{b}$ are vectors defined by $\mathbf{a} = 8\mathbf{i} + 2\mathbf{j} - 3\mathbf{k}$ and $\mathbf{b} = 3\mathbf{i} - 6\mathbf{j} + 4\mathbf{k}$, where $\mathbf{i}$, $\mathbf{j}$, $\mathbf{k}$ are mutually perpendicular unit vectors.

    (a) Calculate $\mathbf{a}.\mathbf{b}$ and show that $\mathbf{a}$ and $\mathbf{b}$ are perpendicular to each other.

    (b) Find the magnitude and the direction cosines of the product vector $\mathbf{a} \times \mathbf{b}$.

12   If the position vectors of P and Q are $\mathbf{i} + 3\mathbf{j} - 7\mathbf{k}$ and $5\mathbf{i} - 2\mathbf{j} + 4\mathbf{k}$ respectively, find $\overline{PQ}$ and determine its direction cosines.

13   If position vectors, $\overline{OA}$, $\overline{OB}$, $\overline{OC}$, are defined by $\overline{OA} = 2\mathbf{i} - \mathbf{j} + 3\mathbf{k}$, $\overline{OB} = 3\mathbf{i} + 2\mathbf{j} - 4\mathbf{k}$, $\overline{OC} = -\mathbf{i} + 3\mathbf{j} - 2\mathbf{k}$, determine:

    (a) the vector $\overline{AB}$

    (b) the vector $\overline{BC}$

    (c) the vector product $\overline{AB} \times \overline{BC}$

    (d) the unit vector perpendicular to the plane ABC.

14   Given that $\mathbf{a} = 2\mathbf{i} - 3\mathbf{j} + \mathbf{k}$, $\mathbf{b} = 4\mathbf{i} + \mathbf{j} - 3\mathbf{k}$ and $\mathbf{c} = -\mathbf{i} + \mathbf{j} - 3\mathbf{k}$ find:

    (a) $(\mathbf{a} \cdot \mathbf{b})\mathbf{c}$

    (b) $\mathbf{a} \cdot (\mathbf{b} \times \mathbf{c})$ (called the scalar triple product)

    (c) $\mathbf{a} \times (\mathbf{b} \times \mathbf{c})$ (called the vector triple product).

15   Given that $\mathbf{a} = 5\mathbf{i} + 2\mathbf{j} - \mathbf{k}$ and $\mathbf{b} = \mathbf{i} - 3\mathbf{j} + \mathbf{k}$ find $(\mathbf{a} + \mathbf{b}) \times (\mathbf{a} - \mathbf{b})$.

16   Prove that the area of the parallelogram with adjacent sides $\mathbf{a}$ and $\mathbf{b}$ is given as $|\mathbf{a} \times \mathbf{b}|$.

17   Show that if $\mathbf{a}$, $\mathbf{b}$ and $\mathbf{c}$ lie in the same plane then $\mathbf{a} \cdot (\mathbf{b} \times \mathbf{c}) = 0$.

18   Show that the volume V of the parallelepiped with adjacent edges $\mathbf{a}$, $\mathbf{b}$ and $\mathbf{c}$ is given as the magnitude of the scalar triple product. That is:
$$V = |\mathbf{a} \cdot (\mathbf{b} \times \mathbf{c})| = |\mathbf{b} \cdot (\mathbf{c} \times \mathbf{a})| = |\mathbf{c} \cdot (\mathbf{a} \times \mathbf{b})|.$$

19   Given the triangle of vectors $\mathbf{a}$, $\mathbf{b}$ and $\mathbf{c}$ where $\mathbf{c} = \mathbf{b} - \mathbf{a}$ derive the cosine law for plane triangles $c^2 = a^2 + b^2 - 2ab\cos\theta$ where $\mathbf{a} \cdot \mathbf{b} = ab\cos\theta$.

20   The point $(1, 2, 4)$ lies in plane P which is perpendicular to the vector $\mathbf{a} = 2\mathbf{i} - 3\mathbf{j} + \mathbf{k}$. Given $\mathbf{r} = x\mathbf{i} + y\mathbf{j} + z\mathbf{k}$ is the position vector of another point in the plane find the equation of the plane.

Binomial series

Learning outcomes

When you have completed this Module you will be able to:

- Define $n!$ and recognize that there are $n!$ different combinations of n different items
- Evaluate $n!$ for moderately sized n using a calculator
- Manipulate expressions containing factorials
- Recognize that there are $\dfrac{n!}{(n-r)!r!}$ combinations of r identical items in n locations
- Recognize simple properties of combinatorial coefficients
- Construct Pascal's triangle
- Write down the binomial expansion for natural number powers
- Obtain specific terms in the binomial expansion using the general term
- Use the sigma notation

Units

Factorials and combinations

1   Factorials

Answer this question. How many different three-digit numbers can you construct using the three numerals 5, 7 and 8 once each?

The answer is in the next frame

2

6

They are: 578   587
             758   785
             857   875

Instead of listing them like this you can work it out. There are 3 choices for the first numeral and for each choice there are a further 2 choices for the second numeral. That is, there are:

$3 \times 2 = 6$ choices of first and second numeral combined

The third numeral is then the one that is left so there is only 1 choice for that. Therefore, there are:

$3 \times 2 \times 1$ choices of first, second and third numeral combined

So, how many four-digit numbers can be constructed using the numerals 3, 5, 7 and 9 once each?

Answer in the next frame

3

$4 \times 3 \times 2 \times 1$

Because

The first numeral can be selected in one of 4 ways, each selection leaving 3 ways to select the second numeral. So there are:

$4 \times 3 = 12$ ways of selecting the first two numerals

Each combination of the first two numerals leaves 2 ways to select the third numeral. So there are:

$4 \times 3 \times 2 = 24$ ways of selecting the first three numerals

The last numeral is the one that is left so there is only 1 choice for that. Therefore, there are:

$4 \times 3 \times 2 \times 1 = 24$ choices of first, second, third and fourth numeral combined.

Can you see the pattern here? If you have n different items then you can form

$n \times (n - 1) \times (n - 2) \times \ldots \times 2 \times 1$

different arrangements, or *combinations*, using each item just once.

▶

This type of product of decreasing natural numbers occurs quite often in mathematics so a general notation has been devised. For example, the product:

$3 \times 2 \times 1$

is called *3-factorial* and is written as 3!

So the value of 5! is . . . .

Next frame

$$\boxed{120}$$

4

Because

$5! = 5 \times 4 \times 3 \times 2 \times 1 = 120$

The factorial expression can be generalized to the factorial of an arbitrary natural number n as:

$n! = n \times (n - 1) \times (n - 2) \times \ldots \times 2 \times 1$

To save time a calculator can be used to evaluate $n!$ for moderately sized n. For example, on your calculator:

Enter the number 9
Press the ! key
The display changes to 362880. that is:

$9! = 9 \times 8 \times 7 \times 6 \times 5 \times 4 \times 3 \times 2 \times 1 = 362\,880$

So the value of 11! is . . . .

$$\boxed{39\,916\,800}$$

5

For reasons that will soon become clear the value of 0! is defined to be 1. Try it on your calculator.

Try a few examples. Evaluate each of the following:

(a) 6!   (b) $\dfrac{8!}{3!}$ (c) $(7 - 2)!$ (d) $\dfrac{3!}{0!}$ (e) $\dfrac{5!}{(5 - 2)!2!}$

$$\boxed{\text{(a) } 720 \quad \text{(b) } 6720 \quad \text{(c) } 120 \quad \text{(d) } 6 \quad \text{(e) } 10}$$

6

Because

(a) $6! = 6 \times 5 \times 4 \times 3 \times 2 \times 1 = 720$

(b) $\dfrac{8!}{3!} = \dfrac{8 \times 7 \times 6 \times 5 \times 4 \times 3 \times 2 \times 1}{3 \times 2 \times 1} = 8 \times 7 \times 6 \times 5 \times 4 = 6720$

(c) $(7 - 2)! = 5! = 120$

(d) $\dfrac{3!}{0!} = \dfrac{6}{1} = 6$ Notice that we have used the fact that $0! = 1$

(e) $\dfrac{5!}{(5 - 2)!2!} = \dfrac{5!}{3!2!} = \dfrac{5 \times 4 \times 3 \times 2 \times 1}{(3 \times 2 \times 1) \times (2 \times 1)} = \dfrac{5 \times 4}{(2 \times 1)} = 10$

▶

Now try some for general n. Because

$$n! = n \times (n-1) \times (n-2) \times \ldots \times 2 \times 1$$
$$= n \times (n-1)!$$

then:

$$\frac{n!}{(n-1)!} = \frac{n \times (n-1)!}{(n-1)!} = n$$

Also note that while

$$2n! = 2 \times n!, (2n)! = (2n) \times (2n-1) \times (2n-2) \times \ldots \times 2 \times 1$$

so that, for example:

$$\frac{(2n+1)!}{(2n-1)!} = \frac{(2n+1) \times (2n+1-1) \times (2n+1-2) \times \ldots \times 2 \times 1}{(2n-1) \times (2n-1-1) \times (2n-1-2) \times \ldots \times 2 \times 1}$$
$$= \frac{(2n+1) \times (2n) \times (2n-1) \times \ldots \times 2 \times 1}{(2n-1) \times (2n-2) \times (2n-3) \times \ldots \times 2 \times 1}$$
$$= (2n+1) \times (2n)$$

So simplify each of these:

(a) $\dfrac{n!}{(n+1)!}$ (b) $\dfrac{(n+1)!}{(n-1)!}$ (c) $\dfrac{(2n)!}{(2n+2)!}$

Answers in the next frame

7

> (a) $\dfrac{1}{n+1}$
>
> (b) $n(n+1)$
>
> (c) $\dfrac{1}{(2n+2) \times (2n+1)}$

Because

(a) $\dfrac{n!}{(n+1)!} = \dfrac{n \times (n-1) \times (n-2) \times \ldots \times 2 \times 1}{(n+1) \times n \times (n-1) \times (n-2) \times \ldots \times 2 \times 1} = \dfrac{1}{(n+1)}$

(b) $\dfrac{(n+1)!}{(n-1)!} = \dfrac{(n+1) \times (n) \times (n-1)!}{(n-1)!} = (n+1) \times (n) = n(n+1)$

(c) $\dfrac{(2n)!}{(2n+2)!} = \dfrac{(2n)!}{(2n+2) \times (2n+1) \times (2n)!} = \dfrac{1}{(2n+2) \times (2n+1)}$

---

8   Try some more. Write each of the following in factorial form:

(a) $4 \times 3 \times 2 \times 1$

(b) $6 \times 5 \times 4$

(c) $\dfrac{(7 \times 6) \times (3 \times 2 \times 1)}{2}$

Next frame for the answers

$$\boxed{\text{(a) } 4! \qquad \text{(b) } \frac{6!}{3!} \qquad \text{(c) } \frac{7! \times 3!}{5! \times 2!}}$$

Because

(a) $4 \times 3 \times 2 \times 1 = 4!$ by the definition of the factorial

(b) $6 \times 5 \times 4 = \dfrac{6 \times 5 \times 4 \times 3 \times 2 \times 1}{3 \times 2 \times 1} = \dfrac{6!}{3!}$

(c) $\dfrac{(7 \times 6) \times (3 \times 2 \times 1)}{2} = \dfrac{(7 \times 6 \times 5 \times 4 \times 3 \times 2 \times 1) \times (3 \times 2 \times 1)}{(5 \times 4 \times 3 \times 2 \times 1) \times (2 \times 1)}$

$$= \frac{7! \times 3!}{5! \times 2!}$$

Now let's use these ideas. Next frame

Combinations

10

Let's assume that you have a part-time job in the weekday evenings where you have to be at work just two evenings out of the five. Let's also assume that your employer is very flexible and allows you to choose which evenings you work provided you ring him up on Sunday and tell him. One possible selection could be:

Mon	Tue	Wed	Thu	Fri
W	W	–	–	–

another selection could be:

Mon	Tue	Wed	Thu	Fri
–	–	W	–	W

How many arrangements are there of two working evenings among the five days?

The answer is in the next frame

11

$$\boxed{5 \times 4 = 20}$$

Because

There are 5 weekdays from which to make a first selection and for each such selection there are 4 days left from which to make the second selection. This gives a total of $5 \times 4 = 20$ possible arrangements.

However, not all arrangements are *different*. For example, if, on the Sunday, you made

your first choice as Friday and your second choice as Wednesday,

this would be the same arrangement as making

your first choice as Wednesday and your second choice as Friday.

So every arrangement is duplicated.

How many *different* arrangements are there?

12

$$\boxed{\frac{5 \times 4}{2} = 10}$$

Because each arrangement is duplicated. List them:

Mon, Tue	Mon, Wed	Mon, Thu	Mon, Fri
	Tue, Wed	Tue, Thu	Tue, Fri
		Wed, Thu	Wed, Fri
			Thu, Fri

There are 10 different ways of combining two identical items in five different places.

The expression $\dfrac{5 \times 4}{2}$ can be written in factorial form as follows:

$$\frac{5 \times 4}{2} = \frac{5 \times 4}{2 \times 1} = \frac{5 \times 4 \times 3 \times 2 \times 1}{(3 \times 2 \times 1)(2 \times 1)} = \frac{5!}{3!2!} \text{ or better still:}$$

$\dfrac{5 \times 4}{2} = \dfrac{5!}{3!2!} = \dfrac{5!}{(5-2)!2!}$ because this just contains the numbers 5 and 2 and will link to a general notation to be introduced in Frame 14.

There are $\dfrac{5!}{(5-2)!2!}$ *combinations of* **two** *identical items in* **five** *different places.*

So, if your employer asked you to work 3 evenings out of the 5, how many arrangements of three days are there? (Some arrangements will contain the same days but in a different order.)

The answer is in the next frame

13

$$\boxed{5 \times 4 \times 3}$$

Because

There are 5 weekdays from which to make a first selection and for each such selection there are 4 days left from which to make the second selection and then 3 days from which to make the third selection. This gives a total of $5 \times 4 \times 3 = 60$ possible arrangements.

However, not all arrangements are different. Some arrangemens will contain the same days but in a different order. Any one arrangement can be rearranged within itself $3 \times 2 \times 1 = 6$ times.

The total number of different arrangements (that is combinations) is then:

. . . .

Next frame

$$\frac{5 \times 4 \times 3}{3 \times 2 \times 1} = 10$$

Because

Any one arrangement of three specific days can be rearranged within itself $3 \times 2 \times 1 = 6$ times. This means that every combination of three specific days appears 6 times in the list of 60 possible arrangements. Therefore, there are:

$$\frac{5 \times 4 \times 3}{3 \times 2 \times 1} = \frac{60}{6} = 10 \quad \text{combinations of 3 days out of the 5}$$

Written in factorial notation $\dfrac{5 \times 4 \times 3}{3 \times 2 \times 1} = \dots\dots\dots$

The answer is in the next frame

$$\frac{5!}{2!3!} = \frac{5!}{(5-3)!3!}$$

Because

$$\frac{5 \times 4 \times 3}{3 \times 2 \times 1} = \frac{5 \times 4 \times 3 \times 2 \times 1}{(2 \times 1)(3 \times 2 \times 1)} = \frac{5!}{2!3!} = \frac{5!}{(5-3)!3!}$$

Can you see the pattern emerging here?

There are $\dfrac{5!}{(5-2)!2!}$ combinations of *two* identical items in *five* different places.

There are $\dfrac{5!}{(5-3)!3!}$ combinations of *three* identical items in *five* different places.

So if you have r identical items to be located in n different places where $n \geq r$, the number of combinations is $\dots\dots\dots$

The answer is in the next frame

16

$$\boxed{\dfrac{n!}{(n-r)!r!}}$$

Because

The first item can be located in any one of n places.
The second item can be located in any one of the remaining $n-1$ places.
The third item can be located in any one of the remaining $n-2$ places.

⋮

The rth item can be located in any one of the remaining $n-(r-1)$ places.

This means that there are $n-(r-1) = \dfrac{n!}{(n-r)!}$ arrangements. However, every arrangement is repeated $r!$ times. (Any one arrangement can be rearranged within itself $r!$ times.) So the total number of combinations is given as:

$$\dfrac{n!}{(n-r)!r!}$$

This particular ratio of factorials is called a *combinatorial coefficient* and is denoted by ^nC_r:

$$^nC_r = \dfrac{n!}{(n-r)!r!} \quad \text{where } 0 \le r \le n$$

So, evaluate each of the following:

(a) $^6C_3$ (b) $^7C_2$ (c) $^4C_4$ (d) $^3C_0$ (e) $^5C_1$

17

$$\boxed{\text{(a) } 20 \quad \text{(b) } 21 \quad \text{(c) } 1 \quad \text{(d) } 1 \quad \text{(e) } 5}$$

Because

(a) $^6C_3 = \dfrac{6!}{(6-3)!3!} = \dfrac{6!}{3!3!} = \dfrac{720}{36} = 20$

(b) $^7C_2 = \dfrac{7!}{(7-2)!2!} = \dfrac{7!}{5!2!} = \dfrac{5040}{120 \times 2} = 21$

(c) $^4C_4 = \dfrac{4!}{(4-4)!4!} = \dfrac{4!}{0!4!} = \dfrac{4!}{4!} = 1$ Remember $0! = 1$

(d) $^3C_0 = \dfrac{3!}{(3-0)!0!} = \dfrac{3!}{3!0!} = 1$

(e) $^5C_1 = \dfrac{5!}{(5-1)!1!} = \dfrac{5!}{4!1!} = \dfrac{5!}{4!} = 5$

Some properties of combinatorial coefficients   **18**

1 $^nC_n = {}^nC_0 = 1$ and $^nC_1 = n$

These are quite straightforward to prove:

$$^nC_n = \frac{n!}{(n-n)!n!} = \frac{n!}{0!n!} = \frac{n!}{n!} = 1 \text{ and } {}^nC_0 = \frac{n!}{(n-0)!0!} = \frac{n!}{n!0!} = \frac{n!}{n!} = 1$$

$$^nC_1 = \frac{n!}{(n-1)!1!} = \frac{n(n-1)!}{(n-1)!} = n$$

2 $^nC_{n-r} = {}^nC_r$. This is a little more involved:

$$^nC_{n-r} = \frac{n!}{(n-[n-r])!(n-r)!} = \frac{n!}{(n-n+r)!(n-r)!} = \frac{n!}{r!(n-r)!} = {}^nC_r$$

3 $^nC_r + {}^nC_{r+1} = {}^{n+1}C_{r+1}$

This requires some care to prove:

$$^nC_r + {}^nC_{r+1} = \frac{n!}{(n-r)!r!} + \frac{n!}{(n-[r+1])!(r+1)!}$$

$$= \frac{n!}{(n-r)!r!} + \frac{n!}{(n-r-1)!(r+1)!}$$

$$= \frac{n!}{(n-r-1)!r!}\left(\frac{1}{n-r} + \frac{1}{r+1}\right)$$
Taking out common factors where
$(n-r)! = (n-r)(n-r-1)!$
and $(r+1)! = (r+1)r!$

$$= \frac{n!}{(n-r-1)!r!}\left(\frac{r+1+n-r}{(n-r)(r+1)}\right)$$
Adding the two fractions together

$$= \frac{n!}{(n-r)!(r+1)!}(n+1)$$
Because $(n-r-1)!(n-r) = (n-r)!$ and $r!(r+1) = (r+1)!$

$$= \frac{(n+1)!}{(n-r)!(r+1)!}$$
Because $n!(n+1) = (n+1)!$

$$= \frac{(n+1)!}{([n+1]-[r+1])!(r+1)!}$$
Because $[n+1] - [r+1] = n - r$

$$= {}^{n+1}C_{r+1}$$

At this point let us pause and summarize the main facts on factorials and combinations

Review summary   Unit 36

19

1   The product $n \times (n-1) \times (n-2) \times \ldots \times 2 \times 1$ is called n-factorial and is denoted by $n!$

2   The number of different arrangements of n different items is $n!$

3   The number of combinations of r identical items among n locations is given by the combinatorial coefficient $^nC_r = \dfrac{n!}{(n-r)!r!}$ where $0 \le r \le n$

4   $^nC_n = {}^nC_0 = 1$ and $^nC_1 = n$

5   $^nC_{n-r} = {}^nC_r$

6   $^nC_r + {}^nC_{r+1} = {}^{n+1}C_{r+1}$

 # Review exercise   Unit 36

20

1   Find the value of:

(a) $5!$ (b) $10!$ (c) $\dfrac{13!}{8!}$ (d) $(9-3)!$ (e) $0!$

2   How many different arrangements are there of 9 identical umbrellas on a rack of 15 coat hooks?

3   Evaluate each of the following:

(a) $^3C_1$ (b) $^9C_4$ (c) $^{100}C_0$

4   Without evaluating, explain why each of the following equations is true:

(a) $^{15}C_5 = {}^{15}C_{10}$ (b) $^5C_3 + {}^5C_4 = {}^6C_4$

Complete all four questions. Take your time, there is no need to rush.
If necessary, look back at the Unit.
The answers and working are in the next frame.

21

1   (a) $5! = 5 \times 4 \times 3 \times 2 \times 1 = 120$

(b) $10! = 3\,628\,800$ (use a calculator)

(c) $\dfrac{13!}{8!} = 13 \times 12 \times 11 \times 10 \times 9 = 154\,440$

(d) $(9-3)! = 6! = 720$ (e) $0! = 1$

2   There are $^{15}C_9$ different arrangements of 9 identical umbrellas on a rack of 15 coat hooks, where:

$$^{15}C_9 = \frac{15!}{(15-9)!9!} = \frac{15!}{6!9!} = 5005$$

3 (a) $^3C_1 = \dfrac{3!}{(3-1)!1!} = \dfrac{3!}{2!1!} = 3$ (b) $^9C_4 = \dfrac{9!}{(9-4)!4!} = \dfrac{9!}{5!4!} = 126$

    (c) $^{100}C_0 = 1$

4 (a) $^{15}C_5 = {}^{15}C_{10}$ because $^nC_{n-r} = {}^nC_r$ where $n = 15$ and $r = 10$

    (b) $^5C_3 + {}^5C_4 = {}^6C_4$ because $^nC_r + {}^nC_{r+1} = {}^{n+1}C_{r+1}$ where $n = 5$ and $r = 3$

Now for the Review test

Review test   **Unit 36**

---

1 In how many different ways can 7 identical bottles of wine be arranged in a wine rack with spaces for 12 bottles?   **22**

2 Find the value of:

    (a) $6!$ (b) $11!$ (c) $\dfrac{15!}{12!}$ (d) $(13-7)!$ (e) $\dfrac{0!}{3!}$

3 Evaluate each of the following:

    (a) $^4C_2$ (b) $^{10}C_6$ (c) $^{89}C_0$ (d) $^{36}C_{35}$

Binomial series   **Unit 37**

Pascal's triangle   **1**

The following triangular array of combinatorial coefficients can be constructed where the *superscript* to the left of each coefficient indicates the row number and the *subscript* to the right indicates the column number:

Row			*Column*		
	0	*1*	*2*	*3*	*4*
0	$^0C_0$				
1	$^1C_0$	$^1C_1$			
2	$^2C_0$	$^2C_1$	$^2C_2$		
3	$^3C_0$	$^3C_1$	$^3C_2$	$^3C_3$	
4	. . . . .	. . . . .	. . . . .	. . . . .	. . . . .

Follow the pattern and fill in the next row.

The answer is in the following frame

2

$$^4C_0, {}^4C_1, {}^4C_2, {}^4C_3 \text{ and } {}^4C_4$$

Because

The superscript indicates row 4 and the subscripts indicate columns 0 to 4.

The pattern devised in this array can be used to demonstrate the third property of combinatorial coefficients that you considered in Frame 18 of Unit 36, namely that:

$$^nC_r + {}^nC_{r+1} = {}^{n+1}C_{r+1}$$

In the following array, arrows have been inserted to indicate that any coefficient is equal to the coefficient immediately above it added to the one above it and to the left.

Row   Column
        0   1   2   3   4

0     ${}^0C_0$

1     ${}^1C_0$ ${}^1C_1$

2     ${}^2C_0$ ${}^2C_1$ ${}^2C_2$ ${}^1C_0 + {}^1C_1 = {}^{1+1}C_{0+1} = {}^2C_1$

3     ${}^3C_0$ ${}^3C_1$ ${}^3C_2$ ${}^3C_3$ ${}^2C_0 + {}^2C_1 = {}^{2+1}C_{0+1} = {}^3C_1$ etc.

4     ${}^4C_0$ ${}^4C_1$ ${}^4C_2$ ${}^4C_3$ ${}^4C_4$ ${}^3C_0 + {}^3C_1 = {}^{3+1}C_{0+1} = {}^4C_1$ etc.

Now you have already seen from the first property of the combinatorial coefficients in Frame 18 of Unit 36 that ${}^nC_n = {}^nC_0 = 1$ so the values of some of these coefficients can be filled in immediately:

Row   Column
        0   1   2   3   4

0   1
1   1   1
2   1 ${}^2C_1$ 1
3   1 ${}^3C_1$ ${}^3C_2$ 1
4   1 ${}^4C_1$ ${}^4C_2$ ${}^4C_3$ 1

Fill in the numerical values of the remaining combinatorial coefficients using the fact that ${}^nC_r + {}^nC_{r+1} = {}^{n+1}C_{r+1}$: that is, any coefficient is equal to the number immediately above it added to the one above it and to the left.

The answer is in Frame 3

Row		Column				
	0	*1*	*2*	*3*	*4*	
0	1					
1	1	1				
2	1	2	1			$1 + 1 = 2$
3	1	3	3	1		$1 + 2 = 3, \, 2 + 1 = 3$
4	1	4	6	4	1	$1 + 3 = 4, \, 3 + 3 = 6, \, 3 + 1 = 4$

This is called *Pascal's triangle*.

The numbers on row 5, reading from left to right would be ....

Answers are in the next frame

1, 5, 10, 10, 5, 1

4

Because

The first number is $^5C_0 = 1$ and then $1 + 4 = 5$, $4 + 6 = 10$, $6 + 4 = 10$, $4 + 1 = 5$. Finally, $^5C_5 = 1$.

Now let's move on to a related topic in the next frame

Binomial expansions

5

A *binomial* is the sum of a pair of numbers raised to a power. In this Module we shall only consider whole number powers, namely binomials of the form:

$(a + b)^n$

where n is a natural number (that is, a whole number). In particular, look at the following expansions:

$(a + b)^1 = 1a + 1b = a + b$ Note the coefficients 1, 1

$(a + b)^2 = 1a^2 + 2ab + 1b^2$

$\qquad\quad = a^2 + 2ab + b^2$ Note the coefficients 1, 2, 1

$(a + b)^3 = 1a^3 + 3a^2b + 3ab^2 + 1b^3$

$\qquad\quad = a^3 + 3a^2b + 3ab^2 + b^3$ Note the coefficients 1, 3, 3, 1

So what is the expansion of $(a + b)^4$ and what are the coefficients?

Next frame for the answer

6

$$(a + b)^4 = a^4 + 4a^3b + 6a^2b^2 + 4ab^3 + b^4$$
Coefficients 1, 4, 6, 4, 1

Because

$$(a + b)^4 = (a + b)^3(a + b)$$
$$= (a^3 + 3a^2b + 3ab^2 + b^3)(a + b)$$
$$= a^4 + 3a^3b + 3a^2b^2 + ab^3 + a^3b + 3a^2b^2 + 3ab^3 + b^4$$
$$= a^4 + 4a^3b + 6a^2b^2 + 4ab^3 + b^4$$

Can you see the connection with Pascal's triangle?

Row			Column			
	0	*1*	*2*	*3*	*4*	
0	1					
1	1	1				Coefficients of $(a + b)^1$
2	1	2	1			Coefficients of $(a + b)^2$
3	1	3	3	1		Coefficients of $(a + b)^3$
4	1	4	6	4	1	Coefficients of $(a + b)^4$

So what are the coefficients of $(a + b)^5$?

Next frame for the answer

7

1, 5, 10, 10, 5, 1

Because

The values in the next row of Pascal's triangle are 1, 5, 10, 10, 5, 1 and the numbers in row 5 are also the values of the coefficients of the binomial expansion:

$$(a + b)^5 = a^5 + 5a^4b + 10a^3b^2 + 10a^2b^3 + 5ab^4 + b^5$$
$$= 1a^5b^0 + 5a^4b^1 + 10a^3b^2 + 10a^2b^3 + 5a^1b^4 + 1a^0b^5$$

Notice how in this expansion, as the power of a decreases, the power of b increases so that in each term the two powers add up to 5.

Because the numbers in row 5 are also the values of the appropriate combinatorial coefficients this expansion can be written as:

$$(a + b)^5 = {}^5C_0a^5b^0 + {}^5C_1a^4b^1 + {}^5C_2a^3b^2 + {}^5C_3a^2b^3 + {}^5C_4a^1b^4 + {}^5C_5a^0b^5$$

The general binomial expansion for natural number n is then given as:

$$(a + b)^n = {}^nC_0a^nb^0 + {}^nC_1a^{n-1}b^1 + {}^nC_2a^{n-2}b^2 + \ldots + {}^nC_na^{n-n}b^n$$
$$= a^n + {}^nC_1a^{n-1}b + {}^nC_2a^{n-2}b^2 + \ldots + b^n$$

Now ${}^nC_1 = \dfrac{n!}{(n-1)!1!} = n$ and ${}^nC_2 = \dfrac{n!}{(n-2)!2!} = \dfrac{n(n-1)}{2!}$ so ${}^nC_3 = \ldots\ldots\ldots$

Next frame for the answer

$$\boxed{\dfrac{n(n-1)(n-2)}{3!}}$$

Because

$$^{n}C_3 = \dfrac{n!}{(n-3)!3!} = \dfrac{n(n-1)(n-2)}{3!}$$

From this, the binomial expansion can be written as:

$$(a+b)^n = a^n + na^{n-1}b + \dfrac{n(n-1)}{2!}a^{n-2}b^2 + \dfrac{n(n-1)(n-2)}{3!}a^{n-3}b^3 + \ldots + b^n$$

So use this form of the binomial expansion to expand $(a+b)^6$.

Next frame for the answer

$$\boxed{(a+b)^6 = a^6 + 6a^5b + 15a^4b^2 + 20a^3b^3 + 15a^2b^4 + 6ab^5 + b^6}$$

Because

$$(a+b)^6 = a^6 + 6a^5b + \dfrac{6\times5}{2!}a^4b^2 + \dfrac{6\times5\times4}{3!}a^3b^3$$
$$+ \dfrac{6\times5\times4\times3}{4!}a^2b^4 + \dfrac{6\times5\times4\times3\times2}{5!}ab^5$$
$$+ \dfrac{6\times5\times4\times3\times2\times1}{6!}b^6$$
$$= a^6 + 6a^5b + \dfrac{6\times5}{2}a^4b^2 + \dfrac{6\times5\times4}{6}a^3b^3$$
$$+ \dfrac{6\times5\times4\times3}{4\times3\times2\times1}a^2b^4 + \dfrac{6\times5\times4\times3\times2}{5\times4\times3\times2\times1}ab^5 + \dfrac{6!}{6!}b^6$$
$$= a^6 + 6a^5b + 15a^4b^2 + 20a^3b^3 + 15a^2b^4 + 6ab^5 + b^6$$

Notice that 1, 6, 15, 20, 15, 6 and 1 are the numbers in row 6 of Pascal's triangle.

There are now two ways of obtaining the binomial expansion of $(a+b)^n$:

1 Use Pascal's triangle.   This is appropriate when n is small
2 Use the combinatorial coefficients.   This is appropriate when n is large

So, expand each of the following binomials:

   (a) $(1+x)^7$ using Pascal's triangle
   (b) $(3-2x)^4$ using the combinatorial coefficients

Next frame for the answer

10

> (a) $(1+x)^7 = 1 + 7x + 21x^2 + 35x^3 + 35x^4 + 21x^5 + 7x^6 + x^7$
>
> (b) $(3 - 2x)^4 = 81 - 216x + 216x^2 - 96x^3 + 16x^4$

Because

(a) Using Pascal's triangle:

$$(1+x)^7 = 1 + (1+6)x + (6+15)x^2 + (15+20)x^3$$
$$+ (20+15)x^4 + (15+6)x^5 + (6+1)x^6 + x^7$$
$$= 1 + 7x + 21x^2 + 35x^3 + 35x^4 + 21x^5 + 7x^6 + x^7$$

(b) Using the general form of the binomial expansion:

$$(3 - 2x)^4 = (3 + [-2x])^4$$

$$= 3^4 + 4 \times 3^3 \times (-2x) + \frac{4 \times 3}{2!} \times 3^2 \times (-2x)^2$$

$$+ \frac{4 \times 3 \times 2}{3!} \times 3 \times (-2x)^3 + \frac{4 \times 3 \times 2 \times 1}{4!} \times (-2x)^4$$

$$= 3^4 + 3^3 \times (-8x) + 6 \times 3^2 \times (4x^2) + 4 \times 3 \times (-8x^3) + (16x^4)$$

$$= 81 - 216x + 216x^2 - 96x^3 + 16x^4$$

11 ## The general term of the binomial expansion

In Frame 7 we found that the binomial expansion of $(a + b)^n$ is given as:

$$(a + b)^n = {}^nC_0 a^n b^0 + {}^nC_1 a^{n-1} b^1 + {}^nC_2 a^{n-2} b^2 + \ldots + {}^nC_n a^{n-n} b^n$$

Each term of this expansion looks like ${}^nC_r a^{n-r} b^r$ where the value of r ranges progressively from $r = 0$ to $r = n$ (there are $n+1$ terms in the expansion). Because the expression ${}^nC_r a^{n-r} b^r$ is typical of each and every term in the expansion we call it the *general term* of the expansion.

Any *specific* term can be derived from the general term. For example, consider the case when $r = 2$. The general term then becomes:

$${}^nC_2 a^{n-2} b^2 = \frac{n!}{(n-2)!2!} a^{n-2} b^2 = \frac{n(n-1)}{2!} a^{n-2} b^2$$

and this is the *third* term of the expansion.

The 3rd term is obtained from the general term by letting $r = 2$.

Consequently, we can say that:

${}^nC_r a^{n-r} b^r$ represents the $(r + 1)$th term in the expansion for $0 \leq r \leq n$

Let's look at an example. To find the 10th term in the binomial expansion of $(1 + x)^{15}$ written in ascending powers of x, we note that $a = 1$, $b = x$, $n = 15$ and $r + 1 = 10$ so $r = 9$. This gives the 10th term as:

$$^{15}C_9 1^{15-9} x^9 = \frac{15!}{(15-9)!9!} 1^{15-9} x^9$$

$$= \frac{15!}{6!9!} x^9$$

$$= 5005x^9 \text{ obtained using a calculator}$$

Try this one yourself. The 8th term in the binomial expansion of

$$\left(2 - \frac{x}{3}\right)^{12} \text{ is}\ldots\ldots\ldots$$

The answer is in the next frame

$$\boxed{-\frac{2816}{243} x^7}$$

12

Because

Here $a = 2$, $b = -x/3$, $n = 12$ and $r + 1 = 8$ so $r = 7$. The 8th term is:

$$^{12}C_7 2^{12-7} (-x/3)^7 = \frac{12!}{(12-7)!7!} 2^5 (-x/3)^7$$

$$= \frac{12!}{5!7!} 32 \times \frac{x^7}{(-3)^7}$$

$$= \frac{792 \times 32}{-2187} x^7$$

$$= -\frac{25\,344}{2187} x^7$$

$$= -\frac{2816}{243} x^7$$

At this point let us pause and summarize the main facts on the binomial series

Review summary Unit 37

13

1 Row n in Pascal's triangle contains the coefficients of the binomial expansion of $(a+b)^n$.

2 An alternative form of the binomial expansion of $(a+b)^n$ is given in terms of combinatorial coefficients as:

$$(a+b)^n = {}^nC_0a^nb^0 + {}^nC_1a^{n-1}b^1 + {}^nC_2a^{n-2}b^2 + \ldots + {}^nC_na^{n-n}b^n$$

$$= a^n + na^{n-1}b + \frac{n(n-1)}{2!}a^{n-2}b^2 + \frac{n(n-1)(n-2)}{3!}a^{n-3}b^3 \ldots + b^n$$

3 There are $n+1$ terms in the binomial expansion of $(a+b)^n$ and the $(r+1)$th term is given by ${}^nC_ra^{n-r}b^r$ where $0 \le r \le n$.

 Review exercise Unit 37

14

1 Using Pascal's triangle write down the binomial expansion of:

(a) $(a+b)^6$ (b) $(a+b)^7$

2 Write down the binomial expansion of each of the following:

(a) $(1+x)^5$ (b) $(2+3x)^4$ (c) $(2-x/2)^3$

3 In the binomial expansion of $\left(2-\dfrac{3}{x}\right)^8$ written in terms of descending powers of x, find:

(a) the 4th term   (b) the coefficient of x^{-4}

Complete all three questions.
Look back at the Unit if necessary but don't rush.
The answers and working are in the next frame.

15

1 From Pascal's triangle:

(a) $(a+b)^6 = a^6b^0 + 6a^5b^1 + 15a^4b^2 + 20a^3b^3 + 15a^2b^4 + 6a^1b^5 + a^0b^6$

$\qquad = a^6 + 6a^5b + 15a^4b^2 + 20a^3b^3 + 15a^2b^4 + 6ab^5 + b^6$

(b) $(a+b)^7 = a^7 + 7a^6b + 21a^5b^2 + 35a^4b^3 + 35a^3b^4 + 21a^2b^5 + 7ab^6 + b^7$

2 (a) $(1+x)^5 = 1^5 + 5 \times 1^4x^1 + 10 \times 1^3x^2 + 10 \times 1^2x^3 + 5 \times 1^1x^4 + x^5$

$\qquad = 1 + 5x + 10x^2 + 10x^3 + 5x^4 + x^5$

(b) $(2+3x)^4 = 2^4 + 4 \times 2^3(3x)^1 + 6 \times 2^2(3x)^2 + 4 \times 2^1(3x)^3 + (3x)^4$

$\qquad = 16 + 96x + 216x^2 + 216x^3 + 81x^4$

(c) $(2-x/2)^3 = 2^3 + 3 \times 2^2(-x/2)^1 + 3 \times 2^1(-x/2)^2 + (-x/2)^3$

$\qquad = 8 - 6x + 3x^2/2 - x^3/8$

3 In the binomial expansion of $\left(2 - \dfrac{3}{x}\right)^8$:

(a) The 4th term is derived from the general term $^nC_r a^{n-r} b^r$ where $a = 2$, $b = -3/x$, $n = 8$ and $r + 1 = 4$ so $r = 3$. That is, the 4th term is:

$$^8C_3 2^{8-3}(-3/x)^3 = \frac{8!}{(8-3)!3!} 2^5 (-3/x)^3$$

$$= \frac{8!}{5!3!} \times 32 \times (-27/x^3)$$

$$= -\frac{56}{x^3} \times 864 = -\frac{48\,384}{x^3}$$

(b) The coefficient of x^{-4} is derived from the general term when $r = 4$. That is: $^8C_4 2^{8-4}(-3/x)^4$, giving the coefficient as:

$$^8C_4 2^{8-4}(-3)^4 = \frac{8!}{(8-4)!4!} 2^4 (-3)^4$$

$$= 70 \times 16 \times 81 = 90\,720$$

Now for the Review test

Review test

Unit 37

1 Expand $(3a + 4b)^4$ as a binomial series.

16

2 In the binomial expansion of $(1 - x/2)^9$ written in terms of ascending powers of x, find:

(a) the 4th term

(b) the coefficient of x^5.

The $\sum$ (sigma) notation

Unit 38

The binomial expansion of $(a + b)^n$ is given as a sum of terms:

1

$$(a + b)^n = {}^nC_0 a^n b^0 + {}^nC_1 a^{n-1} b^1 + {}^nC_2 a^{n-2} b^2 + \ldots + {}^nC_n a^{n-n} b^n$$

Instead of writing down each term in the sum in this way a shorthand notation has been devised. We write down the general term and then use the Greek letter $\sum$ (sigma) to denote the sum. That is:

$$(a + b)^n = {}^nC_0 a^n b^0 + {}^nC_1 a^{n-1} b^1 + {}^nC_2 a^{n-2} b^2 + \ldots + {}^nC_n a^{n-n} b^n$$

$$= \sum_{r=0}^{n} {}^nC_r a^{n-r} b^r$$

Here the Greek letter $\sum$ denotes a *sum of terms* where the typical term is $^nC_r a^{n-r} b^r$ and where the value of r ranges in integer steps from $r = 0$, as indicated at the bottom of the sigma, to $r = n$, as indicated on the top of the sigma.

One immediate benefit of this notation is that it permits further properties of the combinatorial coefficients to be proved. For example:

$$\sum_{r=0}^{n} {}^nC_r = 2^n$$

The sum of the numbers in any row of Pascal's triangle is equal to 2 raised to the power of the row number

This is easily proved using the fact that:

$$\sum_{r=0}^{n} {}^nC_r a^{n-r} b^r = (a+b)^n$$

By choosing $a = 1$ and $b = 1$ and substituting into this equation we find that:

$$\sum_{r=0}^{n} {}^nC_r 1^{n-r} 1^r = (1+1)^n.$$ That is $\sum_{r=0}^{n} {}^nC_r = 2^n$ which proves our assertion.

2 General terms

The sigma notation is widely used because it provides a more compact and convenient notation. Consequently, it is necessary for you to acquire the ability to form the general term from a sum of specific terms and so write the sum of specific terms using the sigma notation. To begin, consider the sum of the first n even numbers:

$2 + 4 + 6 + 8 + \ldots\ldots\ldots$

Every even integer is divisible by 2 so every even integer can be written in the form $2r$ where r is some integer. For example:

$8 = 2 \times 4$ so here $8 = 2r$ where $r = 4$

Every odd integer can be written in the form $2r - 1$ or as $2r + 1$, that is as an even integer minus or plus 1. For example:

$13 = 14 - 1 = 2 \times 7 - 1$ so that $13 = 2r - 1$ where $r = 7$

Alternatively:

$13 = 12 + 1 = 2 \times 6 + 1$ so that $13 = 2r + 1$ where $r = 6$

Writing (a) 16, 248, −32 each in the form $2r$, give the value of r in each case.
(b) 21, 197, −23 each in the form $2r - 1$, give the value of r in each case.

The answers are in the next frame

3

$$
\begin{array}{ll}
\text{(a) } r = 8 & \text{(b) } r = 11 \\
\quad\; r = 124 & \quad\; r = 99 \\
\quad\; r = -16 & \quad\; r = -11
\end{array}
$$

Because

(a) $16 = 2 \times 8 = 2r$ where $r = 8$

$\quad\;\; 248 = 2 \times 124 = 2r$ where $r = 124$

$\quad\; -32 = 2 \times (-16)$ where $r = -16$

(b) $21 = 22 - 1 = 2 \times 11 - 1 = 2r - 1$ where $r = 11$

$\quad\;\; 197 = 198 - 1 = 2 \times 99 - 1 = 2r - 1$ where $r = 99$

$\quad\; -23 = -22 - 1 = 2 \times (-11) - 1 = 2r - 1$ where $r = -11$

Next frame

4

We saw in Frame 1 of this Unit that we can use the sigma notation to denote sums of general terms. We shall now use the notation to denote sums of terms involving integers. For example, in the sum of the odd natural numbers:

$$1 + 3 + 5 + 7 + 9 + \ldots\ldots\ldots$$

the general term can now be denoted by $2r - 1$ where $r \geq 1$. The symbol $\sum$ can then be used to denote a sum of terms of which the general term is typical:

$$1 + 3 + 5 + 7 + 9 + \ldots\ldots\ldots = \sum (2r - 1)$$

We can now also denote the range of terms over which we wish to extend the sum by inserting the appropriate values of the *counting number r* below and above the sigma sign. For example:

$$\sum_{r=1}^{7} (2r - 1) \quad \text{indicates the sum of 7 terms where } r \text{ ranges from } r = 1 \text{ to } r = 7.$$

That is:

$$\sum_{r=1}^{7} (2r - 1) = 1 + 3 + 5 + 7 + 9 + 11 + 13 = 49$$

Now you try one. Write down the general term and then write down the sum of the first 10 terms using the sigma notation:

$$2 + 4 + 6 + 8 + \ldots\ldots\ldots$$

The answer is in the next frame

5

$$2r, \sum_{r=1}^{10} 2r$$

Because

This is the sum of the first 10 even numbers and every even number is divisible by $2r$, giving the sum as $\sum_{r=1}^{10} 2r$.

Now try some more. Write down the general term and then write down the sum of the first 10 terms using the sigma notation of:

(a) $1 + \dfrac{1}{2} + \dfrac{1}{3} + \dfrac{1}{4} + \ldots$ (b) $1 + 8 + 27 + 64 + \ldots$

(c) $-1 + 2 - 3 + 4 - 5 + \ldots$ (d) $\dfrac{1}{0!} + \dfrac{1}{1!} + \dfrac{1}{2!} + \dfrac{1}{3!} + \ldots$

Answers in the next frame

6

(a) $\dfrac{1}{r}, \displaystyle\sum_{r=1}^{10} \dfrac{1}{r}$ (b) $r^3, \displaystyle\sum_{r=1}^{10} r^3$

(c) $(-1)^r r, \displaystyle\sum_{r=1}^{10} (-1)^r r$ (d) $\dfrac{1}{r!}, \displaystyle\sum_{r=0}^{9} \dfrac{1}{r!}$

Because

(a) This is the sum of the first 10 reciprocals and the general reciprocal can be denoted by $\dfrac{1}{r}$ where $r \neq 0$. This gives the sum as $\displaystyle\sum_{r=1}^{10} \dfrac{1}{r}$.

(b) This is the sum of the first 10 numbers cubed and every number can be denoted by r^3, giving the sum as $\displaystyle\sum_{r=1}^{10} r^3$.

(c) Here, every odd number is preceded by a minus sign. This can be denoted by $(-1)^r$ because when r is even $(-1)^r = 1$ and when r is odd $(-1)^r = -1$. This permits the general term to be written as $(-1)^r r$ and the sum is $\displaystyle\sum_{r=1}^{10} (-1)^r r$.

(d) This is the sum of the first 10 reciprocal factorials and the general reciprocal factorial can be denoted by $\dfrac{1}{r!}$, giving the sum as $\displaystyle\sum_{r=0}^{9} \dfrac{1}{r!}$. Notice that the sum of the first 10 terms starts with $r = 0$ and ends with $r = 9$.

7

If the sum of terms is required up to some final but unspecified value of the counting variable r, say $r = n$, then the symbol n is placed on the top of the sigma. For example, the sum of the first n terms of the series with general term r^2 is given by:

$$\sum_{r=1}^{n} r^2 = 1^2 + 2^2 + 3^2 + \ldots + n^2$$

Try some examples. In each of the following write down the general term and then write down the sum of the first n terms using the sigma notation:

(a) $\dfrac{5}{2} + \dfrac{5}{4} + \dfrac{5}{6} + \dfrac{5}{8} + \ldots$ (b) $1 + \dfrac{1}{3^2} + \dfrac{1}{5^2} + \dfrac{1}{7^2} + \ldots$

(c) $2 - 4 + 6 - 8 + \ldots$ (d) $1 - 3 + 9 - 27 + \ldots$

(e) $1 - 1 + 1 - 1 + \ldots$

Answers in the next frame

8

(a) $\dfrac{5}{2r}, \displaystyle\sum_{r=1}^{n} \dfrac{5}{2r}$ (b) $\dfrac{1}{(2r-1)^2}, \displaystyle\sum_{r=1}^{n} \dfrac{1}{(2r-1)^2}$

(c) $(-1)^{r+1} 2r, \displaystyle\sum_{r=1}^{n} (-1)^{r+1} 2r$ (d) $(-1)^r 3^r, \displaystyle\sum_{r=0}^{n-1} (-1)^r 3^r$

(e) $(-1)^r, \displaystyle\sum_{r=0}^{n-1} (-1)^r$

Because

(a) Each term is of the form of 5 divided by an even number.

(b) Each term is of the form of the reciprocal of an odd number squared. Notice that the first term could be written as $\dfrac{1}{1^2}$ to maintain the pattern.

(c) Here the alternating sign is positive for every odd term (r odd) and negative for every even term (r even). Consequently, to force a positive sign for r odd we must raise -1 to an even power – hence $r + 1$ which is even when r is odd and odd when r is even.

(d) Here the counting starts at $r = 0$ for the first term so while odd terms are preceded by a minus sign the value of r is even. Also the nth term corresponds to $r = n - 1$.

(e) Again, the nth term corresponds to $r = n - 1$ as the first term corresponds to $r = 0$.

---

9   **The sum of the first *n* natural numbers**

Consider the sum of the first *n* non-zero natural numbers:

$$\sum_{r=1}^{n} r = 1 + 2 + 3 + \ldots + n. \text{ This can equally well be written as:}$$

$$\sum_{r=1}^{n} r = n + (n-1) + (n-2) + \ldots + 1 \qquad \begin{array}{l}\text{starting with } n \text{ and working} \\ \text{backwards.}\end{array}$$

If these two are added together term by term then:

$$2\sum_{r=1}^{n} r = (1+n) + (2+n-1) + (3+n-2) + \ldots + (n+1)$$

That is:

$$2\sum_{r=1}^{n} r = (n+1) + (n+1) + (n+1) + \ldots + (n+1) \qquad (n+1) \text{ added } n \text{ times}$$

That is:

$$2\sum_{r=1}^{n} r = n(n+1) \text{ so that } \sum_{r=1}^{n} r = \frac{n(n+1)}{2} \qquad \begin{array}{l}\text{the sum of the first } n \text{ non-zero} \\ \text{natural numbers.}\end{array}$$

This is a useful formula to remember so make a note of it.

Example

Find the value of the first 100 natural numbers (excluding zero).

Solution

$$1 + 2 + 3 + \ldots + 100 = \sum_{r=1}^{100} r$$

$$= \frac{100(100+1)}{2} \qquad \text{using the formula}$$

$$= 5050$$

Now for two rules to be used when manipulating sums. Next frame

Rules for manipulating sums

Rule 1

If $f(r)$ is some general term and k is a constant then:

$$\sum_{r=1}^{n} kf(r) = kf(1) + kf(2) + kf(3) + \ldots + kf(n)$$

$$= k(f(1) + f(2) + f(3) + \ldots + f(n))$$

$$= k\sum_{r=1}^{n} f(r) \qquad \textit{Common constants can be factored out of the sigma.}$$

In particular, when $f(r) = 1$ for all values of r:

$$\sum_{r=1}^{n} k = k\sum_{r=1}^{n} 1$$

$$= k(1 + 1 + \ldots + 1) \qquad k \text{ multiplied by 1 added to itself } n \text{ times}$$

$$= kn$$

Rule 2

If $f(r)$ and $g(r)$ are two general terms then:

$$\sum_{r=1}^{n} (f(r) + g(r)) = \{f(1) + g(1) + f(2) + g(2) + \ldots\}$$

$$= \{f(1) + f(2) + \ldots\} + \{g(1) + g(2) + \ldots\}$$

$$= \sum_{r=1}^{n} f(r) + \sum_{r=1}^{n} g(r)$$

Now for a worked example

Example

Find the value of $\displaystyle\sum_{r=1}^{n} (6r + 5)$

Solution

$$\sum_{r=1}^{n} (6r + 5) = \sum_{r=1}^{n} 6r + \sum_{r=1}^{n} 5 \qquad \text{by Rule 2}$$

$$= 6\sum_{r=1}^{n} r + \sum_{r=1}^{n} 5 \qquad \text{by Rule 1}$$

$$= 6\frac{n(n+1)}{2} + 5n \qquad \text{using the formulas in Frames 9 and 10}$$

$$= 3(n^2 + n) + 5n$$

$$= 3n^2 + 8n$$

$$= n(3n + 8)$$

So the values of:

(a) $\displaystyle\sum_{r=1}^{50} r$

(b) $\displaystyle\sum_{r=1}^{n} (8r - 7)$ are ....

Answers in next frame

12

> (a) 1275
> (b) $n(4n - 3)$

Because

(a) $\displaystyle\sum_{r=1}^{50} r = \frac{50 \times 51}{2} = 1275$

(b) $\displaystyle\sum_{r=1}^{n} (8r - 7) = 8\sum_{r=1}^{n} r - \sum_{r=1}^{n} 7 = \frac{8n(n + 1)}{2} - 7n = 4n^2 - 3n = n(4n - 3)$

At this point let us pause and summarize the main facts on the sigma notation

---

 # Review summary   Unit 38

13

1 The sigma notation is used as a shorthand notation for the sum of a number of terms, each term typified by a general term:

$$\sum_{r=1}^{n} f(r) = f(1) + f(2) + f(3) + \ldots + f(n)$$

2 The binomial expansion can be written using the sigma notation as:

$$(a + b)^n = \sum_{r=0}^{n} {}^nC_r a^{n-r} b^r$$

3 Two properties of sums are:

(a) $\displaystyle\sum_{r=0}^{n} {}^nC_r = 2^n$

(b) $\displaystyle\sum_{r=1}^{n} r = \frac{n(n + 1)}{2}$

Two rules for manipulating sums are:

(a) $\displaystyle\sum_{r=1}^{n} k = kn$

(b) $\displaystyle\sum_{r=1}^{n} [f(r) + g(r)] = \sum_{r=1}^{n} f(r) + \sum_{r=1}^{n} g(r)$

Review exercise   Unit 38

1 Evaluate: **14**

(a) $\displaystyle\sum_{r=1}^{35} r$ (b) $\displaystyle\sum_{r=1}^{n}(4r+2)$

2 Determine the 5th term and the sum of the first 20 terms of the series:
$3+6+9+12+\ldots$

> *Complete both questions, looking back at the Unit if you need to.*
> *Don't rush, take your time.*
> *The answers and working are in the next frame.*

---

1 (a) $\displaystyle\sum_{r=1}^{n} r = \frac{n(n+1)}{2}$ so that $\displaystyle\sum_{r=1}^{35} r = \frac{35(36)}{2} = 630$ **15**

(b) $\displaystyle\sum_{r=1}^{n}(4r+2) = \sum_{r=1}^{n} 4r + \sum_{r=1}^{n} 2$

$$= 4\sum_{r=1}^{n} r + (2+2+\ldots+2) \quad \text{2 added to itself } n \text{ times}$$

$$= 4\frac{n(n+1)}{2} + 2n$$

$$= 2n^2 + 4n = 2n(n+2)$$

2 The 5th term of $3+6+9+12+\ldots$ is 15 because the general term is $3r$. The sum of the first 20 terms of the series is then:

$$\sum_{r=1}^{20} 3r = 3\sum_{r=1}^{20} r = 3\frac{20(21)}{2} = 630$$

> *Now for the Review test*

 # Review test   Unit 38

1 Evaluate: **16**

(a) $\displaystyle\sum_{r=1}^{20} r$ (b) $\displaystyle\sum_{r=1}^{n}(2r+3)$

2 Determine the 6th term and the sum of the first 10 terms of the series:
$2+4+6+8+\ldots$

17   You have now come to the end of this Module. A list of **Can You?** questions
follows for you to gauge your understanding of the material in the Module.
You will notice that these questions match the **Learning outcomes** listed at
the beginning of the Module. Now try the **Test exercise**. *Work through the
questions at your own pace, there is no need to hurry.* A set of **Further problems**
provides additional valuable practice.

 # Can You?

1   Checklist: Module 12

Check this list before and after you try the end of Module test.

On a scale of 1 to 5 how confident are you that you can:

- Define $n!$ and recognize that there are $n!$ different combinations of n
  different items?
  Yes ☐ ☐ ☐ ☐ ☐ *No*

- Evaluate $n!$ for moderately sized n using a calculator?
  Yes ☐ ☐ ☐ ☐ ☐ *No*

- Manipulate expressions containing factorials?
  Yes ☐ ☐ ☐ ☐ ☐ *No*

- Recognize that there are $\dfrac{n!}{(n-r)!r!}$ combinations of r identical items in
  n locations?
  Yes ☐ ☐ ☐ ☐ ☐ *No*

- Recognize simple properties of combinatorial coefficients?
  Yes ☐ ☐ ☐ ☐ ☐ *No*

- Construct Pascal's triangle?
  Yes ☐ ☐ ☐ ☐ ☐ *No*

- Write down the binomial expansion for natural number powers?
  Yes ☐ ☐ ☐ ☐ ☐ *No*

- Obtain specific terms in the binomial expansion using the general
  term?
  Yes ☐ ☐ ☐ ☐ ☐ *No*

- Use the sigma notation?
  Yes ☐ ☐ ☐ ☐ ☐ *No*

Test exercise 12

1 In how many different ways can 6 different numbers be selected from the numbers 1 to 49 if the order in which the selection is made does not matter?

2 Find the value of:

(a) $8!$ (b) $10!$ (c) $\dfrac{17!}{14!}$ (d) $(15-11)!$ (e) $\dfrac{4!}{0!}$

3 Evaluate each of the following:

(a) $^8C_3$ (b) $^{15}C_{12}$ (c) $^{159}C_{158}$ (d) $^{204}C_0$

4 Expand $(2a-5b)^7$ as a binomial series.

5 In the binomial expansion of $(1+10/x)^{10}$ written in terms of descending powers of x, find:

(a) the 8th term   (b) the coefficient of x^{-8}

6 Evaluate:

(a) $\displaystyle\sum_{r=1}^{45} r$ (b) $\displaystyle\sum_{r=1}^{n}(9-3r)$

7 Determine the 5th term and the sum of the first 20 terms of the series: $1+3+5+7+\ldots$

 # Further problems 12

1 Given a row of 12 hat pegs, in how many different ways can:

(a) 5 identical red hardhats be hung?

(b) 5 identical red and 4 identical blue hardhats be hung?

(c) 5 identical red, 4 identical blue and 2 identical white hardhats be hung?

2 Show that:

(a) $^{n+1}C_1 - {}^nC_1 = 1$

(b) $\displaystyle\sum_{r=0}^{n}(-1)^r\,{}^nC_r = 0$

(c) $\displaystyle\sum_{r=0}^{n}{}^nC_r\,2^r = 3^n$

3 Write out the binomial expansions of:

(a) $(1-3x)^4$ (b) $(2+x/2)^5$ (c) $\left(1-\dfrac{1}{x}\right)^5$ (d) $(x+1/x)^6$

4 Evaluate:

(a) $\displaystyle\sum_{r=1}^{16}(5r-7)$ (b) $\displaystyle\sum_{r=1}^{n}(4-5r)$

Sets

Learning outcomes

When you have completed this Module you will be able to:

- Define a set and to describe a set using set notation
- Use a Venn diagram to illustrate set properties
- Derive the intersection and union of sets
- Derive the complement of a set
- Compute the number of elements in the union, intersection and complements of sets

Units

Sets and subsets

Unit 39

Definitions

1

A set is a collection of distinct objects be they tools, numbers or ideas; we can even have a set of sets. The notation used to denote a set consists of a pair of curly brackets:

$\{\ldots\}$

where the individual objects that comprise the set are described inside the brackets; these individual objects being called the *elements* of the set.

For example, the set consisting of the whole numbers 1 to 5 can be denoted by:

$\{1, 2, 3, 4, 5\}$

where the elements are listed individually. Alternatively, the set can be denoted by:

$\{n : n$ is a whole number and $1 \leq n \leq 5\}$

This is read as:

The set of n where n is a whole number and n is greater than or equal to one and less than or equal to 5 (the colon (:) stands for the word 'where').

In this description of the set the elements are prescribed by describing those properties that they have in common that are sufficient to generate the list. Notice that if two sets contain the same elements then they are the same set so that:

$\{1, 2, 3, 4, 5\} = \{3, 1, 5, 2, 4\}$

The order in which the elements are listed does not matter. Also the elements of a set are distinct from one another and so repetition of elements is of no account. So, example:

$\{1, 1, 1, 2, 2, 3, 4, 5\} = \{1, 2, 3, 4, 5\}$

Try these. In each of the following give an alternative description of the set:

(a) $\{1, 2, 3, 4, 5, 6\}$

(b) $\{n : n$ is a whole number and $4 \leq n < 8\}$

(c) $\{t : t$ is the result of tossing a coin$\}$

(d) $\{d : d$ is a day of the week$\}$

(e) $\{a, e, i, o, u\}$

(f) $\{p, p, p, p, p, q, r, r, r, s, s\}$

The answers are in the following frame

2

> (a) $\{n : n \text{ is a whole number and } 1 \leq n \leq 6\}$
> (b) $\{4, 5, 6, 7\}$
> (c) {Head, Tail}
> (d) {Monday, Tuesday, Wednesday, Thursday, Friday, Saturday, Sunday}
> (e) $\{v : v \text{ is a vowel}\}$
> (f) $\{p, q, r, s\}$

Because

(a) The statement 'n is a whole number and $1 \leq n \leq 6$' describes a common property of each element and is sufficient for the list of numbers to be generated.

(b) The listing is obtained by selecting those numbers that satisfy the common property.

(c) The two possible outcomes from tossing a coin are that it shows a Head or a Tail.

(d) The list of days of the week.

(e) The list gives all the vowels.

(f) Repetitions are not permitted in a listing of elements.

Next frame

3   **Further set notation**

It is often more convenient to refer to a set without continually describing its elements, in which case we use capital letters to denote the set. For example:

$A = \{\text{Head, Tail}\}$

Here the set consisting of all possible outcomes from tossing a coin has been called set A and can now be referred to in future by simply using the letter A. Membership of a set is denoted by using the symbol $\in$ for inclusion and $\notin$ for exclusion. For example, if now A is the set containing the letters a, b and c and B is the set containing the letters c, d and e:

$A = \{a, b, c\}$ and $B = \{c, d, e\}$

then:

$a \in A$ and $d \in B$, that is, a is an element of A and d is an element of B.

Also

$d \notin A$ and $a \notin B$, that is, d is not an element of A and a is not an element of B.

Note that a strike through a symbol means the negative of the symbol such as $=$ and $\neq$.

▶

So if $A = \{2, 4, 6, 8\}$ place $\in$ or $\notin$ in each of the following:

(a) 2 ... A

(b) 5 ... A

(c) a ... A

The answers are in the next frame

4

(a) $2 \in A$

(b) $5 \notin A$

(c) $a \notin A$

Because

(a) 2 is an element of A so $2 \in A$

(b) 5 is not an element of A so $5 \notin A$

(c) a is not an element of A so $a \notin A$

Move to the next frame

Subsets

5

If all the elements of set B are contained in set A then we say that set B is a **subset** of set A. So if A is the set of all red cars, blue cars and white cars and B is the set of all red cars and blue cars then set B is a subset of set A and we write:

$B \subset A$ because all the elements of B are contained in A

or, alternatively:

$A \supset B$ because A contains all the elements of B

For example, if $A = \{a, b, c, d, e\}$, $B = \{c, d, e\}$ and $C = \{d, e, f\}$ then:

$B \subset A$ because all the elements of B are contained in A

but

$C \not\subset A$ because not all the elements of C are contained in A

Try these. Which of the following are true statements?

(a) $\{1\} \subset \{1, 2, 3\}$

(b) $1 \subset \{1, 2, 3\}$

(c) $\{1\} \subset \{\{1\}, \{2\}, \{3\}\}$

(d) $\{x : x \text{ is an integer and } -2 < x < 4\} \supset \{-1, 0, 1, 2, 3, 4\}$

(e) $\{x : x \text{ is an integer and } -2 < x < 4\} \subset \{-1, 0, 1, 2, 3, 4\}$

Next frame

6

(a) True
(b) False
(c) False
(d) False
(e) True

Because

(a) {1} is a set whose element 1 is contained in the set {1, 2, 3} so set {1} is a subset of the set {1, 2, 3}

(b) 1 is an element of the set {1, 2, 3} so $1 \in \{1, 2, 3\}$, it is not a subset.

(c) {1} is an element of the set {{1}, {2}, {3}} that contains three sets. That is {1} ∈ {{1}, {2}, {3}}, it is not a subset.

(d) The set on the left is {−1, 0, 1, 2, 3} which does not contain all the elements of the set {−1, 0, 1, 2, 3, 4}

(e) The set on the left is {−1, 0, 1, 2, 3} all of whose elements are contained in {−1, 0, 1, 2, 3, 4} therefore it is a subset of the set {−1, 0, 1, 2, 3, 4}

Since it is an obvious truism to say that:

Every element of A is an element of A

the definition of a subset means that set A is a subset of itself – indeed, every set is a subset of itself.

To cater for subsets of A that are either not A or A itself we use the notation $\subseteq$. So if:

$B \subseteq A$ then this means $B \subset A$ or $B = A$

And if $B \subset A$ then we call B a *proper* subset of A. Therefore, given:

$A = \{a, b, c\}$

the proper subsets of A are:

{a}, {b}, {c}, {a, b}, {a, c}, {b, c}

whereas the subsets of A are:

{a}, {b}, {c}, {a, b}, {a, c}, {b, c}, {a, b, c}

– or nearly so!

Move now to the next frame

The empty set   7

The set that contains no elements at all is called the **empty set** and is denoted either by a pair of empty brackets { } or by the symbol Ø which is the 28th letter of the Danish alphabet and is pronounced like the vowel in 'third'.

There is only one empty set but it has a property that links it to every set.

Recall that if it is possible to find an element of set B that is not in set A then set B cannot be a subset of set A and that if it is not so possible then set B is a subset of set A. Now, it is not possible to find an element of Ø that is not in set A so Ø is a subset of A. Indeed:

There is only one empty set and it is a subset of every set.

This now brings us to the complete list of subsets of the set $\{a, b, c\}$ which is:

   Ø, $\{a\}$, $\{b\}$, $\{c\}$, $\{a, b\}$, $\{a, c\}$, $\{b, c\}$, $\{a, b, c\}$

The total number of subsets of $\{a, b, c\}$ is, therefore, 8. It is no accident that $8 = 2^3$ where 3 is the number of elements in the set. Indeed, a set with n elements has 2^n subsets.

Because the number of subsets of a set with n elements is 2 to the *power n* the set that contains all the subsets of set A is called the power set of set A, denoted by $P(A)$. So that the power set of $\{a, b, c\}$ is:

   $P(\{a, b, c\}) = \{$Ø, $\{a\}$, $\{b\}$, $\{c\}$, $\{a, b\}$, $\{a, c\}$, $\{b, c\}$, $\{a, b, c\}\}$

You try one. The power set of the set $A = \{2, 4, 6, 8\}$ is ....

The answer is in the next frame

                                                                          8

   $\{$Ø, $\{2\}$, $\{4\}$, $\{6\}$, $\{8\}$, $\{2, 4\}$, $\{2, 6\}$, $\{2, 8\}$, $\{4, 6\}$, $\{4, 8\}$,
   $\{6, 8\}$, $\{2, 4, 6\}$, $\{2, 4, 8\}$, $\{2, 6, 8\}$, $\{4, 6, 8\}$, $\{2, 4, 6, 8\}\}$

Because

The power set of set A contains all the subsets of A and there are:

1 subset with no elements { } or Ø

4 subsets with 1 element, namely $\{2\}$, $\{4\}$, $\{6\}$, $\{8\}$

6 subsets with 2 elements, namely $\{2, 4\}$, $\{2, 6\}$, $\{2, 8\}$, $\{4, 6\}$,
$\qquad\qquad\qquad\qquad\qquad\qquad\qquad\qquad\qquad$ $\{4, 8\}$, $\{6, 8\}$

4 subsets with 3 elements, namely $\{2, 4, 6\}$, $\{2, 4, 8\}$, $\{2, 6, 8\}$, $\{4, 6, 8\}$

and 1 subset with 4 elements $\{2, 4, 6, 8\}$

a total of $16 = 2^4$ where 4 is the number of elements in A.

Move to the next frame

9 The universal set

The set that contains all the elements under discussion is called the **universal set** and is denoted by the symbol U. For example, if

$$A = \{x : x \text{ is a person with blue eyes}\}$$
$$\text{and } B = \{x : x \text{ is a person with green eyes}\}$$

Then the universal set could be:

$$U = \{x : x \text{ is a person with blue or green eyes}\}$$

Notice the words 'could be'. The universal set is not unique. In the example just given the universal set could have been defined as:

$$U = \{x : x \text{ is a person}\}$$

The definition of the universal set depends upon the context of the problem. The overriding condition is that all the sets that are going to be discussed *must be subsets of the universal set.*

So, define suitable universal sets for the discussions involving each of the following:

(a) the colours of the rainbow

(b) rolling a six-sided die

(c) the setting of three on/off switches A, B and C in the form of ordered triples of values (a, b, c) where 0 represents 'off' and 1 represents 'on'

The answer is in the next frame

10

(a) {red, orange, yellow, green, blue, indigo, violet}
(b) $\{x : x \text{ is whole number and } 1 \le x \le 6\}$
(c) $\{(0, 0, 0), (0, 0, 1), (0, 1, 0), (0, 1, 1), (1, 0, 0), (1, 0, 1), (1, 1, 0), (1, 1, 1)\}$

Because

(a) there are seven colours traditionally associated with a rainbow and they are red, orange, yellow, green, blue, indigo and violet

(b) the outcome from rolling a six-sided die is the upper face showing a 1, 2, 3, 4, 5 or 6

(c) if the switches are labelled as A, B and C with values a, b and c respectively then these values can form the ordered triple (a, b, c). So, if they are all off their values are 0, 0 and 0 which forms the ordered triple $(0, 0, 0)$. If A and B are off and C is on then their values are 0, 0 and 1 respectively which forms the ordered triple $(0, 0, 1)$ and so on.

Review summary   Unit 39

11

1 *Set:*   A collection of distinct elements denoted by either a description or a listing of elements within curly brackets. A set can also be referred to by using a capital letter.

2 *Subset:*   If set B is formed from the elements of set A then B is a subset of set A, denoted by $B \subseteq A$. If B is a proper subset of A then this is denoted by $B \subset A$. Every set is a subset of itself.

3 *The empty set:*   The set with no elements at all is called the empty set and is denoted by $\emptyset$ or empty brackets { }. The empty set is a subset of every set.

4 *The universal set:*   The set that contains all the elements under discussion is called a universal set and is denoted by U. The universal set is not unique.

 # Review exercise   Unit 39

1 Which of the following are true statements:

12

   (a) $1 \in \{1, 2, 3\}$ (b) $\{3\} \in \{1, 2, 3\}$ (c) $\{A, B, C\} \subseteq \{B, A, C\}$

   (d) $\emptyset \in \{-1, 0, 1\}$ (e) $\emptyset \subset \{-1, 0, 1\}$ (f) $\emptyset \in \{\{\emptyset\}, \emptyset\}$

   (g) $\emptyset \subset \{\{\emptyset\}, \emptyset\}$

2 How many subsets are there of the set $A = \{1, 2, 3, a, b, c\}$?

3 Define a suitable universal set for discussions involving each of the following:

   (a) choosing a number between 1 and 7

   (b) selecting a working or a defective component from an assembly line.

Complete all three questions, referring to the Unit if necessary. You can check your answers and working with the next frame

1 (a) True. 1 is an element of the set

13

   (b) False. $\{3\}$ is a set and the elements are numbers.

   (c) True. $\{A, B, C\} \subseteq \{B, A, C\}$ means $\{A, B, C\} \subset \{B, A, C\}$ or $\{A, B, C\} = \{B, A, C\}$

   (d) False. The empty set is not an element of $\{-1, 0, 1\}$ which contains only the elements -1, 0 and 1.

   (e) True. The empty set is a subset of every set.

   (f) True. The set on the right-hand side contains the empty set $\emptyset$ and the set containing the empty set $\{\emptyset\}$ so the empty set is an element of the set.

   (g) True. The empty set is a subset of every set.

2 There are 6 elements in set A so the number of subsets is $2^6 = 64$. There are:

1 subset with 0 elements	{ }
6 subsets with 1 element	$\{1\}$, $\{2\}$, $\{3\}$, $\{a\}$, $\{b\}$, $\{c\}$
15 subsets with 2 elements	$\{1, 2\}$, $\{1, 3\}$, $\{1, a\}$, $\{1, b\}$, $\{1, c\}$
	$\{2, 3\}$, $\{2, a\}$, $\{2, b\}$, $\{2, c\}$
	$\{3, a\}$, $\{3, b\}$, $\{3, c\}$
	$\{a, b\}$, $\{a, c\}$
	$\{b, c\}$
20 subsets with 3 elements	$\{1, 2, 3\}$, $\{1, 2, a\}$, $\{1, 2, b\}$, $\{1, 2, c\}$,
	$\{1, 3, a\}$, $\{1, 3, b\}$, $\{1, 3, c\}$, $\{1, a, b\}$,
	$\{1, a, c\}$, $\{1, b, c\}$, $\{2, 3, a\}$, $\{2, 3, b\}$,
	$\{2, 3, c\}$, $\{2, a, b\}$, $\{2, a, c\}$, $\{2, b, c\}$,
	$\{3, a, b\}$, $\{3, a, c\}$, $\{3, b, c\}$, $\{a, b, c\}$
15 subsets with 4 elements	$\{1, 2, 3, a\}$, $\{1, 2, 3, b\}$, $\{1, 2, 3, c\}$,
	$\{2, 3, a, b\}$, $\{2, 3, a, c\}$, $\{2, 3, b, c\}$,
	$\{1, 3, a, b\}$, $\{1, 3, a, c\}$, $\{1, 3, b, c\}$,
	$\{1, 2, a, b\}$, $\{1, 2, a, c\}$, $\{1, 2, b, c\}$,
	$\{1, a, b, c\}$, $\{2, a, b, c\}$, $\{3, a, b, c\}$
6 subsets with 5 elements	$\{1, 2, 3, a, b\}$, $\{1, 2, 3, a, c\}$,
	$\{1, 2, 3, b, c\}$, $\{1, 2, a, b, c\}$,
	$\{1, 3, a, b, c\}$, $\{2, 3, a, b, c\}$
1 subset with 6 elements	$\{1, 2, 3, a, b, c\}$

3 (a) The universal set $\{1, 2, 3, 4, 5, 6, 7\}$ contains all the possible outcomes which, in this case, are the numbers 1 to 7.

(b) There are only two possible outcomes: so the universal set is {working component, defective component}.

Now for the Review test

 # Review test   Unit 39

14 **1** Which of the following are true statements:

(a) $e \in \{a, b, c, d\}$

(b) $\{c\} \in \{a, b, c, d\}$

(c) {Red, Blue, Green} $\subseteq$ {Green, Orange, Blue, Red}

(d) $\varnothing \notin \{A, B, C\}$

(e) $\varnothing \not\subset \{A, B, C\}$

(f) $\varnothing \notin \{\{\varnothing\}\}$

(g) $\varnothing \subset \{\{\varnothing\}\}$

2 How many subsets are there of the set $A = \{$red, white, blue$\}$?

3 Define a suitable universal set for discussions involving each of the following:

(a) tossing a fair coin

(b) selecting a red card from a deck of playing cards.

Set operations **Unit 40**

Venn diagrams **1**

A Venn diagram is a useful way of illustrating set properties. It is quite a simple device and consists of using a circle to represent a set:

Here the interior of the circle represents the set A. Elements can be represented by labelled points such as:

 here $A = \{a, b, c, d\}$

If set B is a subset of set A this can be illustrated by one circle inside another:

For either of the two following diagrams set B is not a subset of set A:

 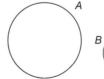

In the first diagram there are elements in B that are not in A. In the second diagram no elements of B are in A.

Move to the next frame

2 Intersection

Consider the two sets:

$A = \{a : a$ is a member of a walking club$\}$

and $B = \{b : b$ is a member of a golf club$\}$

Then by taking those elements that are common to both sets we can construct:

$C = \{c : c$ is a member of both a walking club and a golf club$\}$

Set C is called the **intersection** of sets A and B and is denoted:

$C = A \cap B$

The symbol $\cap$ represents the operation of intersection – the operation of selecting the elements that are common to both the two sets.

The Venn diagram for the intersection of the two sets A and B consists of two overlapping circles with the intersection of the two sets shaded in:

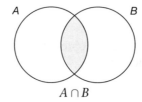

$A \cap B$

Find the intersection $A \cap B$ for each of the following:

(a) $A = \{1, 2, 3, 4\}$ and $B = \{3, 4, 5, 6\}$

(b) $A = \{2, 4, 6, 8\}$ and $B = \{2, 6\}$

(c) $A = \{a : a$ is a red court card from a deck of playing cards$\}$ and
   $B = \{b : b$ is a Jack from a deck of playing cards$\}$

(d) $A = \{c : c$ is a cubist oil painting$\}$ and $B = \{p : p$ is a painting by Picasso$\}$

The answers are in the next frame

3

> (a) {3, 4}
> (b) {2, 6}
> (c) $\{r : r$ is a red Jack$\}$ or {Jack of hearts, Jack of diamonds}
> (d) $\{x : x$ is a cubist oil painting by Picasso$\}$

Because

(a) $A = \{1, 2, 3, 4\}$ and $B = \{3, 4, 5, 6\}$, so
$$A \cap B = \{1, 2, 3, 4\} \cap \{3, 4, 5, 6\} = \{3, 4\}$$
and those elements are common to both sets.

(b) $A = \{2, 4, 6, 8\}$ and $B = \{2, 6\}$, so
$$A \cap B = \{2, 4, 6, 8\} \cap \{2, 6\} = \{2, 6\}$$
and those elements are common to both sets. Notice that $A \cap B = B$ so B is a subset of A.

(c) $A = \{a : a$ is a red court card from a deck of playing cards$\}$ and $B = \{b : b$ is a Jack from a deck of playing cards$\}$, so
$$A \cap B = \{a : a \text{ is a red court card}\} \cap \{b : b \text{ is a Jack}\}$$
$$= \{l : l \text{ is a red Jack}\}$$
$$= \{\text{Jack of hearts, Jack of diamonds}\}$$

(d) $A = \{c : c$ is a cubist oil painting$\}$ and $B = \{p : p$ is a painting by Picasso$\}$, so
$$A \cap B = \{c : c \text{ is a cubist oil painting}\} \cap \{p : p \text{ is a painting by Picasso}\}$$
$$= \{x : x \text{ is a cubist oil painting by Picasso}\}$$

Move to the next frame

Union

4

Consider the two sets:

$$A = \{a : a \text{ is a person with green eyes}\}$$
$$\text{and} \quad B = \{b : b \text{ is a person with brown eyes}\}$$

Then, by combining the elements of both sets, we can construct:

$$C = \{c : c \text{ is a person with green eyes or brown eyes}\}$$

Set C is called the **union** of sets A and B and is denoted:

$$C = A \cup B$$

The symbol $\cup$ represents the operation of union – the operation of combining the elements of two sets.

The Venn diagram for the union of the two sets A and B consists of two overlapping circles with the union of the two sets shaded in:

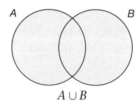

$$A \cup B$$

Find the union $A \cup B$ for each of the following:

(a) $A = \{1, 2, 3, 4\}$ and $B = \{5, 6, 7\}$

(b) $A = \{2, 4, 6, 8\}$ and $B = \{6, 8, 10, 12\}$

(c) $A = \{a : a \text{ is a vowel}\}$ and $B = \{b : b \text{ is a consonant}\}$

(d) $A = \{r : r \text{ is a red Ford car}\}$ and $B = \{g : g \text{ is a green Honda car}\}$

The answers are in the next frame

5

(a) $\{1, 2, 3, 4, 5, 6, 7\}$

(b) $\{2, 4, 6, 8, 10, 12\}$

(c) $\{l : l \text{ is a letter of the alphabet}\}$

(d) $\{c : c \text{ is either a red Ford car or a green Honda car}\}$

Because

(a) $A = \{1, 2, 3, 4\}$ and $B = \{5, 6, 7\}$ and

   $A \cup B = \{1, 2, 3, 4\} \cup \{5, 6, 7\} = \{1, 2, 3, 4, 5, 6, 7\}$

   by combining the elements of the two sets.

(b) $A = \{2, 4, 6, 8\}$ and $B = \{6, 8, 10, 12\}$ and

   $A \cup B = \{2, 4, 6, 8\} \cup \{6, 8, 10, 12\} = \{2, 4, 6, 8, 10, 12\}$

   Repetitions are not permitted.

(c) $A = \{a : a \text{ is a vowel}\}$ and $B = \{b : b \text{ is a consonant}\}$ and

   $A \cup B = \{a : a \text{ is a vowel}\} \cup \{b : b \text{ is a consonant}\}$

   $\qquad = \{l : l \text{ is a letter of the alphabet}\}$

   since the alphabet consists of letters that are either consonants or vowels.

(d) $A = \{r : r \text{ is a red Ford car}\}$ and $B = \{g : g \text{ is a green Honda car}\}$ and

   $A \cup B = \{r : r \text{ is a red Ford car}\} \cup \{g : g \text{ is a green Honda car}\}$

   $\qquad = \{c : c \text{ is either a red Ford car or a green Honda car}\}$

Move to the next frame

Complement

Consider the set:

$A = \{2, 4, 6\}$

and the universal set

$U = \{u : u$ is a number on a six-sided die$\}$

By taking those elements that are within the universal set but not in A we can form the set:

$B = \{1, 3, 5\}$

The set B is called the **complement** of A and is denoted by $\bar{A}$.

The bar over the A indicates the complement of A – those elements in the universal set that are not in A.

The Venn diagram for the complement of set A requires a diagram of the universal set which is represented by a rectangle containing A:

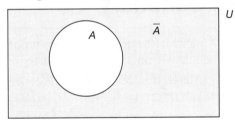

Find the complement $\bar{A}$ of A for each of the following:

(a) $A = \{2, 4, 6\}$ and $U = \{-2, -1, 0, 1, 2, 3, 4, 5, 6, 7\}$
(b) $A = \{1, 3, 5, 7\}$ and $U = \{n : n$ is a whole number and $1 \leq n < 8\}$
(c) $A = \{-1, 1\}$ and $U = \{x : x$ is a solution of the equation $x(x^2 - 1) = 0\}$
(d) $A = \{a : a$ is a mathematics text book$\}$ and $U = \{b : b$ is a text book$\}$

The answers are in the next frame

(a) $\{-2, -1, 0, 1, 3, 5, 7\}$
(b) $\{2, 4, 6\}$
(c) $\{0\}$
(d) $\{x : x$ is a text book that is not a mathematics text book$\}$

Because

(a) $A = \{2, 4, 6\}$ and $U = \{-2, -1, 0, 1, 2, 3, 4, 5, 6, 7\}$ and so

$\bar{A} = \overline{\{2, 4, 6\}} = \{-2, -1, 0, 1, 3, 5, 7\}$

that is, those elements in U but not in A.

(b) $A = \{1, 3, 5, 7\}$ and $U = \{n : n$ is a whole number and $1 \leq n < 8\}$
$$= \{1, 2, 3, 4, 5, 6, 7\} \text{ and so}$$

$\bar{A} = \overline{\{1, 3, 5, 7\}} = \{2, 4, 6\}$

that is, those elements in U but not in A.

(c) $A = \{-1, 1\}$ and $U = \{x : x$ is a solution of the equation $x(x^2 - 1) = 0\}$
$$= \{-1, 0, 1\} \text{ and so}$$

$\bar{A} = \overline{\{-1, 1\}} = \{0\}$

(d) $A = \{a : a$ is a mathematics text book$\}$ and $U = \{b : b$ is a text book$\}$
and so

$\bar{A} = \overline{\{b : b \text{ is a mathematics text book}\}}$
$$= \{x : x \text{ is a text book that is not a mathematics text book}\}$$

Move to the next frame

Review summary   Unit 40

8

1 *Venn diagram:* A pictorial device to illustrate properties of sets. Each set is represented by a circle within a rectangle that represents the universal set.

2 *Intersection:* The intersection of sets A and B consists of those elements in both A and B and is written $A \cap B$.

3 *Union:* The union of sets A and B consists of those elements in either A or B or both A and B and is written $A \cup B$.

4 *Complement:* The complement of set A consists of those elements of the universal set U that are not in A and is denoted by $\bar{A}$.

Review exercise   Unit 40

9

1 Find the intersection $A \cap B$ for each of the following:
   (a) $A = \{a, e, i\}$ and $B = \{i, o, u\}$
   (b) $A = \{3, 6, 9, 12\}$ and $B = \{6, 12\}$
   (c) $A = \{a : a$ is a black card from a deck of playing cards$\}$ and
       $B = \{b : b$ is a king from a deck of playing cards$\}$
   (d) $A = \{s : s$ is a bronze statue$\}$ and $B = \{d : d$ is a bronze statue by Degas$\}$

2 Find the union $A \cup B$ for each of the following:

(a) $A = \{a, e, i, o\}$ and $B = \{i, o, u\}$

(b) $A = \{3, 6, 9, 12\}$ and $B = \{6, 12\}$

(c) $A = \{l : l$ is a straight line segment$\}$ and $B = \{p : p$ is a parabola$\}$

(d) $A = \{r : r$ is a red playing card$\}$ and $B = \{c : c$ is an ace$\}$

3 Find the complement $\bar{A}$ of A for each of the following:

(a) $A = \{3, 6, 9\}$ and $U = \{-6, -3, 0, 3, 6, 9, 12, 15\}$

(b) $A = \{a, o, u\}$ and $U = \{l : l$ is a vowel$\}$

(c) $A = \{0, 2\}$ and
$U = \{x : x$ is a solution of the equation $x(x-1)(x+1)(x-2)(x+2) = 0\}$

(d) $A = \{g : g$ is a green bicycle$\}$ and $U = \{b : b$ is a bicycle$\}$

Complete all three and then check your answers and working with the next frame

10

1 (a) $A = \{a, e, i\}$ and $B = \{i, o, u\}$ so
$A \cap B = \{a, e, i\} \cap \{i, o, u\} = \{i\}$
because i is the only element common to both A and B.

(b) $A = \{3, 6, 9, 12\}$ and $B = \{6, 12\}$ so
$A \cap B = \{3, 6, 9, 12\} \cap \{6, 12\} = \{6, 12\}$
because $B \subset A$.

(c) $A = \{a : a$ is a black card from a deck of playing cards$\}$ and
$B = \{b : b$ is a king from a deck of playing cards$\}$ so
$A \cap B = \{c : c$ is a black king from a deck of playing cards$\}$
because black kings are the only common elements of the two sets.

(d) $A = \{s : s$ is a bronze statue$\}$ and $B = \{d : d$ is a bronze statue by Degas$\}$
so
$A \cap B = B = \{d : d$ is a bronze statue by Degas$\}$ because $B \subset A$.

2 (a) $A = \{a, e, i, o\}$ and $B = \{i, o, u\}$ so

$A \cup B = \{a, e, i\} \cup \{i, o, u\} = \{a, e, i, o, u\}$

because all the elements of A combined with all the elements of B form
the complete collection of vowels.

(b) $A = \{3, 6, 9, 12\}$ and $B = \{6, 12\}$ so
$A \cup B = \{3, 6, 9, 12\} \cup \{6, 12\} = \{3, 6, 9, 12\} = A$ because $B \subset A$.

(c) $A = \{l : l$ is a straight line segment$\}$ and $B = \{p : p$ is a parabola$\}$ so
$A \cup B = \{l : l$ is a straight line segment$\} \cup \{p : p$ is a parabola$\}$
$= \{x : x$ is either a straight line segment or a parabola$\}$

because the union unites all the straight line segments with all the
parabolas.

(d) $A = \{r : r$ is a red playing card$\}$ and $B = \{c : c$ is an ace$\}$ so
$A \cup B = \{r : r$ is either a red playing card or a black ace$\}$
because the union unites all the red cards with the black aces.

3 (a) $A = \{3, 6, 9\}$ and $U = \{-6, -3, 0, 3, 6, 9, 12, 15\}$ so
   $\bar{A} = \{-6, -3, 0, 12, 15\}$ because $-6, -3, 0, 12$ and 15 are those elements
   of the universal set that are not in A.

   (b) $A = \{a, o, u\}$ and $U = \{l : l \text{ is a vowel}\}$ so $\bar{A} = \{e, i\}$ because e and i are
   those vowels that are not in A.

   (c) $A = \{0, 2\}$ and
   $U = \{x : x \text{ is a solution of the equation } x(x-1)(x+1)(x-2)(x+2) = 0\}$
   so $\bar{A} = \{-2, -1, 1\}$ because $U = \{-2, -1, 0, 1, 2\}$.

   (d) $A = \{g : g \text{ is a green bicycle}\}$ and $U = \{b : b \text{ is a bicycle}\}$
   so $\bar{A} = \{x : x \text{ is a bicycle that is not green}\}$ because non-green bicycles,
   whilst being members of U are not members of A.

Now for the Review test

 # Review test

Unit 40

11 **1** Find the intersection $A \cap B$ for each of the following:

   (a) $A = \{11, 12, 13, 14, 15, 16, 17\}$ and $B = \{15, 16, 17, 18, 19\}$
   (b) $A = \{u, v, w, x, y, z\}$ and $B = \{v, z\}$
   (c) $A = \{e : e \text{ is an even-numbered card from a deck of playing cards}\}$ and
   $B = \{b : b \text{ is a black card from a deck of playing cards}\}$
   (d) $A = \{p : p \text{ is a paper-back book}\}$ and
   $B = \{c : c \text{ is } A \text{ Christmas Carol by Charles Dickens}\}$

 2 Find the union $A \cup B$ for each of the following:

   (a) $A = \{\text{red, yellow, indigo}\}$ and $B = \{\text{orange, green, blue, violet}\}$
   (b) $A = \{l, n, p, r, t\}$ and $B = \{r, s, t, u, v, w\}$
   (c) $A = \{b : b \text{ is a black car}\}$ and
   $B = \{a : a \text{ is a car with automatic transmission}\}$
   (d) $A = \{d : d \text{ is a day of the week}\}$ and $B = \{\text{Friday, Saturday}\}$

 3 Find the complement $\bar{A}$ of A for each of the following:

   (a) $A = \{\text{orange, green, blue, violet}\}$ and
   $U = \{c : c \text{ is a colour of the rainbow}\}$
   (b) $A = \{1, 2, 3, 4, 5, 6, 7\}$ and $U = \{p : p \text{ is a prime number and } p \le 19\}$
   (c) $A = \{-1, 1\}$ where $U = \{-3, -2, -1, 0, 1, 2, 3\}$
   (d) $A = \{l : l \text{ is a sheet of lined A4 paper}\}$ and
   $U = \{p : p \text{ is a sheet of A4 paper}\}$

Set properties

Union, intersection and complement

1

Just to recap, the Venn diagram is used to illustrate the union and the intersection of two sets and the complement of a single set in the following ways:

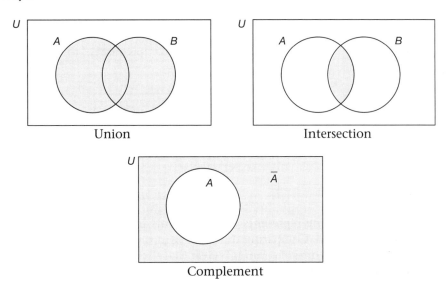

Union   Intersection

Complement

Using these simple ideas we can now illustrate various properties of sets. For example, to show that the complement of a union is the intersection of the complements, namely:

$$\overline{A \cup B} = \bar{A} \cap \bar{B}$$

we can use the following diagrams.

In the first diagram we shade in the $\overline{A \cup B}$ (the complement of the union of A and B) which is the area within the universal set but outside the two circles that represent A and B, thus:

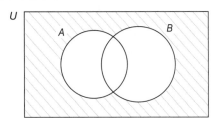

In the second diagram we shade in the $\bar{A}$ (the complement of A) with left to right shading thus:

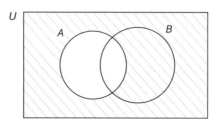

In the same diagram we shade in the $\bar{B}$ (the complement of B) with right to left shading thus:

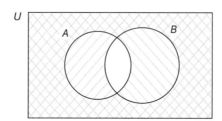

The cross-hatching represents those elements in both $\bar{A}$ and $\bar{B}$, that is those elements in $\bar{A} \cap \bar{B}$. Comparing the shading in the first diagram with the cross-hatching in the third diagram it is clear that they cover the same areas. In this way we have illustrated the fact that:

$$\overline{A \cup B} = \bar{A} \cap \bar{B}$$

Now you try one. Prove that $\overline{A \cap B} = \bar{A} \cup \bar{B}$.

To start, the Venn diagram that illustrates $\bar{A} \cup \bar{B}$ is ....

The answer is in the next frame

2

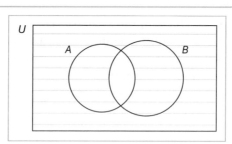

Because

The first diagram represents $\bar{A}$, the second diagram represents $\bar{B}$ and the third diagram represents the union $\bar{A} \cup \bar{B}$:

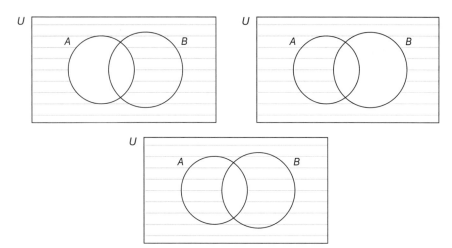

Now, the Venn diagram that illustrates $\overline{A \cap B}$ is . . . .

The answer is in the next frame

3

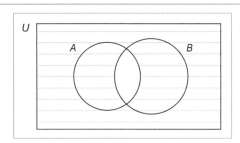

Because

The first diagram represents $A \cap B$ and the second diagram the complement $\overline{A \cap B}$:

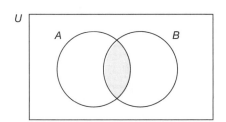
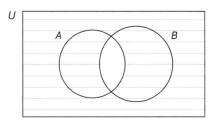

Comparing the results of these last two frames we can conclude that:

$$\overline{A \cap B} = \ldots \ldots \ldots$$

The answer is in the next frame

4

$$\overline{A \cap B} = \bar{A} \cup \bar{B}$$

Because

The shading that represents $\overline{A \cap B}$ covers the same area as the shading that represents $\bar{A} \cup \bar{B}$.

If we wish to consider properties that involve three sets A, B and C then we must ensure that in the Venn diagram each set intersects with the other two as in the following diagram:

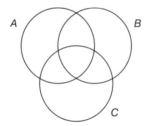

Move to the next frame

5   Distributivity rules

To demonstrate the property of sets that intersection is distributive over union, namely:

$A \cap (B \cup C) = (A \cap B) \cup (A \cap C)$

we consider each side of the equation separately. Firstly the left-hand side:

$A \cap (B \cup C)$:

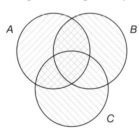

Here the cross-hatching represents the intersection of A with $B \cup C$. Next we consider the right-hand side of the equation:

$(A \cap B) \cup (A \cap C)$:

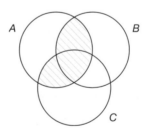

Here the shading represents the union of $A \cap B$ with $A \cap C$.

By comparing the two diagrams we have illustrated that:

$$A(\cap(B \cup C) = \ldots\ldots\ldots$$

The answer is in the next frame

6

$$\boxed{A \cap (B \cup C) = (A \cap B) \cup (A \cap C)}$$

Because

The shading that represents $A \cap (B \cup C)$ covers the same area as the shading that represents $(A \cap B) \cup (A \cap C)$ showing that intersection distributes over union, that is:

$$A \cap (B \cup C) = (A \cap B) \cup (A \cap C)$$

Now you try one. Show that union distributes over intersection, that is:

$$A \cup (B \cap C) = (A \cup B) \cap (A \cup C)$$

To start, the Venn diagram that illustrates $A \cup (B \cap C)$ is $\ldots\ldots\ldots$

The answer is in the next frame

7

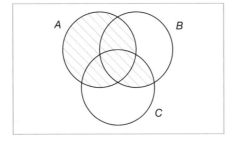

Because

The shading of $B \cap C$ combined with the shading of A clearly represents $A \cup (B \cap C)$.

Considering the right-hand side of the equation, the Venn diagram that illustrates $(A \cup B) \cap (A \cup C)$ is $\ldots\ldots\ldots$

The answer is in the next frame

8

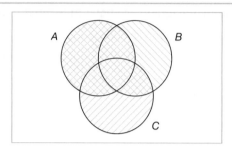

Because

The shading that represents $A \cup B$ and the shading of $A \cup C$ forms a cross-hatching that represents their intersection, namely $(A \cup B) \cap (A \cup C)$.

By comparing the two diagrams we have illustrated that:

$$A \cup (B \cap C) = \ldots\ldots\ldots\ldots$$

The answer is in the next frame

9

$$A \cup (B \cap C) = (A \cup B) \cap (A \cup C)$$

Because

The shading that represents $A \cup (B \cap C)$ covers the same area as the shading that represents $(A \cup B) \cap (A \cup C)$ showing that union distributes over intersection, that is:

$$A \cup (B \cap C) = (A \cup B) \cap (A \cup C)$$

Now move to the next frame

10   The number of elements in a set

The number of elements in set A is denoted by $n(A)$ and the number of elements in the union of sets A and B will depend upon the number of elements in their intersection. If A and B have no elements in common then $n(A \cup B) = n(A) + n(B)$:

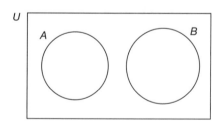

▷

However, if the two sets do have elements in common then adding $n(A)$ to $n(B)$ means that those common elements have been counted twice. In which case the second count must be subtracted. That is:

$$n(A \cup B) = n(A) + n(B) - n(A \cap B)$$

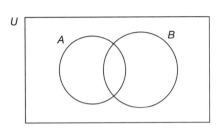

So if $A = \{a, b, c, d, e, f\}$ and $B = \{d, e, f, g, h, i\}$ then $n(A \cup B) = \ldots\ldots\ldots\ldots$

The answer is in the next frame

9

11

Because

$n(A) = 6$, $n(B) = 6$ and $n(A \cap B) = 3$ so $n(A \cup B) = 6 + 6 - 3 = 9$.
Indeed $A \cup B = \{a, b, c, d, e, f, g, h, i\}$

Let's try another.

Of 150 people questioned, 135 claimed they could write with their right hand and 29 claimed they could write with their left hand.

The number of ambidextrous people amongst the 150 was $\ldots\ldots\ldots\ldots$

Next frame

14

12

Because

If R is the set of people who can write with their right hand and L is the set of people who can write with their left hand, $n(R) = 135$, $n(L) = 29$ and $n(L \cup R) = 150$ – the number of people questioned. Also:

$$n(L \cup R) = n(L) + n(R) - n(L \cap R)$$

That is:

$$150 = 29 + 135 - n(L \cap R)$$

therefore:

$$n(L \cap R) = 14$$

And another.

Of 1500 students in the Faculty of Engineering, Mathematics and Science of a university, 500 studied mathematics and science but no engineering, 450 studied mathematics and engineering but no science and 230 studied all three subjects. If mathematics was compulsory for engineering and science, how many studied mathematics by itself?

In this problem all the students studied mathematics so that forms the universal set. If *E* is the set of those students who studied engineering and *S* is the set of those who studied science then the Venn diagram that represents this problem is ....

The answer is in the next frame

13

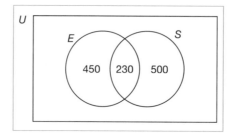

Because

450 studied mathematics and engineering but no science
500 studied mathematics and science but no engineering
230 studied mathematics, engineering and science

The number that studied mathematics by itself is ....

The answer is in the next frame

14

320

Because

$n(U) = 1500$ where $n(E) = 450$, $n(S) = 500$ and $n(E \cap S) = 230$. Therefore:

$$n(\overline{E \cup S}) = n(U) - n(E \cup S) = 1500 - (450 + 230 + 500)$$
$$= 1500 - 1180$$
$$= 320$$

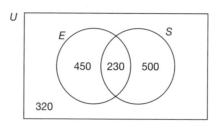

Move to the next frame

Three sets

15

If A, B and C are three sets where:

$n(A)$ is the number of elements in set A
$n(B)$ is the number of elements in set B
$n(C)$ is the number of elements in set C

To find the number of elements in the union of A, B and C, that is:

$n(A \cup B \cup C)$

we begin by adding the number of elements in each of the three sets together to produce a Venn diagram of the form:

$n(A) + n(B) + n(C)$

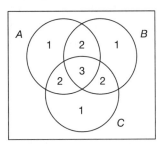

Each number shown in the various regions of the Venn diagram of the three sets A, B and C represents the number of times the number of elements in that region has been added to form the total of:

$$n(A) + n(B) + n(C)$$

The numbers 2 and 3 indicate that we have added the number of elements in those regions too many times. To begin compensating for this we first subtract the number of elements in the intersection of A and B to give a Venn diagram of the form:

$n(A) + n(B) + n(C) - n(A \cap B)$

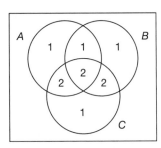

Subtracting the elements in the intersection of A and B reduces the count in the appropriate regions by 1. Hence the 2 and 3 reduce to 1 and 2 respectively because each number shown in the various regions of the Venn diagram represents the number of times the number of elements in that region has been added to form the total of:

$$n(A) + n(B) + n(C) - n(A \cap B)$$

Now subtract the number of elements in the intersection of A and C to give a Venn diagram of the form:

. . . .

The answer is in the next frame

16

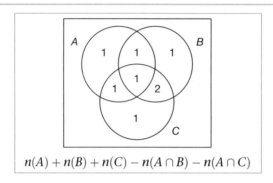

$$n(A) + n(B) + n(C) - n(A \cap B) - n(A \cap C)$$

Because

Subtracting the elements in the intersection of A and C reduces the count in the appropriate regions by 1. Hence the 2 and 2 reduce to 1 and 1. Again, this is because each number shown in the various regions of the Venn diagram represents the number of times the number of elements in that region has been added to form the total of

$$n(A) + n(B) + n(C) - n(A \cap B) - n(A \cap C)$$

Now subtract the number of elements in the intersection of B and C to give a Venn diagram of the form ....

The answer is in the next frame

17

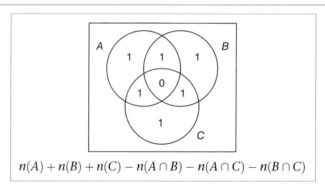

$$n(A) + n(B) + n(C) - n(A \cap B) - n(A \cap C) - n(B \cap C)$$

Because

Subtracting the elements in the intersection of B and C reduces the count in the appropriate regions by 1. Hence the 1 and 2 reduce to 0 and 1 respectively. Again, this is because each number shown in the various regions of the Venn diagram represents the number of times the number of elements in that region has been added to form the total of

$$n(A) + n(B) + n(C) - n(A \cap B) - n(A \cap C) - n(B \cap C)$$

The number of elements in the region containing the number 0 is given in terms of sets A, B and C as: ....

The answer is in the next frame

$$\boxed{n(A \cap B \cap C)}$$

18

Because

That region of the Venn diagram is the intersection of all three sets A, B and C, denoted by $A \cap B \cap C$ and so the number of elements that it contains is $n(A \cap B \cap C)$.

Finally, we can see that the number of elements in the union of all three sets is given as:

$$n(A \cup B \cup C) = \ldots\ldots\ldots\ldots$$

Next frame

$$\boxed{\begin{aligned} n(A \cup B \cup C) = {} & n(A) + n(B) + n(C) - n(A \cap B) - n(A \cap C) - n(B \cap C) \\ & + n(A \cap B \cap C) \end{aligned}}$$

19

Because

From the Venn diagram of the union of all three sets A, B and C it can be seen that the number of elements in the union is equal to the number of elements in each individual region shown. That is:

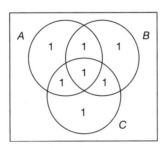

$$n(A \cup B \cup C) = n(A) + n(B) + n(C) - n(A \cap B) - n(A \cap C) - n(B \cap C)$$
$$+ n(A \cap B \cap C)$$

Let's try using that equation.

To find $n(A \cup B \cup C)$ given that $A = \{1, 2, 3, 4, 5\}$, $B = \{1, 3, 5, 7\}$ and $C = \{2, 3, 4, 6, 7, 8, 10\}$ we begin by counting the number of elements in each set A, B and C. That is:

$$n(A) = \ldots\ldots\ldots\ldots$$
$$n(B) = \ldots\ldots\ldots\ldots$$
$$n(C) = \ldots\ldots\ldots\ldots$$

The answers are in the next frame

20

$$n(A) = 3$$
$$n(B) = 4$$
$$n(C) = 7$$

Because

$n(A)$, $n(B)$ and $n(C)$ represent the number of elements in sets A, B and C respectively.

Now let's find the numbers of elements in the intersection of each of the pairs of sets, then:

$$n(A \cap B) = \dots\dots\dots$$
$$n(A \cap C) = \dots\dots\dots$$
$$n(B \cap C) = \dots\dots\dots$$

The answers are in the next frame

21

$$n(A \cap B) = 3$$
$$n(A \cap C) = 3$$
$$n(B \cap C) = 2$$

Because

$A = \{1, 2, 3, 4, 5\}$, $B = \{1, 3, 5, 7\}$ and $C = \{2, 3, 4, 6, 7, 8, 10\}$ so that:

$A \cap B = \{1, 2, 3, 4, 5\} \cap \{1, 3, 5, 7\} = \{1, 3, 5\}$ so $n(A \cap B) = 3$
$A \cap C = \{1, 2, 3, 4, 5\} \cap \{2, 3, 4, 6, 7, 8, 10\} = \{2, 3, 4\}$ so $n(A \cap C) = 3$
$B \cap C = \{1, 3, 5, 7\} \cap \{2, 3, 4, 6, 7, 8, 10\} = \{3, 7\}$ so $n(B \cap C) = 2$

Now we need the number of elements in the intersection of all three sets, so:

$$n(A \cap B \cap C) = \dots\dots\dots$$

The answer is in the next frame

$$\boxed{1}$$

22

Because

$A = \{1, 2, 3, 4, 5\}$, $B = \{1, 3, 5, 7\}$ and $C = \{2, 3, 4, 6, 7, 8, 10\}$ so that

$$A \cap B \cap C = \{1, 2, 3, 4, 5\} \cap \{1, 3, 5, 7\} \cap \{2, 3, 4, 6, 7, 8, 10\}$$
$$= \{1, 3, 5\} \cap \{2, 3, 4, 6, 7, 8, 10\}$$
$$= \{3\}$$

Therefore $n(A \cap B \cap C) = 1$.

Now we can work out the number of elements in the union of all three sets:

$$n(A \cup B \cup C) = \ldots\ldots\ldots\ldots$$

Next frame

$$\boxed{9}$$

23

Because

$$n(A \cup B \cup C) = n(A) + n(B) + n(C) - n(A \cap B) - n(A \cap C) - n(B \cap C)$$
$$+ n(A \cap B \cap C)$$
$$= 5 + 4 + 7 - 3 - 3 - 2 + 1$$
$$= 9$$

Furthermore:

$$A \cup B \cup C = \{1, 2, 3, 4, 5\} \cup \{1, 3, 5, 7\} \cup \{2, 3, 4, 6, 7, 8, 10\}$$
$$= \{1, 2, 3, 4, 5, 7\} \cup \{2, 3, 4, 6, 7, 8, 10\}$$
$$= \{1, 2, 3, 4, 5, 6, 7, 8, 10\}$$

confirming that $n(A \cup B \cup C) = 9$.

Review summary

24

1 *Union, intersection and complement:* The Venn diagram is used to illustrate properties involving the union, intersection and complement of sets by shading.

2 *Number of elements:* The number of elements in set A is denoted by $n(A)$. For two sets A and B the number in their union is given as:

$$n(A \cup B) = n(A) + n(B) - n(A \cap B)$$

For three sets A, B and C the number in their union is given as:

$$n(A \cup B \cup C) = n(A) + n(B) + n(C) - n(A \cap B) - n(A \cap C) - n(B \cap C)$$
$$+ n(A \cap B \cap C)$$

 # Review exercise

25

1 Use Venn diagrams to illustrate that:
   (a) $(A \cap B) \cup \bar{A} = B \cup \bar{A}$
   (b) $(\bar{A} \cap B) \cup C \neq \bar{A} \cap (B \cup C)$ unless $A \cap C = \emptyset$

2 Show that $(\bar{A} \cup B) \cap A = B \cap A$ for $A = \{a, b, c, d, e\}$ and $B = \{d, e, f, g, h\}$ where the universal set $U = \{a, b, c, d, e, f, g, h, i, j, k, l\}$.

3 During the course of one week 132 people visited a health spa of whom 75 had a herbal bath but no massage, 28 had a massage but no herbal bath and 15 had both. How many had neither?

4 In an athletics club there are 87 members, 38 of whom run the 400 m event, 45 of whom run the 200 m event, and 15 members who run neither event. How many of the athletes run both events?

5 A sports club for walkers, cyclists and golfers has 38 members of whom 29 are walkers, 30 are cyclists and 15 play golf. Of those who walk, 12 also cycle and 9 also play golf. And of those who play golf 20 also cycle. How many walk, play golf and cycle?

Complete all five questions, looking back at the Unit if necessary. You can check your answers and working with the next frame

1 (a) $(A \cap B) \cup \bar{A}$ $\qquad\qquad\qquad\qquad$ $B \cup \bar{A}$ $\qquad$ **26**

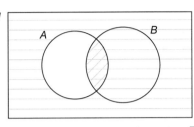 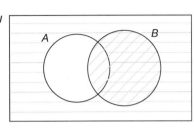

$$(A \cap B) \cup \bar{A} = B \cup \bar{A}$$

(b) $(\bar{A} \cap B) \cup C$ where $A \cap C \neq \emptyset$ $\qquad$ $(\bar{A} \cap B) \cup C$ where $A \cap C = \emptyset$

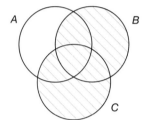

$\bar{A} \cap (B \cup C)$ where $A \cap C \neq \emptyset$ $\qquad$ $\bar{A} \cap (B \cup C)$ where $A \cap C = \emptyset$

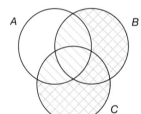 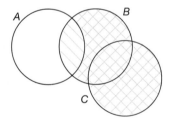

$$(\bar{A} \cap B) \cup C \neq \bar{A} \cap (B \cup C) \text{ unless } A \cap C = \emptyset$$

2 Since $A = \{a, b, c, d, e\}$ and $B = \{d, e, f, g, h\}$ where the universal set $U = \{a, b, c, d, e, f, g, h, i, j, k, l\}$ then:

$\bar{A} = \{f, g, h, i, j, k, l\}$ and so $\bar{A} \cup B = \{d, e, f, g, h, i, j, k, l\}$

so that:

$(\bar{A} \cup B) \cap A = \{d, e\}$. Also $B \cap A = \{d, e\}$, therefore $(\bar{A} \cup B) \cap A = B \cap A$

3 If H is the set of people who visited the health spa and M is the set of those who had a massage, the Venn diagram for this problem is:

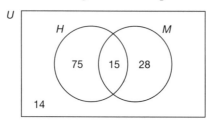

$n(H \cup M) = 75 + 15 + 28 = 118$ and $n(U) = 132$ therefore:

$$n(\overline{H \cup M}) = n(U) - n(H \cup M)$$
$$= 132 - 118$$
$$= 14$$

4 If A is the set of members who run the 400 m event and B is the set of members who run the 200 m event then:

$n(A) = 38$, $n(B) = 45$ and $n(\overline{A \cup B}) = 15$ therefore:

$$n(A \cup B) = n(U) - n(\overline{A \cup B})$$
$$= 87 - 15$$
$$= 72$$

So that:

$$n(A \cup B) = n(A) + n(B) - n(A \cap B) \text{ giving}$$
$$72 = 38 + 45 - n(A \cap B) \text{ hence } n(A \cap B) = 11$$

5 If W is the set of members who walk, C is the set of cyclists and G is the set of golfers then from the information given in the question:

29 are walkers	$n(W) = 29$
30 are cyclists	$n(C) = 30$
15 are golfers	$n(G) = 15$
12 walk and cycle	$n(W \cap C) = 12$
9 walk and play golf	$n(W \cap G) = 9$
20 cycle and play golf	$n(C \cap G) = 20$
There are 38 members	$n(W \cup C \cup G) = 38$

To find $n(W \cap C \cap G)$, since:

$$n(W \cup C \cup G) = n(W) + n(C) + n(G) - n(W \cap C) - n(W \cap G) - n(C \cap G)$$
$$+ n(W \cap C \cap G)$$

then:

$$38 = 29 + 30 + 15 - 12 - 9 - 20 + n(W \cap C \cap G) \text{ so}$$
$$38 = 74 - 41 + n(W \cap C \cap G) \text{ that is}$$
$$38 - 33 = 5 = n(W \cap C \cap G)$$

Now for the Review test

Review test

1 Use Venn diagrams to illustrate that:
    (a) $(A \cap \bar{B}) \cup B = A \cup B$ (b) $(A \cap \bar{B}) \cup (\bar{A} \cap B) = (A \cup B) \cap (\bar{A} \cup \bar{B})$

27

2 Show that $(A \cup \bar{B}) \cap (\bar{A} \cup B) = (A \cap B) \cup (\overline{B \cup A})$ for $A = \{a, b, c, d, e\}$, $B = \{d, e, f, g, h\}$ where the universal set $U = \{a, b, c, d, e, f, g, h, i, j, k, l\}$.

3 Of 1000 shoppers in a supermarket 425 bought white bread but no brown bread, 357 bought brown bread but no white bread, and 135 bought both white and brown bread. How many bought neither?

4 On a bookshelf there are 43 books, 27 of which have a contents page, 18 of which have an index, and 12 which have neither. How many books have both a contents page and an index?

5 Of 150 households, 74 had electric heating, 79 had gas heating, and 12 had solid fuel heating. Of the 74 with electric heating, 22 also had gas heating and 9 also had solid fuel heating, and of those who had gas heating 5 also had solid fuel heating. How many had all three forms of heating?

You have now come to the end of this Module. A list of **Can You?** questions follows for you to gauge your understanding of the material in the Module. These questions match the **Learning outcomes** listed at the beginning of the Module. Now try the **Test exercise**. *Work through the questions at your own pace, there is no need to hurry.* A set of **Further problems** provides additional valuable practice.

28

Can You?

Test exercise 13

2

1 Which of the following are true statements:
   (a) $\{p\} \in \{p, q, r\}$ (b) $s \in \{r, s, t\}$ (c) $\{L, M, N\} \supseteq \{M, N, L\}$
   (d) $\varnothing \notin \{A, B, C\}$ (e) $\varnothing \not\subset \{A, B, C\}$ (f) $\varnothing \in \{\{\varnothing\}, 0\}$
   (g) $\varnothing \not\subset \{\{\varnothing\}, \varnothing\}$

2 How many subsets are there of the set $A = \{$blue, pink, green, yellow$\}$?

3 Find the intersection $A \cap B$ for each of the following:
   (a) $A = \{p, q, r\}$ and $B = \{q, r, s\}$
   (b) $A = \{-8, -6, -4, -2, 0, 2, 4\}$ and $B = \{-2, 2\}$
   (c) $A = \{a : a$ is a red card from a deck of playing cards$\}$ and
        $B = \{b : b$ is an ace or a 2 of hearts from a deck of playing cards$\}$
   (d) $A = \{s : s$ is a cotton bed sheet$\}$ and $B = \{p : p$ is a white bed sheet$\}$

4 Find the union $A \cup B$ for each of the following:
   (a) $A = \{w, x, y, z\}$ and $B = \{u, v, w\}$
   (b) $A = \{-8, -6, -4, -2, 0, 2, 4\}$ and $B = \{2, 6\}$
   (c) $A = \{c : c$ is a circle$\}$ and $B = \{e : e$ is an ellipse$\}$
   (d) $A = \{c : c$ is a court playing card$\}$ and $B = \{t : t$ is the three of diamonds$\}$

5 Find the complement $\bar{A}$ of A for each of the following:
   (a) $A = \{-6, 0, 6\}$ and $U = \{-8, -6, -4, -2, 0, 2, 4, 6, 8\}$
   (b) $A = \{0, 2, 4, 6\}$ and $U = \{n : n$ is an integer less than 8$\}$
   (c) $A = \{0, \pi, 2\pi, 3\pi\}$ and
        $U = \{x : x$ is a solution of the equation $\sin x = 0$ for $-3\pi \le x \le 3\pi\}$
   (d) $A = \{p : p$ is a computer printer$\}$ and $U = \{l : l$ is a laser printer$\}$

6 Use Venn diagrams to illustrate:
   (a) $(\bar{A} \cap B) \cap \bar{C}$
   (b) $(\overline{A \cap B}) \cap (A \cup B)$

7 Show that $(A \cap \bar{B}) \cup B = A \cup B$ for $A = \{a, b, c, d, e\}$, $B = \{d, e, f, g, h\}$ where the universal set $U = \{a, b, c, d, e, f, g, h, i, j, k, l\}$.

8 Of 38 ready made meals, 16 had low salt but high fat content, 12 had high salt content but low fat content, and 4 had both high salt and high fat content. How many had both low salt and low fat content?

9 In a national survey of kennels and catteries it was found that of the 139 approached that 78 boarded dogs and 87 boarded cats. How many boarded both dogs and cats?

10 Of 87 car hire firms, 76 hired petrol cars, 36 hired diesel cars, and 10 hired electric cars. Of the 76 hiring petrol cars, 23 also hired diesel cars and 7 also hired electric cars, and of those who hired diesel cars 7 also hired electric cars. How many hired all three types of car?

Further problems 13

3

1 Describe each of the following sets in an alternative way:
  (a) {0, 1}
  (b) $\{x : x^2 - 5x + 6 = 0\}$

2 From the set of employees from among the three departments of sales, management and finance, describe the union, intersection and complements in words of the sets:

  $F = \{f : f$ is a female employee$\}$ and
  $M = \{m : m$ is a management employee$\}$

3 Which of the following is the empty set?
  (a) $\{x : x \neq x\}$
  (b) $\{x : x + 1 = x^2\}$
  (c) $\{x : x + 1 = x - 1\}$

4 Which of the following are true statements?
  (a) $\{1\} \in \{1, \{1\}, \{\{1\}\}\}$
  (b) $\{\emptyset\} \subseteq \{\emptyset, \{\emptyset\}\}$
  (c) $\{1\} \subset \{1, \{1\}\}$
  (d) $\{1\} \in \{1, \{1\}\}$

5 List all the subsets of $\{\emptyset, \{\emptyset\}\}$

6 Find the power set of $A = \{1, \{1\}\}$

7 Given $U = \{1, 2, 3, 4, 5, 6, 7, 8, 9, 10\}$, $A = \{1, 3, 5, 7\}$ and $B = \{1, 2, 3, 4, 7, 8\}$ find:
  (a) $A \cap B$ (b) $\bar{A} \cap B$ (c) $A \cap \bar{B}$ (d) $\bar{A} \cap \bar{B}$ (e) $A \cup B$
  (f) $\bar{A} \cup B$ (g) $A \cup \bar{B}$ (h) $\bar{A} \cup \bar{B}$ (i) $\overline{A \cap B}$ (j) $\overline{A \cup B}$

8 Draw a Venn diagram to illustrate each of the following:
  (a) $(\bar{A} \cap B) \cup (B \cap \bar{C})$
  (b) $(\bar{A} \cup \bar{B}) \cap C$

9 Use a Venn diagram to show the validity of:
  (a) $(A \cap B) \cup (\bar{A} \cap \bar{B}) = (\bar{A} \cup B) \cap (A \cup \bar{B})$
  (b) $(A \cap \bar{B}) \cup (\bar{A} \cap C) \cup (A \cap B) = A \cup C$
  (c) $(A \cap B \cap C) \cup (A \cap B \cap \bar{C}) = A \cap B$

10 Show that $(A \cup B) \cap C = (A \cap C) \cup (B \cap C)$ for $A = \{a, b, c, d, e\}$, $B = \{d, e, f, g, h\}$ and $C = \{h, j\}$ where the universal set is $U = \{a, b, c, d, e, f, g, h, j, k, l\}$

11 The difference of two sets is defined by the following:

$$A - B = A \cap \bar{B}$$

and the symmetric difference is defined as:

$$A \triangle B = (A - B) \cup (B - A) = (A \cup \bar{B}) \cup (B \cap \bar{A})$$

(a) Draw the Venn diagram of the difference and the symmetric difference.

(b) Show that:

(i) $(A \triangle B) \triangle C = A \triangle (B \triangle C)$

(ii) $A \cup (B \triangle C) \neq (A \cup B) \triangle (A \cup C)$

12 In a school of 1200 pupils, 650 were boys, 400 stayed for lunch and 380 were juniors. Of the juniors, 75 were boys who stayed for lunch and 85 were girls who did not stay for lunch. 405 of the pupils were senior boys of whom 150 stayed for lunch. How many senior girls did not stay for lunch?

13 In an election 10,564 voters were eligible to vote. Of those who voted, 5 476 voted for, 4 326 voted against, while 75 spoiled their ballot papers by voting for and against. Draw the Venn diagram for this problem indicating the number of voters who did not vote.

14 Of the 38 diners using a restaurant, 25 had water with their meal and 23 had wine with their meal. How many diners had both wine and water with their meal?

15 On one summer Sunday a theme park had 150 000 visitors. Of the 150 000, 94 000 were under 15 years of age and were admitted at half price. 86 500 went on the thrill rides and 84 000 ate a meal in one of the twelve restaurants. Of the under 15s, 28 000 went on the thrill rides and ate in the restaurants and of all those who ate in the restaurants 43 500 went on the thrill rides. Of the over 15s, 12 000 ate in a restaurant only and 24 000 went on a thrill ride only. How many of the over 15s neither ate in a restaurant nor went on a thrill ride?

16 One hundred people applied for a job in a computer service bureau. Of the 100, 15 were left-handed, 48 were male and 63 had previous experience in information technology. Of the 15 left-handers, 3 were women with past experience of IT, 6 were men without experience of IT, and 2 were men with past experience of IT. Of the women a total of 30 had past experience of IT. How many women were there that were right-handed with no experience of information technology?

17 Out of 673 sports fans, 325 regularly go to a rugby match, 415 regularly go to a football match and 137 regularly attend cricket matches. Of those that regularly attend rugby matches, 196 also regularly attend cricket matches and 53 also regularly attend football matches. Of those that regularly attend football matches, 49 also regularly attend cricket matches. A further 15 fans attend no sporting events on a regular basis, so how many regularly attend all three sports matches?

Probability

Empirical probability

Introduction

1

In very general terms, *probability* is a measure of the likelihood that a particular *event* will occur in any one *trial*, or experiment, carried out in prescribed conditions. Each separate possible result from a trial is called an *outcome*.

The ability to predict likely occurrences has obvious applications, e.g. in insurance matters and in industrial quality control and the efficient use of resources.

Notation

The probability that a certain event A will occur is denoted by $P(A)$. For example, if A represents the event that a component, picked at random from stock, is faulty, written $A = \{$faulty component$\}$, then

$P(A)$ denotes ....

2

> the probability of picking a faulty component

Sampling

In a manufacturing run, it would be both time-consuming and uneconomical to subject every single component produced to full inspection. It is usual, therefore, to examine a sample batch of components, taken at random, as being representative of the whole output. The larger the *random sample*, the more nearly representative of the whole *population* (total output of components) is it likely to be.

Types of probability

The determination of probability may be undertaken from two approaches: (a) *empirical* (or experimental) probability, and (b) *classical* (or theoretical) probability. Let us look at each of these in turn.

So move on

Empirical probability

Empirical probability is based on previous known results. The relative frequency of the number of times the event has previously occurred is taken as the indication of likely occurrences in the future. **3**

Take an example. A random batch of 240 components is subjected to strict inspection and 20 items are found to be defective. Therefore, if we pick any one component at random from this sample, the chance of its being faulty is '20 in 240', i.e. '1 in 12'.

So, if $A = \{$faulty component$\}$

then $P(A) = 1$ in $12 = \dfrac{1}{12} = 0{\cdot}0833 = 8{\cdot}33$ per cent

The most usual forms are $P(A) = \dfrac{1}{12}$ or $P(A) = 0{\cdot}0833$

Therefore, a run of 600 components from the same machine would be likely to contain .... defectives.

$$\boxed{50}$$ **4**

Because $P($faulty component$) = P(A) = \dfrac{1}{12}$

Therefore, in 600 components, the likely number of defectives x is simply

$x = 600 \times \dfrac{1}{12} = 50$

Expectation

The result does not assert that there will be exactly 50 defectives in any run of 600 components, but having found the probability of the event occurring in any one trial $\left(\dfrac{1}{12}\right)$, we can use it to predict the likely number of times the event will occur in N similar trials. This *expectation E* is defined by the product of the number of trials N and the probability $P(A)$ that the event A will occur in any one trial, i.e.

$\qquad E = N \times P(A)$ $\hfill (1)$

On to the next frame

---

5   Success or failure

Throughout, we are concerned with the probability of the occurrence of a particular event. When it does occur in any one trail, we record a *success*: when it fails to occur, we record a *failure* – whatever the defined event may be.

If, in N trials, there are x successes, there will also be $(N - x)$ failures, so that:

$$x + (N - x) = N$$

$$\therefore \ \frac{x}{N} + \frac{N - x}{N} = 1$$

But, $\dfrac{x}{N} = P(\text{success}) = P(A)$ and $\dfrac{N - x}{N} = P(\text{failure}) = P(\text{not } A)$

$$\therefore \ P(A) + P(\text{not } A) = 1$$

The event not A is called the complement of event A and is often written $\bar{A}$, i.e.

$$P(A) + P(\bar{A}) = 1 \tag{2}$$

$\therefore$ If the probability of picking a defective in any trial is $\dfrac{1}{5}$, the probability of not picking a defective is ....

6

$$\frac{4}{5}$$

Sample size

The size of the original sample from which the probability figure was established affects the reliability of the result.

Probabilities derived from small samples seldom reflect the probabilities associated with the whole population. The larger the sample, the more reliable are the results obtained.

For example, 15 per cent of castings are found to be outside prescribed tolerances. Determine the number of acceptable items likely to be present in a batch of 120 such castings.

That is easy enough. $E = $ ....

7

$$102$$

Because 15 per cent are rejects. $\therefore$ 85 per cent are acceptable.

$\therefore$ If event $A = \{$acceptable casting$\}$, $P(A) = 85$ per cent $= \dfrac{17}{20}$

$N = 120$ and $E = N \times P(A) = 120 \times \dfrac{17}{20} = 102$

$\therefore$ Expected number acceptable $= 102$

Multiple samples

A single random sample of *n* components taken from a whole population of *N* components is not necessarily representative of the parent population. Another random sample of *n* components from the same population could well include a different number of defectives from that in the first sample.

We have not got a production line on hand, but we can simulate the same kind of problem with the aid of a pack of playing cards (jokers removed). Then, the total population *N* is ....

> the number of cards in the whole pack, i.e. 52

8

If we shuffle well and then deal out a random sample of 12 cards (*n*), we can count how many cards in the sample are (say) spades (*x*). These take the place of the defective components.

Replace the 12-card sample in the pack; shuffle well and then take a second sample of 12 cards. Counting the spades in this second sample will most likely give a different total from the result of the first trial.

If we average the two results, dividing the combined number of spades in the two trials by the number of trials, i.e. 2, we are effectively considering a larger sample.

Let us extend this into a useful experiment,
which we can now carry out – so move on to the next frame

Experiment

9

To determine the number of spades in each of 40 trials of a random 12-card sample from a full deck of playing cards and to compile a cumulative proportion (running average) of the results.

Procedure

Start with a full deck of playing cards, excluding the jokers:

(a) Shuffle the cards thoroughly.

(b) Deal a random sample of 12 cards.

(c) From the sample, count and record the number of cards that are spades.

(d) Return the sample cards to the pack and shuffle thoroughly.

(e) Repeat the process for a total of 40 such trials.

(f) Compile a table showing

(i) the number of the trial, *r* (1 to 40)

(ii) the number of spade cards in each trial, *x*

(iii) the cumulative total of spade cards, cum *x*

(iv) the cumulative proportion (running average) of the results.

$$\text{Cum prop} = \frac{\text{cumulative total of spade cards in } r \text{ trials}}{\text{number of trials to date, i.e. } r}$$

(g) Display graphically the distribution of the number of spades at each trial (1 to 40) and the running average for the 40 trials.

Your table of results will look like this, but with different entries:

Trial r	Spades x	Cum x	Running average
1	2	2	2·00
2	5	7	3·50
3	0	7	2·33
⋮	⋮	⋮	⋮

Carry on and see what you get

10

The results from one such experiment gave the following distribution of spades. Yours will be different, but will eventually lead to the same conclusion.

Number of spades in 40 trials of a 12-card sample

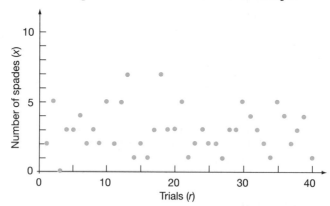

Plotting the running average $\dfrac{\text{cum } x}{r}$ against r, we have shown the result in the next frame. Complete yours likewise.

11

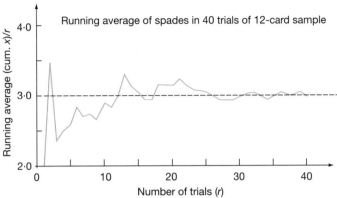

Running average of spades in 40 trials of 12-card sample

We see that:

(a) the running average fluctuates in the early stages

(b) as the number of trials is increased, the updated running average settles much more closely to . . . .

12

$$\boxed{\text{the value 3}}$$

Of course, we happen to know in this case that the pack of 52 cards in fact contains 13 spades, giving a probability that any one card is a spade as . . . .

13

$$\boxed{\dfrac{13}{52} \text{ i.e. } \dfrac{1}{4}; \quad P(\text{spade}) = \dfrac{1}{4}}$$

Therefore, in a random sample of 12 cards, the expectation of spades is
$E = \ldots \ldots \ldots$

14

$$\boxed{E = n \times P(\text{spades}) = 12 \times \dfrac{1}{4} = 3}$$

which is, in fact, the value that the graph settles down to as the number of trials increases.

So the main points so far are as follows:

(a) The empirical probability of an event A occurring is the number, x, of successes experienced in n previous trials, divided by n

i.e. $P(A) = \dfrac{x}{n}$ (the relative frequency) $\qquad$ (3)

(b) The number of successes E expected in a sample of N trials is

$$E = N \times P(A)$$

i.e. Expectation = (Number of trials) × (Probability of success in any one trial)

Example

It is known from past records that 8 per cent of moulded plastic items are defective. Determine:

(a) the probability that any one item is: (i) defective, (ii) acceptable

(b) the number of acceptable items likely to be found in a sample batch of 4500.

Complete it. No snags.

15

(a) (i) $\dfrac{2}{25}$	(ii) $\dfrac{23}{25}$	(b) 4140

Because let $A = \{$defective$\}$ and $B = \{$acceptable$\}$
then:

(a) (i) $P(A) = \dfrac{8}{100} = \dfrac{2}{25}$; (ii) $P(B) = \dfrac{92}{100} = \dfrac{23}{25}$

(b) $E = N \times P(B) = 4500 \times \dfrac{23}{25} = 4140$

Note that, as always: $P(A) + P(B) = P(A) + P(\bar{A}) = 1$

At this point let's just pause and summarize the work so far on empirical probability

 # Review summary   **Unit 42**

16 **1** *Empirical probability*: A probability based on previous known results. The relative frequency of the number of times an event has previously occurred is taken as the indication of likely occurrences in the future. If 20 items out of 240 are found to be defective then the likelihood of a defective item being found in the future is taken to be $P(A)$ where:

$$P(A) = \frac{20}{240} = \frac{1}{12}$$

2 *Expectation*: If the probability of event A occurring in any one trial is $P(A)$ then in N trials the expectation is:
$$E = N \times P(A)$$

3 *Success or failure*: If the occurrence of a particular event, denoted by A, is classified as *success* and its non-occurrence, denoted by $\bar{A}$, is *failure* then:
$$P(A) + P(\bar{A}) = 1$$

Review exercise

Unit 42

1 Fifteen per cent of a collection of dye-stamped washers are defective due to wear of the stamping machine. Determine: **17**

   (a) the probability that any washer drawn at random from a batch is:
   (i) defective   (ii) acceptable

   (b) the number of acceptable washers likely to be found in a batch of 1500.

> *Complete the question. Take your time, there is no need to rush.*
> *If necessary, refer back to the Unit.*
> *The answers and working are in the next frame.*

---

1 (a) (i) Because 15 out of 100 washers (15%) are found to be defective the **18**
   probability $P(A)$ of selecting a defective in the future is $\dfrac{15}{100} = \dfrac{3}{20}$

   (ii) If we treat the selection of a defective as a success with a probability of $P(A)$ and the selection of a non-defective as a failure with a probability of $P(\bar{A})$ then $P(A) + P(\bar{A}) = 1$. That is
   $$P(\bar{A}) = 1 - P(A)$$
   $$= 1 - \frac{3}{20}$$
   $$= \frac{17}{20}$$

   (b) The number of acceptable washers likely to be found in a batch of 1500 is the expectation E, where:
   $$E = 1500 \times \frac{17}{20} = 1275$$

> *Now for the Review test*

 # Review test

Unit 42

1 In an ice cream production run it is found that, due to a fault in the packaging machine, 12 out of 150 cartons of ice cream are damaged and unsaleable. Determine: **19**

   (a) the probability that any carton drawn at random from a batch is:
   (i) saleable   (i) unsaleable

   (b) the number of unsaleable cartons of ice cream likely to be found in a batch of 275.

Classical probability   Unit 43

Definition

1

The classical approach to probability is based on a consideration of the theoretical number of ways in which it is possible for an event A to occur. As before:

(a) In any trial, each separate possible result is called an ....

(b) The particular occurrence being looked for in a trial is the ....

(c) When it occurs, we have a ....; when it does not occur, we have a ....

2

> (a) outcome
> (b) event
> (c) success; failure

To this list, we now add:

(d) The classical probability P of an event A occurring is defined by:

$$P(A) = \frac{\text{number of ways in which event } A \text{ can occur}}{\text{total number of all possible outcomes}} \qquad (4)$$

Example

If we consider the chance result of rolling a normal unbiased die, the total number of possible outcomes is ....

3

> six, i.e. 1, 2, 3, 4, 5, 6

If the event we are considering is the throwing of a 6, this can occur only one way out of the six possible outcomes.

$$\therefore P(\text{six}) = \text{...........} \quad \text{and} \quad P(\text{not six}) = \text{...........}$$

4

> $$P(\text{six}) = \frac{1}{6}; \quad P(\text{not six}) = \frac{5}{6}$$

If, out of n possible outcomes of a trial, it is possible for an event A to occur in x ways and for the event A not to occur in y ways, then

$$n = x + y \quad \text{and} \quad P(A) = \frac{x}{n} = \frac{x}{x+y} \qquad (5)$$

Certain and impossible events

5

(a) If an event A is certain to occur every time, then

$$x = n, \quad y = 0 \qquad \therefore P(A) = \frac{n}{n} = 1$$

(b) If an event A cannot possibly occur at any time, then

$$x = 0, \quad y = n \qquad \therefore P(A) = \frac{0}{n} = 0$$

$$\therefore P(\text{certainty}) = 1; \qquad P(\text{impossibility}) = 0 \tag{6}$$

In most cases, probability values lie between the extreme values.
So:

If one card is drawn at random from a full pack of playing cards, the probability of obtaining

(a) a heart is ....,   (b) a king is ....,

(c) a card other than a king is ....,   (d) a black card is ....

6

$$
\begin{array}{ll}
\text{(a)} \quad \dfrac{13}{52} = \dfrac{1}{4} & \text{(b)} \quad \dfrac{4}{52} = \dfrac{1}{13} \\[2ex]
\text{(c)} \quad \dfrac{48}{52} = \dfrac{12}{13} & \text{(d)} \quad \dfrac{26}{52} = \dfrac{1}{2}
\end{array}
$$

Mutually exclusive and mutually non-exclusive events

7

(a) *Mutually exclusive events* are events which cannot occur together. For example, in rolling a die, the event of throwing a six and that of throwing a five cannot occur at the same time. Similarly, when drawing a card from a pack, the event of drawing an ace and also that of drawing a king cannot occur in the same single trial.

(b) *Mutually non-exclusive events* are events that occur simultaneously. For example, in rolling a die, the event of obtaining a multiple of 3 and the event of obtaining a multiple of 2 can occur together if a 6 is thrown. Similarly, with cards, the event of drawing a black suit and that of drawing a queen can occur together if the .... or the .... is drawn.

8

> queen of clubs; queen of spades

Exercise

State whether the following pairs of events are mutually exclusive or non-exclusive.

(a) $A = \{\text{ace}\}$; $B = \{\text{black card}\}$;
(b) $A = \{\text{heart}\}$; $B = \{\text{ace of spades}\}$;
(c) $A = \{\text{red card}\}$; $B = \{\text{jack}\}$;
(d) $A = \{\text{5 of clubs}\}$; $B = \{\text{10 of clubs}\}$;
(e) $A = \{\text{card} < 10\}$; $B = \{\text{king of diamonds}\}$.

9

> (a) and (c) non-exclusive; (b), (d) (e) exclusive

Let's just pause and summarize the work so far on classical probability

 # Review summary   **Unit 43**

10   **1** *Classical probability*:   A probability based on the number of ways an event can occur divided by the number of possible outcomes:

$$P(A) = \frac{\text{number of ways in which event } A \text{ can occur}}{\text{total number of possible outcomes}}$$

2 *Certain and impossible events*: $P(\text{certainty}) = 1$ and $P(\text{impossibility}) = 0$

3 *Mutually exclusive events*:   These are events that cannot occur together.

4 *Mutually non-exclusive events*:   These are events that can occur simultaneously.

 # Review exercise   **Unit 43**

11   **1** A bag contains 3 red balls, 6 blue balls and 8 white balls. A ball is drawn at random from the bag. Determine the probability that the colour of the ball is:

(a) blue

(b) not white.

> *Complete the question. Take one step at a time, there is no need to rush.*
> *If you need to, look back at the Unit.*
> *The answers and working are in the next frame.*

1   A bag containing 3 red balls, 6 blue balls and 8 white balls has 17 in total. **12** A ball is drawn at random from the bag and every ball has an equal chance of

$$\frac{1}{17}$$

of being drawn. Therefore the probability that the colour of the ball is:

(a) blue is $\frac{6}{17}$ because there are 6 blue balls

(b) not white is $\frac{9}{17}$ because there are 9 balls that are not white.

Now for the Review test

 # Review test   **Unit 43**

1   A bowl contains 25 red sweets, 16 green sweets, 14 yellow sweets and **13** 10 orange sweets. A sweet is drawn at random from the bowl. Determine the probability that the colour of the sweet is:

(a) red
(b) not yellow
(c) neither green nor orange.

Addition law of probability   **Unit 44**

If there are n possible outcomes to a trial, of which x give an event A and y give **1** an event B, then, provided the events A and B are *mutually exclusive*, the probability of either event A or event B occurring – but clearly not both – is:

$$P(A \text{ or } B) = \frac{x+y}{n}$$
$$= \frac{x}{n} + \frac{y}{n} = P(A) + P(B)$$
$$\therefore P(A \text{ or } B) = P(A) + P(B) \tag{7}$$

If the events A and B are *non-exclusive*, so that events A and B can occur together, then the probability of A or B occurring is given by:

$$P(A \text{ or } B) = P(A) + P(B) - P(A \text{ and } B) \tag{8}$$

For example, in rolling a die:

the probability of scoring a multiple of 3 (i.e. 3 or 6) $= \frac{2}{6} = P(A)$

the probability of scoring a multiple of 2 (i.e. 2, 4, 6) $= \frac{3}{6} = P(B)$

So the probability of scoring a multiple of 3 or a multiple of 2 would seem to be . . . .

2

$$P(A) + P(B) = \frac{2}{6} + \frac{3}{6} = \frac{5}{6}$$

But wait. One outcome (6) appears in both sets of outcomes, i.e. the events are not mutually exclusive, and this outcome has been counted twice. The probability of scoring one of the sixes must therefore be removed.

$$P(\text{six}) = \frac{1}{6} = P(A \text{ and } B)$$

Therefore, in this example $P(A \text{ or } B) = \ldots\ldots\ldots\ldots$

3

$$P(A \text{ or } B) = \frac{2}{3}$$

Because $P(A) = \frac{2}{6}$; $\quad P(B) = \frac{3}{6}$; $\quad P(A \text{ and } B) = \frac{1}{6}$

$$\therefore P(A \text{ or } B) = P(A) + P(B) - P(A \text{ and } B) = \frac{2}{6} + \frac{3}{6} - \frac{1}{6} = \frac{2}{3}$$

Example

A single card is drawn from a pack of 52 playing cards.

(a) If event $A = $ {drawing an ace} and event $B = $ {drawing a seven}, the probability of drawing either an ace or a seven, i.e.

$\quad P(A \text{ or } B) = \ldots\ldots\ldots\ldots$

4

$$P(A \text{ or } B) = P(\text{ace or seven}) = \frac{2}{13}$$

Because these are mutually exclusive events in any one trial

$\quad \therefore P(A \text{ or } B) = P(A) + P(B)$

$\quad$ 4 aces in the pack $\qquad \therefore P(A) = P(\text{ace}) = \frac{4}{52} = \frac{1}{13}$

$\quad$ 4 sevens in the pack $\qquad \therefore P(B) = P(\text{seven}) = \frac{4}{52} = \frac{1}{13}$

$\quad \therefore P(A \text{ or } B) = P(\text{ace or seven}) = \frac{1}{13} + \frac{1}{13} = \frac{2}{13}$

(b) If event $A = $ {drawing a king} and event $B = $ {drawing a red card}, then
$\quad P(A \text{ or } B) = \ldots\ldots\ldots\ldots$

5

$$P(A \text{ or } B) = P(\text{king or red card}) = \frac{7}{13}$$

These are non-exclusive events since we could draw the king of hearts or the king of diamonds.

$\therefore P(A \text{ or } B) = P(A) + P(B) - P(A \text{ and } B)$

4 kings in the pack $\qquad \therefore P(A) = P(\text{king}) = \dfrac{4}{52} = \dfrac{1}{13}$

26 red cards in the pack $\qquad \therefore P(B) = P(\text{red card}) = \dfrac{26}{52} = \dfrac{1}{2}$

2 red kings in the pack $\qquad \therefore P(A \text{ and } B) = P(\text{red king}) = \dfrac{2}{52} = \dfrac{1}{26}$

$\therefore P(A \text{ or } B) = \dfrac{1}{13} + \dfrac{1}{2} - \dfrac{1}{26} = \dfrac{2 + 13 - 1}{26} = \dfrac{14}{26} = \dfrac{7}{13}$

Take it in steps, then all is well.

Next frame

So, for *mutually exclusive* events A and B, i.e. A or B can occur, but not both at the same time:

$P(A \text{ or } B) = \dots\dots\dots$

6

$$P(A \text{ or } B) = P(A) + P(B)$$

7

and for *mutually non-exclusive* events A and B, i.e. A or B or both can occur in any one trial:

$P(A \text{ or } B) = \dots\dots\dots$

$$P(A \text{ or } B) = P(A) + P(B) - P(A \text{ and } B)$$

8

Next frame

 # Review summary $\qquad$ Unit 44 $\qquad$ **9**

1 *Addition law of probability:* If there are n possible outcomes to a trial, of which x give an event A and y give an event B then, provided A and B are *mutually exclusive events*, the probability of either event A or event B occurring is:

$P(A \text{ or } B) = P(A) + P(B)$

2 For *mutually non-exclusive events* (these are events that can occur simultaneously) then:

$P(A \text{ or } B) = P(A) + P(B) - P(A \text{ and } B)$

Review exercise   Unit 44

10 1 (a) Determine the probability of obtaining a court card or a 3 from a deck of
            52 playing cards in a single draw.

        (b) Determine the probability of drawing a court card or a heart from a deck
            on 52 playing cards in a single draw.

Complete the question. Take your time, there is no need to rush.
If necessary, refer back to the Unit.
The answers and working are in the next frame.

11 1 (a) In a deck of 52 playing cards there are 12 court cards – a jack, a queen and
            a king in each of the four suits. Let the drawing of a court card be event A
            and the drawing of a 3 be event B. Events A and B are mutually exclusive
            because only one of the events can occur at any one time.

            The probability of drawing a court card is $P(A) = 12/52$ and the
            probability of drawing a 3 is $P(B) = 4/52$ so the probability of drawing
            either a court card or a 3 is:

$$P(A \text{ and } B) = P(A) + P(B) = \frac{12}{52} + \frac{4}{52} = \frac{16}{52} = \frac{4}{13}$$

        (b) Let the drawing of a court card be event A and the drawing of a heart be
            event C. Events A and C are mutually non-exclusive because both events
            can occur at the same time, for example drawing a king of hearts. The
            probability of drawing a court card is $P(A) = 12/52$ and the probability of
            drawing a heart is $P(C) = 13/52$. Since there are 3 court cards which are
            hearts the probability of drawing a court card which is a heart is
            $P(A \text{ and } C) = 3/52$. Therefore the probability of drawing either a court
            card or a heart is:

$$P(A \text{ or } C) = P(A) + P(C) - P(A \text{ and } C) = \frac{12}{52} + \frac{13}{52} - \frac{3}{52} = \frac{22}{52} = \frac{11}{26}$$

Now for the Review test

 # Review test   Unit 44

12 1 A mass-produced electronic unit is susceptible to one of two faults: a solder
           fault with a probability of 0·04 and a component fault with a probability
           of 0·06. A unit is selected at random from a batch and tested. What is the
           probability of the unit having either a solder fault or a component fault if:

        (a) the two faults are mutually exclusive
        (b) the probability of the unit having both faults is 0·005.

Multiplication law and conditional probability

Independent events and dependent events

1

Make a note of these results, if you have not already done so:

(a) Events are *independent* when the occurrence of one event does not affect the probability of the occurrence of the second event. For example, in rolling a die on two occasions, the outcome of the first throw will not affect the probability of throwing a six on the second throw.

(b) Events are *dependent* when one event does affect the probability of the occurrence of the second.

Thus, the probability of drawing an ace from a pack of cards is $\frac{4}{52} = \frac{1}{13}$.

If the card is replaced so that the pack is complete and shuffled, the probability of drawing an ace on the second occasion is similarly $\frac{1}{13}$ (independent events).

However, if the ace is drawn on the first cutting and *not* replaced, the probability of drawing an ace on the second occasion is now ....

2

$$\boxed{\frac{3}{51}}$$

Because there are now only 3 remaining aces in the incomplete pack of 51 cards (dependent events).

Multiplication law of probabilities

3

Let us consider the probability of the occurrence of both events A and B

where event $A = \{$throwing a six$\}$ when rolling a die and
event $B = \{$drawing an ace$\}$ from a pack of cards.

These are clearly independent events:

$$P(A) = P(\text{six}) = \frac{1}{6}; \quad P(B) = P(\text{ace}) = \frac{4}{52} = \frac{1}{13}$$

If we now both roll the die and draw a card as one trial, then there are 6 possible outcomes from the die and for each one of these there are 52 possible outcomes from the cards, giving (6×52) outcomes altogether.

There are 4 possibilities of obtaining a six and an ace together, so the probability of event A and event B occurring is

$$P(A \text{ and } B) = \frac{4}{6 \times 52} = \frac{1 \times 4}{6 \times 52} = \frac{1}{6} \times \frac{4}{52} = P(A) \times P(B)$$

So, when A and B are *independent events*:

$$P(A \text{ and } B) = P(A) \times P(B) \tag{9}$$

Note this result

Conditional probability

4

We are concerned here with the probability of an event B occurring, given that an event A has already taken place. This is denoted by the symbol $P(B|A)$. If A and B are *independent* events, the fact that event A has already occurred will not affect the probability of event B. In that case:

$$P(B|A) = \ldots\ldots\ldots\ldots$$

5

$$\boxed{P(B|A) = P(B)} \tag{10}$$

If A and B are *dependent* events, then event A having occurred will affect the probability of the occurrence of event B. Let us see an example.

Example

A box contains five 10 ohm resistors and twelve 30 ohm resistors. The resistors are all unmarked and of the same physical size.

(a) If one resistor is picked out at random, determine the probability of its resistance being 10 ohms.

(b) If this first resistor is found to be 10 ohms and it is retained on one side, find the probability that a second selected resistor will be of resistance 30 ohms.

Let $A = \{10 \text{ ohm resistor}\}$ and $B = \{30 \text{ ohm resistor}\}$

(a) $N = 5 + 12 = 17$ Then $P(A) = \ldots\ldots\ldots\ldots$

6

$$\boxed{P(A) = \frac{5}{17}}$$

(b) The box now contains four 10 ohm resistors and twelve 30 ohm resistors. Then the probability of B, A having occurred

$$= P(B|A) = \ldots\ldots\ldots\ldots$$

7

$$P(B|A) = \frac{12}{16} = \frac{3}{4}$$

So the probability of getting a 10 ohm resistor at the first selection, retaining it, and getting a 30 ohm resistor at the second selection is

$P(A \text{ and } B|A) = \ldots\ldots\ldots$

8

$$\frac{15}{68}$$

Because $P(A \text{ and } B|A) = P(A) \times P(B|A) = \frac{5}{17} \times \frac{3}{4} = \frac{15}{68}$

So, if A and B are independent events:

$P(A \text{ and } B) = P(A) \times P(B)$

and if A and B are dependent events:

$P(A \text{ and } B) = P(A) \times P(B|A)$ \hfill (11)

Make a note of these results. Then on to the example in the next frame

---

9

A box contains 100 copper plugs, 27 of which are oversize and 16 undersize. A plug is taken from the box, tested and replaced: a second plug is then similarly treated. Determine the probability that (a) both plugs are acceptable, (b) the first is oversize and the second undersize, (c) one is oversize and the other undersize.

Let $A = \{\text{oversize plug}\}; \quad B = \{\text{undersize plug}\}$

$N = 100; \quad 27 \text{ oversize}; \quad 16 \text{ undersize}; \quad \therefore 57 \text{ acceptable}$

(a) $P_1(\text{first plug acceptable}) = \dfrac{57}{100}$

$P_2(\text{second acceptable}) = \dfrac{57}{100}$

$\therefore P_{12}(\text{first acceptable and second acceptable}) = \ldots\ldots\ldots$

10

$$P_1 \times P_2 = \frac{57}{100} \times \frac{57}{100} = \frac{3249}{10\,000} = 0{\cdot}3249$$

(b) $P_1(\text{first oversize}) = \dfrac{27}{100}$

$P_2(\text{second undersize}) = \dfrac{16}{100}$

$\therefore P_{12}(\text{first oversize and second undersize}) = \ldots\ldots\ldots$

11

$$P_1 \times P_2 = \frac{27}{100} \times \frac{16}{100} = \frac{432}{10\,000} = 0.0432$$

(c) This section, of course, includes part (b) of the problem, but also covers the case when the first is undersize and the second oversize.

$$P_3(\text{first undersize}) = \frac{16}{100}; \quad P_4(\text{second oversize}) = \frac{27}{100}$$

∴ P_{34}(first undersize and second oversize)

$$= \frac{16}{100} \times \frac{27}{100} = \frac{432}{10\,000} = 0.0432$$

∴ P(one oversize and one undersize)

$$= P\{(\text{first oversize and second undersize}) \text{ or (first undersize and}$$
$$\text{second oversize})\} = \ldots\ldots\ldots\ldots$$

12

$$P_{12} + P_{34} = \frac{432}{10\,000} + \frac{432}{10\,000} = \frac{864}{10\,000} = 0.0864$$

One must be careful to read the precise requirements of the problem. Then the solution is straightforward, the main tools being:

Independent events

$P(A \text{ or } B) = P(A) + P(B)$ i.e. 'or' is associated with $+$
$P(A \text{ and } B) = P(A) \times P(B)$ i.e. 'and' is associated with $\times$

Dependent events

$P(A \text{ or } B) = P(A) + P(B); \quad P(A \text{ and } B) = P(A) \times P(B|A)$

Now one more on your own. It is the previous example repeated with one important variation.

A box contains 100 copper plugs, 27 oversize and 16 undersize. A plug is taken, tested but *not* replaced: a second plug is then treated similarly. Determine the probability that (a) both plugs are acceptable, (b) the first is oversize and the second undersize, (c) one is oversize and the other undersize.

Complete all three sections and then check with the next frame

See if you agree.

13

Let $A = \{\text{oversize plug}\}$; $B = \{\text{undersize plug}\}$

$N = 100$; 27 oversize; 16 undersize; $\therefore$ 57 acceptable

(a) $P_1(\text{first plug acceptable}) = \dfrac{57}{100}$ Plug *not* replaced

$P_2(\text{second acceptable}) = \dfrac{56}{99}$

$\therefore P_{12}(\text{first acceptable and second acceptable})$

$= \dfrac{57}{100} \times \dfrac{56}{99} = \dfrac{3192}{9900} = 0 \cdot 3224$

(b) $P_1(\text{first oversize}) = \dfrac{27}{100}$ Plug *not* replaced

$P_2(\text{second undersize}) = \dfrac{16}{99}$

$\therefore P_{12}(\text{first oversize and second undersize})$

$= \dfrac{27}{100} \times \dfrac{16}{99} = \dfrac{432}{9900} = 0 \cdot 0436$

(c) Here again, we must include the two cases of either (first oversize and second undersize) or (first undersize and second oversize). The first of these we have already covered above:

$$P_{12} = \dfrac{432}{9900} = 0 \cdot 0436$$

Now we have:

$P_3(\text{first undersize}) = \dfrac{16}{100}$ Plug *not* replaced

$P_4(\text{second oversize}) = \dfrac{27}{99}$

$\therefore P_{34}(\text{first undersize and second oversize})$

$= \dfrac{16}{100} \times \dfrac{27}{99} = \dfrac{432}{9900} = 0 \cdot 0436$

$\therefore P(\text{one oversize and one undersize}) =$

$P\{(\text{first oversize and second undersize}) \text{ or } (\text{first undersize and second}$

$$\text{oversize})\} = P_{12} + P_{34}$$

$$= 0 \cdot 0436 + 0 \cdot 0436 = 0 \cdot 0872$$

Let us pause and summarize the work on multiplication of probabilities and conditional probability

Review summary

Unit 45

14

1 *Independent events*: A and B are independent events if the occurrence of one does not depend on the occurrence of the other.

2 *Dependent events*: A and B are dependent events if the occurrence of one depends upon the occurrence of the other.

3 *Multiplication law of probability*: If A and B are independent events the probability of either event A or event B occurring is:

$P(A \text{ and } B) = P(A)P(B)$

4 *Conditional probability*: The probability of event B occurring given that event A has already occurred is denoted by:

$P(B|A)$

If A and B are independent events then:

$P(B|A) = P(B)$

If A and B are dependent events then:

$P(A \text{ and } B) = P(A)P(B|A)$

 # Review exercise

Unit 45

15

1 If the probability of student A passing an examination is 0·8 and the probability of student B passing the same examination is 0·6, what is the probability of both students passing the examination?

2 A card is drawn from a deck of 52 playing cards. The card selected is not replaced before a second card is selected. What is the probability that the two cards selected are both aces?

Complete both questions. Take care, there is no need to rush.
If you need to, look back at the Unit.
The answers and working are in the next frame.

16

1 If the probability of student A passing an examination is $P(A) = 0·8$ and the probability of student B passing the same examination is $P(B) = 0·6$ then because these two events are independent of each other and the probability of both students passing the examination is:

$P(A \text{ and } B) = P(A)P(B) = 0·8 \times 0·6 = 0·48$

2 The probability of drawing an ace from a deck of 52 playing cards is $P(A) = 4/52$. The probability of drawing a second ace is then $P(B|A) = 3/51$ because there are only 3 aces and 51 cards left from which to draw. Therefore:

$$P(A \text{ and } B) = P(A)P(B|A) = \frac{4}{52} \times \frac{3}{51} = \frac{12}{2652} = \frac{1}{221}$$

Now for the Review test

Review test

Unit 45

17

1   If the probability of a bus being late is 0·2 and the probability of a train being late is 0·3, what is the probability of both the bus and the train being late?

2   From a committee of 4 men and 5 women two people are selected at random to form a sub-committee. What is the probability that the sub-committee is composed of:

   (a) 2 women   (b) 1 man and 1 woman

18

You have now come to the end of this Module. A list of **Can You?** questions follows for you to gauge your understanding of the material in the Module. You will notice that these questions match the **Learning outcomes** listed at the beginning of the Module. Now try the **Test exercise**. *Work through the questions at your own pace, there is no need to hurry.* A set of **Further problems** provides valuable additional practice.

 # Can You?

Checklist: Module 14

1

Check this list before and after you try the end of Module test.

On a scale of 1 to 5 how confident are you that you can:

• Understand the nature of probability as a measure of chance?
   Yes ☐ ☐ ☐ ☐ ☐ *No*

• Compute expectations of events arising from an experiment with a number of possible outcomes?
   Yes ☐ ☐ ☐ ☐ ☐ *No*

• Assign classical measures to the probability of an event and be able to define the probabilities of both certainty and impossibility?
   Yes ☐ ☐ ☐ ☐ ☐ *No*

• Distinguish between mutually exclusive and mutually non-exclusive events and compute their probabilities?
   Yes ☐ ☐ ☐ ☐ ☐ *No*

• Distinguish between dependent and independent events and apply the multiplication law of probabilities?
   Yes ☐ ☐ ☐ ☐ ☐ *No*

• Compute conditional probabilities?
   Yes ☐ ☐ ☐ ☐ ☐ *No*

Test exercise 14

2 There are no tricks here, so you will have no difficulty.

1 Twelve per cent of a type of plastic bushes are rejects. Determine:
(a) the probability that any one item drawn at random is
   (i) defective   (ii) acceptable
(b) the number of acceptable bushes likely to be found in a sample batch of 4000.

2 A box contains 12 transistors of type *A* and 18 of type *B*, all identical in appearance. If one transistor is taken at random, tested and returned to the box, and a second transistor then treated in the same manner, determine the probability that:
(a) the first is type *A* and the second type *B*
(b) both are type *A*
(c) neither is type *A*.

3 A packet contains 100 washers, 24 of which are brass, 36 copper and the remainder steel. One washer is taken at random, retained, and a second washer similarly drawn. Determine the probability that:
(a) both washers are steel
(b) the first is brass and the second copper
(c) one is brass and one is steel.

Further problems 14

3 **1** One card is taken from each of two packs. What is the probability that one is a diamond and the other is not a diamond?

2 Again, one card is taken from each of two packs. What is the probability that:
(a) at least one of the cards is a diamond
(b) at most one of the cards is a diamond
(c) neither card is a diamond?

3 If I travel to my destination by bus there is a 2% chance of being late whereas if I travel by car there is a 4% chance of being late. There is a 40% likelihood of me travelling by bus and a 60% likelihood of me taking the car. What is the chance of me arriving at my destination on time?

4 When rolling two dice simultaneously what is the probability of rolling:
(a) a 7
(b) an 11
(c) a 7 or an 11?

5 What is the probability of taking three cards off the top of a deck of playing cards and that they are an ace, a two and a three?

6 What is the probability of taking three cards off the top of a deck of playing cards and that they are the ace, the two and the three of spades?

7 A sub-committee of three is selected from six committee members A, B, C, D, E and F. What is the probability that:

(a) C and E are selected
(b) D is not selected?

8 What is the probability of tossing exactly two heads when three coins are tossed?

9 What is the probability of tossing n tails when n coins are tossed simultaneously?

10 A student applies to two universities for admission to study history. The probability of being accepted at A is 0·8 and of being rejected at B is 0·3. If the probability of being rejected by at least one university is 0·2 what is the probability of being accepted by at least one university?

11 An integer is chosen at random from the first 42 positive integers. What is the probability that it can be divided by 7 or 3?

12 On an engineering course, 4% failed the Principles course, 5% failed the Mathematics course and 2% failed both courses. What percentage:

(a) passed Principles and failed Mathematics
(b) failed Principles and passed Mathematics
(c) passed both Principles and Mathematics?

13 (a) A block of seats at an arena is arranged in a rectangle of 10 rows and 9 columns – the first row being the front row. A spectator buys a ticket. What is the probability that the seat is:

(i) on the front row
(ii) in the centre of the 5th row?

(b) A second spectator subsequently buys another ticket. What is the probability that the seat is also:

(i) on the front row
(ii) in the centre of the 5th row?

(c) What is the probability that the second spectator is sitting directly behind the first spectator but with three rows between them?

14 An insecticide has 60% success rate on first application. The remaining 40% of insects develop a resistance so that the proportion killed is only one-half of the remaining insects in the next application. What is the probability of an insect surviving 4 applications of the insecticide, and of 1000 insects how many will survive?

15 100 tickets have been sold at a raffle and the tickets are numbered 1 to 100. Three prizes are available and so three tickets are drawn at random. What is the probability that the drawn tickets are:

(a) alternately odd, even and odd
(b) all odd?

Statistics

When you have completed this Module you will be able to:

- Distinguish between discrete and continuous data
- Construct frequency and relative frequency tables for grouped and ungrouped discrete data
- Determine class boundaries, class intervals and central values for discrete and continuous data
- Construct a histogram and a frequency polygon
- Determine the mean, median and mode of grouped and ungrouped data
- Determine the range, variance and standard deviation of discrete data
- Measure the dispersion of data using the normal and standard normal curves

Units

Data

Unit 46

1

Statistics is concerned with the collection, ordering and analysis of data. *Data* consist of sets of recorded observations or values. Any quantity that can have a number of values is a *variable*. A variable may be one of two kinds:

(a) *Discrete* – a variable that can be counted, or for which there is a fixed set of values.

(b) *Continuous* – a variable that can be measured on a continuous scale, the result depending on the precision of the measuring instrument, or the accuracy of the observer.

A statistical exercise normally consists of four stages:

(a) collection of data by counting or measuring
(b) ordering and presentation of the data in a convenient form
(c) analysis of the collected data
(d) interpretation of the results and conclusions formulated.

State whether each of the following is a discrete or continuous variable:

(a) the number of components in a machine
(b) the capacity of a container
(c) the size of workforce in a factory
(d) the speed of rotation of a shaft
(e) the temperature of a coolant.

2

(a) and (c) discrete; (b), (d) and (e) continuous

Next frame

Arrangement of data

3

The contents of each of 30 packets of washers are recorded:

```
28   31   29   27   30   29   29   26   30   28
28   29   27   26   32   28   32   31   25   30
27   30   29   30   28   29   31   27   28   28
```

We can appreciate this set of numbers better if we now arrange the values in ascending order, writing them still in 3 lines of 10. If we do this, we get . . . .

4

25	26	26	27	27	27	27	28	28	28
28	28	28	28	29	29	29	29	29	29
30	30	30	30	30	31	31	31	32	32

Some values occur more than once. Therefore, we can form a table showing how many times each value occurs:

Value	Number of times
25	1
26	2
27	4
28	7
29	6
30	5
31	3
32	2

The number of occasions on which any particular value occurs is called its *frequency*, denoted by the symbol f.

The total frequency is therefore ....

5

30, the total number of readings

Tally diagram

When dealing with large numbers of readings, instead of writing all the values in ascending order, it is more convenient to compile a *tally diagram*, recording the range of values of the variable and adding a stroke for each occurrence of that reading, thus:

Variable (x)	Tally marks	Frequency (f)
25	/	1
26	//	2
27	////	4
28	̶H̶H̶ //	7
29	̶H̶H̶ /	6
30	̶H̶H̶	5
31	///	3
32	//	2

It is usual to denote the variable by x and the frequency by f. The right-hand column gives the *frequency distribution* of the values of the variable.

▶

Exercise

The number of components per hour turned out on a lathe was measured on 40 occasions:

18	17	21	18	19	17	18	20	16	17
19	19	16	17	15	19	17	17	20	18
17	18	19	19	18	19	18	18	19	20
18	15	18	17	20	18	16	17	18	17

Compile a tally diagram and so determine the frequency distribution of the values. This gives . . . .

Variable (x)	Tally marks	Frequency (f)
15	/	2
16	///	3
17	ⅠⅠⅠ ⅠⅠⅠ	10
18	ⅠⅠⅠ ⅠⅠⅠ //	12
19	ⅠⅠⅠ ///	8
20	////	4
21	/	1
	$n = \sum f = 40$	

6

Grouped data

If the range of values of the variable is large, it is often helpful to consider these values arranged in regular groups, or *classes*.

For example, the numbers of the overtime hours per week worked by employees at a factory are as follows:

45	31	46	25	57	39	42	55	20	37
40	59	11	38	34	22	62	33	48	43
57	37	43	51	29	41	35	66	45	32
44	47	42	46	54	65	17	35	53	27
38	22	33	39	45	32	43	41	57	45

Lowest value of the variable $= 11$
Highest value of the variable $= 66$ $\Bigg\}$ $\therefore$ Arrange 6 classes of 10 h each.

To determine the frequency distribution, we set up a table as follows:

Overtime hours (x)	Tally marks	Frequency (f)
10–19		
20–29		
30–39		
etc.		

Complete the frequency distribution.

7

Overtime hours (x)	Tally marks	Frequency (f)
10–19	//	2
20–29	⊬⊬ /	6
30–39	⊬⊬ ⊬⊬ ////	14
40–49	⊬⊬ ⊬⊬ ⊬⊬ //	17
50–59	⊬⊬ ///	8
60–70	///	3
	$n = \sum f = 50$	

Grouping with continuous data

In the previous example using discrete data, there is no difficulty in allocating any given value to its appropriate group, since, for example, there is no value between, say, 29 and 30. However, with continuous data, the variable is measured on .... and may well have values lying between 29 and 30, e.g. 29·7, 29·8 etc.

8

a continuous scale

In practice, where the values of the variable are all given to the same number of significant figures or decimal places, there is no trouble and we form the groups accordingly.

For example, the lengths (in mm) of 40 spindles were measured with the following results:

20·90 20·57 20·86 20·74 20·82 20·63 20·53 20·89 20·75 20·65
20·71 21·03 20·72 20·41 20·94 20·75 20·79 20·65 21·08 20·89
20·50 20·88 20·97 20·78 20·61 20·92 21·07 21·16 20·80 20·77
20·82 20·72 20·60 20·90 20·86 20·68 20·75 20·88 20·56 20·94

Lowest value = 20·41 ⎱ ∴ Form classes from 20·40 to 21·20
Highest value = 21·16 ⎰   at 0·10 intervals.

▶

Length (mm) (x)	Tally marks	Frequency (f)
20·40–20·49		
20·50–20·59		
20·60–20·69		
etc.		

Complete the table and so determine the frequency distribution . . . .

9

Length (mm) (x)	Tally marks	Frequency (f)
20·40–20·49	/	1
20·50–20·59	////	4
20·60–20·69	̶H̶H̶ /	6
20·70–20·79	̶H̶H̶ ̶H̶H̶	10
20·80–20·89	̶H̶H̶ ////	9
20·90–20·99	̶H̶H̶ /	6
21·00–21·09	///	3
21·10–21·20	/	1
	$n = \sum f = 40$	

Note that the last class is slightly larger than the others, but this has negligible effect, since there are very few entries in the end classes.

Relative frequency

In the frequency distribution just determined, if the frequency of any one class is compared with the sum of the frequencies of all classes (i.e. the total frequency), the ratio is the *relative frequency* of that class. The result is generally expressed as a percentage.

Add a fourth column to the table above showing the relative frequency of each class expressed as a percentage.

Check with the next frame

10

Length (mm) (x)	Frequency (f)	Relative frequency (%)
20·40–20·49	1	2·5
20·50–20·59	4	10·0
20·60–20·69	6	15·0
20·70–20·79	10	25·0
20·80–20·89	9	22·5
20·90–20·99	6	15·0
21·00–21·09	3	7·5
21·10–21·20	1	2·5
	$n = \sum f = 40$	100·0%

The sum of the relative frequencies of all classes must add up to the whole and therefore has a value 1 or 100 per cent.

Rounding off data

If the value 21·7 is expressed to two significant figures, the result is, of course, 22. Similarly, 21·4 is rounded off to 21. In order to maintain consistency of class boundaries, 'middle values' will, in what follows, always be rounded up. For example, 21·5 is rounded up to 22 and 42·5 is rounded up to 43.

Therefore, when a result is quoted to two significant figures as 37 on a continuous scale, this includes all possible values between 36·500000... and 37·499999... (37·5 itself would be rounded off to 38). This could be expressed by saying that 37 included all values between 36·5 and 37·5⁻, the small negative sign in the index position indicating that the value is just under 37·5 without actually reaching it.

So   42   includes all values between .... and ....

      31·4   includes all values between .... and ....

      17·63   includes all values between .... and ....

11

42:	41·5	and 42·5⁻
31·4:	31·35	and 31·45⁻
17·63:	17·625	and 17·635⁻

Exercise

The thicknesses of 20 samples of steel plate are measured and the results (in millimetres) to two significant figures are as follows:

7·3	7·1	6·6	7·0	7·8	7·3	7·5	6·2	6·9	6·7
6·5	6·8	7·2	7·4	6·5	6·9	7·2	7·6	7·0	6·8

Compile a table showing the frequency distribution and the relative frequency distribution for regular classes of 0·2 mm from 6·2 mm to 7·9 mm.

12

Length (mm) (x)	Frequency (f)	Relative frequency (%)
6·2–6·4	1	5
6·5–6·7	4	20
6·8–7·0	6	30
7·1–7·3	5	25
7·4–7·6	3	15
7·7–7·9	1	5
	$n = \sum f = 20$	100%

Class boundaries

In the example above, the values of the variable are given to 2 significant figures. With the usual rounding-off procedure, each class in effect extends from 0·05 below the first stated value of the class to just under 0·05 above the second stated value of the class. So, the class

7·1–7·3 includes all values between 7·05 and 7·35⁻
7·4–7·6 includes all values between 7·35 and 7·65⁻ etc.

We can use this example to define a number of terms we shall need. Let us consider in particular the class labelled 7·1–7·3.

(a) The class values stated in the table are the *lower* and *upper limits* of the class and their difference gives the *class width*.

(b) The *class boundaries* are 0·05 below the lower class limit and 0·05 above the upper class limit, that is:

the lower class boundary is $7·1 - 0·05 = 7·05$

the upper class boundary is $7·3 + 0·05 = 7·35$.

(c) The *class interval* is the difference between the upper and lower class boundaries.

$$\text{class interval} = \text{upper class boundary} - \text{lower class boundary}$$
$$= 7·35 - 7·05 = 0·30$$

Where the classes are regular, the class interval can also be found by subtracting any lower class limit from the lower class limit of the following class.

(d) The *central value* (or mid-value) of the class is the average of the upper and lower class boundaries. So, in the particular class we are considering, the central value is . . . .

13

$$\boxed{7·20}$$

For the central value $= \dfrac{1}{2}(7·05 + 7·35) = 7·20$.

We can summarize these terms in the following diagram, using the class 7·1–7·3 (inclusive) as our example.

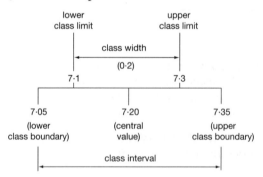

So now you can complete the following table:

Thickness (mm) (x)	Central value	Lower class boundary	Upper class boundary
6·2–6·4			
6·5–6·7			
6·8–7·0			
7·1–7·3			
7·4–7·6			
7·7–7·9			

14

Thickness (mm) (x)	Central value	Lower class boundary	Upper class boundary
6·2–6·4	6·3	6·15	6·45
6·5–6·7	6·6	6·45	6·75
6·8–7·0	6·9	6·75	7·05
7·1–7·3	7·2	7·05	7·35
7·4–7·6	7·5	7·35	7·65
7·7–7·9	7·8	7·65	7·95

Exercise

A machine is set to produce metal washers of nominal diameter 20·0 mm. The diameters of 34 samples are measured and the following results in millimetres obtained:

19·63 19·82 19·96 19·75 19·86 19·82 19·61 19·97 20·07
19·89 20·16 19·56 20·05 19·72 19·96 19·68 19·87 19·90
19·73 19·93 20·03 19·86 19·81 19·77 19·78 19·75 19·87
19·66 19·77 19·99 20·00 20·11 20·01 19·84

Arrange the values into 7 equal classes of width 0·09 mm for the range 19·50 mm to 20·19 mm and determine the frequency distribution.

Check your result with the next frame

15

Diameter (mm) (x)	Frequency (f)
19·50–19·59	1
19·60–19·69	4
19·70–19·79	7
19·80–19·89	9
19·90–19·99	6
20·00–20·09	5
20·10–20·19	2
$n = \sum f = 34$	

So (a) the class having the highest frequency is ....

(b) the lower class boundary of the third class is ....

(c) the upper class boundary of the seventh class is ....

(d) the central value of the fifth class is ....

16

(a) 19·80 – 19·89	(b) 19·695
(c) 20·195	(d) 19·945

Next frame

Histograms

17   Frequency histogram

A histogram is a graphical representation of a frequency distribution, in which vertical rectangular blocks are drawn so that:

(a) the centre of the base indicates the central value of the class and

(b) the area of the rectangle represents the class frequency.

If the class intervals are regular, the frequency is then denoted by the height of the rectangle.

For example, measurement of the lengths of 50 brass rods gave the following frequency distribution:

Length (mm) (x)	Lower class boundary	Upper class boundary	Central value	Frequency (f)
3·45–3·47	3·445	3·475	3·460	2
3·48–3·50	3·475	3·505	3·490	6
3·51–3·53	3·505	3·535	3·520	12
3·54–3·56	3·535	3·565	3·550	14
3·57–3·59	3·565	3·595	3·580	10
3·60–3·62	3·595	3·625	3·610	5
3·63–3·65	3·625	3·655	3·640	1

First, we draw a base line and on it mark a scale of x on which we can indicate the central values of the classes. Do that for a start.

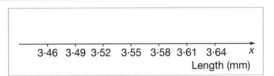

18

Since the classes are of regular class interval, the class boundaries will coincide with the points mid-way between the central values, thus:

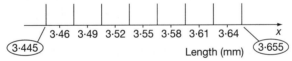

Note that the lower boundary of the first class extends to 3·445 and that the upper boundary of the seventh class extends to 3·655. Because the class intervals are regular, we can now erect a vertical scale to represent class frequencies and rectangles can be drawn to the appropriate height. Complete the work and give it a title.

19   **Relative frequency histogram**

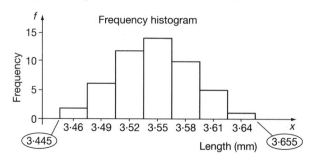

The same diagram can be made to represent the relative frequency distribution by replacing the scale of frequency along the y-axis with a scale of relative frequencies (percentages).

Considerable information can be gleaned from a histogram on sight. For instance, we see that the class having the highest frequency is the .... class.

20

> fourth

and the class with the lowest frequency is the .... class.

21

> seventh

Most of the whole range of values is clustered within the middle classes and knowledge of the centre region of the histogram is important. We can put a numerical value on this by determining a *measure of central tendency* which we shall deal with next.

For now, let's pause and summarize the work on data

Review summary

1 *Data* (a) Discrete data – values that can be precisely counted.
(b) Continuous data – values measured on a continuous scale.

22

2 *Grouped data*

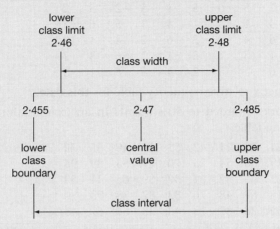

3 *Frequency (f)* – the number of occasions on which each value, or class, occurs.

Relative frequency – each frequency expressed as a percentage or fraction of the total frequency.

4 *Histogram* – graphical representation of a frequency or relative frequency distribution.

The frequency of any one class is given by the *area* of its column. If the class intervals are constant, the height of the rectangle indicates the frequency on the vertical scale.

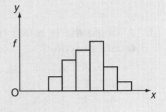

5 *Frequency polygon* – the figure formed by joining the centre points of the tops of the rectangles of a frequency histogram with straight lines and extended to include the two zero frequency columns on the sides.

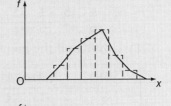

6 *Frequency curve* – obtained by 'smoothing' the boundary of the frequency polygon, or by plotting centre values and joining with a smooth curve.

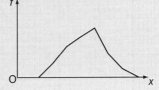

Review exercise   **Unit 46**

23   **1** A single die is rolled 60 times. The score on each roll was recorded and the
following table contains the complete record:

1	2	6	1	4	3	6	4	1	3
1	1	5	6	5	1	5	2	5	2
4	3	1	3	1	4	3	5	6	4
1	6	4	4	2	6	2	4	1	5
2	1	3	1	5	1	2	6	3	2
3	6	1	5	2	4	5	6	1	3

Construct a relative frequency table for this data.

2 The marks awarded to 50 students in an examination are recorded in the
following table.

4	21	31	11	42	51	19	40	55	72
38	49	60	54	23	70	7	55	89	95
23	62	5	32	61	37	59	65	41	53
51	18	59	48	76	83	82	78	93	80
42	69	28	57	15	44	33	66	57	45

Group this data and construct a relative frequency table. Draw the histogram
for this data.

Complete both questions. Take care, there is no need to rush.
If you need to, look back at the Unit.
The answers and working are in the next frame.

24   **1** A single die is rolled 60 times. The following table contains the complete
record of scores:

1̶	2̶	6	1̶	4	3	6	4	1̶	3
1̶	1̶	5	6	5	1̶	5	2̶	5	2̶
4	3	1̶	3	1̶	4	3	5	6	4
1̶	6	4	4	2̶	6	2̶	4	1̶	5
2̶	1̶	3	1̶	5	1̶	2̶	6	3	2̶
3	6	1̶	5	2̶	4	5	6	1̶	3

By crossing off the numbers it is possible to make an accurate count.

Score	Frequency	Relative frequency
1	15	0·25
2	9	0·15
3	9	0·15
4	9	0·15
5	9	0·15
6	9	0·15
$\sum_{1}^{6} =$	60	1

2 The marks awarded to 50 students in an examination are recorded in the following table.

```
 4  21  31  11  42  51  19  40  55  72
38  49  60  54  23  70   7  55  89  95
23  62   5  32  61  37  59  65  41  53
51  18  59  48  76  83  82  78  93  80
42  69  28  57  15  44  33  66  57  45
```

By crossing off the numbers it is possible to make an accurate count.

Marks	Central value	Frequency	Relative frequency
0–9	4·5	3	0·06
10–19	14·5	4	0·08
20–29	24·5	4	0·08
30–39	34·5	5	0·10
40–49	44·5	8	0·16
50–59	54·5	10	0·20
60–69	64·5	6	0·12
70–79	74·5	4	0·08
80–89	84·5	4	0·08
90–100	95·0	2	0·04
$\sum\limits_{1}^{10} =$		50	1

The histogram can then be drawn:

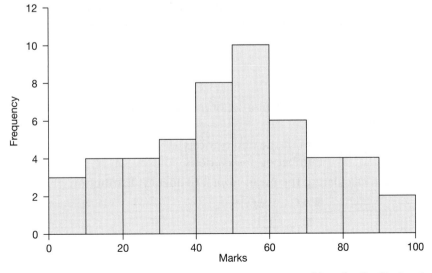

Now for the Review test

Review test   Unit 46

25

1 Given the collection of numbers in the following table:

5	3	4	1	4	3	2	4	1	3
1	1	5	2	5	1	5	2	5	2
4	4	4	3	1	4	3	4	3	4
1	3	4	4	2	3	4	4	3	5
2	1	3	3	4	5	2	2	4	2
3	2	1	5	2	4	4	5	3	3

Construct a relative frequency table for this data.

2 A factory makes pots of mustard that are supposed to weigh 95 gm. A sample of 50 pots from one day's production had the weights (in gm) recorded in the following table:

94·9	95·1	95·0	95·1	95·0	95·1	95·2	94·9	95·3	95·4
95·2	95·0	95·3	95·3	95·0	94·9	95·0	95·1	94·9	95·0
95·1	95·2	94·9	95·0	95·1	95·4	95·1	95·0	94·9	95·3
95·3	94·9	95·0	94·9	95·2	95·0	95·3	95·2	95·1	95·2
95·1	95·2	95·1	95·1	94·9	95·0	95·4	94·9	95·1	95·0

Group this data and construct a relative frequency table, then draw the histogram for this data.

Measures of central tendency   Unit 47

1

There are three common measures of central tendency, the *mean*, *mode* and *median*, of a set of observations and we shall discuss each of them in turn.

Mean

The arithmetic mean $\bar{x}$ of a set of n observations x is simply their average,

i.e. mean $= \dfrac{\text{sum of the observations}}{\text{number of observations}}$ $\therefore \ \bar{x} = \dfrac{\sum x}{n}$

When calculating the mean from a frequency distribution, this becomes

mean $= \bar{x} = \dfrac{\sum xf}{n} = \dfrac{\sum xf}{\sum f}$

For example, for the following frequency distribution, we need to add a third column showing the values of the product $x \times f$, after which the mean can be found:

Variable (x)	Frequency (f)	Product (xf)
15	1	15
16	4	64
17	9	153
18	10	180
19	6	114
20	2	40

So, $\bar{x} = \ldots\ldots\ldots\ldots$

$$\boxed{17\cdot69}$$

2

Because

$$n = \sum f = 32 \text{ and } \sum xf = 566 \qquad \therefore \ \bar{x} = \frac{\sum xf}{n} = \frac{566}{32} = 17\cdot69$$

When calculating the mean from a frequency distribution with grouped data, the central value, x_m, of the class is taken as the x-value in forming the product xf. So, for the frequency distribution

Variable (x)	12–14	15–17	18–20	21–23	24–26	27–29
Frequency (f)	2	6	9	8	4	1

$\bar{x} = \ldots\ldots\ldots\ldots$

$$\boxed{\bar{x} = 19\cdot9}$$

3

Because

$$n = \sum f = 30 \text{ and } \sum x_m f = 597 \qquad \therefore \ \bar{x} = \frac{\sum x_m f}{n} = \frac{597}{30} = 19\cdot9$$

Here is one more.

Measurement in millimetres of 60 bolts gave the following frequency distribution:

Length x (mm)	30·2	30·4	30·6	30·8	31·0	31·2	31·4
Frequency f	3	7	12	17	11	8	2

The mean $\bar{x} = \ldots\ldots\ldots\ldots$

4

$$\bar{x} = 30{\cdot}79$$

Length (mm) (x)	Frequency (f)	Product (xf)
30·2	3	90·6
30·4	7	212·8
30·6	12	367·2
30·8	17	523·6
31·0	11	341·0
31·2	8	249·6
31·4	2	62·8

$$\therefore \bar{x} = \frac{\sum xf}{n} = \frac{1847{\cdot}6}{60}$$
$$= 30{\cdot}79$$
$$\therefore \bar{x} = 30{\cdot}79$$

$$n = \sum f = 60$$

$$\downarrow$$
$$\sum xf = 1847{\cdot}6$$

5   Coding for calculating the mean

We can save ourselves some of the tedious work by using a system of *coding* which involves converting the x-values into simpler values for the calculation and then converting back again for the final result. An example will show the method in detail: we will use the same frequency distribution as above.

First, we choose a convenient value of x (near the middle of the range) and subtract this from every other value of x to give the second column.

Length (mm) (x)	Deviation from chosen value $(x - 30{\cdot}8)$	In units of 0·2 mm $\left(x_c = \dfrac{x - 30{\cdot}8}{0{\cdot}2}\right)$	
30·2	−0·6		Then we change the values into an even simpler form by dividing the values in column 2 by 0·2 mm. These are entered in column 3 and give the coded values of x, i.e. x_c
30·4	−0·4		
30·6	−0·2		
30·8	0		
31·0	0·2		
31·2	0·4		
31·4	0·6		

All we have to do then is to add column 4 showing the class frequencies (from the table above) and then column 5 containing values of the product $x_c f$. Using the last two columns, the mean value of x_c can be found as before.
$$\bar{x}_c = \ldots\ldots\ldots\ldots$$

$$\bar{x}_c = -0.0333$$

6

Length (mm) (x)	Deviation from chosen value $(x - 30.8)$	In units of 0.2 mm $\left(x_c = \dfrac{x - 30.8}{0.2}\right)$	Frequency (f)	Product $(x_c f)$
30.2	−0.6	−3	3	−9
30.4	−0.4	−2	7	−14
30.6	−0.2	−1	12	−12
30.8	0	0	17	0
31.0	0.2	1	11	11
31.2	0.4	2	8	16
31.4	0.6	3	2	6

$$n = \sum f = 60; \qquad \sum x_c f = -2.0$$

$$\therefore \ \bar{x}_c = \frac{-2}{60} = -0.0333$$

So we have $\bar{x}_c = -0.0333$. Now we have to retrace our steps back to the original units of x.

7

Decoding

In the coding procedure, our last step was to divide by 0.2. We therefore now multiply by 0.2 to return to the correct units of $(x - 30.8)$:

$$\therefore \ \bar{x} - 30.8 = -0.0333 \times 0.2 = -0.00667$$

We now add the 30.8 to both sides:

$$\therefore \ \bar{x} = 30.8 - 0.00667 = 30.79333 \qquad \therefore \ \bar{x} = 30.79$$

Note that in decoding, we reverse the operations used in the original coding process. The value (30.8) which was subtracted from all values of x is near the centre of the range of x-values and is therefore sometimes referred to as a *false mean*.

8   Coding with a grouped frequency distribution

The method is precisely the same, except that we work with the centre values of the classes of x, i.e. x_m, for the calculation purposes.

Here is an exercise:

The thicknesses of 50 spacing pieces were measured:

Thickness x (mm)	2·20–2·22	2·23–2·25	2·26–2·28	2·29–2·31	2·32–2·34
Frequency f	1	5	8	15	12

Thickness x (mm)	2·35–2·37	2·38–2·40
Frequency f	7	2

Using coding, determine the mean value of the thickness. $\bar{x} = \ldots\ldots\ldots$

9

$$\boxed{\bar{x} = 2\cdot307 \text{ mm}}$$

Thickness (mm) (x)	Central value (x_m)	Deviation from chosen value $(x_m - 2\cdot30)$	In units of 0·03 mm $\left(x_c = \dfrac{x_m - 2\cdot30}{0\cdot03}\right)$	Frequency (f)	Product $(x_c f)$
2·20–2·22	2·21	−0·09	−3	1	−3
2·23–2·25	2·24	−0·06	−2	5	−10
2·26–2·28	2·27	−0·03	−1	8	−8
2·29–2·31	2·30	0	0	15	0
2·32–2·34	2·33	0·03	1	12	12
2·35–2·37	2·36	0·06	2	7	14
2·38–2·40	2·39	0·09	3	2	6

$$n = \sum f = 50$$
$$\downarrow$$
$$\sum x_c f = 11$$

$$\therefore \bar{x}_c = \frac{\sum x_c f}{n} = \frac{11}{50} = 0\cdot22$$
$$\therefore \bar{x}_m - 2\cdot30 = \bar{x}_c \times 0\cdot03 = 0\cdot22 \times 0\cdot03 = 0\cdot0066$$
$$\therefore \bar{x}_m = 2\cdot30 + 0\cdot0066 = 2\cdot3067 \quad \therefore \bar{x} = 2\cdot307$$

Now let us deal with the mode, so move on to the next frame

Mode

The *mode* of a set of data is that value of the variable that occurs most often. For instance, in the set of values 2, 2, 6, 7, 7, 7, 10, 13, the mode is clearly 7. There could, of course, be more than one mode in a set of observations, e.g. 23, 25, 25, 25, 27, 27, 28, 28, 28, has two modes, 25 and 28, each of which appears three times.

10

Mode of a grouped frequency distribution

11

The masses of 50 castings gave the following frequency distribution:

Mass x (kg)	10–12	13–15	16–18	19–21	22–24	25–27	28–30
Frequency f	3	7	16	10	8	5	1

If we draw the histogram, using the central values as the mid-points of the bases of the rectangles, we obtain ....

12

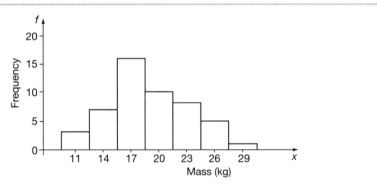

The class having the highest frequency is the .... class

$\boxed{\text{third}}$

13

The modal class is therefore the third class, with boundaries 15·5 and 18·5 kg. The value of the mode itself lies somewhere within that range and its value can be found by a simple construction.

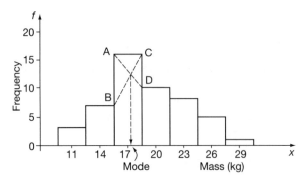

The two diagonal lines AD and BC are drawn as shown. The *x*-value of their point of intersection is taken as the mode of the set of observations.

Carry out the construction on your histogram and you will find that, for this set of observations, the mode = ...

14

$$\text{mode} = 17\cdot3$$

The value of the mode can also be calculated.

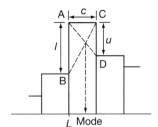

If L = lower boundary value
l = AB = difference in frequency on the lower boundary
u = CD = difference in frequency on the upper boundary
c = class interval

then the mode $= L + \left(\dfrac{l}{l+u}\right)c$

In the example we have just considered:

$L =$ .... $l =$ ....
$u =$ .... $c =$ ....

15

$$L = 15\cdot5; \quad l = 16 - 7 = 9; \quad u = 16 - 10 = 6; \quad c = 3$$

$$\text{Then the mode} = 15\cdot5 + \left(\frac{9}{9+6}\right) \times 3$$
$$= 15\cdot5 + 1\cdot8 = 17\cdot3 \quad \therefore \ \text{Mode} = 17\cdot3 \text{ kg}$$

This result agrees with the graphical method we used before.

Median

The third measure of central tendency is the *median*, which is the value of the middle term when all the observations are arranged in ascending or descending order.

For example, with 4, 7, 8, 9, 12, 15, 26, the median = 9.

$$\uparrow$$
$$\text{median}$$

Where there is an even number of values, the median is then the average of the two middle terms, e.g. 5, 6, 10, $\underbrace{12, 14,}$ 17, 23, 30 has a median of

$$\frac{12 + 14}{2} = 13$$

Similarly, for the set of values 13, 4, 18, 23, 9, 16, 18, 10, 20, 6, the median is ....

$$\boxed{14 \cdot 5}$$

Because arranging the terms in order, we have:

$$4, 6, 9, 10, 13, 16, 18, 18, 20, 23 \text{ and median} = \frac{13 + 16}{2} = 14 \cdot 5$$

Now we must see how we can get the median
with grouped data, so move on to Frame 18

Median with grouped data

18

Since the median is the value of the middle term, it divides the frequency histogram into two equal areas. This fact gives us a method for determining the median.

For example, the temperature of a component was monitored at regular intervals on 80 occasions. The frequency distribution was as follows:

Temperature x (°C)	30·0–30·2	30·3–30·5	30·6–30·8	30·9–31·1
Frequency f	6	12	15	20

Temperature x (°C)	31·2–31·4	31·5–31·7	31·8–32·0
Frequency f	13	9	5

First we draw the frequency histogram.

19

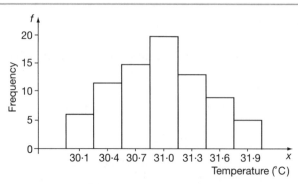

The median is the average of the 40th and 41st terms and, if we count up the frequencies of the rectangles, these terms are included in the .... class.

20

| fourth |

If we insert a dashed line to represent the value of the median, it will divide the area of the histogram into two equal parts.

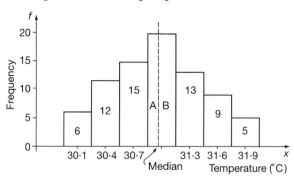

$6 + 12 + 15 + A = B + 13 + 9 + 5$

$\therefore\ 33 + A = B + 37$ But $A + B = 20$ $\therefore\ B = 20 - A$

$\therefore\ 33 + A = 20 - A + 27$ $\therefore\ 2A = 14$ $\therefore\ A = 7$

$\therefore$ Width of $A = \dfrac{7}{20} \times$ class interval

$\qquad\qquad = 0{\cdot}35 \times 0{\cdot}3 = 0{\cdot}105$ $\therefore$ Median $= 30{\cdot}85 + 0{\cdot}105$

$\qquad\qquad = 30{\cdot}96°C$

Now, by way of revision, here is a short exercise.

Exercise

Determine (a) the mean, (b) the mode and (c) the median of the following:

x	5–9	10–14	15–19	20–24	25–29	30–34
f	4	9	16	12	6	3

21

> (a) mean = 18·6   (b) mode = 17·7   (c) median = 18·25

At this point let's just pause and summarize the work so far on measures of central tendency

 # Review summary   Unit 47

22

1 *Mean* (arithmetic mean)

$$\bar{x} = \frac{\sum xf}{n} = \frac{\sum xf}{\sum f}$$

2 *Mode* – the value of the variable that occurs most often. For grouped distribution:

$$\text{mode} = L + \left(\frac{l}{l+u}\right)c$$

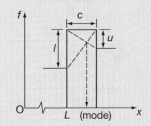

3 *Median* – the value of the middle term when all values are put in ascending or descending order. With an even number of terms, the median is the average of the two middle terms.

 # Review exercise   Unit 47

23

1 The lengths of 60 samples of blue steel cables were recorded in the following frequency table:

Length *l* (m)	24·8	24·9	25·0	25·1	25·2	25·3
Frequency *f*	4	10	16	19	8	3

Find the mode, median and mean of this data.

2 The weights of 40 samples of lapped steel collars were recorded in the following frequency table:

Weight w (kg)	3·82–3·86	3·87–3·91	3·92–3·96	3·97–4·01	4·02–4·06	4·07–4·11
Frequency f	2	5	10	12	8	3

Find the mode, median and mean of this data.

Complete both questions. Take one step at a time, there is no need to rush.
If you need to, look back at the Unit.
The answers and working are in the next frame.

---

24 **1** The lengths of 60 samples of blue steel cables are as recorded in the following frequency table:

Length l (m)	24·8	24·9	25·0	25·1	25·2	25·3
Frequency f	4	10	16	19	8	3

Mode

The mode is the value of the length which occurs most often, that is 25·1 m.

Median

The median is the length that splits the population into two equal halves. If the data were arranged in order of increasing length then (because there is an even number of items) the middle two are 25·0 and 25·1. The median is then the average of these two lengths which is 25·05 m.

Mean

To find the mean we extend the data table:

Length l (m)	24·8	24·9	25·0	25·1	25·2	25·3
Frequency f	4	10	16	19	8	3
$l \times f$	99·2	249·0	400·0	476·9	201·6	75·9

From this table we can now calculate the mean length which is given by:

$$\bar{l} = \frac{\sum l \times f}{\sum f} = \frac{1502 \cdot 6}{60} = 25 \cdot 04 \, \text{m}$$

▶

2 The weights of 40 samples of lapped steel collars are as recorded in the following frequency table:

Weight w (kg)	3·82– 3·86	3·87– 3·91	3·92– 3·96	3·97– 4·01	4·02– 4·06	4·07– 4·11
Frequency f	2	5	10	12	8	3

Mode

The group which occurs most often is 3·97–4·01 where:

L = lower boundary value = 3·965
l = difference in the frequency on the lower boundary = $12 - 10 = 2$
u = difference in the frequency on the upper boundary = $12 - 8 = 4$
c = the class interval = 0·05 (for example $3·865 - 3·815 = 0·05$)

Therefore:

$$\text{mode} = L + \left(\frac{l}{l+u}\right)c$$
$$= 3·965 + \left(\frac{2}{2+4}\right)0·05$$
$$= 3·976 \, \text{kg}$$

Median

Constructing the histogram:

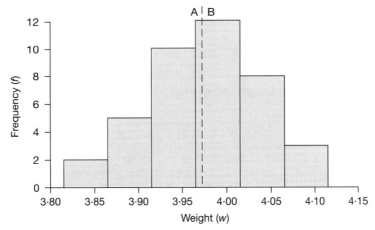

The dashed line divides the area of the histogram into two equal parts and so represents the median. Then:

$2 + 5 + 10 + A = B + 8 + 3$ so that
$A - B = -6$ but $A + B = 12$ so that $A = 3$ and $B = 9$

Therefore, width of $A = (3/20) \times$ class interval $= (3/20) \times 0·05 = 0·0075$
Giving the median as $3·965 + 0.0075 = 3·9725$ kg.

Mean

To calculate the mean we construct the following table:

Weight w	Central value w_m	Deviation from 3·94 $w_m - 3·94$	In units of w_c 0·05	Frequency f	Product $w_c f$
3·82–3·86	3·84	−0·10	−2	2	−4
3·87–3·91	3·89	−0·05	−1	5	−5
3·92–3·96	3·94	0·00	0	10	0
3·97–4·01	3·99	0·05	1	12	12
4·02–4·06	4·04	0·10	2	8	16
4·07–4·11	4·09	0·15	3	3	9

$$n = \sum f = 40$$
$$\downarrow$$
$$\sum w_c f = 28$$

Therefore:

$$\bar{w}_c = \frac{\sum w_c f}{n} = \frac{28}{40} = 0·7 \text{ so that}$$

$$\bar{w}_m - 3·94 = \bar{w}_c \times 0·05 = 0·7 \times 0·05 = 0·035 \text{ therefore}$$

$$\bar{w}_m = 3·94 + 0·035 = 3·975 \text{ kg}$$

Now for the Review test

Review test   Unit 47

25

1 The weights of a sample of 35 small loaves were recorded in the following frequency table:

Weight w (gm)	120	121	122	123	124	125
Frequency f	1	6	10	9	6	3

Find the mode, median and mean of the loaves.

2 The lengths of 22 samples of rolled steel joists were recorded in the following frequency table:

Length l (m)	17·1–17·2	17·2–17·3	17·3–17·4	17·4–17·5
Frequency f	1	7	12	2

Find the mode, median and mean of the lengths.

Dispersion

Unit 48

1

The mean, mode and median give important information regarding the general mass of the observations recorded. They do not, however, tell us anything about the dispersion of the values around the central value.

The set   26, 27, 28, 29, 30 has a mean of 28
and   5, 19, 20, 36, 60 also has a mean of 28

These two sets have the same mean, but clearly the first is more tightly arranged around the mean than is the second. We therefore need a measure to indicate the spread of the values about the mean.

Range

The simplest value to indicate the spread or dispersion is the *range* which is merely the difference between the highest and lowest values in the set of observations. In the two cases quoted above, the range of set 1 is $30 - 26 = 4$, while that of set 2 is $60 - 5 = 55$. The disadvantage of the range is that it deals only with the extreme values: it does not take into account the behaviour of the intermediate values.

Standard deviation

The *standard deviation from the mean* is used widely in statistics to indicate the degree of dispersion. It takes into account the deviation of every value from the mean and it is found as follows:

(a) The mean $\bar{x}$ of the set of n values is first calculated.

(b) The deviation of each of these n values, $x_1, x_2, x_3, \ldots, x_n$ from the mean is calculated and the results squared, i.e. $(x_1 - \bar{x})^2$; $(x_2 - \bar{x})^2$; $(x_3 - \bar{x})^2$; $\ldots$; $(x_n - \bar{x})^2$.

(c) The average of these results is then found and the result is called the *variance* of the set of observations:

i.e. variance $= \dfrac{(x_1 - \bar{x})^2 + (x_2 - \bar{x})^2 + \ldots + (x_n - \bar{x})^2}{n}$

(d) The square root of the variance gives the standard deviation, denoted by the Greek letter 'sigma':

standard deviation $= \sigma = \sqrt{\dfrac{\sum(x - \bar{x})^2}{n}}$

So, for the set of values 3, 6, 7, 8, 11, 13, the mean $\bar{x} = \ldots\ldots\ldots\ldots$, the variance $= \ldots\ldots\ldots\ldots$ and the standard deviation $= \ldots\ldots\ldots\ldots$

2

$$\bar{x} = 8 \cdot 0; \text{ variance} = 10 \cdot 67; \sigma = 3 \cdot 27$$

Because

x	f
3	1
6	1
7	1
8	1
11	1
13	1

$$\sum x = 48$$
$$n = 6$$
$$\therefore \bar{x} = 8$$

$x - 8$	$(x - 8)^2$
-5	25
-2	4
-1	1
0	0
3	9
5	25

$$\therefore \sum (x - \bar{x})^2 = 64$$

$$\text{variance} = \frac{\sum (x - \bar{x})^2}{n} = \frac{64}{6} = 10.67$$

$$\therefore \text{ standard deviation} = \sigma = \sqrt{10 \cdot 67} = 3 \cdot 266 \quad \therefore \sigma = 3 \cdot 27$$

Alternative formula for the standard deviation

The formula $\sigma = \sqrt{\dfrac{\sum (x - \bar{x})^2}{n}}$ can also be written in another and more convenient form, $\sigma = \sqrt{\dfrac{\sum x^2}{n} - (\bar{x})^2}$ which requires only the mean and the squares of the values of x. For grouped data, this then becomes $\sigma = \sqrt{\dfrac{\sum x^2 f}{n} - (\bar{x})^2}$ and the working is even more simplified if we use the coding procedure as we did earlier in the Module.

Move on to the next frame for a typical example

3

Here is an example. The lengths of 70 bars were measured and the following frequency distribution obtained:

Length x (mm)	21·2–21·4	21·5–21·7	21·8–22·0	22·1–22·3
Frequency f	3	5	10	16

Length x (mm)	22·4–22·6	22·7–22·9	23·0–23·2
Frequency f	18	12	6

First we prepare a table with the following headings:

Length (mm) (x)	Central value (x_m)	Deviation from chosen value ($x_m - 22.2$)	Units of 0·3 $\left(x_c = \dfrac{x_m - 22.2}{0.3}\right)$	Freq. f	$x_c f$		

Leave room on the right-hand side for two more columns yet to come. Now you can complete the six columns shown and from the values entered, determine the coded value of the mean and also the actual mean.

Do that and then move on

So far, the work looks like this:

4

Length (mm) (x)	Central value (x_m)	Deviation from chosen value ($x_m - 22.2$)	Units of 0·3 $\left(x_c = \dfrac{x_m - 22.2}{0.3}\right)$	Freq. f	$x_c f$		
21·2–21·4	21·3	−0·9	−3	3	−9		
21·5–21·7	21·6	−0·6	−2	5	−10		
21·8–22·0	21·9	−0·3	−1	10	−10		
22·1–22·3	22·2	0	0	16	0		
22·4–22·6	22·5	0·3	1	18	18		
22·7–22·9	22·8	0·6	2	12	24		
23·0–23·2	23·1	0·9	3	6	18		

$$n = \sum f = 70 \qquad \downarrow$$
$$\sum x_c f = 31$$

$$\text{Coded mean} = \bar{x}_c = \frac{\sum x_c f}{n} = \frac{31}{70} = 0.4429 \text{ (in units of 0·3)}$$

$$\therefore \ \bar{x}_m - 22.2 = 0.4429 \times 0.3 = 0.1329$$

$$\bar{x}_m = 22.2 + 0.1329 = 22.333 \qquad \therefore \ \bar{x} = 22.33$$

Now we can complete the remaining two columns. Head these x_c^2 and $x_c^2 f$.

Fill in the appropriate values and using $\sigma_c = \sqrt{\dfrac{\sum x_c^2 f}{n} - (\bar{x}_c)^2}$ we can find the coded value of the standard deviation (σ_c).

Complete the table then and determine the coded standard deviation on your own

5

Finally the table now becomes:

Length (mm) (x)	Central value (x_m)	Deviation from chosen value ($x_m - 22 \cdot 2$)	Units of 0·3 $\left(x_c = \dfrac{x_m - 22 \cdot 2}{0 \cdot 3}\right)$	Freq. f	$x_c f$	x_c^2	$x_c^2 f$
21·2–21·4	21·3	−0·9	−3	3	−9	9	27
21·5–21·7	21·6	−0·6	−2	5	−10	4	20
21·8–22·0	21·9	−0·3	−1	10	−10	1	10
22·1–22·3	22·2	0	0	16	0	0	0
22·4–22·6	22·5	0·3	1	18	18	1	18
22·7–22·9	22·8	0·6	2	12	24	4	48
23·0–23·2	23·1	0·9	3	6	18	9	54
				70	31		177

Now $\sigma_c = \sqrt{\dfrac{\sum x_c^2 f}{n} - (\bar{x}_c)^2}$ and from the table, $n = 70$ and $\sum x_c^2 f = 177$.

Also, from the previous work, $\bar{x}_c = 0 \cdot 443$. $\therefore$ $\sigma_c = \ldots\ldots\ldots\ldots$

6

$$\therefore \; \sigma_c = 1 \cdot 527 \; \text{(in units of 0·3)}$$

$\therefore \; \sigma = 1 \cdot 527 \times 0 \cdot 3 = 0 \cdot 4581$ $\therefore \; \sigma = 0 \cdot 458$

Note: In calculating the standard deviation, we do not restore the 'false mean' subtracted from the original values, since the standard deviation is relative to the mean and not to the zero or origin of the set of observations.

Let's just pause and summarize the work so far on dispersion

Review summary

Unit 48

7

1 *Standard deviation*

$$\sigma = \sqrt{\dfrac{\sum x^2 f}{n} - (\bar{x})^2}$$

$n = $ number of observations
$\bar{x} = $ mean
$\sigma = $ standard deviation from the mean.

With coding, $\bar{x}_c = $ coded mean

$$\sigma_c = \sqrt{\dfrac{\sum x_c^2 f}{n} - (\bar{x}_c)^2}$$

Review exercise

1 Find the standard deviation for the following distribution of numbers:

8

Number n	2	4	6	8	10	12	14	16	18
Frequency f	1	5	10	15	14	12	8	4	1

2 The weights of 40 samples of lapped steel collars were recorded in the following frequency table:

Weight w (kg)	3·82– 3·86	3·87– 3·91	3·92– 3·96	3·97– 4·01	4·02– 4·06	4·07– 4·11
Frequency f	2	5	10	12	8	3

Find the standard deviation of the weights.

> *Complete both questions. Take one step at a time, there is no need to rush.*
> *If you need to, look back at the Unit.*
> *The answers and working are in the next frame.*

1 To find the standard deviation for the distribution of numbers:

9

Number n	2	4	6	8	10	12	14	16	18
Frequency f	1	5	10	15	14	12	8	4	1

we construct the following table:

Number n	Deviation from $n = 10$	In units of 2 n_c	Frequency f	Product $n_c f$	n_c^2	$n_c^2 f$
2	−8	−4	1	−4	16	16
4	−6	−3	5	−15	9	45
6	−4	−2	10	−20	4	40
8	−2	−1	15	−15	1	15
10	0	0	14	0	0	0
12	2	1	12	12	1	12
14	4	2	8	16	4	32
16	6	3	4	12	9	36
18	8	4	1	4	16	16

$$\sum f = 70 \qquad \qquad \downarrow$$

$$\sum n_c f = -10 \qquad \sum n_c^2 f = 212$$

$$\bar{n}_c = \frac{\sum n_c f}{\sum f} = -0.1429$$

Therefore: $\sigma_c = \sqrt{\dfrac{\sum n_c^2 f}{\sum f} - (\bar{n}_c)^2} = 1.7344$ to 4 dp

and so: $\sigma = 1.7344 \times 2 = 3.47$ to 2 dp

2 The weights of 40 samples of lapped steel collars are as recorded in the following frequency table:

Weight w (kg)	3·82– 3·86	3·87– 3·91	3·92– 3·96	3·97– 4·01	4·02– 4·06	4·07– 4·11
Frequency f	2	5	10	12	8	3

To calculate the standard deviation we construct the following table:

Weight w	Central value w_m	Deviation from 3·94 $w_m - 3.94$	In units of w_c 0·05	Frequency f	Product $w_c f$	w_c^2	$w_c^2 f$
3·82–3·86	3·84	−0·10	−2	2	−4	4	8
3·87–3·91	3·89	−0·05	−1	5	−5	1	5
3·92–3·96	3·94	0·00	0	10	0	0	0
3·97–4·01	3·99	0·05	1	12	12	1	12
4·02–4·06	4·04	0·10	2	8	16	4	32
4·07–4·11	4·09	0·15	3	3	9	9	27

$$n = \sum f = 40 \qquad\qquad \downarrow$$
$$\sum w_c^2 f = 84$$

Therefore: $\sigma_c = \sqrt{\sum \dfrac{w_c^2 f}{n} - (\bar{w}_c)^2} = \sqrt{\dfrac{84}{40} - (0.7)^2} = 1.27$ and so $\sigma = 0.064$

Now for the Review test

 # Review test   **Unit 48**

10   **1** Find the standard deviation for the following distribution of weights:

Weight w (kg)	1	3	5	7	9	11	13	15	17
Frequency f	1	2	7	25	44	22	5	3	1

2 The lengths of a sample of 47 units were recorded in the following frequency table:

Length l (m)	22·1–22·2	22·2–22·3	22·3–22·4	22·4–22·5	22·5–22·6	22·6–22·7
Frequency f	1	7	18	14	5	2

Find the standard deviation of the lengths.

The normal distribution   Unit 49

Frequency polygons   1

If the centre points of the tops of the rectangular blocks of a frequency histogram are joined, the resulting figure is a *frequency polygon*. If the polygon is extended to include the mid-points of the zero frequency classes at each end of the histogram, then the area of the complete polygon is equal to the area of the histogram and therefore represents the total frequency of the variable.

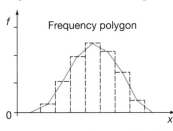 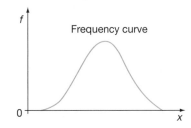

Frequency curves   2

If the frequency polygon is 'smoothed out', or if we plot the frequency against the central value of each class and draw a smooth curve, the result is a frequency curve.

Normal distribution curve   3

When very large numbers of observations are made and the range is divided into a very large number of 'narrow' classes, the resulting frequency curve, in many cases, approximates closely to a standard curve known as the *normal distribution curve*, which has a characteristic bell-shaped formation.

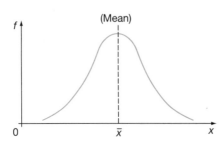

The normal distribution curve (or normal curve) is symmetrical about its centre line which coincides with the mean $\bar{x}$ of the observations.

There is, in fact, a close connection between the standard deviation (sd) from the mean of a set of values and the normal curve.

Values within 1 sd of the mean

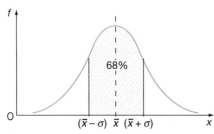

The normal curve has a complicated equation, but it can be shown that the shaded area is 68 per cent of the total area under the normal curve, i.e. 68 per cent of the observations occur within the range $(\bar{x} - \sigma)$ to $(\bar{x} + \sigma)$.

For example, on a manufacturing run to produce 1000 bolts of nominal length 32·5 mm, sampling gave a mean of 32·58 and a standard deviation of 0·06 mm. From this information, $\bar{x} = 32.58$ mm and $\sigma = 0.06$ mm.

$$\left. \begin{aligned} \bar{x} - \sigma = 32.58 - 0.06 = 32.52 \\ \bar{x} + \sigma = 32.58 + 0.06 = 32.64 \end{aligned} \right\}$$ $\therefore$ 68 per cent of the bolts, i.e. 680, are likely to have lengths between 32·52 mm and 32·64 mm.

Values within 2 sd of the mean

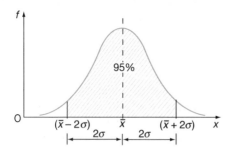

Between $(\bar{x} - 2\sigma)$ and $(\bar{x} + 2\sigma)$ the shaded area accounts for 95 per cent of the area of the whole figure, i.e. 95 per cent of the observations occur between these two values.

Values within 3 sd of the mean

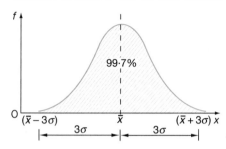

Between $(\bar{x} - 3\sigma)$ and $(\bar{x} + 3\sigma)$ the shaded area is 99·7 per cent of the total area under the normal curve and therefore 99·7 per cent of the observations occur within this range, i.e. almost all the values occur within 3σ or 3 sd of the mean.

Therefore, in our previous example where $\bar{x} = 32{\cdot}58$ and $\sigma = 0{\cdot}06$ mm:

68 per cent of the bolts are likely to have lengths between 32·52 and 32·64 mm,

95 per cent of the bolts are likely to have lengths between $32{\cdot}5 - 0{\cdot}12$ and $32{\cdot}5 + 0{\cdot}12$, i.e. between 32·38 and 32·62 mm,

99·7 per cent, i.e. almost all, are likely to have lengths between $32{\cdot}5 - 0{\cdot}18$ and $32{\cdot}5 + 0{\cdot}18$, i.e. between 32·32 and 32·68 mm.

We can enter the same information in a slightly different manner, dividing the figure into columns of 1σ width on each side of the mean.

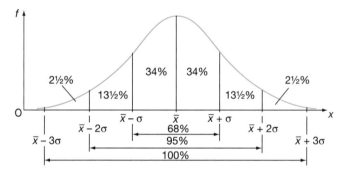

Exercise

Measurement of the diameters of 1600 plugs gave a mean of 74·82 mm and a standard deviation of 0·14 mm. Calculate:

(a) the number of plugs likely to have diameters less than 74·54 mm

(b) the number of plugs likely to have diameters between 74·68 mm and 75·10 mm.

(a) 40	(b) 1304

4

Because

(a) $74{\cdot}54 = 74{\cdot}82 - 2(0{\cdot}14) = \bar{x} - 2\sigma$ ∴ number of plugs with diameters less than 74·54 mm $= 2\frac{1}{2}$ per cent $\times 1600 = 40$

(b) $74\cdot68 = 74\cdot82 - 0\cdot14 = \bar{x} - \sigma$

$\quad 75\cdot10 = 74\cdot82 + 2(0\cdot14) = \bar{x} + 2\sigma$

$\quad \therefore$ Numbers of plugs between these values $= (34 + 34 + 13\frac{1}{2})$ per cent

$$= 81\frac{1}{2} \text{ per cent}$$

$$81\frac{1}{2} \text{ per cent of } 1600 = 1304$$

---

5 **Standard normal curve**

The conversion from normal distribution to standard normal distribution is achieved by the substitution $z = \dfrac{x - \mu}{\sigma}$ which effectively moves the distribution curve along the x-axis and reduces the scale of the horizontal units by dividing by σ. To keep the total area under the curve at unity, we multiply the y-values by σ. The equation of the standardized normal curve then becomes

$$y = \phi(z) = \frac{1}{\sqrt{2\pi}}e^{-z^2/2}$$

$z = \dfrac{x - \mu}{\sigma}$ is called the *standard normal variable*,

$\phi(z)$ is the *probability density function*.

Make a note of these

---

6   Standard normal curve:

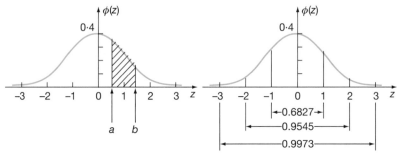

Note the following:

(a) Mean $\mu = 0$.

(b) z-values are in standard deviation units.

(c) Total area under the curve from $z = -\infty$ to $z = +\infty = 1$.

(d) Area between $z = a$ and $z = b$ represents the probability that z lies between the values $z = a$ and $z = b$, i.e. $P(a \leq z \leq b) = $ area shaded.

(e) The probability of a value of z being

between $z = -1$ and $z = 1$ is $68\cdot27$ per cent $= 0\cdot6827$

between $z = -2$ and $z = 2$ is $95\cdot45$ per cent $= 0\cdot9545$

between $z = -3$ and $z = 3$ is $99\cdot73$ per cent $= 0\cdot9973$

In each case the probability is given by . . . .

the area under the curve between the stated limits.

Similarly, the probability of a randomly selected value of z lying between $z = 0\cdot5$ and $z = 1\cdot5$ is given by the area shaded.

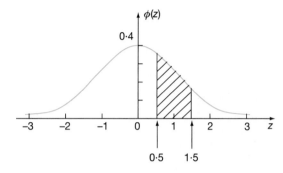

That is $P(0\cdot5 \leq z \leq 1\cdot5) = \displaystyle\int_{0\cdot5}^{1\cdot5} \frac{1}{\sqrt{2\pi}} e^{-z^2/2} \mathrm{d}z$

This integral cannot be evaluated by ordinary means, so we use a table giving the area under the standard normal curve from $z = 0$ to $z = z_1$.

Move to Frame 8

8

Area under the standard normal curve

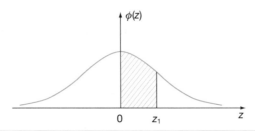

z_1	0·00	0·01	0·02	0·03	0·04	0·05	0·06	0·07	0·08	0·09
0·0	0·0000	0·0040	0·0080	0·0120	0·0160	0·0199	0·0239	0·0279	0·0319	0·0359
0·1	0·0398	0·0438	0·0478	0·0517	0·0557	0·0596	0·0636	0·0675	0·0714	0·0753
0·2	0·0793	0·0832	0·0871	0·0910	0·0948	0·0987	0·1026	0·1064	0·1103	0·1141
0·3	0·1179	0·1217	0·1255	0·1293	0·1331	0·1368	0·1406	0·1443	0·1480	0·1517
0·4	0·1554	0·1591	0·1628	0·1664	0·1700	0·1736	0·1772	0·1808	0·1844	0·1879
0·5	0·1915	0·1950	0·1985	0·2019	0·2054	0·2088	0·2123	0·2157	0·2190	0·2224
0·6	0·2257	0·2291	0·2324	0·2357	0·2389	0·2422	0·2454	0·2486	0·2517	0·2549
0·7	0·2580	0·2611	0·2642	0·2673	0·2704	0·2734	0·2764	0·2794	0·2823	0·2852
0·8	0·2881	0·2910	0·2939	0·2967	0·2995	0·3023	0·3051	0·3078	0·3106	0·3133
0·9	0·3159	0·3186	0·3212	0·3238	0·3264	0·3289	0·3315	0·3340	0·3365	0·3389
1·0	0·3413	0·3438	0·3461	0·3485	0·3508	0·3531	0·3554	0·3577	0·3599	0·3621
1·1	0·3643	0·3665	0·3686	0·3708	0·3729	0·3749	0·3770	0·3790	0·3810	0·3830
1·2	0·3849	0·3869	0·3888	0·3907	0·3925	0·3944	0·3962	0·3980	0·3997	0·4015
1·3	0·4032	0·4049	0·4066	0·4082	0·4099	0·4115	0·4131	0·4147	0·4162	0·4177
1·4	0·4192	0·4207	0·4222	0·4236	0·4251	0·4265	0·4279	0·4292	0·4306	0·4319
1·5	0·4332	0·4345	0·4357	0·4370	0·4382	0·4394	0·4406	0·4418	0·4429	0·4441
1·6	0·4452	0·4463	0·4474	0·4484	0·4495	0·4505	0·4515	0·4525	0·4535	0·4545
1·7	0·4554	0·4564	0·4573	0·4582	0·4591	0·4599	0·4608	0·4616	0·4625	0·4633
1·8	0·4641	0·4649	0·4656	0·4664	0·4671	0·4678	0·4686	0·4693	0·4699	0·4706
1·9	0·4713	0·4719	0·4726	0·4732	0·4738	0·4744	0·4750	0·4756	0·4761	0·4767
2·0	0·4773	0·4778	0·4783	0·4788	0·4793	0·4798	0·4803	0·4808	0·4812	0·4817
2·1	0·4821	0·4826	0·4830	0·4834	0·4838	0·4842	0·4846	0·4850	0·4854	0·4857
2·2	0·4861	0·4864	0·4868	0·4871	0·4875	0·4878	0·4881	0·4884	0·4887	0·4890
2·3	0·4893	0·4896	0·4898	0·4901	0·4904	0·4906	0·4909	0·4911	0·4913	0·4916
2·4	0·4918	0·4920	0·4922	0·4925	0·4927	0·4929	0·4931	0·4932	0·4934	0·4936
2·5	0·4938	0·4940	0·4941	0·4943	0·4945	0·4946	0·4948	0·4949	0·4951	0·4952
2·6	0·4953	0·4955	0·4956	0·4957	0·4959	0·4960	0·4961	0·4962	0·4963	0·4964
2·7	0·4965	0·4966	0·4967	0·4968	0·4969	0·4970	0·4971	0·4972	0·4973	0·4974
2·8	0·4974	0·4975	0·4976	0·4977	0·4977	0·4978	0·4979	0·4979	0·4980	0·4981
2·9	0·4981	0·4982	0·4983	0·4983	0·4984	0·4984	0·4985	0·4985	0·4986	0·4986
3·0	0·4987	0·4987	0·4987	0·4988	0·4988	0·4989	0·4989	0·4989	0·4989	0·4990
3·1	0·4990	0·4991	0·4991	0·4991	0·4992	0·4992	0·4992	0·4992	0·4993	0·4993
3·2	0·4993	0·4993	0·4994	0·4994	0·4994	0·4994	0·4994	0·4995	0·4995	0·4995

Now move on to Frame 9

9

We need to find the area between $z = 0.5$ and $z = 1.5$:

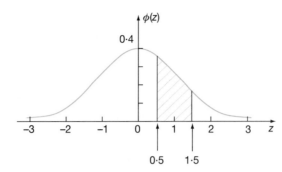

From the table:

area from $z = 0$ to $z = 1.5 = 0.4332$

area from $z = 0$ to $z = 0.5 = 0.1915$

∴   area from $z = 0.5$ to $z = 1.5 = 0.2417$

∴ $P(0.5 \leq z \leq 1.5) = 0.2417 = 24.17$ per cent

Although the table gives areas for only positive values of z, the symmetry of the curve enables us to deal equally well with negative values. Now for some examples.

Example 1

Determine the probability that a random value of z lies between $z = -1.4$ and $z = 0.7$.

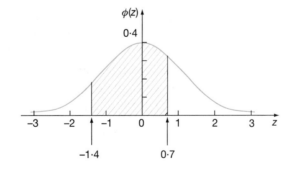

area from $z = -1.4$ to $z = 0 =$ area from $z = 0$ to $z = 1.4$

$= 0.4192$ (from the table)

area from $z = 0$ to $z = 0.7 = 0.2580$ (from the table)

∴   area from $z = -1.4$ to $z = 0.7 = \ldots\ldots\ldots\ldots$

10

$$0{\cdot}4192 + 0{\cdot}2580 = 0{\cdot}6772$$

$\therefore$ $P(-1{\cdot}4 \leq z \leq 0{\cdot}7) = 0{\cdot}6772 = 67{\cdot}72$ per cent

Example 2

Determine the probability that a value of z is greater than $2{\cdot}5$.
Draw a diagram: then it is easy enough.

Finish it off

11

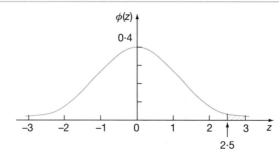

The total area from $z = 0$ to $z = \infty$ is $0{\cdot}5000$.
The total area from $z = 0$ to $z = 2{\cdot}5 = 0{\cdot}4938$ (from the table).
$\therefore$ The area for $z \geq 2{\cdot}5 = 0{\cdot}5000 - 0{\cdot}4938 = 0{\cdot}0062$
$\therefore$ $P(z \geq 2{\cdot}5) = 0{\cdot}0062 = 0{\cdot}62$ per cent

Example 3

The mean diameter of a sample of 400 rollers is $22{\cdot}50$ mm and the standard
deviation $0{\cdot}50$ mm. Rollers are acceptable with diameters $22{\cdot}36 \pm 0{\cdot}53$ mm.
Determine the probability of any one roller being within the acceptable limits.

We have $\mu = 22{\cdot}50$ mm; $\sigma = 0{\cdot}50$ mm

Limits of $x_1 = 22{\cdot}36 - 0{\cdot}53 = 21{\cdot}83$ mm

$x_2 = 22{\cdot}36 + 0{\cdot}53 = 22{\cdot}89$ mm

Using $z = \dfrac{x - \mu}{\sigma}$ we convert x_1 and x_2 into z_1 and z_2

$z_1 = \ldots\ldots\ldots\ldots;$ $z_2 = \ldots\ldots\ldots\ldots$

12

$$z_1 = -1{\cdot}34;\quad z_2 = 0{\cdot}78$$

Now, using the table, we can find the area under the normal curve between
$z = -1{\cdot}34$ and $z = 0{\cdot}78$ which gives us the required result.

$P(21{\cdot}83 \leq x \leq 22{\cdot}89) = P(-1{\cdot}34 \leq z \leq 0{\cdot}78) = \ldots\ldots\ldots\ldots$

$$\boxed{0.6922}$$

So that is it. Convert the given values into z-values and apply the table as required. They are all done in much the same way.

Finally one more entirely on your own.

14

Example 4

A thermostat set to switch at 20°C operates at a range of temperatures having a mean of 20·4°C and a standard deviation of 1·3°C. Determine the probability of its opening at temperatures between 19·5°C and 20·5°C.
  Complete it and then check your working with the next frame.

$$\boxed{P(19\cdot5 \leq x \leq 20\cdot5) = 0\cdot2868}$$

Here it is:

$$x_1 = 19\cdot5 \quad \therefore \quad z_1 = \frac{19\cdot5 - 20\cdot4}{1\cdot3} = -\frac{0\cdot9}{1\cdot3} = -0\cdot692$$

$$x_2 = 20\cdot5 \quad \therefore \quad z_2 = \frac{20\cdot5 - 20\cdot4}{1\cdot3} = \frac{0\cdot1}{1\cdot3} = 0\cdot077$$

$$\text{Area } (z = -0\cdot69 \text{ to } z = 0) = \text{ area } (z = 0 \text{ to } z = 0\cdot69)$$
$$= 0\cdot2549$$
$$\text{Area } (z = 0 \text{ to } z = 0\cdot08) = 0\cdot0319$$
$$\therefore \quad \text{Area } (z = -0\cdot69 \text{ to } z = 0\cdot08) = 0\cdot2868$$
$$\therefore \quad P(19\cdot5 \leq x \leq 20\cdot5) = P(-0\cdot692 \leq z \leq 0\cdot077) = 0\cdot2868$$

For now, let's pause and summarize the work on normal distributions

Review summary

16   **1**   *Normal distribution curve* – large numbers of observations.

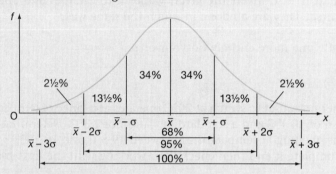

Symmetrical about the mean.
68 per cent of observations lie within ± 1 sd of the mean.
95 per cent of observations lie within ± 2 sd of the mean.
99·7 per cent of observations lie within ± 3 sd of the mean.

2   *Standardized normal curve* – the axis of symmetry of the normal curve becomes the vertical axis with a scale of relative frequency. The horizontal axis carries a scale of z-values indicated as multiples of the standard deviation. The curve therefore represents a distribution with zero mean and unit standard deviation.

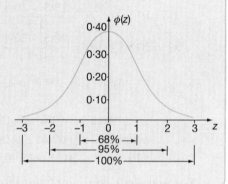

 Review exercise

17   **1**   A normally distributed variable x has a mean $\bar{x} = 34$ and a standard deviation $\sigma = 2\cdot5$. Find the z-values of the standard normal distribution that correspond to:

(a) $x = 37$

(b) $x = 30$

(c) $29 \leq x \leq 36$

Complete the question. Take your time, there is no need to rush.
Refer back to the Unit if necessary.
The answers and working are in the next frame.

18

1 (a) $x = 37$ and so $z_{37} = \dfrac{37 - 34}{2 \cdot 5} = \dfrac{3}{2 \cdot 5} = 1 \cdot 2$

   (b) $x = 30$ and so $z_{30} = \dfrac{30 - 34}{2 \cdot 5} = \dfrac{-4}{2 \cdot 5} = -1 \cdot 6$

   (c) Since $29 \leq x \leq 36$ then:

$$z_{29} = \frac{29 - 34}{2 \cdot 5} = \frac{-5}{2 \cdot 5} = -2 \cdot 0 \text{ and } z_{36} = \frac{36 - 34}{2 \cdot 5} = \frac{2}{2 \cdot 5} = 0 \cdot 8$$

The area beneath the standard normal curve between $z = -2 \cdot 0$ and $z = 0$ is the same as the area between $z = 0$ and $z = 2$ by the symmetry of the curve and from the table this is $0 \cdot 4773$.

The area between $z = 0$ and $z = 0 \cdot 8$ is, from the table, $0 \cdot 2881$.

Therefore the area beneath the standard normal curve between:

$$z = -2 \cdot 0 \text{ and } z = 0 \cdot 8 \text{ is } 0 \cdot 4773 + 0 \cdot 2881 = 0 \cdot 7654$$

That is, $76 \cdot 54\%$ of the area under the standard curve lies between $z = -2 \cdot 0$ and $z = 0 \cdot 8$ which is the same as the area beneath the normal curve for $29 \leq x \leq 36$.

Now for the Review test

Review test

Unit 49

1 A normally distributed variable x has a mean $\bar{x} = 165$ and a standard deviation $\sigma = 24 \cdot 8$. Find the z-values of the standard normal distribution that correspond to:

   (a) $x = 176$

   (b) $x = 158$

   (c) $140 \leq x \leq 180$

19

You have now come to the end of this Module. A list of **Can You?** questions follows for you to gauge your understanding of the material in the Module. You will notice that these questions match the **Learning outcomes** listed at the beginning of the Module. Now try the **Test exercise**. *Work through the questions at your own pace, there is no need to hurry.* A set of **Further problems** provides valuable additional practice.

20

Can You?

1   Checklist: Module 15

Check this list before and after you try the end of Module test.

On a scale of 1 to 5 how confident are you that you can:

• Distinguish between discrete and continuous data?
   Yes □ □ □ □ □ *No*

• Construct frequency and relative frequency tables for grouped and
   ungrouped discrete data?
   Yes □ □ □ □ □ *No*

• Determine class boundaries, class intervals and central values for
   discrete and continuous data?
   Yes □ □ □ □ □ *No*

• Construct a histogram and a frequency polygon?
   Yes □ □ □ □ □ *No*

• Determine the mean, median and mode of grouped and ungrouped
   data?
   Yes □ □ □ □ □ *No*

• Determine the range, variance and standard deviation of discrete data?
   Yes □ □ □ □ □ *No*

• Measure the dispersion of data using the normal and standard normal
   curves?
   Yes □ □ □ □ □ *No*

Test exercise 15

2   **1** The masses of 50 castings were measured. The results in kilograms were as
        follows:

        4·6   4·7   4·5   4·6   4·7   4·4   4·8   4·3   4·2   4·8
        4·7   4·5   4·7   4·4   4·5   4·5   4·6   4·4   4·6   4·6
        4·8   4·3   4·8   4·5   4·5   4·6   4·6   4·7   4·6   4·7
        4·4   4·6   4·5   4·4   4·3   4·7   4·7   4·6   4·6   4·8
        4·9   4·4   4·5   4·7   4·4   4·5   4·9   4·7   4·5   4·6

        (a) Arrange the data in 8 equal classes between 4·2 and 4·9 mm.

        (b) Determine the frequency distribution.

        (c) Draw the frequency histogram.

2 The diameters of 75 rollers gave the following frequency distribution:

Diameter x (mm)	8·82–8·86	8·87–8·91	8·92–8·96	8·97–9·01
Frequency f	1	8	16	18

Diameter x (mm)	9·02–9·06	9·07–9·11	9·12–9·16	9·17–9·21
Frequency f	15	10	5	2

(a) For each class, calculate (i) the central value, (ii) the relative frequency.
(b) Draw the relative frequency histogram.
(c) State (i) the lower boundary of the third class, (ii) the upper boundary of the sixth class, (iii) the class interval.

3 The thicknesses of 40 samples of steel plate were measured:

Thickness x (mm)	9·60–9·80	9·90–10·1	10·2–10·4	10·5–10·7
Frequency f	1	4	10	11

Thickness x (mm)	10·8–11·0	11·1–11·3	11·4–11·6
Frequency f	7	4	3

Using coding procedure, calculate:
(a) the mean
(b) the standard deviation
(c) the mode
(d) the median of the set of values given.

4 The lengths of 50 copper plugs gave the following frequency distribution:

Length x (mm)	14·0–14·2	14·3–14·5	14·6–14·8	14·9–15·1
Frequency f	2	4	9	15

Length x (mm)	15·2–15·4	15·5–15·7	15·8–16·00
Frequency f	11	6	3

(a) Calculate the mean and the standard deviation.
(b) For a full batch of 2400 plugs, calculate (i) the limits between which all the lengths are likely to occur, (ii) the number of plugs with lengths greater than 15·09 mm.

5 A machine delivers rods having a mean length of 18·0 mm and a standard deviation of 1·33 mm. If the lengths are normally distributed, determine the number of rods between 16·0 mm and 19·0 mm long likely to occur in a run of 300.

Further problems 15

3

1 The number of components processed in one hour on a new machine was recorded on 40 occasions:

66	87	79	74	84	72	81	78	68	74
80	71	91	62	77	86	87	72	80	77
76	83	75	71	83	67	94	64	82	78
77	67	76	82	78	88	66	79	74	64

     (a) Divide the set of values into seven equal width classes from 60 to 94.

     (b) Calculate (i) the frequency distribution, (ii) the mean, (iii) the standard deviation.

2 The lengths, in millimetres, of 40 bearings were determined with the following results:

16·6	15·3	16·3	14·2	16·7	17·3	18·2	15·6	14·9	17·2
18·7	16·4	19·0	15·8	18·4	15·1	17·0	18·9	18·3	15·9
13·6	18·3	17·2	18·0	15·8	19·3	16·8	17·7	16·8	17·9
17·3	16·6	15·3	16·4	17·3	16·9	14·7	16·2	17·4	15·6

     (a) Group the data into six equal width classes between 13·5 and 19·4 mm.

     (b) Obtain the frequency distribution.

     (c) Calculate (i) the mean, (ii) the standard deviation.

3 Masses of 80 brass junctions gave the following frequency distribution:

Mass x (kg)	4·12–4·16	4·17–4·21	4·22–4·26	4·27–4·31
Frequency f	5	12	16	20

Mass x (kg)	4·32–4·36	4·37–4·41	4·42–4·46
Frequency f	14	9	4

     (a) Calculate (i) the mean, (ii) the standard deviation.

     (b) For a batch of 1800 such components, calculate (i) the limits between which all the masses are likely to lie, (ii) the number of junctions with masses greater than 4·36 kg.

4 The values of the resistance of 90 carbon resistors were determined:

Resistance x (MΩ)	2·35	2·36	2·37	2·38	2·39	2·40	2·41
Frequency f	3	10	19	20	18	13	7

Calculate (a) the mean, (b) the standard deviation, (c) the mode and (d) the median of the set of values.

▶

5 Forty concrete cubes were subjected to failure tests in a crushing machine. Failure loads were as follows:

Load x (kN)	30·6	30·8	31·0	31·2	31·4	31·6
Frequency f	2	8	14	10	4	2

Calculate (a) the mean, (b) the standard deviation, (c) the mode and (d) the median of the set of results.

6 The time taken by employees to complete an operation was recorded on 80 occasions:

Time (min)	10·0	10·5	11·0	11·5	12·0	12·5	13·0
Frequency f	4	8	14	22	19	10	3

(a) Determine (i) the mean, (ii) the standard deviation, (iii) the mode and (iv) the median of the set of observations.

(b) State (i) the class interval, (ii) the lower boundary of the third class, (iii) the upper boundary of the seventh class.

7 Components are machined to a nominal diameter of 32·65 mm. A sample batch of 400 components gave a mean diameter of 32·66 mm with a standard deviation of 0·02 mm. For a production total of 2400 components, calculate:

(a) the limits between which all the diameters are likely to lie

(b) the number of acceptable components if those with diameters less than 32·62 mm or greater than 32·68 mm are rejected.

8 The masses of 80 castings were determined with the following results:

Mass x (kg)	7·3	7·4	7·5	7·6	7·7	7·8
Frequency f	4	13	21	23	14	5

(a) Calculate (i) the mean mass, (ii) the standard deviation from the mean.

(b) For a batch of 2000 such castings, determine (i) the likely limits of all the masses and (ii) the number of castings likely to have a mass greater than 7·43 kg.

9 The heights of 120 pivot blocks were measured:

Height x (mm)	29·4	29·5	29·6	29·7	29·8	29·9
Frequency f	6	25	34	32	18	5

(a) Calculate (i) the mean height and (ii) the standard deviation.

(b) For a batch of 2500 such blocks, calculate (i) the limits between which all the heights are likely to lie and (ii) the number of blocks with height greater than 29·52 mm.

10 A machine is set to produce bolts of nominal diameter 25·0 mm. Measurement of the diameters of 60 bolts gave the following frequency distribution:

Diameter x (mm)	23·3–23·7	23·8–24·2	24·3–24·7	24·8–25·2
Frequency f	2	4	10	17

Diameter x (mm)	25·3–25·7	25·8–26·2	26·3–26·7
Frequency f	16	8	3

(a) Calculate (i) the mean diameter and (ii) the standard deviation from the mean.

(b) For a full run of 3000, calculate (i) the limits between which all the diameters are likely to lie, (ii) the number of bolts with diameters less than 24·45 mm.

11 Samples of 10 A fuses have a mean fusing current of 9·9 A and a standard deviation of 1·2 A. Determine the probability of a fuse blowing with a current:

(a) less than 7·0 A

(b) between 8·0 A and 12·0 A.

12 Resistors of a certain type have a mean resistance of 420 ohms with a standard deviation of 12 ohms. Determine the percentage of resistors having resistance values:

(a) between 400 ohms and 430 ohms

(b) equal to 450 ohms.

13 Washers formed on a machine have a mean diameter of 12·60 mm with a standard deviation of 0·52 mm. Determine the number of washers in a random sample of 400 likely to have diameters between 12·00 mm and 13·50 mm.

14 The life of a drill bit has a mean of 16 hours and a standard deviation of 2·6 hours. Assuming a normal distribution, determine the probability of a sample bit lasting for:

(a) more than 20 hours

(b) fewer than 14 hours.

15 A type of bearing has an average life of 1500 hours and a standard deviation of 40 hours. Assuming a normal distribution, determine the number of bearings in a batch of 1200 likely:

(a) to fail before 1400 hours

(b) to last for more than 1550 hours.

16 Telephone calls from an office are monitored and found to have a mean duration of 452 s and a standard deviation of 123 s. Determine:

(a) the probability of the length of a call being between 300 s and 480 s

(b) the proportion of calls likely to last for more than 720 s.

17 Light bulbs, having a mean life of 2400 hours and standard deviation of 62 hours, are used for a consignment of 4000 bulbs. Determine:

(a) the number of bulbs likely to have a life in excess of 2500 hours

(b) the percentage of bulbs with a life length between 2300 hours and 2500 hours

(c) the probability of any one bulb having a life of 2500 hours (to the nearest hour).

Regression and correlation

Learning outcomes

When you have completed this Module you will be able to:

- Draw by eye a straight line of best fit to a graph of distinct data points
- Use least squares analysis to find the equation of the straight line of best fit to a graph of distinct data points
- Understand what is meant by regression
- Understand what is meant by the correlation of two variables
- Calculate the Pearson product-moment correlation coefficient
- Calculate Spearman's rank correlation coefficient

Units

Regression

Unit 50

Explicit linear variation of two variables

1

You are familiar with the equation of a straight line:

$$y = a + bx$$

where the value of the dependent variable y varies *linearly* with the value of the independent variable x. What is meant by this statement is that when pairs of corresponding x- and y-values are plotted on a graph the plotted points are found to lie on a straight line and the equation of the straight line is given explicitly as $y = a + bx$, a being the vertical intercept (the y-value when $x = 0$) and b the gradient or slope of the line.

For example, the temperature of a body can be measured in degrees Celsius (°C) or degrees Fahrenheit (°F). The relationship between these two measuring systems is given as:

$$y = 32 + 1 \cdot 8x$$

Here y represents the temperature in degrees Fahrenheit and x the temperature in degrees Celsius. From this equation we can compute the Fahrenheit value of a given Celsius value. That is:

$$20°C = \ldots \ldots \ldots °F$$
$$0°C = \ldots \ldots \ldots °F$$
$$100°C = \ldots \ldots \ldots °F$$
$$-40°C = \ldots \ldots \ldots °F$$

The answers are given in the next frame

2

$$20°C = 68°F$$
$$0°C = 32°F$$
$$100°C = 212°F$$
$$-40°C = -40°F$$

Because

The relationship between x and y is given explicitly as $y = 32 + 1 \cdot 8x$ so that:

$$y_1 = 32 + 1 \cdot 8 \times 20 = 68$$
$$y_2 = 32 + 1 \cdot 8 \times 0 = 32 \text{ the freezing point of water}$$
$$y_3 = 32 + 1 \cdot 8 \times 100 = 212 \text{ the boiling point of water}$$
$$y_4 = 32 + 1 \cdot 8 \times (-40) = -40$$

Move to the next frame

3 Implicit linear variation of two variables

Many times it is known that two variables are linearly related but the explicit linear equation that links them together is not given. Instead, pairs of *x*- and *y*-values are given, possibly from experimental data, and it is from these values that the relationship must be found and made explicit. For example, the following table records measurements made of exact values of *x* and experimentally observed, corresponding values of *y*:

x	−2·4	−0·8	0·3	1·9	3·2
y	−5·0	−1·5	2·5	6·4	11·0

If we plot these pairs of *x*- and *y* values on a graph called a **scattergraph** because the points are scattered we obtain the following graph:

. . . .

Next frame

4

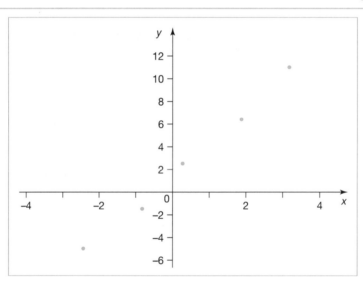

Because

Each individual pair of *x*- and corresponding *y*-values contributes a single point on the scattergraph.

We can now draw a straight line that best fits all these plotted points called the line of best fit and from that line we can deduce the linear relationship.

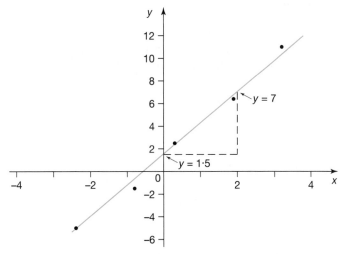

The vertical intercept is approximately 1·5 and the gradient is approximately $(7 - 1·5)/2 = 2·75$. This gives the equation of the line as:

. . . .

Next frame

$$y = 1·5 + 2·75x$$

5

Because

The equation of a straight line is $y = a + bx$ where a is the vertical intercept (in this case 1·5) and b is the gradient (in this case 2·75).

This is a rather crude way of determining the equation of the straight line but fortunately there is a better way. The method is known as **regression analysis** and it employs a method known as **least squares fit** to find by algebraic means the line of best fit. It is called regression analysis because it finds the line to which the plotted points fall back to – that is, regress to.

Move to the next frame

Regression – method of least squares

6

The method of least squares calculates the sum of the squares of the vertical distances of the plotted points from the regression line and then adjusts the line until this sum takes on its smallest value. We shall not work through the least squares process here but instead just give the result.

Given a collection of n pairs of values of x_i and the corresponding y_i for $i = 1$ (that is the first pair) ... n (the last pair) and given the equation of the line of best fit as:

$$y = a + bx$$

then, according to the method of least squares, the vertical intercept a and the gradient b of the line satisfy the two equations:

$$an + b\sum_{i=1}^{n} x_i = \sum_{i=1}^{n} y_i$$

and

$$a\sum_{i=1}^{n} x_i + b\sum_{i=1}^{n} x_i^2 = \sum_{i=1}^{n} x_i y_i$$

By solving these equations we can find the line of best fit. For example, given the data:

x	$-2\cdot4$	$-0\cdot8$	$0\cdot3$	$1\cdot9$	$3\cdot2$
y	$-5\cdot0$	$-1\cdot5$	$2\cdot5$	$6\cdot4$	$11\cdot0$

We construct the following table:

i	x_i	y_i	x_i^2	$x_i y_i$
1	$-2\cdot4$	$-5\cdot0$	$5\cdot76$	$12\cdot00$
2	$-0\cdot8$	$-1\cdot5$	....	....
3	$0\cdot3$	$2\cdot5$	....	....
4	$1\cdot9$	$6\cdot4$	....	....
5	$3\cdot2$	$11\cdot0$	....	....
$\sum_{i=1}^{5}$	$2\cdot2$	$13\cdot4$	....	....

Complete the table.
The answer is in the next frame

i	x_i	y_i	x_i^2	$x_i y_i$
1	$-2 \cdot 4$	$-5 \cdot 0$	$5 \cdot 76$	$12 \cdot 00$
2	$-0 \cdot 8$	$-1 \cdot 5$	$0 \cdot 64$	$1 \cdot 20$
3	$0 \cdot 3$	$2 \cdot 5$	$0 \cdot 09$	$0 \cdot 75$
4	$1 \cdot 9$	$6 \cdot 4$	$3 \cdot 61$	$12 \cdot 16$
5	$3 \cdot 2$	$11 \cdot 0$	$10 \cdot 24$	$35 \cdot 20$
$\sum\limits_{i=1}^{5}$	$2 \cdot 2$	$13 \cdot 4$	$20 \cdot 34$	$61 \cdot 31$

The two equations now become:

....

The answer is in the next frame

8

$$5a + 2 \cdot 2b = 13 \cdot 4$$
$$2 \cdot 2a + 20 \cdot 34b = 61 \cdot 31$$

Because

$$an + b \sum_{i=1}^{n} x_i = \sum_{i=1}^{n} y_i \quad \text{and} \quad a \sum_{i=1}^{n} x_i + b \sum_{i=1}^{n} x_i^2 = \sum_{i=1}^{n} x_i y_i.$$

That is, from the table in Frame 7:

$$5a + 2 \cdot 2b = 13 \cdot 4$$

$$2 \cdot 2a + 20 \cdot 34b = 61 \cdot 31$$

Solving these two equations for a and b yields:

$$a = \ldots \ldots \ldots$$
$$b = \ldots \ldots \ldots$$

Next frame

9

$$a = 1 \cdot 421$$
$$b = 2 \cdot 861$$

Because

Given:

$$5a + 2 \cdot 2b = 13 \cdot 4$$
$$2 \cdot 2a + 20 \cdot 34b = 61 \cdot 31$$

dividing by the coefficients of a gives:

$$a + 0 \cdot 440b = 2 \cdot 68$$
$$\underline{a + 9 \cdot 245b = 27 \cdot 87}$$
$$\therefore 8 \cdot 805b = 25 \cdot 19$$

Therefore:

$$8 \cdot 805b = 25 \cdot 19 \qquad \therefore b = 2 \cdot 861$$
$$\therefore a = 2 \cdot 68 - 1 \cdot 2588 \qquad \therefore a = 1 \cdot 421$$

So that the best straight line for the given values is:

$$y = 1.42 + 2.86x$$

Try another, just to make sure – it's all very straightforward.

To find the line of best fit $y = b + bx$ for the data values:

x	1·792	1·398	0·8573	0·4472
y	2·079	2·114	2·230	2·301

We construct the following table:

i	x_i	y_i	x_i^2	x_iy_i
1	1·792	2·079	3·211	3·726
2	....	....	....	....
3	....	....	....	....
4	....	....	....	....
$\sum_{i=1}^{4}$	....	....	....	....

Complete the table.
The answer is in the next frame

i	x_i	y_i	x_i^2	$x_i y_i$
1	1·792	2·079	3·211	3·726
2	1·398	2·114	1·954	2·955
3	0·8573	2·230	0·735	1·912
4	0·4472	2·301	0·200	1·029
$\sum_{i=1}^{4}$	4·4945	8·724	6·100	9·622

The two equations now become:

. . . .

The answer is in the next frame

$$4a + 4·495b = 8·724$$
$$4·495a + 6·100b = 9·622$$

Because

$$an + b\sum_{i=1}^{n} x_i = \sum_{i=1}^{n} y_i \quad \text{and} \quad a\sum_{i=1}^{n} x_i + b\sum_{i=1}^{n} x_i^2 = \sum_{i=1}^{n} x_i y_i.$$

That is, from the table in Frame 7:

$$4a + 4·495b = 8·724$$
$$4·495a + 6·100b = 9·622$$

Solving these two equations for a and b yields:

$$a = \ldots \ldots \ldots$$
$$b = \ldots \ldots \ldots$$

Next frame

12

$$a = 2 \cdot 379$$
$$b = -0 \cdot 176$$

Because

Given:

$$4a + 4 \cdot 495b = 8 \cdot 724$$
$$4 \cdot 495a + 6 \cdot 100b = 9 \cdot 622$$

dividing by the coefficients of b gives:

$$0 \cdot 890a + b = 1 \cdot 941$$
$$\underline{0 \cdot 737a + b = 1 \cdot 577}$$
$$\therefore 0 \cdot 153a = 0 \cdot 364$$

Therefore:

$$0 \cdot 153a = 0 \cdot 364 \qquad \therefore a = 2 \cdot 379$$
$$\therefore b = 1 \cdot 577 - 1 \cdot 753 \qquad \therefore b = -0 \cdot 176$$

So that the best straight line for the given values is:

$$y = 2 \cdot 38 - 0 \cdot 18x$$

At this point let us pause and summarize the main facts so far on regression

 # Review summary   **Unit 50**

13   1   *Explicit linear variation of two variables*:   The equation of the straight line:
$$y = a + bx$$
displays an explicit linear relationship between the two variables x and y. When pairs of corresponding x- and y-values are plotted on a graph they will be found to lie on a straight line.

   2   *Implicit linear variation of two variables*:   Many times it is known that two variables are linearly related but the exact form of the relationship is unknown. Instead, pairs of values of the two variables are given and it is from these values that the explicit form of the relationship can be found.

   3   *Scattergraph*:   A plot of isolated pairs of data values which are plotted on a Cartesian graph.

   4   *Line of best fit by eye*:   The straight line drawn against a collection of points drawn on a scattergraph. From the point where the line intersects the vertical axis and from the gradient of the line the explicit relationship between the two variables is estimated.

5　*Line of best fit by regression analysis*:　The equation of the straight line is made explicit by using a least squares fit to find the straight line to which the plotted points regress.

6　*Method of least squares*:　The line of best fit is given as:

$$y = a + bx$$

where:

$$an + b\sum_{i=1}^{n} x_i = \sum_{i=1}^{n} y_i \quad \text{and} \quad a\sum_{i=1}^{n} x_i + b\sum_{i=1}^{n} x_i^2 = \sum_{i=1}^{n} x_i y_i$$

and where n is the number of data pairs (x_i, y_i), $i = 1 \ldots n$.

 # Review exercise　　　　　　　　　**Unit 50**

1　Variables x and y are thought to be related by the law $y = a + bx$. Determine the values of a and b that best fit the set of values given in the table: **14**

x	25	56	144	225	625
y	13·1	28·1	70·2	109	301

Complete this question, looking back at the Unit if you need to.
The answers and working are in the next frame.

1　Given a collection of pairs of values x_i and the corresponding y_i for $i = 1 \ldots n$ and given the equation of the line of best fit as: **15**

$$y = a + bx$$

then the vertical intercept a and the gradient b of the line satisfy the two equations:

$$an + b\sum_{i=1}^{n} x_i = \sum_{i=1}^{n} y_i$$

and

$$a\sum_{i=1}^{n} x_i + b\sum_{i=1}^{n} x_i^2 = \sum_{i=1}^{n} x_i y_i$$

So, given the data:

x	25	56	144	225	625
y	13·1	28·1	70·2	109	301

we construct the following table:

i	x_i	y_i	x_i^2	x_iy_i
1	25	13·1	625	327·5
2	56	28·1	3136	1573·6
3	144	70·2	20736	10108·8
4	225	109	50625	24525
5	625	301	390625	188125
$\sum_{i=1}^{5}$	1075	521·4	465747	224659·9

The two equations now become:

$$an + b\sum_{i=1}^{n}x_i = \sum_{i=1}^{n}y_i \text{ and } a\sum_{i=1}^{n}x_i + b\sum_{i=1}^{n}x_i^2 = \sum_{i=1}^{n}x_iy_i.$$

That is, from the table:

$$5a + 1075b = 521\cdot4$$
$$1075a + 465747b = 224659\cdot9$$

and dividing by the coefficients of a gives:

$$a + 215b = 104\cdot28$$
$$\underline{a + 433.253b = 208\cdot986}$$
$$\therefore 218\cdot253b = 104\cdot706$$

Therefore:

$$218\cdot253b = 104\cdot706 \qquad \therefore b = 0\cdot480$$
$$\therefore a = 104\cdot28 - 103\cdot2 \qquad \therefore a = 1\cdot08$$

So that the best straight line for the given values is:

$$y = 1\cdot08 + 0\cdot48x$$

Now for the Review test

 # Review test   **Unit 50**

16 **1** Variables x and y are thought to be related by the law $y = a + bx$. Determine the values of a and b that best fit the set of values given in the table:

x	10	13	15	17	21
y	31·03	−40·5	−46·95	−53·45	−66·37

Correlation

Correlation

If two variables vary together they are said to be **correlated**. For example, the fuel consumed by a car is correlated to the number of miles travelled. The exact relationship is not easily found because so many other factors are involved such as speed of travel, condition of the engine, the tyre pressures and so on. What is important is that as the mileage increases so does the fuel used – there is a correlation.

It is most important to note that whilst cause-and-effect does display correlation between the cause and the effect as in the case of mileage travelled (cause) and petrol consumed (effect) the converse does not; the correlation of variables x and y does not imply cause-and-effect between x and y. For example, if it were noted that during the month of July in a particular year the temperature rose week by week whereas the average share price fell week by week we would say that there was a correlation between the temperature and the share price but clearly there was no cause-and-effect; the rising temperature did not cause the share price to fall and the falling share price did not cause the temperature to rise!

You may think that these are trite observations but it is surprising how many statistical surveys do imply a cause-and-effect when in fact only a correlation is observed.

So how do we measure correlation?

Move to the next frame

Measures of correlation

The strength of the correlation between two variables is given by a **correlation coefficient** – a number whose value ranges from -1 to $+1$. The strongest *positive* correlation is when the correlation coefficient is $+1$; in this case the two variables increase and decrease together in perfect unison. For example, the distance s travelled by a car moving at a constant velocity v it given by the equation:

$$s = vt$$

where t is the time taken to cover the distance. A plot of s values against the corresponding t values gives a straight line of gradient v passing through the origin. If distances s_i were measured and the corresponding times t_i recorded then they would exhibit perfect correlation with a correlation coefficient of 1.

The strongest *negative* correlation is when the correlation coefficient is -1; in this case one variable will increase or decrease as the other decreases or increases respectively, again in perfect unison. An example of such perfect negative correlation would be the lengths of adjacent sides a and b of rectangles of a given constant area A. In this case:

$$ab = A \text{ so that } a = \frac{A}{b}$$

So as one side increased or decreased the adjacent side would decrease or increase in perfect unison. If adjacent lengths a and b were recorded for different rectangles all with the same area then they would exhibit perfect correlation with a correlation coefficient of -1.

When the correlation coefficient is zero then there is no correlation between the two variables whatsoever.

Move to the next frame

3 The Pearson product-moment correlation coefficient

The *Pearson product-moment correlation coefficient* r gives the strength of a linear relationship between the n values of two variables x_i and y_i for $i = 1 \ldots n$, where r is given by the rather daunting equation:

$$r = \frac{n \sum_{i=1}^{n} x_i y_i - \left(\sum_{i=1}^{n} x_i \right) \left(\sum_{i=1}^{n} y_i \right)}{\sqrt{\left[n \sum_{i=1}^{n} x_i^2 - \left(\sum_{i=1}^{n} x_i \right)^2 \right] \left[n \sum_{i=1}^{n} y_i^2 - \left(\sum_{i=1}^{n} y_i \right)^2 \right]}}$$

However, by taking things one step at a time it is not as bad as it first appears. For example, a factory produces engine components and finds that its profit profile can be given as:

Units sold x	100	200	300	400	500	600
Profit y (£1000s)	1·4	3·1	4·8	5·5	6·4	8·1

To find the Pearson product-moment correlation coefficient r from this data we must first construct the following table:

i	x_i	y_i	x_iy_i	x_i^2	y_i^2
1	100	1·4	140	10000	1·96
2	200	3·1	....	....	....
3	300	4·8	....	....	....
4	400	5·5	....	....	....
5	500	6·4	....	....	....
6	600	8·1	....	....	....
	$\sum_{i=1}^{6} x_i = \ldots$	$\sum_{i=1}^{6} y_i = \ldots$	$\sum_{i=1}^{6} x_iy_i = \ldots$	$\sum_{i=1}^{6} x_i^2 = \ldots$	$\sum_{i=1}^{6} y_i^2 = \ldots$

Complete the table.
The answer is in the next frame

4

i	x_i	y_i	x_iy_i	x_i^2	y_i^2
1	100	1·4	140	10000	1·96
2	200	3·1	620	40000	9·61
3	300	4·8	1440	90000	23·04
4	400	5·5	2200	160000	30·25
5	500	6·4	3200	250000	40·96
6	600	8·1	4860	360000	65·61
	$\sum_{i=1}^{6} x_i$ $= 2100$	$\sum_{i=1}^{6} y_i$ $= 29\cdot3$	$\sum_{i=1}^{6} x_iy_i$ $= 12460$	$\sum_{i=1}^{6} x_i^2$ $= 910000$	$\sum_{i=1}^{6} y_i^2$ $= 171\cdot43$

It is a straightforward matter to substitute these numbers into:

$$r = \frac{6\sum_{i=1}^{6} x_iy_i - \left(\sum_{i=1}^{6} x_i\right)\left(\sum_{i=1}^{6} y_i\right)}{\sqrt{\left[6\sum_{i=1}^{6} x_i^2 - \left(\sum_{i=1}^{6} x_i\right)^2\right]\left[6\sum_{i=1}^{6} y_i^2 - \left(\sum_{i=1}^{6} y_i\right)^2\right]}}$$

$$= \ldots\ldots\ldots$$

Next frame

5

$$r = 0 \cdot 99$$

Because

$$r = \frac{6(12460) - (2100)(29 \cdot 3)}{\sqrt{\left[6(910000) - (2100)^2\right]\left[6(171 \cdot 43) - (29 \cdot 3)^2\right]}}$$

$$= \frac{13230}{\sqrt{[1050000][170 \cdot 09]}}$$

$$= \frac{13230}{13363 \cdot 9}$$

$$= 0 \cdot 99$$

There is a strong correlation between the profit and the number of units sold, as one would hope to be the case.

As another example, consider the following table recording the unit output and the size of the workforce involved in production from seven different manufacturing plants.

Number of staff x	6	11	12	9	11	7	9
Units of production y	21	18	33	39	25	27	21

To calculate the Pearson product-moment correlation coefficient the following table is constructed:

i	x_i	y_i	$x_i y_i$	x_i^2	y_i^2
1	6	21	126	36	441
2	11	18	. . . .	. . . .	. . . .
3	12	33	. . . .	. . . .	. . . .
4	9	39	. . . .	. . . .	. . . .
5	11	25	. . . .	. . . .	. . . .
6	7	27	. . . .	. . . .	. . . .
7	9	21	. . . .	. . . .	. . . .
	$\displaystyle\sum_{i=1}^{7} x_i = \ldots$	$\displaystyle\sum_{i=1}^{7} y_i = \ldots$	$\displaystyle\sum_{i=1}^{7} x_i y_i = \ldots$	$\displaystyle\sum_{i=1}^{7} x_i^2 = \ldots$	$\displaystyle\sum_{i=1}^{7} y_i^2 = \ldots$

Complete the table.
The answer is in the next frame

6

i	x_i	y_i	$x_i y_i$	x_i^2	y_i^2
1	6	21	126	36	441
2	11	18	198	121	324
3	12	33	396	144	1089
4	9	39	351	81	1521
5	11	25	275	121	625
6	7	27	189	49	729
7	9	21	189	81	441
	$\sum_{i=1}^{7} x_i$ $= 65$	$\sum_{i=1}^{7} y_i$ $= 184$	$\sum_{i=1}^{7} x_i y_i$ $= 1724$	$\sum_{i=1}^{7} x_i^2$ $= 633$	$\sum_{i=1}^{7} y_i^2$ $= 5170$

It is a straightforward matter to substitute these numbers into:

$$r = \frac{n\sum_{i=1}^{n} x_i y_i - \left(\sum_{i=1}^{n} x_i\right)\left(\sum_{i=1}^{n} y_i\right)}{\sqrt{\left[n\sum_{i=1}^{n} x_i^2 - \left(\sum_{i=1}^{n} x_i\right)^2\right]\left[n\sum_{i=1}^{n} y_i^2 - \left(\sum_{i=1}^{n} y_i\right)^2\right]}} = \ldots\ldots\ldots$$

Insert the correct value for n.
The answer is in the next frame

7

$$r = \frac{7\sum_{i=1}^{7} x_i y_i - \left(\sum_{i=1}^{7} x_i\right)\left(\sum_{i=1}^{7} y_i\right)}{\sqrt{\left[7\sum_{i=1}^{7} x_i^2 - \left(\sum_{i=1}^{7} x_i\right)^2\right]\left[7\sum_{i=1}^{7} y_i^2 - \left(\sum_{i=1}^{7} y_i\right)^2\right]}}$$

$$= 0\cdot16$$

Because

The number of pairs of data is $n = 7$, so that:

$$r = \frac{7\sum_{i=1}^{7} x_i y_i - \left(\sum_{i=1}^{7} x_i\right)\left(\sum_{i=1}^{7} y_i\right)}{\sqrt{\left[7\sum_{i=1}^{7} x_i^2 - \left(\sum_{i=1}^{7} x_i\right)^2\right]\left[7\sum_{i=1}^{7} y_i^2 - \left(\sum_{i=1}^{7} y_i\right)^2\right]}}$$

Furthermore:

$$r = \frac{7(1724) - (65)(184)}{\sqrt{\left[7(633) - (65)^2\right]\left[7(5170) - (184)^2\right]}}$$

$$= \frac{108}{\sqrt{[206][2334]}}$$

$$= \frac{108}{693 \cdot 4}$$

$$= 0 \cdot 16$$

So a weak correlation between the size of the workforce at any given site and the number of units manufactured is exhibited. This could be taken to imply that productivity is more dependent upon location than upon the size of the workforce.

Move to the next frame

8   Spearman's rank correlation coefficient

Another method of measuring correlation that does not use the actual values of the data but rather the rankings of the data values is *Spearman's rank correlation coefficient*. For example, five colleges had their mathematics courses ranked according to their quality by two outside academics. Their rankings of the quality scores is as follows where 1 represents the highest quality and 5 the lowest:

College		1	2	3	4	5
Academic	A	2	3	1	4	5
	B	1	4	2	3	5

The difference in the rankings and their square is tabulated below:

College	Academic		d_i	d_i^2
	A	B		
1	2	1	1	1
2	3	4	-1	1
3	1	2	-1	1
4	4	3	1	1
5	5	5	0	0
			$\sum_{i=1}^{5} d_i^2 = 4$	

The Spearman's rank correlation coefficient is then given as:

$$r_S = 1 - \frac{6\sum_{i=1}^{n} d_i^2}{n(n^2 - 1)}$$

so that:

$$r_S = \ldots\ldots\ldots$$

Next frame

$$\boxed{r_S = 0{\cdot}8}$$

9

Because

$$r_S = 1 - \frac{6\sum_{i=1}^{5} d_i^2}{5(5^2 - 1)}$$
$$= 1 - \frac{6 \times 4}{5 \times 24}$$
$$= 1 - 0{\cdot}2$$
$$= 0{\cdot}8$$

There is a reasonably high correlation between the two academics and this indicates that their judgments have a good measure of agreement.

Now you try another one. The following table records the sales of two employees of an office supply company:

Office supply sales	Value of sales	
	A	B
A4 paper	1200	400
Ballpoint pens	600	1000
Folders	800	700
Printer ink units	2100	1800
Miscellaneous item packs	750	1200

By ranking this data the Spearman's rank correlation coefficient is found to be:

$$r_S = \ldots\ldots\ldots$$

Next frame

10

$$r_S = 0.1$$

Because

In the table that now follows, the sales values are ranked according to size where 1 represents the highest sale, 2 the next highest and so on. Also tabulated is the difference in the ranking and its square:

Office supply sales	Value of sales			
	A	**B**	d_i	d_i^2
A4 paper	2	5	-3	9
Ballpoint pens	5	3	2	4
Folders	3	4	-1	1
Printer ink units	1	1	0	0
Miscellaneous item packs	4	2	2	4
			$\sum_{i=1}^{5} d_i^2 = 18$	

The Spearman's rank correlation coefficient is then given as:

$$r_S = 1 - \frac{6\sum_{i=1}^{n} d_i^2}{n(n^2 - 1)}$$

so that:

$$r_S = 1 - \frac{6\sum_{i=1}^{5} d_i^2}{5(5^2 - 1)}$$

$$= 1 - \frac{6 \times 18}{5 \times 24}$$

$$= 1 - 0.9$$

$$= 0.1$$

There is very little correlation between the two employee's sales – just because A sells a lot of one particular item does not mean that B will also.

Let's now pause and summarize the work on correlation

Review summary

Unit 51

1 *Correlation*: If two variables vary together they are said to be correlated. Cause-and-effect may be the reason for the correlation but it may also be coincidence.

11

2 *Measures of correlation*: The strength of the correlation between two variables is given by a correlation coefficient whose value ranges from -1 for perfect negative correlation to $+1$ for perfect positive correlation.

3 *Pearson product-moment correlation coefficient r*: This gives the strength of a linear relationship between the n values of two variables x_i and y_i for $i = 1 \ldots n$, where r is given by the rather daunting equation:

$$r = \frac{n\sum_{i=1}^{n} x_i y_i - \left(\sum_{i=1}^{n} x_i\right)\left(\sum_{i=1}^{n} y_i\right)}{\sqrt{\left[n\sum_{i=1}^{n} x_i^2 - \left(\sum_{i=1}^{n} x_i\right)^2\right]\left[n\sum_{i=1}^{n} y_i^2 - \left(\sum_{i=1}^{n} y_i\right)^2\right]}}$$

4 *Spearman's rank correlation coefficient r_S*: By taking the ranking of the values of two variables the Spearman's rank correlation coefficient measures the correlation of the ranking. The coefficient is given as:

$$r_S = 1 - \frac{6\sum_{i=1}^{n} d_i^2}{n(n^2 - 1)}$$

where the n values of the two variables are ranked and d_i is the difference in ranking of the ith pair.

Review exercise

Unit 51

1 Find the Pearson product-moment correlation coefficient for the data in the Unit 50 Review exercise to find the strength of the linear relationship between the two variables x and y. The data was:

12

x	25	56	144	225	625
y	13·1	28·1	70·2	109	301

2 Five different manufacturers produce their own brand of pain-killer. Two individuals were then asked to rank the effectiveness of each product with the following results:

Make	1	2	3	4	5
Trial 1	5	3	4	1	2
Trial 2	2	3	5	4	1

Calculate Spearman's rank correlation coefficient for this data.

Complete both questions. Take your time.
Whilst the formulae may seem daunting, remember that the mathematics is quite
straightforward. If necessary, look back at the Unit.
The answers and working are in the next frame.

13   1 To find the Pearson product-moment correlation coefficient for the data:

x	25	56	144	225	625
y	13·1	28·1	70·2	109	301

we construct the following table:

i	x_i	y_i	x_i^2	$x_i y_i$	y_i^2
1	25	13·1	625	327·5	171·61
2	56	28·1	3136	1573·6	789·61
3	144	70·2	20736	10108·8	4928·04
4	225	109	50625	24525	11881
5	625	301	390625	188125	90601
$\sum_{i=1}^{5}$	1075	521·4	465747	224659·9	108371·26

Substituting these numbers into:

$$r = \frac{n\sum_{i=1}^{n} x_i y_i - \left(\sum_{i=1}^{n} x_i\right)\left(\sum_{i=1}^{n} y_i\right)}{\sqrt{\left[n\sum_{i=1}^{n} x_i^2 - \left(\sum_{i=1}^{n} x_i\right)^2\right]\left[n\sum_{i=1}^{n} y_i^2 - \left(\sum_{i=1}^{n} y_i\right)^2\right]}}$$

where the number of pairs of data is $n = 5$, gives:

$$r = \frac{5\sum_{i=1}^{5} x_i y_i - \left(\sum_{i=1}^{5} x_i\right)\left(\sum_{i=1}^{5} y_i\right)}{\sqrt{\left[5\sum_{i=1}^{5} x_i^2 - \left(\sum_{i=1}^{5} x_i\right)^2\right]\left[5\sum_{i=1}^{5} y_i^2 - \left(\sum_{i=1}^{5} y_i\right)^2\right]}}$$

$$= \frac{5(224659 \cdot 9) - (1075)(521 \cdot 4)}{\sqrt{\left[5(465747) - (1075)^2\right]\left[5(108371.26) - (521 \cdot 4)^2\right]}}$$

$$= \frac{562794 \cdot 5}{\sqrt{[1173110][269998 \cdot 34]}}$$

$$= \frac{562794 \cdot 5}{562794 \cdot 592}$$

$$= 0 \cdot 99999984$$

This indicates practically perfect positive correlation indicating that the assumption that the data lies on a straight line is correct.

2   The table of data:

Make	1	2	3	4	5
Trial 1	5	3	4	1	2
Trial 2	2	3	5	4	1

can be extended to cater for differences in ranking as follows:

Make	1	2	3	4	5
Trial 1	5	3	4	1	2
Trial 2	2	3	5	4	1
d_i	3	0	−1	−3	1
d_i^2	9	0	1	9	1

The Spearman's rank correlation coefficient is then given as:

$$r_S = 1 - \frac{6\sum\limits_{i=1}^{n} d_i^2}{n(n^2 - 1)}$$

$$= 1 - \frac{6\sum\limits_{i=1}^{5} d_i^2}{5(5^2 - 1)}$$

$$= 1 - \frac{6 \times 20}{5 \times 24}$$

$$= 1 - 1$$

$$= 0$$

There is no correlation between the two judges indicating that any beneficial effects were probably due to personal factors.

Now for the Review test

 # Review test   **Unit 51**

14

1   Find the Pearson product-moment correlation coefficient for the data in the Unit 50 Review test to find the strength of the linear relationship between the two variables x and y. The data was:

x	10	13	15	17	21
y	31·03	−40·5	−46·95	−53·45	−66·37

2   In a pub quiz the average age of a team was related to the percentage score attained. The data was then ranked with 1 representing the oldest average age and 1 representing the highest score to produce the following table:

Team	1	2	3	4	5
Average age	5	3	4	1	2
Score	4	2	5	1	3

Calculate Spearman's rank correlation coefficient for this data.

15

You have now come to the end of this Module. A list of **Can You?** questions follows for you to gauge your understanding of the material in the Module. You will notice that these questions match the **Learning outcomes** listed at the beginning of the Module. Now try the **Test exercise**. *Work through the questions at your own pace, there is no need to hurry.* A set of **Further problems** provides valuable additional practice.

Can You?

 # Test exercise 16

1 In an experiment the corresponding values of the two variables x and y were recorded as follows: **2**

x	2·100	2·250	2·420	2·600	2·720	2·860
y	0·270	0·369	0·439	0·560	0·620	0·680

It is assumed that the relationship between x and y is of the form $y = a + bx$.

(a) Plot the data pairs on a scattergraph.

(b) Draw the line of best fit by eye and estimate the values of a and b.

(c) Use the method of least squares to find more accurate values of a and b.

2   The price of a particular commodity was reduced by different percentages in different branches of a store to determine the effect of price change on sales. The percentage changes in subsequent sales were recorded in the following table:

Price cut %	5	7	10	14	20
Sales change %	3	4	8	8	10

Calculate the Pearson product-moment correlation coefficient and comment on the result.

3   Ten candidates for a teaching post were ranked by two members of the interviewing panel. Their ranking is as follows:

Candidate		1	2	3	4	5	6	7	8	9	10
Panel member	1	8	10	3	9	6	5	1	7	2	4
	2	10	8	7	6	9	4	1	5	2	3

Calculate Spearman's rank correlation coefficient and comment on the measure of agreement between the two panel members.

 # Further problems 16

3

1   In an experiment the corresponding values of the two variables x and y were recorded as follows:

x	9·8	12·2	13·7	16·3	17·9	20·3
y	43·7	52·9	62·1	71·3	80·5	89·7

It is assumed that the relationship between x and y is of the form $y = a + bx$.

(a) Plot the data pairs on a scattergraph.
(b) Draw the line of best fit by eye and estimate the values of a and b.
(c) Use the method of least squares to find more accurate values of a and b.

2 The two variables x and y were assumed to be linked via the equation $y = ax^n$ where a and n are constants. The corresponding values of the two variables x and y were recorded as follows:

x	2	5	12	25	32	40
y	5·62	13·8	52·5	112	160	200

(a) By defining $Y = \log y$ and $X = \log x$ write the equation above as a linear equation and extend the table to include values of X and Y.
(b) Plot the new data pairs on a scattergraph.
(c) Draw the line of best fit by eye and estimate the values of a and n.
(d) Use the method of least squares to find more accurate values of a and n.

3 The two variables W and T were assumed to be linked via the equation $W = ae^{nT}$ where a and n are constants. The corresponding values of the two variables W and T were recorded as follows:

T	3·0	10	15	30	50	90
W	3·857	1·974	1·733	0·4966	0·1738	0·0091

(a) By defining $Y = \ln W$ write the equation above as a linear equation and extend the table to include values of Y.
(b) Plot the new data pairs on a scattergraph.
(c) Draw the line of best fit by eye and estimate the values of a and n.
(d) Use the method of least squares to find more accurate values of a and n.

4 Find the Pearson product-moment correlation coefficient for the following data to find the strength of the linear relationship proposed in question **1**:

x	9·8	12·2	13·7	16·3	17·9	20·3
y	43·7	52·9	62·1	71·3	80·5	89·7

It is assumed that the relationship between x and y is of the form $y = a + bx$.

5 Find the Pearson product-moment correlation coefficient for the following data to find the strength of the relationship proposed in question **2**. The two variables x and y were assumed to be linked via the equation $y = ax^n$ where a and n are constants. The corresponding values of the two variables x and y were recorded as follows:

x	2	5	12	25	32	40
y	5·62	13·8	52·5	112	160	200

6   Find the Pearson product-moment correlation coefficient for the following data to find the strength of the linear relationship proposed in question **3**. The two variables W and T were assumed to be linked via the equation $W = ae^{nT}$ where a and n are constants. The corresponding values of the two variables W and T were recorded as follows:

T	3·0	10	15	30	50	90
W	3·857	1·974	1·733	0·4966	0·1738	0·0091

7   Five golfers ranked their handicaps against the ranking of the number of years they have been playing the game. The resulting table of ranks, where 1 represents both the lowest handicap and the highest number of years playing the game, is given as:

Handicap	1	2	3	4	5
Years of play	2	3	4	4	1

Calculate Spearman's rank correlation coefficient and comment on the result.

8   Eight teas were ranked by two professional tea tasters for quality. Their ranking is as follows:

Tea		1	2	3	4	5	6	7	8
Tasters	1	1	4	6	5	7	5	3	2
	2	2	3	5	1	6	2	3	4

Calculate Spearman's rank correlation coefficient and comment on the measure of agreement between the two panel members.

9   The following table shows the number of attempts to correctly guess the suit of a playing card selected in secret and the corresponding number of successes. Rank this data and calculate Spearman's rank correlation coefficient for this data.

Attempts	104	97	101	116	125
Successes	13	11	11	14	16

Introduction to differentiation

When you have completed this Module you will be able to:
Determine the gradient of a straight-line graph

- Evaluate from first principles the gradient at a point on a quadratic curve
- Differentiate powers of x and polynomials
- Evaluate second derivatives and use tables of standard derivatives
- Differentiate products and quotients of expressions
- Differentiate using the chain rule for a 'function of a function'
- Use the Newton–Raphson method to obtain a numerical solution to an equation

Units

Gradients   Unit 52

The gradient of a straight-line graph

1

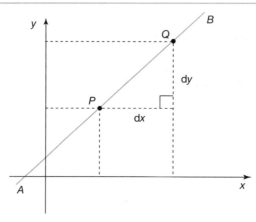

The *gradient* of the sloping straight line shown in the figure is defined as:

$$\frac{\text{the vertical distance the line rises or falls between two points } P \text{ and } Q}{\text{the horizontal distance between } P \text{ and } Q}$$

where P is a point to the left of point Q on the straight line AB which slopes upwards from left to right. The changes in the x- and y-values of the points P and Q are denoted by dx and dy respectively. So the gradient of this line is given as:

$$\frac{dy}{dx}$$

We could have chosen any pair of points on the straight line for P and Q and by similar triangles this ratio would have worked out to the same value:

the gradient of a straight line is constant throughout its length

Its value is denoted by the symbol m.

Therefore $m = \dfrac{dy}{dx}$.

For example, if, for some line (see the figure on the next page), P is the point $(2, 3)$ and Q is the point $(6, 4)$, then P is to the left and *below* the point Q. In this case:

$dy =$ the change in the y-values $= 4 - 3 = 1$ and
$dx =$ the change in the x-values $= 6 - 2 = 4$ so that:

$$m = \frac{dy}{dx} = \frac{1}{4} = 0{\cdot}25. \quad \text{The sloping line } \textit{falls vertically} \text{ from left to right by}$$
$$\text{1 unit for every 1 unit horizontally.}$$

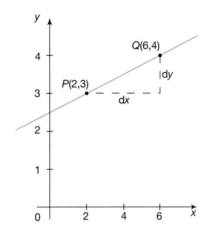

If, for some other line (see the figure below), P is the point $(3, 5)$ and Q is the point $(7, 1)$, then P is to the left and *above* the point Q. In this case:

dy = the change in the y-values $= 1 - 5 = -4$ and
dx = the change in the x-values $= 7 - 3 = 4$ so that:

$m = \dfrac{dy}{dx} = \dfrac{-4}{4} = -1.$ The sloping line *falls vertically* from left to right by 1 unit for every 1 unit horizontally.

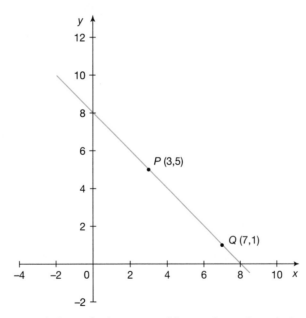

So, lines going *up* to the right have a *positive* gradient, lines going *down* to the right have a *negative* gradient.

Try the following exercises. Determine the gradients of the straight lines joining:

1 P (3, 7) and Q (5, 8)   **4** P (−3, 6) and Q (5, 2)
2 P (2, 4) and Q (6, 9)   **5** P (−2, 4) and Q (3, −2)
3 P (1, 6) and Q (4, 4)

Draw a diagram in each case

2 Here they are:

1

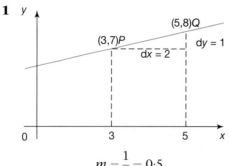

$$m = \frac{1}{2} = 0{\cdot}5$$

4

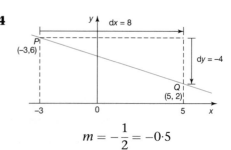

$$m = -\frac{1}{2} = -0{\cdot}5$$

2

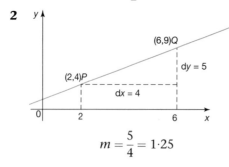

$$m = \frac{5}{4} = 1{\cdot}25$$

5

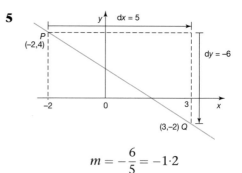

$$m = -\frac{6}{5} = -1{\cdot}2$$

3

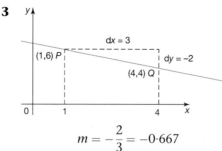

$$m = -\frac{2}{3} = -0{\cdot}667$$

Now let us extend these ideas to graphs that are not straight lines.

On then to the next frame

The gradient of a curve at a given point

3

If we take two points P and Q on a curve and calculate, as we did for the straight line:

$$\frac{\text{the vertical distance the curve rises or falls between two points } P \text{ and } Q}{\text{the horizontal distance between } P \text{ and } Q}$$

the result will depend upon the choice of points P and Q as can be seen from the figure below:

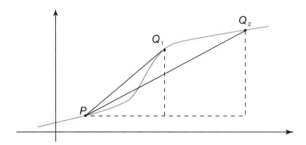

This is because the gradient of the curve varies along its length, as anyone who has climbed a hill will appreciate. Because of this the gradient of a curve is not defined *between two points* as in the case of a straight line but *at a single point*. The gradient of a curve at a given point is defined to be the gradient of the straight line that touches the curve at that point – the *gradient of the tangent*.

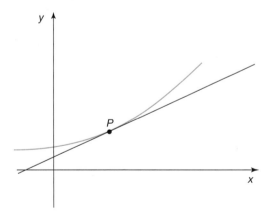

The gradient of the curve at P is equal to the gradient of the tangent to the curve at P.

This is a straightforward but very important definition because all of the differential calculus depends upon it. Make a note of it.

For example, find the gradient of the curve of $y = 2x^2 + 5$ at the point P at which $x = 1\cdot5$.

First we must compile a table giving the y-values of $y = 2x^2 + 5$ at $0\cdot5$ intervals of x between $x = 0$ and $x = 3$.

Complete the table and then move on to the next frame

4

x	0	0·5	1·0	1·5	2·0	2·5	3·0
x^2	0	0·25	1·0	2·25	4·0	6·25	9·0
$2x^2$	0	0·5	2·0	4·5	8·0	12·5	18·0
$y = 2x^2 + 5$	5	5·5	7·0	9·5	13·0	17·5	23·0

Then we can plot the graph accurately, using these results, and mark the point P on the graph at which $x = 1\cdot5$.

So do that

5

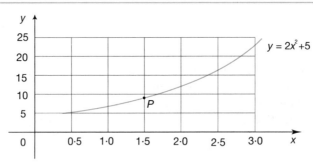

Now we bring a ruler up to the point P and adjust the angle of the ruler by eye until it takes on the position of the tangent to the curve. Then we carefully draw the tangent.

6

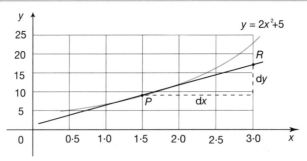

Now select a second point on the tangent at R, for example R at $x = 3\cdot0$ where at P, $x = 1\cdot5$. Determine dx and dy between them and hence calculate the gradient of the tangent, i.e. the gradient of the curve at P.

This gives $m = \ldots\ldots\ldots$

$$\boxed{m = 5{\cdot}9}$$

7

Because, at $x = 1{\cdot}5$, $y = 9{\cdot}3$ (gauged from the graph)
$\quad x = 3{\cdot}0$, $y = 18{\cdot}1$ (again, gauged from the graph) therefore $dx = 1{\cdot}5$
$\quad$ and $dy = 8{\cdot}8$.

Therefore for the tangent $m = \dfrac{dy}{dx} = \dfrac{8{\cdot}8}{1{\cdot}5} = 5{\cdot}9$ to 2 sig fig.

Therefore the gradient of the curve at P is approximately 5·9.

Note: Your results may differ slightly from those given here because the value obtained is the result of practical plotting and construction and may, therefore, contain minor differences. However, the description given is designed to clarify the method.

Algebraic determination of the gradient of a curve

The limitations of the practical construction in the previous method call for a more accurate method of finding the gradient, so let us start afresh from a different viewpoint and prove a general rule.

8

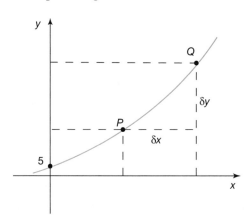

Let P be a fixed point (x, y) on the curve $y = 2x^2 + 5$, and Q be a neighbouring point, with coordinates $(x + \delta x, y + \delta y)$.

Now draw two straight lines, one through the points P and Q and the other tangential to the curve at point P.

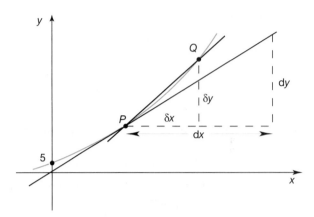

Notice that we use δx and δy to denote the respective differences in the *x*- and *y*-values of points P and Q on the curve. We reserve the notation d*x* and d*y* for the differences in the *x*- and *y*-values of two points on a *straight line* and particularly for a tangent. The quantities d*x* and d*y* are called *differentials*.

Next frame

9

At Q: $y + \delta y = 2(x + \delta x)^2 + 5$

$\qquad\qquad = 2(x^2 + 2x.\delta x + [\delta x]^2) + 5$ Expanding the bracket

$\qquad\qquad = 2x^2 + 4x.\delta x + 2[\delta x]^2 + 5$ Multiplying through by 2

Subtracting *y* from both sides:

$\qquad y + \delta y - y = 2x^2 + 4x.\delta x + 2[\delta x]^2 + 5 - (2x^2 + 5)$

$\qquad\qquad\qquad = 4x.\delta x + 2[\delta x]^2$

Therefore: $\delta y = 4x.\delta x + 2[\delta x]^2$

δy is the *vertical* distance between point P and point Q and this has now been given in terms of *x* and δx where δx is the *horizontal* distance between point P and point Q.

Now, if we divide both sides by δx:

$$\frac{\delta y}{\delta x} = 4x + 2.\delta x$$

This expression, giving the change in vertical distance per unit change in horizontal distance, is the *gradient* of the straight line through P and Q.

If the line through P and Q now rotates clockwise about P, the point Q moves down the curve and approaches P. Also, both δx and δy approach 0 ($\delta x \to 0$ and $\delta y \to 0$). However, *their ratio*, which is the gradient of PQ, approaches the gradient of the tangent at P:

$$\frac{\delta y}{\delta x} \to \text{the gradient of the tangent at } P = \frac{dy}{dx}$$

Therefore $\dfrac{dy}{dx} = \ldots\ldots\ldots\ldots$

$$\frac{dy}{dx} = 4x$$

Because

$$\frac{\delta y}{\delta x} = 4x + 2.\delta x \text{ and as } \delta x \to 0 \text{ so } \frac{\delta y}{\delta x} \to 4x + 2 \times 0 = 4x$$

This is a general result giving the slope of the curve at any point on the curve $y = 2x^2 + 5$.

$$\therefore \text{ At } x = 1 \cdot 5 \quad \frac{dy}{dx} = 4(1 \cdot 5) = 6$$

$\therefore$ The real slope of the curve $y = 2x^2 + 5$ at $x = 1 \cdot 5$ is 6.

The graphical solution previously obtained, i.e. 5·9, is an approximation.

The expression $\frac{dy}{dx}$ is called *the derivative of y with respect to x* because it is *derived* from the expression for y. The process of finding the derivative is called *differentiation* because it involves manipulating differences in coordinate values.

Derivative of powers of *x*

Two straight lines

(a) $y = c$ (constant)
The graph of $y = c$ is a straight line parallel to the *x*-axis. Therefore its gradient is zero.

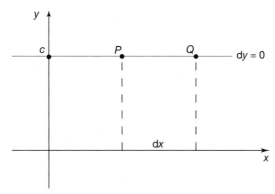

$y = c$, $dy = 0$ therefore $\dfrac{dy}{dx} = \dfrac{0}{dx} = 0$

$$\text{If } y = c, \quad \frac{dy}{dx} = 0$$

Move to the next frame

12

(b) $y = ax$

So for a point further up the line

$$y + dy = a(x + dx)$$
$$= ax + a.dx$$

Subtract y:

$$dy = ax + a.dx - ax \text{ where } y = ax$$
$$= a.dx$$

Therefore:

$$\frac{dy}{dx} = a$$

Meaning that the gradient of a straight line is constant along its length.

$$\text{If } y = ax, \frac{dy}{dx} = a$$

In particular, if $a = 1$ then $y = x$ and $\dfrac{dy}{dx} = 1$

And now for curves

13   **Two curves**

(a) $y = x^2$

At Q, $y + \delta y = (x + \delta x)^2$
$$y + \delta y = x^2 + 2x.\delta x + (\delta x)^2$$
$$y = x^2$$
$$\therefore \delta y = 2x.\delta x + (\delta x)^2$$
$$\therefore \frac{\delta y}{\delta x} = 2x + \delta x$$

If $\delta x \to 0$, $\dfrac{\delta y}{\delta x} \to \dfrac{dy}{dx}$ and $\therefore \dfrac{dy}{dx} = 2x$

$\therefore$ If $y = x^2$, $\dfrac{dy}{dx} = 2x$

Remember that $\dfrac{dy}{dx}$ *is the gradient of the tangent to the curve.*

(b) $y = x^3$

14

At Q, $y + \delta y = (x + \delta x)^3$
$$= x^3 + 3x^2.\delta x + 3x(\delta x)^2 + (\delta x)^3$$

Now subtract $y = x^3$ from each side and finish it off as before.

$\therefore$ If $y = x^3$, $\dfrac{dy}{dx} = \dots\dots\dots\dots$

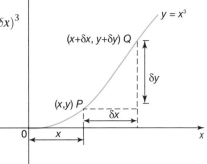

$$3x^2$$

15

Because

$$\delta y = 3x^2.\delta x + 3x(\delta x)^2 + (\delta x)^3$$

$$\therefore \frac{\delta y}{\delta x} = 3x^2 + 3x(\delta x) + (\delta x)^2$$

Then, if $\delta x \to 0$, $\dfrac{\delta y}{\delta x} \to \dfrac{dy}{dx}$ $\therefore \dfrac{dy}{dx} = 3x^2$

$\therefore$ If $y = x^3$, $\dfrac{dy}{dx} = 3x^2$

Continuing in the same manner, we should find that:

$$\text{when } y = x^4, \quad \frac{dy}{dx} = 4x^3$$
$$\text{and when } y = x^5, \quad \frac{dy}{dx} = 5x^4$$

If we now collect these results together and look for a pattern, we have the evidence given in the next frame.

So move on

y	$\dfrac{dy}{dx}$
c	0
x	1
x^2	$2x$
x^3	$3x^2$
x^4	$4x^3$
x^5	$5x^4$

We soon see a clear pattern emerging.

16

For a power of x:

In the derivative, the old index becomes a coefficient and the new index is one less than the index in the original power.

i.e. If $y = x^n$, $\dfrac{dy}{dx} = nx^{n-1}$

While this has only been demonstrated to be valid for $n = 0, 1, 2, 3$ it is in fact true for any value of n.

We have established that if $y = a$ (a constant) then $\dfrac{dy}{dx} = 0$. It can also be established that if $y = ax^n$ then $\dfrac{dy}{dx} = anx^{n-1}$. For example, if $y = ax^4$ then $\dfrac{dy}{dx} = a \times 4x^3 = 4ax^3$.

We can prove these results by using the Binomial theorem:

$$(a + b)^n = a^n + na^{n-1}b + \frac{n(n-1)}{2!}a^{n-2}b^2 + \frac{n(n-1)(n-2)}{3!}a^{n-3}b^3 + \dots$$

If $y = x^n$, $y + \delta y = (x + \delta x)^n$

$$\therefore y + \delta y = x^n + nx^{n-1}(\delta x) + \frac{n(n-1)}{2!}x^{n-2}(\delta x)^2$$
$$+ \frac{n(n-1)(n-2)}{3!}x^{n-3}(\delta x)^3 + \dots$$

$$y = x^n$$

$$\therefore \delta y = nx^{n-1}(\delta x) + \frac{n(n-1)}{2!}x^{n-2}(\delta x)^2 + \frac{n(n-1)(n-2)}{3!}x^{n-3}(\delta x)^3 + \dots$$

$$\frac{\delta y}{\delta x} = nx^{n-1} + \frac{n(n-1)}{2!}x^{n-2}(\delta x) + \frac{n(n-1)(n-2)}{3!}x^{n-3}(\delta x)^2 + \dots$$

If $\delta x \to 0$, $\dfrac{\delta y}{\delta x} \to \dfrac{dy}{dx}$ and all terms on the RHS, except the first $\to 0$.

$$\therefore \text{If } \delta x \to 0, \frac{dy}{dx} = nx^{n-1} + 0 + 0 + 0 + \dots$$

$$\therefore \text{If } y = x^n, \frac{dy}{dx} = nx^{n-1}$$

which is, of course, the general form of the results we obtained in the examples above.

Make a note of this important result

Differentiation of polynomials

17

To *differentiate a polynomial*, we differentiate each term in turn.

e.g. If $y = x^4 + 5x^3 - 4x^2 + 7x - 2$

$$\frac{dy}{dx} = 4x^3 + 5 \times 3x^2 - 4 \times 2x + 7 \times 1 - 0$$

$$\therefore \frac{dy}{dx} = 4x^3 + 15x^2 - 8x + 7$$

Example

If $y = 2x^5 + 4x^4 - x^3 + 3x^2 - 5x + 7$, find an expression for $\dfrac{dy}{dx}$ and the value of $\dfrac{dy}{dx}$ at $x = 2$.

So, first of all, $\dfrac{dy}{dx} = \ldots\ldots\ldots\ldots$

18

$$\frac{dy}{dx} = 10x^4 + 16x^3 - 3x^2 + 6x - 5$$

Then, expressing the RHS in nested form and substituting $x = 2$, we have:

$$\text{At } x = 2, \ \frac{dy}{dx} = \ldots\ldots\ldots\ldots$$

19

283

Because

$$\frac{dy}{dx} = f(x) = \{[(10x + 16)x - 3]x + 6\}x - 5$$

$$\therefore f(2) = 283$$

Now, as an exercise, determine an expression for $\dfrac{dy}{dx}$ in each of the following cases and find the value of $\dfrac{dy}{dx}$ at the stated value of x:

1 $y = 3x^4 - 7x^3 + 4x^2 + 3x - 4$ $\qquad [x = 2]$

2 $y = x^5 + 2x^4 - 4x^3 - 5x^2 + 8x - 3$ $\qquad [x = -1]$

3 $y = 6x^3 - 7x^2 + 4x + 5$ $\qquad [x = 3]$

20

1 $\dfrac{dy}{dx} = 12x^3 - 21x^2 + 8x + 3.$ $\qquad$ At $x = 2$, $\dfrac{dy}{dx} = 31$

2 $\dfrac{dy}{dx} = 5x^4 + 8x^3 - 12x^2 - 10x + 8.$ $\qquad$ At $x = -1$, $\dfrac{dy}{dx} = 3$

3 $\dfrac{dy}{dx} = 18x^2 - 14x + 4.$ $\qquad$ At $x = 3$, $\dfrac{dy}{dx} = 124$

Now on to the next frame

Derivatives – an alternative notation

21

If $y = 2x^2 - 5x + 3$, then $\dfrac{dy}{dx} = 4x - 5$. This double statement can be written as a single statement by putting $2x^2 - 5x + 3$ in place of y in $\dfrac{dy}{dx}$.

i.e. $\dfrac{d}{dx}(2x^2 - 5x + 3) = 4x - 5$

In the same way, $\dfrac{d}{dx}(4x^3 - 7x^2 + 2x - 5) = \ldots\ldots\ldots$

22

$$\boxed{12x^2 - 14x + 2}$$

Either of the two methods is acceptable: it is just a case of which is the more convenient in any situation.

At this point let us pause and summarize derivatives of powers of x and polynomials

 ## Review summary   Unit 52

23

1 *Gradient of a straight line graph* $m = \dfrac{dy}{dx}$

2 *Gradient of a curve at a given point P at* (x, y)

$\dfrac{dy}{dx}$ = gradient of the tangent to the curve at P.

3 *Derivatives of powers of x*

(a) $y = c$ (constant), $\dfrac{dy}{dx} = 0$

(b) $y = x^n$, $\dfrac{dy}{dx} = nx^{n-1}$

(c) $y = ax^n$, $\dfrac{dy}{dx} = anx^{n-1}$

4 *Differentiation of polynomials* – differentiate each term in turn.

Review exercise

Unit 52

1 Calculate the gradient of the straight line joining:

24

   (a) $P(4,0)$ and $Q(7,3)$

   (b) $P(5,6)$ and $Q(9,2)$

   (c) $P(-4,-7)$ and $Q(1,3)$

   (d) $P(0,5)$ and $Q(5,-6)$

2 Determine, algebraically, from first principles, the gradient of the graph of $y = 5x^2 + 2$ at the point P where $x = -1.6$.

3 If $y = -2x^4 - 3x^3 + 4x^2 - x + 5$, obtain an expression for $\dfrac{dy}{dx}$ and hence calculate the value of $\dfrac{dy}{dx}$ at $x = -3$.

Complete all three questions.
Refer back to the Unit if necessary but don't rush.
The answers and working are in the next frame.

---

1 (a) $P(4,0)$ and $Q(7,3)$:

25

$$dy = 3 - 0 = 3, \ dx = 7 - 4 = 3 \text{ so } \frac{dy}{dx} = \frac{3}{3} = 1$$

   (b) $P(5,6)$ and $Q(9,2)$:

$$dy = 2 - 6 = -4, \ dx = 9 - 5 = 4 \text{ so } \frac{dy}{dx} = \frac{-4}{4} = -1$$

   (c) $P(-4,-7)$ and $Q(1,3)$:

$$dy = 3 - (-7) = 10, \ dx = 1 - (-4) = 5 \text{ so } \frac{dy}{dx} = \frac{10}{5} = 2$$

   (d) $P(0,5)$ and $Q(5,-6)$:

$$dy = -6 - 5 = -11, \ dx = 5 - 0 = 5 \text{ so } \frac{dy}{dx} = \frac{-11}{5} = -2.2$$

2 $y = 5x^2 + 2$ so that:

$$y + \delta y = 5(x + \delta x)^2 + 2$$
$$= 5(x^2 + 2x\delta x + [\delta x]^2) + 2$$
$$= 5x^2 + 10x\delta x + 5[\delta x]^2 + 2 \text{ so that:}$$
$$y + \delta y - y = 5x^2 + 10x\delta x + 5[\delta x]^2 + 2 - (5x^2 + 2)$$
$$= 10x\delta x + 5[\delta x]^2$$
$$= \delta y$$

Hence:

$$\frac{\delta y}{\delta x} = \frac{10x\delta x + 5[\delta x]^2}{\delta x} = 10x + 5\delta x \text{ therefore } \frac{dy}{dx} = 10x. \text{ When } x = 1{\cdot}6:$$

$$\frac{dy}{dx} = -16$$

3   If $y = -2x^4 - 3x^3 + 4x^2 - x + 5$ then, differentiating term by term:

$$\frac{dy}{dx} = -2 \times 4x^3 - 3 \times 3x^2 + 4 \times 2x - 1$$
$$= -8x^3 - 9x^2 + 8x - 1 = ((-8x - 9)x + 8)x - 1. \text{ When } x = -3:$$
$$\frac{dy}{dx} = ((-8(-3) - 9)(-3) + 8)(-3) - 1$$
$$= 110$$

Now for the Review test

 # Review test   **Unit 52**

26   **1**   Calculate the gradient of the straight line joining:
  (a) $P(1, 3)$ and $Q(4, 9)$
  (b) $P(-2, 6)$ and $Q(7, -5)$
  (c) $P(0, 0)$ and $Q(4, 8)$

2   Determine, algebraically, from first principles, the gradient of the graph of $y = 3x^2 - 2$ at point P where $x = 2.4$.

3   If $y = 3x^4 - 5x^3 + x^2 + 2x - 4$, obtain an expression for $\dfrac{dy}{dx}$ and hence calculate the value of $\dfrac{dy}{dx}$ at $x = 4$.

Further differentiation

Unit 53

Second derivatives

If $y = 2x^4 - 5x^3 + 3x^2 - 2x + 4$, then, by the previous method:

1

$$\frac{dy}{dx} = \frac{d}{dx}(2x^4 - 5x^3 + 3x^2 - 2x + 4) = 8x^3 - 15x^2 + 6x - 2$$

This expression for $\frac{dy}{dx}$ is itself a polynomial in powers of x and can be differentiated in the same way as before, i.e. we can find the derivative of $\frac{dy}{dx}$.

$\frac{d}{dx}\left(\frac{dy}{dx}\right)$ is written $\frac{d^2y}{dx^2}$ and is the *second derivative of y with respect to x* (spoken as 'dee two y by dee x squared').

So, in this example, we have:

$$y = 2x^4 - 5x^3 + 3x^2 - 2x + 4$$
$$\frac{dy}{dx} = 8x^3 - 15x^2 + 6x - 2$$
$$\frac{d^2y}{dx^2} = 24x^2 - 30x + 6$$

We could, if necessary, find the third derivative of y in the same way:

$$\frac{d^3y}{dx^3} = \dots\dots\dots$$

$$\boxed{48x - 30}$$

2

Similarly, if $y = 3x^4 + 2x^3 - 4x^2 + 5x + 1$

$$\frac{dy}{dx} = \dots\dots\dots$$
$$\frac{d^2y}{dx^2} = \dots\dots\dots$$

3

$$\boxed{\begin{array}{l}\frac{dy}{dx} = 12x^3 + 6x^2 - 8x + 5\\[2mm]\frac{d^2y}{dx^2} = 36x^2 + 12x - 8\end{array}}$$

Let us now establish a limiting value that we shall need in the future.

Limiting value of $\dfrac{\sin \theta}{\theta}$ as $\theta \to 0$

P is a point on the circumference of a circle, centre O and radius r. PT is a tangent to the circle at P.

Use your trigonometry to show that:

$h = r \sin \theta$ and $H = r \tan \theta$

Recollect:

Area of a triangle $= \dfrac{1}{2} \times$ (base) $\times$ (height)

Area of a circle $= \pi r^2 = \dfrac{1}{2} r^2 . 2\pi$, so

Area of a sector $= \dfrac{1}{2} r^2 . \theta$ (θ in radians)

$\triangle POA$: area $= \dfrac{1}{2} r.h = \dfrac{1}{2} r.r \sin \theta = \dfrac{1}{2} r^2 \sin \theta$

Sector POA: area $= \dfrac{1}{2} r^2 \theta$

$\triangle POT$: area $= \dfrac{1}{2} r.H = \dfrac{1}{2} r.r \tan \theta = \dfrac{1}{2} r^2 \tan \theta$

In terms of area: $\triangle POA <$ sector $POA < \triangle POT$

$\therefore \dfrac{1}{2} r^2 \sin \theta < \dfrac{1}{2} r^2 \theta < \dfrac{1}{2} r^2 \tan \theta$

$\therefore \sin \theta < \theta < \tan \theta \qquad \left[\tan \theta = \dfrac{\sin \theta}{\cos \theta} \right] \qquad \therefore \dfrac{1}{\sin \theta} > \dfrac{1}{\theta} > \dfrac{\cos \theta}{\sin \theta}$

Multiplying throughout by $\sin \theta$: $\quad 1 > \dfrac{\sin \theta}{\theta} > \cos \theta$

When $\theta \to 0$, $\cos \theta \to 1$. $\therefore$ The limiting value of $\dfrac{\sin \theta}{\theta}$ is bounded on both sides by the value 1.

$\therefore$ Limiting value of $\dfrac{\sin \theta}{\theta}$ as $\theta \to 0 = 1$

Make a note of this result. We shall certainly meet it again in due course

4   Standard derivatives

So far, we have found derivatives of polynomials using the standard derivative $\dfrac{d}{dx}(x^n) = nx^{n-1}$. Derivatives of trigonometric expressions can be established by using a number of trigonometrical formulas. We shall deal with some of these in the next few frames.

Derivative of $y = \sin x$

If $y = \sin x$, $\quad y + \delta y = \sin(x + \delta x)$ $\quad \therefore$ $\delta y = \sin(x + \delta x) - \sin x$

We now apply the trigonometrical formula:

$$\sin A - \sin B = 2 \cos \frac{A+B}{2} \sin \frac{A-B}{2} \quad \text{where } A = x + \delta x \text{ and } B = x$$

$$\therefore \; \delta y = 2 \cos \left(\frac{2x + \delta x}{2} \right) . \sin \left(\frac{\delta x}{2} \right) = 2 \cos \left(x + \frac{\delta x}{2} \right) . \sin \left(\frac{\delta x}{2} \right)$$

$$\therefore \; \frac{\delta y}{\delta x} = \frac{2 \cos \left(x + \frac{\delta x}{2} \right) . \sin \left(\frac{\delta x}{2} \right)}{\delta x} = \frac{\cos \left(x + \frac{\delta x}{2} \right) . \sin \left(\frac{\delta x}{2} \right)}{\frac{\delta x}{2}}$$

$$= \cos \left(x + \frac{\delta x}{2} \right) . \frac{\sin \left(\frac{\delta x}{2} \right)}{\frac{\delta x}{2}}$$

When $\delta x \to 0$, $\dfrac{\delta y}{\delta x} \to \dfrac{dy}{dx}$ and $\dfrac{dy}{dx} \to \cos x.1$ using the result of Frame 3

$$\therefore \; \text{If } y = \sin x, \; \frac{dy}{dx} = \cos x$$

Derivative of $y = \cos x$

This is obtained in much the same way as for the previous case.

If $y = \cos x$, $y + \delta y = \cos(x + \delta x)$ $\therefore$ $\delta y = \cos(x + \delta x) - \cos x$

Now we use the formula $\cos A - \cos B = -2 \sin \dfrac{A+B}{2} \sin \dfrac{A-B}{2}$

$$\therefore \; \delta y = -2 \sin \left(\frac{2x + \delta x}{2} \right) \cdot \sin \left(\frac{\delta x}{2} \right)$$

$$= -2 \sin \left(x + \frac{\delta x}{2} \right) . \sin \left(\frac{\delta x}{2} \right)$$

$$\therefore \; \frac{\delta y}{\delta x} = \frac{-2 \sin \left(x + \frac{\delta x}{2} \right) . \sin \left(\frac{\delta x}{2} \right)}{\delta x} = -\sin \left(x + \frac{\delta x}{2} \right) . \frac{\sin \left(\frac{\delta x}{2} \right)}{\frac{\delta x}{2}}$$

As $\delta x \to 0$, $\dfrac{\delta y}{\delta x} \to (-\sin x).1$ $\quad \therefore$ If $y = \cos x$, $\dfrac{dy}{dx} = -\sin x$ using the result of Frame 3

5

At this stage, there is one more derivative that we should determine.

Derivative of $y = e^x$

The series representation of e^x is:

$$y = e^x = 1 + x + \frac{x^2}{2!} + \frac{x^3}{3!} + \frac{x^4}{4!} + \dots$$

If we differentiate each power of x on the RHS, this gives

$$\frac{dy}{dx} = 0 + 1 + \frac{2x}{2!} + \frac{3x^2}{3!} + \frac{4x^3}{4!} + \dots$$

$$= 1 + x + \frac{x^2}{2!} + \frac{x^3}{3!} + \dots$$

$$= e^x$$

$$\therefore \text{ If } y = e^x, \ \frac{dy}{dx} = e^x$$

Note that e^x is a special function in which the derivative is equal to the function itself.

So, we have obtained some important standard results:

(a) If $y = x^n$, $\dfrac{dy}{dx} = n.x^{n-1}$

(b) If $y = c$ (a constant), $\dfrac{dy}{dx} = 0$

(c) A constant factor is unchanged, for example:

$$\text{if } y = a.x^n, \ \frac{dy}{dx} = a.n.x^{n-1}$$

(d) If $y = \sin x$, $\dfrac{dy}{dx} = \cos x$

(e) If $y = \cos x$, $\dfrac{dy}{dx} = -\sin x$

(f) if $y = e^x$, $\dfrac{dy}{dx} = e^x$

Now cover up the list of results above and complete the following table:

y	$\dfrac{dy}{dx}$	y	$\dfrac{dy}{dx}$
14		e^x	
$\cos x$		$5x^3$	
x^n		$\sin x$	

You can check these in the next frame

y	$\dfrac{dy}{dx}$	y	$\dfrac{dy}{dx}$
14	0	e^x	e^x
$\cos x$	$-\sin x$	$5x^3$	$15x^2$
x^n	nx^{n-1}	$\sin x$	$\cos x$

6

On to the next frame

Differentiation of products of functions

Let $y = uv$, where u and v are functions of x.
If $x \to x + \delta x$, $u \to u + \delta u$, $v \to v + \delta v$ and, as a result, $y \to y + \delta y$.

7

$$y = uv \quad \therefore\ y + \delta y = (u + \delta u)(v + \delta v)$$
$$= uv + u.\delta v + v.\delta u + \delta u.\delta v$$

Subtract $y = uv$

$$\therefore\ \delta y = u.\delta v + v.\delta u + \delta u.\delta v$$
$$\therefore\ \frac{\delta y}{\delta x} = u\frac{\delta v}{\delta x} + v\frac{\delta u}{\delta x} + \delta u.\frac{\delta v}{\delta x}$$

If $\delta x \to 0$, $\dfrac{\delta y}{\delta x} \to \dfrac{dy}{dx}$, $\dfrac{\delta u}{\delta x} \to \dfrac{du}{dx}$, $\dfrac{\delta v}{\delta x} \to \dfrac{dv}{dx}$, $\delta u \to 0$

$$\therefore\ \frac{dy}{dx} = u\frac{dv}{dx} + v\frac{du}{dx} + 0\frac{dv}{dx} \qquad \therefore\ \text{If } y = uv,\ \frac{dy}{dx} = u\frac{dv}{dx} + v\frac{du}{dx}$$

i.e. To differentiate a product of two functions:

Put down the first (differentiate the second) + put down the second (differentiate the first)

Example 1

$$y = x^3.\sin x$$
$$\frac{dy}{dx} = x^3(\cos x) + \sin x(3x^2)$$
$$= x^3.\cos x + 3x^2.\sin x = x^2(x\cos x + 3\sin x)$$

Example 2

$$y = x^4.\cos x$$
$$\frac{dy}{dx} = x^4(-\sin x) + \cos x.(4x^3)$$
$$= -x^4\sin x + 4x^3\cos x = x^3(4\cos x - x\sin x)$$

Example 3

$$y = x^5 . e^x$$

$$\frac{dy}{dx} = x^5 e^x + e^x 5x^4 = x^4 e^x (x + 5)$$

In the same way, as an exercise, now you can differentiate the following:

1 $y = e^x . \sin x$ 4 $y = \cos x . \sin x$

2 $y = 4x^3 . \sin x$ 5 $y = 3x^3 . e^x$

3 $y = e^x . \cos x$ 6 $y = 2x^5 . \cos x$

Finish all six and then check with the next frame

8

1 $\dfrac{dy}{dx} = e^x . \cos x + e^x . \sin x = e^x (\cos x + \sin x)$

2 $\dfrac{dy}{dx} = 4x^3 . \cos x + 12x^2 . \sin x = 4x^2 (x \cos x + 3 \sin x)$

3 $\dfrac{dy}{dx} = e^x (-\sin x) + e^x \cos x = e^x (\cos x - \sin x)$

4 $\dfrac{dy}{dx} = \cos x . \cos x + \sin x(-\sin x) = \cos^2 x - \sin^2 x$

5 $\dfrac{dy}{dx} = 3x^3 . e^x + 9x^2 . e^x = 3x^2 e^x (x + 3)$

6 $\dfrac{dy}{dx} = 2x^5 (-\sin x) + 10x^4 . \cos x = 2x^4 (5 \cos x - x \sin x)$

Now we will see how to deal with the quotient of two functions

---

Differentiation of a quotient of two functions

9

Let $y = \dfrac{u}{v}$, where u and v are functions of x.

$$\text{Then } y + \delta y = \frac{u + \delta u}{v + \delta v}$$

$$\therefore \delta y = \frac{u + \delta u}{v + \delta v} - \frac{u}{v}$$

$$= \frac{uv + v.\delta u - uv - u.\delta v}{v(v + \delta v)} = \frac{v.\delta u - u.\delta v}{v^2 + v.\delta v}$$

$$\therefore \frac{\delta y}{\delta x} = \frac{v\dfrac{\delta u}{\delta x} - u\dfrac{\delta v}{\delta x}}{v^2 + v.\delta v}$$

If $\delta x \to 0$, $\delta u \to 0$ and $\delta v \to 0$

$$\therefore \frac{dy}{dx} = \frac{v\dfrac{du}{dx} - u\dfrac{dv}{dx}}{v^2}$$

$$\therefore \text{If } y = \frac{u}{v}, \quad \frac{dy}{dx} = \frac{v\dfrac{du}{dx} - u\dfrac{dv}{dx}}{v^2}$$

$\therefore$ To differentiate a quotient of two functions:

[Put down the bottom (differentiate the top) – put down the top (differentiate the bottom)] all over the bottom squared.

$$\therefore \text{If } y = \frac{\sin x}{x^2}, \quad \frac{dy}{dx} = \ldots\ldots\ldots\ldots$$

$$\boxed{\dfrac{x\cos x - 2\sin x}{x^3}}$$

10

Because

$$y = \frac{\sin x}{x^2} \quad \therefore \quad \frac{dy}{dx} = \frac{x^2 \cos x - \sin x.(2x)}{(x^2)^2}$$

$$= \frac{x^2 \cos x - 2x \sin x}{x^4} = \frac{x \cos x - 2 \sin x}{x^3}$$

Another example: $y = \dfrac{5e^x}{\cos x}$

$$y = \frac{u}{v}, \quad \frac{dy}{dx} = \frac{v\dfrac{du}{dx} - u\dfrac{dv}{dx}}{v^2}$$

$$= \frac{\cos x.(5e^x) - 5e^x(-\sin x)}{\cos^2 x}$$

$$\therefore \quad \frac{dy}{dx} = \frac{5e^x(\cos x + \sin x)}{\cos^2 x}$$

Now let us deal with this one:

$$y = \frac{\sin x}{\cos x}$$

$$\frac{dy}{dx} = \frac{\cos x. \cos x - \sin x(-\sin x)}{\cos^2 x}$$

$$= \frac{\cos^2 x + \sin^2 x}{\cos^2 x} \qquad \text{But } \sin^2 x + \cos^2 x = 1$$

$$= \frac{1}{\cos^2 x} \qquad\qquad \text{Also } \frac{1}{\cos x} = \sec x$$

$$= \sec^2 x \qquad\qquad \text{and } \frac{\sin x}{\cos x} = \tan x$$

$$\therefore \text{ If } y = \tan x, \quad \frac{dy}{dx} = \sec^2 x$$

This is another one for our list of standard derivatives, so make a note of it. ▶

We now have the following standard derivatives:

y	$\dfrac{dy}{dx}$
x^n	
c	
$a.x^n$	
$\sin x$	
$\cos x$	
$\tan x$	
e^x	

Fill in the results in the right-hand column and then check with the next frame

11

y	$\dfrac{dy}{dx}$
x^n	$n.x^{n-1}$
c	0
$a.x^n$	$an.x^{n-1}$
$\sin x$	$\cos x$
$\cos x$	$-\sin x$
$\tan x$	$\sec^2 x$
e^x	e^x

Also **1**   If $y = uv$, $\dfrac{dy}{dx} = \ldots\ldots\ldots$

and **2**   If $y = \dfrac{u}{v}$, $\dfrac{dy}{dx} = \ldots\ldots\ldots$

12

$$\mathbf{1}\quad y = uv, \quad \frac{dy}{dx} = u\frac{dv}{dx} + v\frac{du}{dx}$$

$$\mathbf{2}\quad y = \frac{u}{v}, \quad \frac{dy}{dx} = \frac{v\dfrac{du}{dx} - u\dfrac{dv}{dx}}{v^2}$$

Now here is an exercise covering the work we have been doing. In each of the following functions, determine an expression for $\dfrac{dy}{dx}$:

1 $y = x^2.\cos x$ **2** $y = e^x.\sin x$

3 $y = \dfrac{4e^x}{\sin x}$ **4** $y = \dfrac{\cos x}{x^4}$

5 $y = 5x^3.e^x$ **6** $y = 4x^2.\tan x$

7 $y = \dfrac{\cos x}{\sin x}$ **8** $y = \dfrac{\tan x}{e^x}$

9 $y = x^5.\sin x$ **10** $y = \dfrac{3x^2}{\cos x}$

All the solutions are shown in the next frame

13

1 $y = x^2.\cos x$

$\dfrac{dy}{dx} = x^2(-\sin x) + \cos x.(2x)$

$= x(2\cos x - x\sin x)$

2 $y = e^x.\sin x$

$\dfrac{dy}{dx} = e^x.\cos x + \sin x(e^x)$

$= e^x(\sin x + \cos x)$

3 $y = \dfrac{4e^x}{\sin x}$

$\dfrac{dy}{dx} = \dfrac{\sin x.(4e^x) - 4e^x.\cos x}{\sin^2 x}$

$= \dfrac{4e^x(\sin x - \cos x)}{\sin^2 x}$

4 $y = \dfrac{\cos x}{x^4}$

$\dfrac{dy}{dx} = \dfrac{x^4(-\sin x) - \cos x(4x^3)}{x^8}$

$= \dfrac{-x\sin x - 4\cos x}{x^5}$

5 $y = 5x^3.e^x$

$\dfrac{dy}{dx} = 5x^3.e^x + e^x.15x^2$

$= 5x^2 e^x(x + 3)$

6 $y = 4x^2.\tan x$

$\dfrac{dy}{dx} = 4x^2.\sec^2 x + \tan x.(8x)$

$= 4x(x.\sec^2 x + 2\tan x)$

7 $y = \dfrac{\cos x}{\sin x}$

$\dfrac{dy}{dx} = \dfrac{\sin x(-\sin x) - \cos x.\cos x}{\sin^2 x}$

$= \dfrac{-(\sin^2 x + \cos^2 x)}{\sin^2 x}$

$= -\dfrac{1}{\sin^2 x} = -\operatorname{cosec}^2 x$

8 $y = \dfrac{\tan x}{e^x}$

$\dfrac{dy}{dx} = \dfrac{e^x.\sec^2 x - \tan x.e^x}{e^{2x}}$

$= \dfrac{\sec^2 x - \tan x}{e^x}$

9 $y = x^5.\sin x$

$\dfrac{dy}{dx} = x^5.\cos x + \sin x.(5x^4)$

$= x^4(x\cos x + 5\sin x)$

10 $y = \dfrac{3x^2}{\cos x}$

$\dfrac{dy}{dx} = \dfrac{\cos x.6x - 3x^2(-\sin x)}{\cos^2 x}$

$= \dfrac{3x(2\cos x + x\sin x)}{\cos^2 x}$

Check the results and then move on to the next frame

Function of a function

14

If $y = \sin x$, y is a function of the angle x, since the value of y depends on the value given to x.

If $y = \sin(2x - 3)$, y is a function of the angle $(2x - 3)$ which is itself a function of x.

Therefore, y is a function of (a function of x) and is said to be a *function of a function* of x.

Differentiation of a function of a function

To differentiate a function of a function, we must first introduce the *chain rule*.

With the example above, $y = \sin(2x - 3)$, we put $u = (2x - 3)$

i.e. $y = \sin u$ where $u = 2x - 3$.

If x has an increase δx, u will have an increase δu and then y will have an increase δy, i.e. $x \to x + \delta x$, $u \to u + \delta u$ and $y \to y + \delta y$.

At this stage, the increases δx, δu and δy are all finite values and therefore we can say that $\dfrac{\delta y}{\delta x} = \dfrac{\delta y}{\delta u} \times \dfrac{\delta u}{\delta x}$ because the δu in $\dfrac{\delta y}{\delta u}$ cancels the δu in $\dfrac{\delta u}{\delta x}$.

If now $\delta x \to 0$, then $\delta u \to 0$ and $\delta y \to 0$

Also $\dfrac{\delta y}{\delta x} \to \dfrac{dy}{dx}$, $\dfrac{\delta y}{\delta u} \to \dfrac{dy}{du}$ and $\dfrac{\delta u}{\delta x} \to \dfrac{du}{dx}$, and the previous statement now becomes $\dfrac{dy}{dx} = \dfrac{dy}{du} \cdot \dfrac{du}{dx}$

This is the *chain rule* and is particularly useful when determining the derivatives of a function of a function.

Example 1

To differentiate $y = \sin(2x - 3)$

Put $u = (2x - 3)$ $\therefore y = \sin u$

$\therefore \dfrac{du}{dx} = 2$ and $\dfrac{dy}{du} = \cos u$

$\dfrac{dy}{dx} = \dfrac{dy}{du} \cdot \dfrac{du}{dx} = \cos u.(2) = 2\cos u = 2\cos(2x - 3)$

$\therefore$ If $y = \sin(2x - 3)$, $\dfrac{dy}{dx} = 2\cos(2x - 3)$

Further examples follow

Example 2

If $y = (3x + 5)^4$, determine $\dfrac{dy}{dx}$

$y = (3x + 5)^4$. Put $u = (3x + 5)$. $\therefore y = u^4$

$\therefore \dfrac{dy}{du} = 4u^3$ and $\dfrac{du}{dx} = 3$. $\dfrac{dy}{dx} = \dfrac{dy}{du} \cdot \dfrac{du}{dx}$

$\therefore \dfrac{dy}{dx} = 4u^3.(3) = 12u^3 = 12(3x + 5)^3$

$\therefore$ If $y = (3x + 5)^4$, $\dfrac{dy}{dx} = 12(3x + 5)^3$

And in just the same way:

Example 3

If $y = \tan(4x + 1)$, $\dfrac{dy}{dx} = \ldots\ldots\ldots$

16

$$4\sec^2(4x + 1)$$

Because

$y = \tan(4x + 1)$ $\quad \therefore$ Put $u = 4x + 1$ $\quad \therefore \dfrac{du}{dx} = 4$

$y = \tan u$ $\quad \therefore \dfrac{dy}{du} = \sec^2 u$

$\dfrac{dy}{dx} = \dfrac{dy}{du} \cdot \dfrac{du}{dx} = \sec^2 u.(4) = 4\sec^2(4x + 1)$

$\therefore$ If $y = \tan(4x + 1)$, $\dfrac{dy}{dx} = 4\sec^2(4x + 1)$

And now this one:

Example 4

If $y = e^{5x}$, $\dfrac{dy}{dx} = \ldots\ldots\ldots$

17

$$5e^{5x}$$

Because

$y = e^{5x}$ $\quad \therefore$ Put $u = 5x$ $\quad \therefore \dfrac{du}{dx} = 5$

$y = e^u$ $\quad \therefore \dfrac{dy}{du} = e^u$

$\dfrac{dy}{dx} = \dfrac{dy}{du} \cdot \dfrac{du}{dx} = e^u.(5) = 5e^u = 5e^{5x}$

$\therefore$ If $y = e^{5x}$, $\dfrac{dy}{dx} = 5e^{5x}$

Many of these functions can be differentiated at sight by slight modification
to our list of standard derivatives:

F is a function of x			
y	$\dfrac{dy}{dx}$	y	$\dfrac{dy}{dx}$
F^n	$nF^{n-1}.\dfrac{dF}{dx}$	$\cos F$	$-\sin F.\dfrac{dF}{dx}$
$a.F^n$	$a.nF^{n-1}.\dfrac{dF}{dx}$	$\tan F$	$\sec^2 F.\dfrac{dF}{dx}$
$\sin F$	$\cos F.\dfrac{dF}{dx}$	e^F	$e^F.\dfrac{dF}{dx}$

Let us now apply these results

18

Here are four examples:

1 $y = \cos(2x - 1)$, $\dfrac{dy}{dx} = -\sin(2x - 1) \times 2 = -2\sin(2x - 1)$

2 $y = e^{(3x+4)}$, $\dfrac{dy}{dx} = e^{(3x+4)} \times 3 = 3.e^{(3x+4)}$

3 $y = (5x - 2)^3$, $\dfrac{dy}{dx} = 3(5x - 2)^2 \times 5 = 15(5x - 2)^2$

4 $y = 4.e^{\sin x}$, $\dfrac{dy}{dx} = 4.e^{\sin x} \times \cos x = 4\cos x.e^{\sin x}$

In just the same way, as an exercise, differentiate the following:

1 $y = \sin(4x + 3)$ **4** $y = \tan 5x$
2 $y = (2x - 5)^4$ **5** $y = e^{2x-3}$
3 $y = \sin^3 x$ **6** $y = 4\cos(7x + 2)$

19

1 $\dfrac{dy}{dx} = 4.\cos(4x + 3)$		**4** $\dfrac{dy}{dx} = 5.\sec^2 5x$	
2 $\dfrac{dy}{dx} = 8.(2x - 5)^3$		**5** $\dfrac{dy}{dx} = 2.e^{2x-3}$	
3 $\dfrac{dy}{dx} = 3.\sin^2 x \cos x$		**6** $\dfrac{dy}{dx} = -28.\sin(7x + 2)$	

▶

Now let us consider the derivative of $y = \ln x$:

   If $y = \ln x$ then $x = e^y$

Differentiating with respect to x: $\dfrac{d}{dx}(x) = \dfrac{d}{dx}(e^y)$ we find that the derivative on the LHS is equal to 1 and, by the chain rule, the derivative on the RHS is $\dfrac{d}{dy}(e^y)\dfrac{dy}{dx} = e^y \dfrac{dy}{dx}$ so that:

   $1 = e^y \dfrac{dy}{dx}$ and, since $x = e^y$ this can be written as:

   $1 = x\dfrac{dy}{dx}$ therefore $\dfrac{dy}{dx} = \dfrac{1}{x}$. Therefore if $y = \ln x$, $\dfrac{dy}{dx} = \dfrac{1}{x}$

We can add this to our list of standard derivatives. Also, if F is a function of x then:

   if $y = \ln F$ we have $\dfrac{dy}{dx} = \dfrac{1}{F} \cdot \dfrac{dF}{dx}$.

   Here is an example:

   If $y = \ln(3x - 5)$, $\dfrac{dy}{dx} = \dfrac{1}{3x - 5} \cdot 3 = \dfrac{3}{3x - 5}$

   and if $y = \ln(\sin x)$, $\dfrac{dy}{dx} = \dfrac{1}{\sin x} \cdot \cos x = \cot x$

There is one further standard derivative to be established at this stage, so move on to the next frame

---

Derivative of $y = a^x$ **20**

We already know that   if $y = e^x$, $\dfrac{dy}{dx} = e^x$

                  and that   if $y = e^F$, $\dfrac{dy}{dx} = e^F \cdot \dfrac{dF}{dx}$

Then, if $y = a^x$, we can write $a = e^k$ and then $y = a^x = (e^k)^x = e^{kx}$

$\therefore \dfrac{dy}{dx} = e^{kx} \cdot \dfrac{d}{dx}(kx) = e^{kx}(k) = k.e^{kx}$

But, $e^{kx} = a^x$ and $k = \ln a$ $\therefore$ If $y = a^x$, $\dfrac{dy}{dx} = a^x \ln a$

We can add this result to our list for future reference.

Newton–Raphson iterative method

21

Consider the graph of $y = f(x)$ as shown. Then the x-value at the point A, where the graph crosses the x-axis, gives a solution of the equation $f(x) = 0$.

If P is a point on the curve near to A, then $x = x_0$ is an approximate value of the root of $f(x) = 0$, the error of the approximation being given by AB.

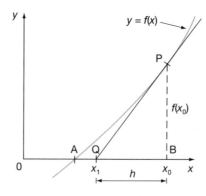

Let PQ be the tangent to the curve at P, crossing the x-axis at Q $(x_1, 0)$. Then $x = x_1$ is a better approximation to the required root.

From the diagram, $\dfrac{\text{PB}}{\text{QB}} = \left[\dfrac{dy}{dx}\right]_\text{P}$ i.e. the value of the derivative of y at the point P, $x = x_0$.

$$\therefore \frac{\text{PB}}{\text{QB}} = f'(x_0) \quad \text{and} \quad \text{PB} = f(x_0)$$

$$\therefore \text{QB} = \frac{\text{PB}}{f'(x_0)} = \frac{f(x_0)}{f'(x_0)} = h \text{ (say)}$$

$$x_1 = x_0 - h \qquad \therefore x_1 = x_0 - \frac{f(x_0)}{f'(x_0)}$$

If we begin, therefore, with an approximate value (x_0) of the root, we can determine a better approximation (x_1). Naturally, the process can be repeated to improve the result still further. Let us see this in operation.

On to the next frame

Example 1

22

The equation $x^3 - 3x - 4 = 0$ is of the form $f(x) = 0$ where $f(1) < 0$ and $f(3) > 0$ so there is a solution to the equation between 1 and 3. We shall take this to be 2, by bisection. Find a better approximation to the root.

We have $f(x) = x^3 - 3x - 4 \quad \therefore f'(x) = 3x^2 - 3$

If the first approximation is $x_0 = 2$, then

$$f(x_0) = f(2) = -2 \quad \text{and} \quad f'(x_0) = f'(2) = 9$$

A better approximation x_1 is given by

$$x_1 = x_0 - \frac{f(x_0)}{f'(x_0)} = x_0 - \frac{x_0{}^3 - 3x_0 - 4}{3x_0{}^2 - 3}$$

$$x_1 = 2 - \frac{(-2)}{9} = 2 \cdot 22$$

$$\therefore x_0 = 2; \quad x_1 = 2 \cdot 22$$

If we now start from x_1 we can get a better approximation still by repeating the process.

$$x_2 = x_1 - \frac{f(x_1)}{f'(x_1)} = x_1 - \frac{x_1{}^3 - 3x_1 - 4}{3x_1{}^2 - 3}$$

Here $x_1 = 2 \cdot 22 \qquad f(x_1) = \ldots\ldots\ldots\ldots ; \quad f'(x_1) = \ldots\ldots\ldots\ldots$

$$\boxed{f(x_1) = 0 \cdot 281; \quad f'(x_1) = 11 \cdot 785}$$

23

Then $x_2 = \ldots\ldots\ldots\ldots$

$$\boxed{x_2 = 2 \cdot 196}$$

24

Because

$$x_2 = 2 \cdot 22 - \frac{0 \cdot 281}{11 \cdot 79} = 2 \cdot 196$$

Using $x_2 = 2 \cdot 196$ as a starter value, we can continue the process until successive results agree to the desired degree of accuracy.

$$x_3 = \ldots\ldots\ldots\ldots$$

25

$$x_3 = 2.196$$

Because

$$f(x_2) = f(2.196) = 0.002026; \qquad f'(x_2) = f'(2.196) = 11.467$$

$$\therefore x_3 = x_2 - \frac{f(x_2)}{f'(x_2)} = 2.196 - \frac{0.00203}{11.467} = 2.196 \text{ (to 4 sig fig)}$$

The process is simple but effective and can be repeated again and again. Each repetition, or *iteration*, usually gives a result nearer to the required root $x = x_A$.

In general $x_{n+1} = \dots\dots\dots$

26

$$x_{n+1} = x_n - \frac{f(x_n)}{f'(x_n)}$$

Tabular display of results

Open your spreadsheet and in cells **A1** to **D1** enter the headings n, x, $f(x)$ and $f'(x)$

Fill cells **A2** to **A6** with the numbers 0 to 4

In cell **B2** enter the value for x_0, namely 2

In cell **C2** enter the formula for $f(x_0)$, namely **= B2^3 – 3*B2 – 4** and copy into cells **C3** to **C6**

In cell **D2** enter the formula for $f'(x_0)$, namely **= 3*B2^2 – 3** and copy into cells **D3** to **D6**

In cell **B3** enter the formula for x_1, namely **= B2 – C2/D2** and copy into cells **B4** to **B6**.

The final display is $\dots\dots\dots$

27

n	x	$f(x)$	$f'(x)$
0	2	−2	9
1	2·222222	0·30727	11·81481
2	2·196215	0·004492	11·47008
3	2·195823	1·01E-06	11·46492
4	2·195823	5·15E-14	11·46492

As soon as the number in the second column is repeated then we know that we have arrived at that particular level of accuracy. The required root is therefore $x = 2.195823$ to 6 dp. Save the spreadsheet so that it can be used as a template for other such problems.

Now let us have another example.

Next frame

Example 2

28

The equation $x^3 + 2x^2 - 5x - 1 = 0$ is of the form $f(x) = 0$ where $f(1) < 0$ and $f(2) > 0$ so there is a solution to the equation between 1 and 2. We shall take this to be $x = 1.5$. Use the Newton–Raphson method to find the root to six decimal places.

Use the previous spreadsheet as a template and make the following amendments:

In cell **B2** enter the number ....

1·5

29

Because

That is the value of x_0 that is used to start the iteration

In cell **C2** enter the formula ....

= B2^3 + 2*B2^2 – 5*B2 – 1

30

Because

That is the value of $f(x_0) = x_0{}^3 + 2x_0{}^2 - 5x_0 - 1$. Copy the contents of cell **C2** into cells **C3** to **C5**.

In cell **D2** enter the formula ....

= 3*B2^2 + 4*B2 – 5

31

Because

That is the value of $f'(x_0) = 3x_0{}^2 + 4x_0 - 5$. Copy the contents of cell **D2** into cells **D3** to **D5**.

In cell **B2** the formula remains the same as ....

= B2 – C2/D2

32

The final display is then ....

33

n	x	$f(x)$	$f'(x)$
0	1·5	−0·625	7·75
1	1·580645	0·042798	8·817898
2	1·575792	0·000159	8·752524
3	1·575773	2·21E-09	8·75228

We cannot be sure that the value 1·575773 is accurate to the sixth decimal place so we must extend the table.

Highlight cells **A5** to **D5**, click **Edit** on the Command bar and select **Copy** from the drop-down menu.

Place the cell highlight in cell **A6**, click **Edit** and then **Paste**.

The sixth row of the spreadsheet then fills to produce the display

n	x	$f(x)$	$f'(x)$
0	1·5	−0·625	7·75
1	1·580645	0·042798	8·817898
2	1·575792	0·000159	8·752524
3	1·575773	2·21E-09	8·75228
4	1·575773	−8·9E-16	8·75228

And the repetition of the x-value ensures that the solution $x = 1·575773$ is indeed accurate to 6 dp.

Now do one completely on your own.

Next frame

34   *Example 3*

The equation $2x^3 - 7x^2 - x + 12 = 0$ has a root near to $x = 1·5$. Use the Newton–Raphson method to find the root to six decimal places.

The spreadsheet solution produces ....

35 $$x = 1·686141 \text{ to 6 dp}$$

Because

Fill cells **A2** to **A6** with the numbers 0 to 4

In cell **B2** enter the value for x_0, namely 1·5

In cell **C2** enter the formula for $f(x_0)$, namely **= 2*B2^3 − 7*B2^2 − B2 + 12** and copy into cells **C3** to **C6**

In cell **D2** enter the formula for $f'(x_0)$, namely **= 6*B2^2 − 14*B2 − 1** and copy into cells **D3** to **D6**

In cell **B3** enter the formula for x_1, namely **= B2 − C2/D2** and copy into cells **B4** to **B6**.

The final display is ....

36

n	x	f(x)	f'(x)
0	1·5	1·5	−8·5
1	1·676471	0·073275	−7·60727
2	1·686103	0·000286	−7·54778
3	1·686141	4·46E-09	−7·54755
4	1·686141	0	−7·54755

As soon as the number in the second column is repeated then we know that we have arrived at that particular level of accuracy. The required root is therefore $x = 1·686141$ to 6 dp.

First approximations

The whole process hinges on knowing a 'starter' value as first approximation. If we are not given a hint, this information can be found by either

(a) applying the remainder theorem if the function is a polynomial
(b) drawing a sketch graph of the function.

Example 4

Find the real root of the equation $x^3 + 5x^2 - 3x - 4 = 0$ correct to six significant figures.

Application of the remainder theorem involves substituting $x = 0$, $x = \pm 1$, $x = \pm 2$, etc. until two adjacent values give a change in sign.

$$f(x) = x^3 + 5x^2 - 3x - 4$$
$$f(0) = -4; \quad f(1) = -1; \quad f(-1) = 3$$

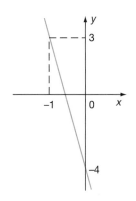

The sign changes from $f(0)$ to $f(-1)$. There is thus a root between $x = 0$ and $x = -1$.

Therefore choose $x = -0·5$ as the first approximation and then proceed as before.

Complete the table and obtain the root
$$x = \ldots\ldots\ldots\ldots$$

$$\boxed{x = -0·675527}$$

37

The final spreadsheet display is

n	x	f(x)	f'(x)
0	−0·5	−1·375	−7·25
1	−0·689655	0·11907	−8·469679
2	−0·675597	0·000582	−8·386675
3	−0·675527	1·43E-08	−8·386262
4	−0·675527	0	−8·386262

38

Example 5

Solve the equation $e^x + x - 2 = 0$ giving the root to 6 significant figures.

It is sometimes more convenient to obtain a first approximation to the required root from a sketch graph of the function, or by some other graphical means.

In this case, the equation can be rewritten as $e^x = 2 - x$ and we therefore sketch graphs of $y = e^x$ and $y = 2 - x$.

x	0·2	0·4	0·6	0·8	1
e^x	1·22	1·49	1·82	2·23	2·72
$2 - x$	1·8	1·6	1·4	1·2	1

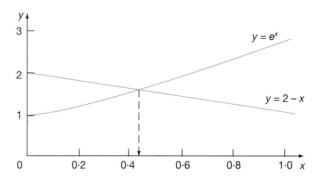

It can be seen that the two curves cross over between $x = 0·4$ and $x = 0·6$.
Approximate root $x = 0·4$

$$f(x) = e^x + x - 2 \qquad f'(x) = e^x + 1$$

$$x = \ldots\ldots\ldots\ldots$$

Finish it off

39

$$x = 0·442854$$

The final spreadsheet display is

n	x	$f(x)$	$f'(x)$
0	0·4	−0·10818	2·491825
1	0·443412	0·001426	2·558014
2	0·442854	2·42E-07	2·557146
3	0·442854	7·11E-15	2·557146

Note: There are times when the normal application of the Newton–Raphson method fails to converge to the required root. This is particularly so when $f'(x_0)$ is very small.

At this point let us pause and summarize the main facts so far on further differentiation

Review summary

1 *Standard derivatives*

40

y	$\dfrac{dy}{dx}$	y	$\dfrac{dy}{dx}$
x^n	$n.x^{n-1}$	$\tan x$	$\sec^2 x$
c	0	e^x	e^x
$\sin x$	$\cos x$	$\ln x$	$\dfrac{1}{x}$
$\cos x$	$-\sin x$	a^x	$a^x \ln a$

2 *Rules of derivatives*

(a) of product $y = uv$, $\dfrac{dy}{dx} = u\dfrac{dv}{dx} + v\dfrac{du}{dx}$

(b) of quotient $y = \dfrac{u}{v}$, $\dfrac{dy}{dx} = \dfrac{v\dfrac{du}{dx} - u\dfrac{dv}{dx}}{v^2}$

3 *Chain rule*

F is a function of x			
y	$\dfrac{dy}{dx}$	y	$\dfrac{dy}{dx}$
F^n	$nF^{n-1}.\dfrac{dF}{dx}$	$\tan F$	$\sec^2 F.\dfrac{dF}{dx}$
$\sin F$	$\cos F.\dfrac{dF}{dx}$	e^F	$e^F.\dfrac{dF}{dx}$
$\cos F$	$-\sin F.\dfrac{dF}{dx}$	$\ln F$	$\dfrac{1}{F}.\dfrac{dF}{dx}$

4 *Newton–Raphson method*

If $x = x_0$ is an approximate solution to the equation $f(x) = 0$ then $x_1 = x_0 + \dfrac{f(x_0)}{f'(x_0)}$ gives a more accurate value. This defines an iterative

procedure $x_{n+1} = x_n + \dfrac{f(x_n)}{f'(x_n)}$ for $n \geq 0$ to find progressively more accurate values.

Review exercise   **Unit 53**

41

1 Determine $\dfrac{dy}{dx}$ in each of the following cases:

(a) $y = x^3 \tan x$ (b) $y = x^2 e^x$

(c) $y = \dfrac{2e^x}{x^2}$ (d) $y = \dfrac{x^3}{\sin x}$

2 Differentiate the following with respect to x:

(a) $y = (4x + 3)^6$ (b) $y = \tan(2x + 3)$

(c) $y = \ln(3x - 4)$ (d) $y = -2e^{(1-3x)}$

(e) $y = e^{-3x} \sin(2x)$ (f) $y = \tan^2 x$

3 (a) If $y = x^3 + 2x^2 - 3x - 4$, determine:

(i) $\dfrac{dy}{dx}$ and $\dfrac{d^2y}{dx^2}$

(ii) the values of x at which $\dfrac{dy}{dx} = 0$.

(b) If $y = 2\cos(x + 1) - 3\sin(x - 1)$, obtain expressions for $\dfrac{dy}{dx}$ and $\dfrac{d^2y}{dx^2}$.

4 Use the Newton–Raphson method to solve the equation $x^2 - 5 = 0$, accurate to 6 decimal places given that $x = 2 \cdot 2$ is the solution accurate to 1 decimal place.

Complete all four questions. Take your time, there is no need to rush.
If necessary, refer back to the Unit.
The answers and working are in the next frame.

42

1 (a) $y = x^3 \tan x$. Applying the product rule we find that:

$$\frac{dy}{dx} = 3x^2 \tan x + x^3 \sec^2 x$$
$$= x^2(3 \tan x + x \sec^2 x)$$

(b) $y = x^2 e^x$. Applying the product rule we find that:

$$\frac{dy}{dx} = 2xe^x + x^2 e^x$$
$$= xe^x(2 + x)$$

(c) $y = \dfrac{2e^x}{x^2}$. Applying the quotient rule we find that:

$$\frac{dy}{dx} = \frac{2e^x x^2 - 2e^x 2x}{[x^2]^2}$$
$$= 2e^x \left(\frac{x^2 - 2x}{x^4} \right)$$
$$= 2e^x \left(\frac{x - 2}{x^3} \right)$$

(d) $y = \dfrac{x^3}{\sin x}$. Applying the quotient rule we find that:

$$\frac{dy}{dx} = \frac{3x^2 \sin x - x^3 \cos x}{[\sin x]^2}$$

$$= x^2 \left(\frac{3 \sin x - x \cos x}{\sin^2 x} \right)$$

2 (a) $y = (4x + 3)^6$. Applying the chain rule we find that:

$$\frac{dy}{dx} = 6(4x + 3)^5 \times 4$$

$$= 24(4x + 3)^5$$

(b) $y = \tan(2x + 3)$. Applying the chain rule we find that:

$$\frac{dy}{dx} = \sec^2(2x + 3) \times 2$$

$$= 2 \sec^2(2x + 3)$$

(c) $y = \ln(3x - 4)$. Applying the chain rule we find that:

$$\frac{dy}{dx} = \frac{1}{3x - 4} \times 3$$

$$= \frac{3}{3x - 4}$$

(d) $y = -2e^{(1-3x)}$. Applying the chain rule we find that:

$$\frac{dy}{dx} = -2e^{(1-3x)} \times (-3)$$

$$= 6e^{(1-3x)}$$

(e) $y = e^{-3x} \sin(2x)$. Applying the chain rule and product rule combined we find that:

$$\frac{dy}{dx} = -3e^{-3x} \times \sin(2x) + e^{-3x} \times 2 \cos(2x)$$

$$= e^{-3x}(2 \cos(2x) - 3 \sin(2x))$$

(f) $y = \tan^2 x$. Applying the chain rule we find that:

$$\frac{dy}{dx} = (2 \tan x) \times \sec^2 x$$

$$= 2 \tan x \sec^2 x$$

3 (a) If $y = x^3 + 2x^2 - 3x - 4$, then:

(i) $\dfrac{dy}{dx} = 3x^2 + 4x - 3$ and $\dfrac{d^2y}{dx^2} = \dfrac{d}{dx}\left(\dfrac{dy}{dx}\right) = 6x + 4.$

(ii) $\dfrac{dy}{dx} = 0$ when $3x^2 + 4x - 3 = 0$. That is when:

$$x = \frac{-4 \pm \sqrt{16 - 4 \times 3 \times (-3)}}{2 \times 3}$$

$$= \frac{-4 \pm \sqrt{52}}{6} = -1{\cdot}869 \text{ or } 0{\cdot}535 \text{ to 3 dp.}$$

(b) If $y = 2\cos(x+1) - 3\sin(x-1)$, then:

$$\frac{dy}{dx} = -2\sin(x+1) - 3\cos(x-1) \text{ and}$$

$$\frac{d^2y}{dx^2} = -2\cos(x+1) + 3\sin(x-1) = -y$$

4   Let $f(x) = x^2 - 5$ so that $f'(x) = 2x$. The iterative solution is then given as:

$$x_{n+1} = x_n + \frac{f(x_n)}{f'(x_n)} \text{ for } n \geq 0$$

We are given that $x = 2 \cdot 2$ is the solution accurate to 1 decimal place so we can set up our tabular display as follows where cell **C2** contains the formula **= B2^-5** and cell **D2** contains the formula **= 2*B2**:

n	x	$f(x)$	$f'(x)$
0	2·2000000	−0·1600000	4·4000000
1	2·2363636	0·0013223	4·4727273
2	2·2360680	0·0000001	4·4721360
3	2·2360680	0.0000000	4·4721360

Giving the solution as 2·236068 accurate to 6 decimal places.

Now for the Review test

 # Review test   **Unit 53**

43   **1**   (a) If $y = 4x^3 + x^2 - 6x - 3$, determine

   (i) $\dfrac{dy}{dx}$ and $\dfrac{d^2y}{dx^2}$

   (ii) the values of x at which $\dfrac{dy}{dx} = 0$.

   (b) If $y = 2\cos(9x - 1) - \sin(3x + 7)$, obtain expressions for $\dfrac{dy}{dx}$ and $\dfrac{d^2y}{dx^2}$

2   Determine $\dfrac{dy}{dx}$ in each of the following cases:

   (a) $y = x^4 \cos x$ (b) $y = x^2 \ln x$

   (c) $y = \dfrac{\sin x}{x}$ (d) $y = \dfrac{3\ln x}{x^4}$

3   Differentiate the following with respect to x:

   (a) $y = (3x - 2)^5$ (b) $y = 2\sin(3x - 1)$
   (c) $y = -e^{(2x+3)}$ (d) $y = \ln(3 - 2x)$
   (e) $y = e^{2x} \cos 3x$ (f) $y = (\ln x)\sin(2 - 5x)$

4   Use the Newton–Raphson method to solve the equation $x^3 - 3 = 0$, accurate to 6 decimal places given that $x = 1 \cdot 4$ is the solution accurate to 1 decimal place.

You have now come to the end of this Module. A list of **Can You?** questions follows for you to gauge your understanding of the material in the Module. You will notice that these questions match the **Learning outcomes** listed at the beginning of the Module. Now try the **Test exercise.** *Work through the questions at your own pace, there is no need to hurry.* A set of **Further problems** provides additional valuable practice.

44

Can You?

Checklist: Module 17

1

Check this list before and after you try the end of Module test.

On a scale of 1 to 5 how confident are you that you can:

- Determine the gradient of a straight-line graph?
  Yes ☐ ☐ ☐ ☐ ☐ No

- Evaluate from first principles the gradient at a point on a quadratic curve?
  Yes ☐ ☐ ☐ ☐ ☐ No

- Differentiate powers of x and polynomials?
  Yes ☐ ☐ ☐ ☐ ☐ No

- Evaluate second derivatives and use tables of standard derivatives?
  Yes ☐ ☐ ☐ ☐ ☐ No

- Differentiate products and quotients of expressions?
  Yes ☐ ☐ ☐ ☐ ☐ No

- Differentiate using the chain rule for a function of a function?
  Yes ☐ ☐ ☐ ☐ ☐ No

- Use the Newton–Raphson method to obtain a numerical solution to an equation?
  Yes ☐ ☐ ☐ ☐ ☐ No

Test exercise 17

2

1 Calculate the slope of the straight line joining:
   (a) P (3, 5) and Q (6, 9)   (c) P (− 3, 8) and Q (4, 2)
   (b) P(2, 6) and Q (7, 4)   (d) P (− 2, 5) and Q (3, − 8)

2 Determine algebraically, from first principles, the slope of the graph of
   $y = 3x^2 + 4$ at the point P where $x = 1\cdot 2$.

3 If $y = x^4 + 5x^3 − 6x^2 + 7x − 3$, obtain an expression for $\dfrac{dy}{dx}$

   and hence calculate the value of $\dfrac{dy}{dx}$ at $x = -2$.

4 (a) If $y = 2x^3 − 11x^2 + 12x − 5$, determine:

   (i) $\dfrac{dy}{dx}$ and $\dfrac{d^2y}{dx^2}$ (ii) the values of x at which $\dfrac{dy}{dx} = 0$.

   (b) If $y = 3\sin(2x + 1) + 4\cos(3x − 1)$, obtain expressions

   for $\dfrac{dy}{dx}$ and $\dfrac{d^2y}{dx^2}$.

5 Determine $\dfrac{dy}{dx}$ in each of the following cases:

   (a) $y = x^2.\sin x$ (b) $y = x^3.e^x$ (c) $y = \dfrac{\cos x}{x^2}$ (d) $y = \dfrac{2e^x}{\tan x}$

6 Differentiate the following with respect to x:
   (a) $y = (5x + 2)^4$ (d) $y = 5\cos(2x + 3)$
   (b) $y = \sin(3x + 2)$ (e) $y = \cos^3 x$
   (c) $y = e^{(4x−1)}$ (f) $y = \ln(4x − 5)$

7 Use the Newton–Raphson method to solve the equation $4 − 2x − x^3 = 0$,
   accurate to 6 decimal places.

Further problems 17

3

1 Determine algebraically from first principles, the slope of the following graphs
   at the value of x indicated:
   (a) $y = 4x^2 − 7$ at $x = -0\cdot 5$ (c) $y = 3x^3 − 2x^2 + x − 4$ at $x = -1$
   (b) $y = 2x^3 + x − 4$ at $x = 2$

2 Differentiate the following and calculate the value of dy/dx at the value of
   x stated:
   (a) $y = 2x^3 + 4x^2 − 2x + 7$ [$x = -2$]
   (b) $y = 3x^4 − 5x^3 + 4x^2 − x + 4$ [$x = 3$]
   (c) $y = 4x^5 + 2x^4 − 3x^3 + 7x^2 − 2x + 3$ [$x = 1$]

Differentiate the functions given in questions **3**, **4** and **5**:

3 (a) $y = x^5 \sin x$ (c) $y = x^3 \tan x$ (e) $y = 5x^2 \sin x$

   (b) $y = e^x \cos x$ (d) $y = x^4 \cos x$ (f) $y = 2e^x \ln x$

4 (a) $y = \dfrac{\cos x}{x^3}$ (c) $y = \dfrac{\cos x}{\tan x}$ (e) $y = \dfrac{\tan x}{e^x}$

   (b) $y = \dfrac{\sin x}{2e^x}$ (d) $y = \dfrac{4x^3}{\cos x}$ (f) $y = \dfrac{\ln x}{x^3}$

5 (a) $y = (2x - 3)^3$ (e) $y = \sin(2x - 3)$ (i) $y = \ln(3x^2)$

   (b) $y = e^{3x+2}$ (f) $y = \tan(x^2 - 3)$ (j) $y = 3\sin(4 - 5x)$

   (c) $y = 4\cos(3x + 1)$ (g) $y = 5(4x + 5)^2$

   (d) $y = \ln(x^2 + 4)$ (h) $y = 6e^{x^2+2}$

6 Use the Newton–Raphson iterative method to solve the following.

(a) Show that a root of the equation $x^3 + 3x^2 + 5x + 9 = 0$ occurs between $x = -2$ and $x = -3$. Evaluate the root to four significant figures.

(b) Show graphically that the equation $e^{2x} = 25x - 10$ has two real roots and find the larger root correct to four significant figures.

(c) Verify that the equation $x - \cos x = 0$ has a root near to $x = 0 \cdot 8$ and determine the root correct to three significant figures.

(d) Obtain graphically an approximate root of the equation $2\ln x = 3 - x$. Evaluate the root correct to four significant figures.

(e) Verify that the equation $x^4 + 5x - 20 = 0$ has a root at approximately $x = 1 \cdot 8$. Determine the root correct to five significant figures.

(f) Show that the equation $x + 3\sin x = 2$ has a root between $x = 0 \cdot 4$ and $x = 0 \cdot 6$. Evaluate the root correct to four significant figures.

(g) The equation $2\cos x = e^x - 1$ has a real root between $x = 0 \cdot 8$ and $x = 0 \cdot 9$. Evaluate the root correct to four significant figures.

(h) The equation $20x^3 - 22x^2 + 5x - 1 = 0$ has a root at approximately $x = 0 \cdot 8$. Determine the value of the root correct to four significant figures.

Module 18

Partial differentiation

Learning outcomes

When you have completed this Module you will be able to:

- Find the first partial derivatives of a function of two real variables
- Find second-order partial derivatives of a function of two real variables
- Calculate errors using partial differentiation

Units

Partial differentiation

Unit 54

Partial derivatives

1

The volume V of a cylinder of radius r and height h is given by

$$V = \pi r^2 h$$

i.e. V depends on two quantities, the values of r and h.

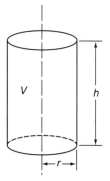

If we keep r constant and increase the height h, the volume V will increase. In these circumstances, we can consider the derivative of V with respect to h – but only if r is kept constant.

i.e. $\left[\dfrac{\mathrm{d}V}{\mathrm{d}h}\right]_{r\ \text{constant}}$ is written $\dfrac{\partial V}{\partial h}$

Notice the new type of 'delta'. We already know the meaning of $\dfrac{\delta y}{\delta x}$ and $\dfrac{\mathrm{d}y}{\mathrm{d}x}$.

Now we have a new one, $\dfrac{\partial V}{\partial h}$. $\dfrac{\partial V}{\partial h}$ is called the *partial derivative* of V with respect to h and implies that for our present purpose, the value of r is considered as being kept . . . .

---

constant

2

$V = \pi r^2 h$. To find $\dfrac{\partial V}{\partial h}$, we differentiate the given expression, taking all symbols

except V and h as being constant $\therefore \dfrac{\partial V}{\partial h} = \pi r^2 . 1 = \pi r^2$

Of course, we could have considered h as being kept constant, in which case, a change in r would also produce a change in V. We can therefore talk about $\dfrac{\partial V}{\partial r}$ which simply means that we now differentiate $V = \pi r^2 h$ with respect to r, taking all symbols except V and r as being constant for the time being.

$$\therefore \dfrac{\partial V}{\partial r} = \pi 2rh = 2\pi rh$$

In the statement $V = \pi r^2 h$, V is expressed as a function of two variables, r and h. It therefore has two partial derivatives, one with respect to . . . . and one with respect to . . . .

3

One with respect to r; one with respect to h

Another example:

Let us consider the area of the curved surface of the cylinder $A = 2\pi rh$

A is a function of r and h, so we can find $\dfrac{\partial A}{\partial r}$ and $\dfrac{\partial A}{\partial h}$

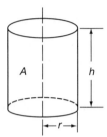

To find $\dfrac{\partial A}{\partial r}$ we differentiate the expression for A with respect to r, keeping all other symbols constant.

To find $\dfrac{\partial A}{\partial h}$ we differentiate the expression for A with respect to h, keeping all other symbols constant.

So, if $A = 2\pi rh$, then $\dfrac{\partial A}{\partial r} = \ldots\ldots\ldots$ and $\dfrac{\partial A}{\partial h} = \ldots\ldots\ldots$

4

$$\dfrac{\partial A}{\partial r} = 2\pi h \text{ and } \dfrac{\partial A}{\partial h} = 2\pi r$$

Of course, we are not restricted to the mensuration of the cylinder. The same will happen with any function which is a function of two independent variables. For example, consider $z = x^2y^3$.

Here z is a function of x and y. We can therefore find $\dfrac{\partial z}{\partial x}$ and $\dfrac{\partial z}{\partial y}$.

(a) To find $\dfrac{\partial z}{\partial x}$, differentiate with respect to x, regarding y as a constant.

$\therefore \dfrac{\partial z}{\partial x} = 2xy^3$

(b) To find $\dfrac{\partial z}{\partial y}$, differentiate with respect to y, regarding x as a constant.

$\dfrac{\partial z}{\partial y} = x^2 3y^2 = 3x^2y^2$

Partial differentiation is easy! For we regard every independent variable, except the one with respect to which we are differentiating, as being for the time being $\ldots\ldots\ldots$

$$\boxed{\text{constant}}$$

Here are some examples. 'With respect to' is abbreviated to w.r.t.

Example 1

$u = x^2 + xy + y^2$

(a) To find $\dfrac{\partial u}{\partial x}$, we regard y as being constant.

        Partial diff w.r.t. x of $x^2 = 2x$

        Partial diff w.r.t. x of $xy = y$ (y is a constant factor)

        Partial diff w.r.t. x of $y^2 = 0$ (y^2 is a constant term)

        $\dfrac{\partial u}{\partial x} = 2x + y$

(b) To find $\dfrac{\partial u}{\partial y}$, we regard x as being constant.

        Partial diff w.r.t. y of $x^2 = 0$ (x^2 is a constant term)

        Partial diff w.r.t. y of $xy = x$ (x is a constant factor)

        Partial diff w.r.t. y of $y^2 = 2y$

        $\dfrac{\partial u}{\partial y} = x + 2y$

Another example in Frame 6

Example 2

$z = x^3 + y^3 - 2x^2 y$

    $\dfrac{\partial z}{\partial x} = 3x^2 + 0 - 4xy = 3x^2 - 4xy$

    $\dfrac{\partial z}{\partial y} = 0 + 3y^2 - 2x^2 = 3y^2 - 2x^2$

And it is all just as easy as that.

Example 3

$z = (2x - y)(x + 3y)$

This is a product, and the usual product rule applies except that we keep y constant when finding $\dfrac{\partial z}{\partial x}$, and x constant when finding $\dfrac{\partial z}{\partial y}$.

    $\dfrac{\partial z}{\partial x} = (2x - y)(1 + 0) + (x + 3y)(2 - 0) = 2x - y + 2x + 6y = 4x + 5y$

    $\dfrac{\partial z}{\partial y} = (2x - y)(0 + 3) + (x + 3y)(0 - 1) = 6x - 3y - x - 3y = 5x - 6y$

Here is one for you to do.

If $z = (4x - 2y)(3x + 5y)$, find $\dfrac{\partial z}{\partial x}$ and $\dfrac{\partial z}{\partial y}$

Find the results and then move on to Frame 7

Find the results and then move on to Frame 7

7

$$\frac{\partial z}{\partial x} = 24x + 14y; \qquad \frac{\partial z}{\partial y} = 14x - 20y$$

Because $z = (4x - 2y)(3x + 5y)$, i.e. product

$$\therefore \frac{\partial z}{\partial x} = (4x - 2y)(3 + 0) + (3x + 5y)(4 - 0)$$

$$= 12x - 6y + 12x + 20y = 24x + 14y$$

$$\frac{\partial z}{\partial y} = (4x - 2y)(0 + 5) + (3x + 5y)(0 - 2)$$

$$= 20x - 10y - 6x - 10y = 14x - 20y$$

There we are. Now what about this one?

Example 4

If $z = \dfrac{2x - y}{x + y}$, find $\dfrac{\partial z}{\partial x}$ and $\dfrac{\partial z}{\partial y}$

Applying the quotient rule, we have:

$$\frac{\partial z}{\partial x} = \frac{(x + y)(2 - 0) - (2x - y)(1 + 0)}{(x + y)^2} = \frac{3y}{(x + y)^2}$$

and $\dfrac{\partial z}{\partial y} = \dfrac{(x + y)(0 - 1) - (2x - y)(0 + 1)}{(x + y)^2} = \dfrac{-3x}{(x + y)^2}$

That was not difficult. Now you do this one:

If $z = \dfrac{5x + y}{x - 2y}$, find $\dfrac{\partial z}{\partial x}$ and $\dfrac{\partial z}{\partial y}$

When you have finished, on to the next frame

8

$$\frac{\partial z}{\partial x}=\frac{-11y}{(x-2y)^2};\ \frac{\partial z}{\partial y}=\frac{11x}{(x-2y)^2}$$

Here is the working:

(a) To find $\dfrac{\partial z}{\partial x}$, we regard y as being constant.

$$\therefore\ \frac{\partial z}{\partial x}=\frac{(x-2y)(5+0)-(5x+y)(1-0)}{(x-2y)^2}$$
$$=\frac{5x-10y-5x-y}{(x-2y)^2}=\frac{-11y}{(x-2y)^2}$$

(b) To find $\dfrac{\partial z}{\partial y}$, we regard x as being constant.

$$\therefore\ \frac{\partial z}{\partial y}=\frac{(x-2y)(0+1)-(5x+y)(0-2)}{(x-2y)^2}$$
$$=\frac{x-2y+10x+2y}{(x-2y)^2}=\frac{11y}{(x-2y)^2}$$

In practice, we do not write down the zeros that occur in the working, but this is how we think.

Let us do one more example, so move on to the next frame

Example 5

9

If $z=\sin(3x+2y)$ find $\dfrac{\partial z}{\partial x}$ and $\dfrac{\partial z}{\partial y}$

Here we have what is clearly a 'function of a function'. So we apply the usual procedure, except to remember that when we are finding:

(a) $\dfrac{\partial z}{\partial x}$, we treat y as constant, and

(b) $\dfrac{\partial z}{\partial y}$, we treat x as constant.

Here goes then.

$$\frac{\partial z}{\partial x}=\cos(3x+2y)\times\frac{\partial}{\partial x}(3x+2y)=\cos(3x+2y)\times 3=3\cos(3x+2y)$$
$$\frac{\partial z}{\partial y}=\cos(3x+2y)\times\frac{\partial}{\partial y}(3x+2y)=\cos(3x+2y)\times 2=2\cos(3x+2y)$$

There it is. So in partial differentiation, we can apply all the ordinary rules of normal differentiation, except that we regard the independent variables, other than the one we are using, for the time being as ....

10   constant

At this point let us pause and summarize the main facts so far on partial differentiation

 # Review summary   Unit 54

11   **1** If dependent variable z is given in terms of two independent variables x and y:

$$z = f(x, y)$$

then we are required to know how z varies as each of the independent variables varies. This give rise to the two partial derivatives:

$$\frac{\partial z}{\partial x} \text{ and } \frac{\partial z}{\partial y}$$

In partial differentiation we can apply all the ordinary rules of normal differentiation, except that we regard the independent variables other than the one we are using as being constant at that time.

 # Review exercise   Unit 54

12 In each of the following cases, find $\dfrac{\partial z}{\partial x}$ and $\dfrac{\partial z}{\partial y}$:

1 $z = 4x^2 + 3xy + 5y^2$ **3** $z = \tan(3x + 4y)$

2 $z = (3x + 2y)(4x - 5y)$ **4** $z = \dfrac{\sin(3x + 2y)}{xy}$

Complete all four questions. Take your time, there is no need to rush.
If necessary, refer back to the Unit.
The answers and working are in the next frame.

13 **1** $z = 4x^2 + 3xy + 5y^2$

To find $\dfrac{\partial z}{\partial x}$, regard y as a constant:

$$\therefore \frac{\partial z}{\partial x} = 8x + 3y + 0, \quad \text{i.e. } 8x + 3y \qquad \therefore \frac{\partial z}{\partial x} = 8x + 3y$$

Similarly, regarding x as constant:

$$\frac{\partial z}{\partial y} = 0 + 3x + 10y, \quad \text{i.e. } 3x + 10y \qquad \therefore \frac{\partial z}{\partial y} = 3x + 10y$$

▶

2 $z = (3x + 2y)(4x - 5y)$ Product rule

$\dfrac{\partial z}{\partial x} = (3x + 2y)(4) + (4x - 5y)(3)$

$\qquad = 12x + 8y + 12x - 15y = 24x - 7y$

$\dfrac{\partial z}{\partial y} = (3x + 2y)(-5) + (4x - 5y)(2)$

$\qquad = -15x - 10y + 8x - 10y = -7x - 20y$

3 $z = \tan(3x + 4y)$

$\dfrac{\partial z}{\partial x} = \sec^2(3x + 4y)(3) = 3\sec^2(3x + 4y)$

$\dfrac{\partial z}{\partial y} = \sec^2(3x + 4y)(4) = 4\sec^2(3x + 4y)$

4 $z = \dfrac{\sin(3x + 2y)}{xy}$

$\dfrac{\partial z}{\partial x} = \dfrac{xy\cos(3x + 2y)(3) - \sin(3x + 2y)(y)}{x^2y^2}$

$\qquad = \dfrac{3x\cos(3x + 2y) - \sin(3x + 2y)}{x^2y}$

$\dfrac{\partial z}{\partial y} = \dfrac{xy\cos(3x + 2y)(2) - \sin(3x + 2y)(x)}{x^2y^2}$

$\qquad = \dfrac{2y\cos(3x + 2y) - \sin(3x + 2y)}{xy^2}$

That should have cleared up any troubles. This business of partial differentiation is perfectly straightforward. All you have to remember is that for the time being, all the independent variables except the one you are using are kept constant – and behave like constant factors or constant terms according to their positions.

Now for the Review test

 # Review test   **Unit 54**

In each of the following cases, find $\dfrac{\partial z}{\partial x}$ and $\dfrac{\partial z}{\partial y}$:

14

1 $z = 2x^3 - 7xy + 9y^3$ **3** $z = \sin(6x + 2y)$

2 $z = (4x - 3y)(6x + 8y)$ **4** $z = \dfrac{\cos(x - 4y)}{xy}$

Further partial differentiation

Unit 55

1 Second derivatives

Consider $z = 3x^2 + 4xy - 5y^2$

Then $\dfrac{\partial z}{\partial x} = 6x + 4y$ and $\dfrac{\partial z}{\partial y} = 4x - 10y$

The expression $\dfrac{\partial z}{\partial x} = 6x + 4y$ is itself a function of x and y. We could therefore find its partial derivatives with respect to x or to y.

(a) If we differentiate it partially w.r.t. x, we get:

$$\frac{\partial}{\partial x}\left\{\frac{\partial z}{\partial x}\right\} \text{ and this is written } \frac{\partial^2 z}{\partial x^2} \text{ (much like an ordinary second derivative,}$$

but with the partial ∂)

$$\therefore \frac{\partial^2 z}{\partial x^2} = \frac{\partial}{\partial x}(6x + 4y) = 6$$

This is called the second partial derivative of z with respect to x.

(b) If we differentiate partially w.r.t. y, we get:

$$\frac{\partial}{\partial y}\left\{\frac{\partial z}{\partial x}\right\} \text{ and this is written } \frac{\partial^2 z}{\partial y.\partial x}$$

Note that the operation now being performed is given by the left-hand of the two symbols in the denominator.

$$\frac{\partial^2 z}{\partial y.\partial x} = \frac{\partial}{\partial y}\left\{\frac{\partial z}{\partial x}\right\} = \frac{\partial}{\partial y}\{6x + 4y\} = 4$$

2 So we have this:

$$z = 3x^2 + 4xy - 5y^2$$

$$\frac{\partial z}{\partial x} = 6x + 4y \qquad \frac{\partial z}{\partial y} = 4x - 10y$$

$$\frac{\partial^2 z}{\partial x^2} = 6$$

$$\frac{\partial^2 z}{\partial y.\partial x} = 4$$

Of course, we could carry out similar steps with the expression for $\dfrac{\partial z}{\partial y}$ on the right. This would give us:

$$\frac{\partial^2 z}{\partial y^2} = -10$$

$$\frac{\partial^2 z}{\partial x . \partial y} = 4$$

Note that $\dfrac{\partial^2 z}{\partial y . \partial x}$ means $\dfrac{\partial}{\partial y}\left\{\dfrac{\partial z}{\partial x}\right\}$ so $\dfrac{\partial^2 z}{\partial x . \partial y}$ means ....

3

$$\boxed{\frac{\partial^2 z}{\partial x . \partial y} \text{ means } \frac{\partial}{\partial x}\left\{\frac{\partial z}{\partial y}\right\}}$$

Collecting our previous results together then, we have:

$$z = 3x^2 + 4xy - 5y^2$$

$$\frac{\partial z}{\partial x} = 6x + 4y \qquad\qquad \frac{\partial z}{\partial y} = 4x - 10y$$

$$\frac{\partial^2 z}{\partial x^2} = 6 \qquad\qquad \frac{\partial^2 z}{\partial y^2} = -10$$

$$\frac{\partial^2 z}{\partial y . \partial x} = 4 \qquad\qquad \frac{\partial^2 z}{\partial x . \partial y} = 4$$

We see in this case, that $\dfrac{\partial^2 z}{\partial y . \partial x} = \dfrac{\partial^2 z}{\partial x . \partial y}$. There are then, *two* first derivatives and *four* second derivatives, though the last two seem to have the same value.

Here is one for you to do.

If $z = 5x^3 + 3x^2y + 4y^3$, find $\dfrac{\partial z}{\partial x}, \dfrac{\partial z}{\partial y}, \dfrac{\partial^2 z}{\partial x^2}, \dfrac{\partial^2 z}{\partial y^2}, \dfrac{\partial^2 z}{\partial x . \partial y}$ and $\dfrac{\partial^2 z}{\partial y . \partial x}$

When you have completed all that, move to Frame 4

Here are the results:

4

$$z = 5x^3 + 3x^2y + 4y^3$$

$$\frac{\partial z}{\partial x} = 15x^2 + 6xy \qquad \frac{\partial z}{\partial y} = 3x^2 + 12y^2$$

$$\frac{\partial^2 z}{\partial x^2} = 30x + 6y \qquad \frac{\partial^2 z}{\partial y^2} = 24y$$

$$\frac{\partial^2 z}{\partial y . \partial x} = 6x \qquad \frac{\partial^2 z}{\partial x . \partial y} = 6x$$

Again in this example also, we see that $\dfrac{\partial^2 z}{\partial y . \partial x} = \dfrac{\partial^2 z}{\partial x . \partial y}$. Now do this one.

▶

It looks more complicated, but it is done in just the same way. Do not rush at it; take your time and all will be well. Here it is. Find all the first and second partial derivatives of $z = x\cos y - y\cos x$.

Then to Frame 5

5

Check your results with these.

$z = x\cos y - y\cos x$

When differentiating w.r.t. x, y is constant (and therefore cos y also).
When differentiating w.r.t. y, x is constant (and therefore cos x also).

So we get:

$$\frac{\partial z}{\partial x} = \cos y + y.\sin x \qquad \frac{\partial z}{\partial y} = -x.\sin y - \cos x$$

$$\frac{\partial^2 z}{\partial x^2} = y.\cos x \qquad \frac{\partial^2 z}{\partial y^2} = -x.\cos y$$

$$\frac{\partial^2 z}{\partial y.\partial x} = -\sin y + \sin x \qquad \frac{\partial^2 z}{\partial x.\partial y} = -\sin y + \sin x$$

And again, $\dfrac{\partial^2 z}{\partial y.\partial x} = \dfrac{\partial^2 z}{\partial x.\partial y}$

In fact this will always be so for the functions you are likely to meet, so that there are really *three* different second partial derivatives (and not four). In practice, if you have found $\dfrac{\partial^2 z}{\partial y.\partial x}$ it is a useful check to find $\dfrac{\partial^2 z}{\partial x.\partial y}$ separately. They should give the same result, of course.

6

What about this one?

If $V = \ln(x^2 + y^2)$, prove that $\dfrac{\partial^2 V}{\partial x^2} + \dfrac{\partial^2 V}{\partial y^2} = 0$

This merely entails finding the two second partial derivatives and substituting them in the left-hand side of the statement. So here goes:

$$V = \ln(x^2 + y^2)$$
$$\frac{\partial V}{\partial x} = \frac{1}{(x^2 + y^2)}2x$$
$$= \frac{2x}{x^2 + y^2}$$
$$\frac{\partial^2 V}{\partial x^2} = \frac{(x^2 + y^2)2 - 2x.2x}{(x^2 + y^2)^2}$$
$$= \frac{2x^2 + 2y^2 - 4x^2}{(x^2 + y^2)^2} = \frac{2y^2 - 2x^2}{(x^2 + y^2)^2} \qquad (a)$$

Now you find $\dfrac{\partial^2 V}{\partial y^2}$ in the same way and hence prove the given identity.

We had found that $\dfrac{\partial^2 V}{\partial x^2} = \dfrac{2y^2 - 2x^2}{(x^2 + y^2)^2}$

So making a fresh start from $V = \ln(x^2 + y^2)$, we get:

$$\frac{\partial V}{\partial y} = \frac{1}{(x^2 + y^2)}.2y = \frac{2y}{x^2 + y^2}$$

$$\frac{\partial^2 V}{\partial y^2} = \frac{(x^2 + y^2)2 - 2y.2y}{(x^2 + y^2)^2}$$

$$= \frac{2x^2 + 2y^2 - 4y^2}{(x^2 + y^2)^2} = \frac{2x^2 - 2y^2}{(x^2 + y^2)^2} \qquad \text{(b)}$$

Substituting now the two results in the identity, gives:

$$\frac{\partial^2 V}{\partial x^2} + \frac{\partial^2 V}{\partial y^2} = \frac{2y^2 - 2x^2}{(x^2 + y^2)^2} + \frac{2x^2 - 2y^2}{(x^2 + y^2)^2}$$

$$= \frac{2y^2 - 2x^2 + 2x^2 - 2y^2}{(x^2 + y^2)^2} = 0$$

Now on to Frame 8

Here is another kind of example that you should see.

Example 1

If $V = f(x^2 + y^2)$, show that $x\dfrac{\partial V}{\partial y} - y\dfrac{\partial V}{\partial x} = 0$

Here we are told that V is a function of $(x^2 + y^2)$ but the precise nature of the function is not given. However, we can treat this as a 'function of a function' and write $f'(x^2 + y^2)$ to represent the derivative of the function w.r.t. its own combined variable $(x^2 + y^2)$.

$$\therefore \quad \frac{\partial V}{\partial x} = f'(x^2 + y^2) \times \frac{\partial}{\partial x}(x^2 + y^2) = f'(x^2 + y^2).2x$$

$$\frac{\partial V}{\partial y} = f'(x^2 + y^2).\frac{\partial}{\partial y}(x^2 + y^2) = f'(x^2 + y^2).2y$$

$$\therefore \quad x\frac{\partial V}{\partial y} - y\frac{\partial V}{\partial x} = x.f'(x^2 + y^2).2y - y.f'(x^2 + y^2).2x$$

$$= 2xy.f'(x^2 + y^2) - 2xy.f'(x^2 + y^2)$$

$$= 0$$

Let us have another one of that kind in the next frame

9

Example 2

If $z = f\left\{\dfrac{y}{x}\right\}$, show that $x\dfrac{\partial z}{\partial x} + y\dfrac{\partial z}{\partial y} = 0$

Much the same as before:

$$\dfrac{\partial z}{\partial x} = f'\left\{\dfrac{y}{x}\right\} \cdot \dfrac{\partial}{\partial x}\left\{\dfrac{y}{x}\right\} = f'\left\{\dfrac{y}{x}\right\}\left(-\dfrac{y}{x^2}\right) = -\dfrac{y}{x^2}f'\left\{\dfrac{y}{x}\right\}$$

$$\dfrac{\partial z}{\partial y} = f'\left\{\dfrac{y}{x}\right\} \cdot \dfrac{\partial}{\partial y}\left\{\dfrac{y}{x}\right\} = f'\left\{\dfrac{y}{x}\right\} \cdot \dfrac{1}{x} = \dfrac{1}{x}f'\left\{\dfrac{y}{x}\right\}$$

$$\therefore x\dfrac{\partial z}{\partial x} + y\dfrac{\partial z}{\partial y} = x\left(-\dfrac{y}{x^2}\right)f'\left\{\dfrac{y}{x}\right\} + y\dfrac{1}{x}f'\left\{\dfrac{y}{x}\right\}$$

$$= -\dfrac{y}{x}f'\left\{\dfrac{y}{x}\right\} + \dfrac{y}{x}f'\left\{\dfrac{y}{x}\right\}$$

$$= 0$$

And one for you, just to get your hand in:

If $V = f(ax + by)$, show that $b\dfrac{\partial V}{\partial x} - a\dfrac{\partial V}{\partial y} = 0$

When you have done it, check your working against that in Frame 10

10

Here is the working; this is how it goes.

$$V = f(ax + by)$$

$$\therefore \dfrac{\partial V}{\partial x} = f'(ax + by) \cdot \dfrac{\partial}{\partial x}(ax + by)$$

$$= f'(ax + by).a = a.f'(ax + by) \qquad \text{(a)}$$

$$\dfrac{\partial z}{\partial y} = f'(ax + by) \cdot \dfrac{\partial}{\partial y}(ax + by)$$

$$= f'(ax + by).b = b.f'(ax + by) \qquad \text{(b)}$$

$$\therefore b\dfrac{\partial V}{\partial x} - a\dfrac{\partial V}{\partial y} = ab.f'(ax + by) - ab.f'(ax + by)$$

$$= 0$$

At this point let us pause and summarize the main facts so far on further partial differentiation

Review summary

Unit 55

1 Partial differentiation is easy, no matter how complicated the expression to be differentiated may seem. **11**

2 To differentiate partially w.r.t. x, all independent variables other than x are constant for the time being.

3 To differentiate partially w.r.t. y, all independent variables other than y are constant for the time being.

4 So that, if z is a function of x and y, i.e. if $z = f(x, y)$, we can find:

$$\frac{\partial z}{\partial x} \qquad \frac{\partial z}{\partial y}$$

$$\frac{\partial^2 z}{\partial x^2} \qquad \frac{\partial^2 z}{\partial y^2}$$

$$\frac{\partial^2 z}{\partial y.\partial x} \qquad \frac{\partial^2 z}{\partial x.\partial y} \qquad \text{And also:} \qquad \frac{\partial^2 z}{\partial y.\partial x} = \frac{\partial^2 z}{\partial x.\partial y}$$

 # Review exercise

Unit 55

1 Find all first and second partial derivatives for each of the following functions: **12**

(a) $z = 3x^2 + 2xy + 4y^2$

(b) $z = \sin xy$

(c) $z = \dfrac{x + y}{x - y}$

2 If $z = \ln(e^x + e^y)$, show that $\dfrac{\partial z}{\partial x} + \dfrac{\partial z}{\partial y} = 1$.

3 If $z = x.f(xy)$, express $x\dfrac{\partial z}{\partial x} - y\dfrac{\partial z}{\partial y}$ in its simplest form.

Complete all three questions.
Look back at the Unit if necessary but don't rush.
The answers and working are in the next frame.

1 (a) $z = 3x^2 + 2xy + 4y^2$ **13**

$$\frac{\partial z}{\partial x} = 6x + 2y \qquad\qquad \frac{\partial z}{\partial y} = 2x + 8y$$

$$\frac{\partial^2 z}{\partial x^2} = 6 \qquad\qquad\qquad \frac{\partial^2 z}{\partial y^2} = 8$$

$$\frac{\partial^2 z}{\partial y.\partial x} = 2 \qquad\qquad\quad \frac{\partial^2 z}{\partial x.\partial y} = 2$$

(b) $z = \sin xy$

$$\frac{\partial z}{\partial x} = y \cos xy \qquad\qquad\qquad \frac{\partial z}{\partial y} = x \cos xy$$

$$\frac{\partial^2 z}{\partial x^2} = -y^2 \sin xy \qquad\qquad \frac{\partial^2 z}{\partial y^2} = -x^2 \sin xy$$

$$\frac{\partial^2 z}{\partial y.\partial x} = y(-x \sin xy) + \cos xy \qquad \frac{\partial^2 z}{\partial x.\partial y} = x(-y \sin xy) + \cos xy$$

$$= \cos xy - xy \sin xy \qquad\qquad = \cos xy - xy \sin xy$$

(c) $z = \dfrac{x+y}{x-y}$

$$\frac{\partial z}{\partial x} = \frac{(x-y)1 - (x+y)1}{(x-y)^2} = \frac{-2y}{(x-y)^2}$$

$$\frac{\partial z}{\partial y} = \frac{(x-y)1 - (x+y)(-1)}{(x-y)^2} = \frac{2x}{(x-y)^2}$$

$$\frac{\partial^2 z}{\partial x^2} = (-2y)\frac{(-2)}{(x-y)^3} = \frac{4y}{(x-y)^3}$$

$$\frac{\partial^2 z}{\partial y^2} = 2x\frac{(-2)}{(x-y)^3}(-1) = \frac{4x}{(x-y)^3}$$

$$\frac{\partial^2 z}{\partial y.\partial x} = \frac{(x-y)^2(-2) - (-2y)2(x-y)(-1)}{(x-y)^4}$$

$$= \frac{-2(x-y)^2 - 4y(x-y)}{(x-y)^4}$$

$$= \frac{-2}{(x-y)^2} - \frac{4y}{(x-y)^3}$$

$$= \frac{-2x+2y-4y}{(x-y)^3} = \frac{-2x-2y}{(x-y)^3}$$

$$\frac{\partial^2 z}{\partial x.\partial y} = \frac{(x-y)^2(2) - 2x.2(x-y)1}{(x-y)^4}$$

$$= \frac{2(x-y)^2 - 4x(x-y)}{(x-y)^4}$$

$$= \frac{2}{(x-y)^2} - \frac{4x}{(x-y)^3}$$

$$= \frac{2x-2y-4x}{(x-y)^3} = \frac{-2x-2y}{(x-y)^3}$$

2 $z = \ln(e^x + e^y)$

$$\frac{\partial z}{\partial x} = \frac{1}{e^x + e^y}.e^x \qquad \frac{\partial z}{\partial y} = \frac{1}{e^x + e^y}.e^y$$

$$\frac{\partial z}{\partial x} + \frac{\partial z}{\partial y} = \frac{e^x}{e^x + e^y} + \frac{e^y}{e^x + e^y}$$

$$= \frac{e^x + e^y}{e^x + e^y} = 1$$

$$\frac{\partial z}{\partial x} + \frac{\partial z}{\partial y} = 1$$

3 $z = x.f(xy)$

$$\frac{\partial z}{\partial x} = x.f'(xy).y + f(xy)$$

$$\frac{\partial z}{\partial y} = x.f'(xy).x$$

$$x\frac{\partial z}{\partial x} - y\frac{\partial z}{\partial y} = x^2yf'(xy) + xf(xy) - x^2yf'(xy)$$

$$x\frac{\partial z}{\partial x} - y\frac{\partial z}{\partial y} = xf(xy) = z$$

Now for the Review test

Review test
Unit 55

1 Find all first and second partial derivatives for each of the following functions: **14**
(a) $z = 5x^3 + 3xy - 2y^3$
(b) $z = \tan xy$
(c) $z = \dfrac{2x + 3y}{3x - 2y}$

2 If $z = \dfrac{1}{\sqrt{x^2 + y^2}}$, show that $x\dfrac{\partial z}{\partial x} + y\dfrac{\partial z}{\partial y} = -z$

3 If $z = f(x - ct)$ where c is a constant, show that:
$$\frac{\partial^2 z}{\partial x^2} = \frac{1}{c^2}\frac{\partial^2 z}{\partial t^2}$$

Calculating errors   Unit 56

Small increments

1

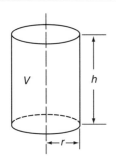

If we return to the volume of the cylinder with which we started this Module, we have once again that $V = \pi r^2 h$. We have seen that we can find $\dfrac{\partial V}{\partial r}$ with h constant, and $\dfrac{\partial V}{\partial h}$ with r constant.

$$\frac{\partial V}{\partial r} = 2\pi rh; \quad \frac{\partial V}{\partial h} = \pi r^2$$

Now let us see what we get if r and h both change simultaneously.

If r becomes $r + \delta r$, and h becomes $h + \delta h$, let V become $V + \delta V$. Then the new volume is given by:

$$V + \delta V = \pi(r + \delta r)^2(h + \delta h)$$
$$= \pi(r^2 + 2r\delta r + [\delta r]^2)(h + \delta h)$$
$$= \pi(r^2 h + 2rh\delta r + h[\delta r]^2 + r^2\delta h + 2r\delta r\delta h + [\delta r]^2\delta h)$$

Subtract $V = \pi r^2 h$ from each side, giving:

$$\delta V = \pi(2rh\delta r + h[\delta r]^2 + r^2\delta h + 2r\delta r\delta h + [\delta r]^2\delta h)$$
$$\approx \pi(2rh\delta r + r^2\delta h) \quad \text{since } \delta r \text{ and } \delta h \text{ are small and all the remaining terms are of a higher degree of smallness.}$$

Therefore

$$\delta V \approx 2\pi rh\delta r + \pi r^2\delta h, \quad \text{that is:}$$
$$\delta V \approx \frac{\partial V}{\partial r}\delta r + \frac{\partial V}{\partial h}\delta h$$

Let us now do a numerical example to see how it all works out.

On to Frame 2

2

A cylinder has dimensions $r = 5$ cm, $h = 10$ cm. Find the approximate increase in volume when r increases by 0·2 cm and h decreases by 0·1 cm.

Well now

$$V = \pi r^2 h \text{ so } \frac{\partial V}{\partial r} = 2\pi rh \qquad \frac{\partial V}{\partial h} = \pi r^2$$

In this case, when $r = 5$ cm, $h = 10$ cm so

$$\frac{\partial V}{\partial r} = 2\pi 5.10 = 100\pi \qquad \frac{\partial V}{\partial h} = \pi r^2 = \pi 5^2 = 25\pi$$

$\delta r = 0·2$ and $\delta h = -0·1$ (minus because h is decreasing)

$$\therefore \delta V \approx \frac{\partial V}{\partial r}.\delta r + \frac{\partial V}{\partial h}.\delta h$$

$$\delta V = 100\pi(0·2) + 25\pi(-0·1)$$

$$= 20\pi - 2·5\pi = 17·5\pi$$

$$\therefore \delta V \approx 54·98 \text{ cm}^3$$

i.e. the volume increases by 54·98 cm^3

Just like that!

3

This kind of result applies not only to the volume of the cylinder, but to any function of two independent variables. Here is an example:

If z is a function of x and y, i.e. $z = f(x, y)$ and if x and y increase by small amounts δx and δy, the increase δz will also be relatively small. If we expand δz in powers of δx and δy, we get:

$$\delta z = A\delta x + B\delta y + \text{higher powers of } \delta x \text{ and } \delta y,$$

where A and B are functions of x and y.

If y remains constant, so that $\delta y = 0$, then:

$$\delta z = A\delta x + \text{higher powers of } \delta x$$

$$\therefore \frac{\delta z}{\delta x} = A. \text{ So that if } \delta x \to 0, \text{ this becomes } A = \frac{\partial z}{\partial x}$$

Similarly, if x remains constant, making $\delta y \to 0$ gives $B = \dfrac{\partial z}{\partial y}$

$$\therefore \delta z = \frac{\partial z}{\partial x}\delta x + \frac{\partial z}{\partial y}\delta y + \text{ higher powers of very small quantities which can be ignored}$$

$$\delta z = \frac{\partial z}{\partial x}\delta x + \frac{\partial z}{\partial y}\delta y$$

4

So, if $z = f(x, y)$

$$\delta z = \frac{\partial z}{\partial x}\delta x + \frac{\partial z}{\partial y}\delta y$$

This is the key to all the forthcoming applications and will be quoted over and over again.

The result is quite general and a similar result applies for a function of three independent variables. For example:

If $z = f(x, y, w)$

then $\delta z = \dfrac{\partial z}{\partial x}\delta x + \dfrac{\partial z}{\partial y}\delta y + \dfrac{\partial z}{\partial w}\delta w$

If we remember the rule for a function of two independent variables, we can easily extend it when necessary.

Here it is once again:

$$\text{If } z = f(x, y) \text{ then } \delta z = \frac{\partial z}{\partial x}\delta x + \frac{\partial z}{\partial y}\delta y$$

Copy this result into your work book in a prominent position, such as it deserves!

5

Now for a couple of examples

Example 1

If $I = \dfrac{V}{R}$, and $V = 250$ volts and $R = 50$ ohms, find the change in I resulting from an increase of 1 volt in V and an increase of 0·5 ohm in R.

$$I = f(V, R) \qquad \therefore \ \delta I = \frac{\partial I}{\partial V}\delta V + \frac{\partial I}{\partial R}\delta R$$

$$\frac{\partial I}{\partial V} = \frac{1}{R} \text{ and } \frac{\partial I}{\partial R} = -\frac{V}{R^2}$$

$$\therefore \ \delta I = \frac{1}{R}\delta V - \frac{V}{R^2}\delta R$$

So when $R = 50$, $V = 250$, $\delta V = 1$ and $\delta R = 0\cdot5$:

$$\delta I = \frac{1}{50}(1) - \frac{250}{2500}(0\cdot5)$$

$$= \frac{1}{50} - \frac{1}{20}$$

$$= 0\cdot02 - 0\cdot05 = -0\cdot03$$

i.e. I decreases by 0·03 amperes

Here is another example.

6

Example 2

If $y = \dfrac{ws^3}{d^4}$, find the percentage increase in y when w increases by 2 per cent, s decreases by 3 per cent and d increases by 1 per cent.

Notice that, in this case, y is a function of three variables, w, s and d. The formula therefore becomes:

$$\delta y = \frac{\partial y}{\partial w}\delta w + \frac{\partial y}{\partial s}\delta s + \frac{\partial y}{\partial d}\delta d$$

We have

$$\frac{\partial y}{\partial w} = \frac{s^3}{d^4}; \quad \frac{\partial y}{\partial s} = \frac{3ws^2}{d^4}; \quad \frac{\partial y}{\partial d} = -\frac{4ws^3}{d^5}$$

$$\therefore \delta y = \frac{s^3}{d^4}\delta w + \frac{3ws^2}{d^4}\delta s + \frac{-4ws^3}{d^5}\delta d$$

Now then, what are the values of $\delta w, \delta s$ and δd?

Is it true to say that $\delta w = \dfrac{2}{100}; \quad \delta s = \dfrac{-3}{100}; \quad \delta d = \dfrac{1}{100}$?

If not, why not?

Next frame

$$\boxed{\text{No. It is not correct}}$$

7

Because δw is not $\dfrac{2}{100}$ of a unit, but 2 per cent of w, i.e. $\delta w = \dfrac{2}{100}$ of $w = \dfrac{2w}{100}$

Similarly, $\delta s = \dfrac{-3}{100}$ of $s = \dfrac{-3s}{100}$ and $\delta d = \dfrac{d}{100}$. Now that we have cleared that point up, we can continue with the problem.

$$\delta y = \frac{s^3}{d^4}\left(\frac{2w}{100}\right) + \frac{3ws^2}{d^4}\left(\frac{-3s}{100}\right) - \frac{4ws^3}{d^5}\left(\frac{d}{100}\right)$$

$$= \frac{ws^3}{d^4}\left(\frac{2}{100}\right) - \frac{ws^3}{d^4}\left(\frac{9}{100}\right) - \frac{ws^3}{d^4}\left(\frac{4}{100}\right)$$

$$= \frac{ws^3}{d^4}\left\{\frac{2}{100} - \frac{9}{100} - \frac{4}{100}\right\}$$

$$= y\left\{-\frac{11}{100}\right\} = -11 \text{ per cent of } y$$

i.e. y decreases by 11 per cent

Remember that where the increment of w is given as 2 per cent, it is *not* $\dfrac{2}{100}$ of a unit, but $\dfrac{2}{100}$ of w, and the symbol w must be included.

Move on to Frame 8

8

Now here is an exercise for you to do.

$P = w^2hd$. If errors of up to 1 per cent (plus or minus) are possible in the measured values of w, h and d, find the maximum possible percentage error in the calculated values of P.

This is very much like the previous example, so you will be able to deal with it without any trouble. Work it right through and then go on to Frame 9 and check your result.

9

$$P = w^2hd \quad \therefore \ \delta P = \frac{\partial P}{\partial w}.\delta w + \frac{\partial P}{\partial h}.\delta h + \frac{\partial P}{\partial d}.\delta d$$

$$\frac{\partial P}{\partial w} = 2whd; \quad \frac{\partial P}{\partial h} = w^2d; \quad \frac{\partial P}{\partial d} = w^2h$$

$$\delta P = 2whd.\delta w + w^2d.\delta h + w^2h.\delta d$$

Now $\quad \delta w = \pm\dfrac{w}{100}; \quad \delta h = \pm\dfrac{h}{100}; \quad \delta d = \pm\dfrac{d}{100}$

$$\delta P = 2whd\left(\pm\frac{w}{100}\right) + w^2d\left(\pm\frac{h}{100}\right) + w^2h\left(\pm\frac{d}{100}\right)$$

$$= \pm\frac{2w^2hd}{100} \pm \frac{w^2dh}{100} \pm \frac{w^2hd}{100}$$

The greatest possible error in P will occur when the signs are chosen so that they are all of the same kind, i.e. all plus or minus. If they were mixed, they would tend to cancel each other out.

$$\therefore \ \delta P = \pm w^2hd\left\{\frac{2}{100} + \frac{1}{100} + \frac{1}{100}\right\} = \pm P\left(\frac{4}{100}\right)$$

$\therefore$ Maximum possible error in P is 4 per cent of P

Finally, here is one last example for you to do. Work right through it and then check your results with those in Frame 10.

The two sides forming the right-angle of a right-angled triangle are denoted by a and b. The hypotenuse is h. If there are possible errors of $\pm0\cdot5$ per cent in measuring a and b, find the maximum possible error in calculating (a) the area of the triangle and (b) the length of h.

10

(a) $\delta A = 1$ per cent of A

(b) $\delta h = 0.5$ per cent of h

Here is the working in detail:

(a) $A = \dfrac{a.b}{2}$ $\delta A = \dfrac{\partial A}{\partial a}.\delta a + \dfrac{\partial A}{\partial b}.\delta b$

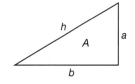

$\dfrac{\partial A}{\partial a} = \dfrac{b}{2};$ $\dfrac{\partial A}{\partial b} = \dfrac{a}{2};$ $\delta a = \pm\dfrac{a}{200};$ $\delta b = \pm\dfrac{b}{200}$

$\delta A = \dfrac{b}{2}\left(\pm\dfrac{a}{200}\right) + \dfrac{a}{2}\left(\pm\dfrac{b}{200}\right)$

$\quad = \pm\dfrac{a.b}{2}\left[\dfrac{1}{200} + \dfrac{1}{200}\right] = \pm A.\dfrac{1}{100}$

$\therefore \delta A = 1$ per cent of A

(b) $h = \sqrt{a^2 + b^2} = (a^2 + b^2)^{\frac{1}{2}}$

$\delta h = \dfrac{\partial h}{\partial a}\delta a + \dfrac{\partial h}{\partial b}\delta b$

$\dfrac{\partial h}{\partial a} = \dfrac{1}{2}(a^2 + b^2)^{-\frac{1}{2}}(2a) = \dfrac{a}{\sqrt{a^2 + b^2}}$

$\dfrac{\partial h}{\partial b} = \dfrac{1}{2}(a^2 + b^2)^{-\frac{1}{2}}(2b) = \dfrac{b}{\sqrt{a^2 + b^2}}$

Also $\delta a = \pm\dfrac{a}{200};$ $\delta b = \pm\dfrac{b}{200}$

$\therefore \delta h = \dfrac{a}{\sqrt{a^2 + b^2}}\left(\pm\dfrac{a}{200}\right) + \dfrac{b}{\sqrt{a^2 + b^2}}\left(\pm\dfrac{b}{200}\right)$

$\quad = \pm\dfrac{1}{200}\dfrac{a^2 + b^2}{\sqrt{a^2 + b^2}}$

$\quad = \pm\dfrac{1}{200}\sqrt{a^2 + b^2} = \pm\dfrac{1}{200}(h)$

$\therefore \delta h = 0.5$ per cent of h

At this point let's pause and summarize the work so far on calculating errors

Review summary

11

1 Given a function of two variables $z = f(x, y)$ then if δx and δy represent small changes in x and y, the corresponding change in z is given by δz where:

$$\delta z = \frac{\partial z}{\partial x}\delta x + \frac{\partial z}{\partial y}\delta y + \text{higher powers of small quantities}$$

So, provided δx and δy are sufficiently small then:

$$\delta z = \frac{\partial z}{\partial x}\delta x + \frac{\partial z}{\partial y}\delta y$$

2 This can be extended to functions of more than two variables where, if $z = f(x_1, x_2, x_3, \ldots)$ then:

$$\delta z = \frac{\partial z}{\partial x_1}\delta x_1 + \frac{\partial z}{\partial x_2}\delta x_2 + \frac{\partial z}{\partial x_3}\delta x_3 + \ldots$$

Review exercise

12

1 The period T of a simple pendulum of length l is given as:

$$T = 2\pi\sqrt{\frac{l}{g}}$$

where g is the acceleration due to gravity. If the length l is measured as 2% too large and the period T is measured as 1% too small calculate the approximate percentage error in the value of g as calculated from the formula.

Complete the question. Take your time, there is no need to rush.
If necessary, look back at the Unit.
The answers and working are in the next frame.

13

1   Given $T = 2\pi\sqrt{\dfrac{l}{g}}$ then $T^2 = 4\pi^2\dfrac{l}{g}$ so $g = 4\pi^2\dfrac{l}{T^2} = g(l,\,T)$. Then:

$$\delta g = \frac{\partial g}{\partial l}\delta l + \frac{\partial g}{\partial T}\delta T$$

$$= 4\pi^2\frac{1}{T^2}\delta l - 8\pi^2\frac{l}{T^3}\delta T$$

and so:

$$\frac{\delta g}{g} = \frac{T^2}{4\pi^2 l}\left(4\pi^2\frac{1}{T^2}\delta l - 8\pi^2\frac{l}{T^3}\delta T\right)$$

$$= \frac{\delta l}{l} - 2\frac{\delta T}{T}$$

Therefore:

$$\frac{\delta g}{g} \times 100 = \left(\frac{\delta l}{l}\right) \times 100 - 2\left(\frac{\delta T}{T}\right) \times 100$$

So the percentage error in $g = 2 - 2(-1) = 4$

The calculated error in g is approximately 4% too large.

Now for the Review test

Review test

Unit 56

14

1   The pressure P, volume V and temperature T of a gas obey the gas law:

$$PV = kT$$

where k is a constant. If, at a certain pressure, volume and temperature, the pressure decreases by 1% and the temperature rises by 2%, show that the approximate change in the volume will be 3%.

15

You have now come to the end of this Module. A list of **Can You?** questions follows for you to gauge your understanding of the material in the Module. You will notice that these questions match the **Learning outcomes** listed at the beginning of the Module. Now try the **Test exercise**. *Work through the questions at your own pace, there is no need to hurry.* A set of **Further problems** provides valuable additional practice.

Can You?

1 Checklist: Module 18

Check this list before and after you try the end of Module test.

On a scale of 1 to 5 how confident are you that you can:

• Find the first partial derivatives of a function of two real variables?

    Yes ☐ ☐ ☐ ☐ ☐   *No*

• Find second-order partial derivatives of a function of two real variables?

    Yes ☐ ☐ ☐ ☐ ☐   *No*

• Calculate errors using partial differentiation?

    Yes ☐ ☐ ☐ ☐ ☐   *No*

 Test exercise 18

2 Take your time over the questions; do them carefully.

1 Find all first and second partial derivatives of the following:

    (a) $z = 4x^3 - 5xy^2 + 3y^3$

    (b) $z = \cos(2x + 3y)$

    (c) $z = e^{x^2 - y^2}$

    (d) $z = x^2 \sin(2x + 3y)$

2 (a) If $V = x^2 + y^2 + z^2$, express in its simplest form

$$x\frac{\partial V}{\partial x} + y\frac{\partial V}{\partial y} + z\frac{\partial V}{\partial z}.$$

    (b) If $z = f(x + ay) + F(x - ay)$, find $\dfrac{\partial^2 z}{\partial x^2}$ and $\dfrac{\partial^2 z}{\partial y^2}$ and hence prove that

$$\frac{\partial^2 z}{\partial y^2} = a^2 \cdot \frac{\partial^2 z}{\partial x^2}.$$

3 The power P dissipated in a resistor is given by $P = \dfrac{E^2}{R}$.

If $E = 200$ volts and $R = 8$ ohms, find the change in P resulting from a drop of 5 volts in E and an increase of 0·2 ohm in R.

4 If $\theta = kHLV^{-\frac{1}{2}}$, where k is a constant, and there are possible errors of ±1 per cent in measuring H, L and V, find the maximum possible error in the calculated value of θ.

That's it

Further problems 18

3

1 If $z = \dfrac{1}{x^2 + y^2 - 1}$, show that $x\dfrac{\partial z}{\partial x} + y\dfrac{\partial z}{\partial y} = -2z(1 + z)$.

2 Prove that, if $V = \ln(x^2 + y^2)$, then $\dfrac{\partial^2 V}{\partial x^2} + \dfrac{\partial^2 V}{\partial y^2} = 0$.

3 If $z = \sin(3x + 2y)$, verify that $3\dfrac{\partial^2 z}{\partial y^2} - 2\dfrac{\partial^2 z}{\partial x^2} = 6z$.

4 If $u = \dfrac{x + y + z}{(x^2 + y^2 + z^2)^{\frac{1}{2}}}$, show that $x\dfrac{\partial u}{\partial x} + y\dfrac{\partial u}{\partial y} + z\dfrac{\partial u}{\partial z} = 0$.

5 Show that the equation $\dfrac{\partial^2 z}{\partial x^2} + \dfrac{\partial^2 z}{\partial y^2} = 0$, is satisfied by

$$z = \ln\sqrt{x^2 + y^2} + \frac{1}{2}\tan^{-1}\left(\frac{y}{x}\right)$$

6 If $z = e^x(x\cos y - y\sin y)$, show that $\dfrac{\partial^2 z}{\partial x^2} + \dfrac{\partial^2 z}{\partial y^2} = 0$.

7 If $u = (1 + x)\sin(5x - 2y)$, verify that $4\dfrac{\partial^2 u}{\partial x^2} + 20\dfrac{\partial^2 u}{\partial x.\partial y} + 25\dfrac{\partial^2 u}{\partial y^2} = 0$.

8 If $z = f\left(\dfrac{y}{x}\right)$, show that $x^2\dfrac{\partial^2 z}{\partial x^2} + 2xy\dfrac{\partial^2 z}{\partial x.\partial y} + y^2\dfrac{\partial^2 z}{\partial y^2} = 0$.

9 If $z = (x + y).f\left(\dfrac{y}{x}\right)$, where f is an arbitrary function, show that

$$x\dfrac{\partial z}{\partial x} + y\dfrac{\partial z}{\partial y} = z.$$

10 In the formula $D = \dfrac{Eh^3}{12(1 - v^2)}$, h is given as $0{\cdot}1 \pm 0{\cdot}002$ and v as $0{\cdot}3 \pm 0{\cdot}02$. Express the approximate maximum error in D in terms of E.

11 The formula $z = \dfrac{a^2}{x^2 + y^2 - a^2}$ is used to calculate z from observed values of x and y. If x and y have the same percentage error p, show that the percentage error in z is approximately $-2p(1 + z)$.

12 In a balanced bridge circuit, $R_1 = R_2R_3/R_4$. If R_2, R_3, R_4 have known tolerances of $\pm x$ per cent, $\pm y$ per cent, $\pm z$ per cent respectively, determine the maximum percentage error in R_1, expressed in terms of x, y and z.

13 The deflection y at the centre of a circular plate suspended at the edge and uniformly loaded is given by $y = \dfrac{kwd^4}{t^3}$, where $w =$ total load, $d =$ diameter of plate, $t =$ thickness and k is a constant.

Calculate the approximate percentage change in y if w is increased by 3 per cent, d is decreased by $2\frac{1}{2}$ per cent and t is increased by 4 per cent.

14 The coefficient of rigidity (n) of a wire of length (L) and uniform diameter (d) is given by $n = \dfrac{AL}{d^4}$, where A is a constant. If errors of $\pm0\cdot25$ per cent and ±1 per cent are possible in measuring L and d respectively, determine the maximum percentage error in the calculated value of n.

15 If $k/k_0 = (T/T_0)^n.p/760$, show that the change in k due to small changes of a per cent in T and b per cent in p is approximately $(na + b)$ per cent.

16 The deflection y at the centre of a rod is known to be given by $y = \dfrac{kwl^3}{d^4}$, where k is a constant. If w increases by 2 per cent, l by 3 per cent, and d decreases by 2 per cent, find the percentage increase in y.

17 The displacement y of a point on a vibrating stretched string, at a distance x from one end, at time t, is given by

$$\frac{\partial^2 y}{\partial t^2} = c^2 . \frac{\partial^2 y}{\partial x^2}$$

Show that one solution of this equation is $y = A \sin\dfrac{px}{c} . \sin(pt + a)$, where A, p, c and a are constants.

18 If $y = A \sin(px + a) \cos(qt + b)$, find the error in y due to small errors δx and δt in x and t respectively.

19 Show that $\phi = Ae^{-kt/2} \sin pt \cos qx$, satisfies the equation

$$\frac{\partial^2 \phi}{\partial x^2} = \frac{1}{c^2}\left\{\frac{\partial^2 \phi}{\partial t^2} + k\frac{\partial \phi}{\partial t}\right\}, \text{ provided that } p^2 = c^2 q^2 - \frac{k^2}{4}.$$

20 Show that (a) the equation $\dfrac{\partial^2 V}{\partial x^2} + \dfrac{\partial^2 V}{\partial y^2} + \dfrac{\partial^2 V}{\partial z^2} = 0$ is satisfied by

$V = \dfrac{1}{\sqrt{x^2 + y^2 + z^2}}$, and that (b) the equation $\dfrac{\partial^2 V}{\partial x^2} + +\dfrac{\partial^2 V}{\partial y^2} = 0$

is satisfied by $V = \tan^{-1}\left(\dfrac{y}{x}\right)$.

Integration

Learning outcomes

When you have completed this Module you will be able to:

- Appreciate that integration is the reverse process of differentiation
- Recognize the need for a constant of integration
- Evaluate indefinite integrals of standard forms
- Evaluate indefinite integrals of polynomials
- Evaluate indefinite integrals of 'functions of a linear function of *x*'
- Integrate by partial fractions
- Integrate by parts
- Appreciate that a definite integral is a measure of the area under a curve
- Evaluate definite integrals of standard forms
- Use Simpson's rule to approximate the area beneath a curve
- Use the definite integral to find areas between a curve and the horizontal axis
- Use the definite integral to find areas between a curve and a given straight line

Units

Integration

1

Integration is the reverse process of differentiation. When we differentiate we start with an expression and proceed to find its derivative. When we integrate, we start with the derivative and then find the expression from which it has been derived.

For example, $\dfrac{d}{dx}(x^4) = 4x^3$. Therefore, the integral of $4x^3$ with respect to x we know to be x^4. This is written:

$$\int 4x^3\, dx = x^4$$

The symbols $\int f(x)\, dx$ denote the *integral of $f(x)$ with respect to the variable x*; the symbol $\int$ was developed from a capital S which was used in the 17th century when the ideas of the calculus were first devised. The expression $f(x)$ to be integrated is called the *integrand* and the differential dx is usefully there to assist in the evaluation of certain integrals.

Constant of integration

So $\dfrac{d}{dx}(x^4) = 4x^3$ $\quad\therefore\quad \int 4x^3.dx = x^4$

Also $\dfrac{d}{dx}(x^4 + 2) = 4x^3$ $\quad\therefore\quad \int 4x^3.dx = x^4 + 2$

and $\dfrac{d}{dx}(x^4 - 5) = 4x^3$ $\quad\therefore\quad \int 4x^3.dx = x^4 - 5$

In these three examples we happen to know the expressions from which the derivative $4x^3$ was derived. But any constant term in the original expression becomes zero in the derivative and all trace of it is lost. So if we do not know the history of the derivative $4x^3$ we have no evidence of the value of the constant term, be it 0, +2, −5 or any other value. We therefore acknowledge the presence of such a constant term of some value by adding a symbol C to the result of the integration:

i.e. $\int 4x^3.dx = x^4 + C$

C is called the *constant of integration* and must always be included.

Such an integral is called an *indefinite integral* since normally we do not know the value of C. In certain circumstances, however, the value of C might be found if further information about the integral is available.

For example, to determine $I = \int 4x^3.dx$, given that $I = 3$ when $x = 2$. As before:

$$I = \int 4x^3\, dx = x^4 + C$$

But $I = 3$ when $x = 2$ so that $3 = 2^4 + C = 16 + C$ $\therefore$ $C = -13$.
So, in this case $I = x^4 - 13$.

Next frame

Standard integrals

Every derivative written in reverse gives an integral.

e.g. $\dfrac{d}{dx}(\sin x) = \cos x$ $\therefore$ $\displaystyle\int \cos x . dx = \sin x + C$

It follows, therefore, that our list of standard derivatives provides a source of standard integrals.

(a) $\dfrac{d}{dx}(x^n) = nx^{n-1}$. Replacing n by $(n+1)$, $\dfrac{d}{dx}(x^{n+1}) = (n+1)x^n$

$\therefore$ $\dfrac{d}{dx}\left(\dfrac{x^{n+1}}{n+1}\right) = x^n$ $\therefore$ $\displaystyle\int x^n . dx = \dfrac{x^{n+1}}{n+1} + C$

This is true except when $n = -1$, for then we should be dividing by 0.

(b) $\dfrac{d}{dx}(\sin x) = \cos x$

$\therefore$ $\displaystyle\int \cos x . dx = \sin x + C$

(c) $\dfrac{d}{dx}(\cos x) = -\sin x$

$\therefore$ $\dfrac{d}{dx}(-\cos x) = \sin x$

$\therefore$ $\displaystyle\int \sin x . dx = -\cos x + C$

(d) $\dfrac{d}{dx}(\tan x) = \sec^2 x$

$\therefore$ $\displaystyle\int \sec^2 x . dx = \tan x + C$

(e) $\dfrac{d}{dx}(e^x) = e^x$

$\therefore$ $\displaystyle\int e^x . dx = e^x + C$

(f) $\dfrac{d}{dx}(\ln x) = \dfrac{1}{x}$

$\therefore$ $\displaystyle\int \dfrac{1}{x} dx = \ln x + C$

(g) $\dfrac{d}{dx}(a^x) = a^x . \ln a$

$\therefore$ $\displaystyle\int a^x . dx = \dfrac{a^x}{\ln a} + C$

As with differentiation, a constant coefficient remains unchanged

e.g. $\displaystyle\int 5 . \cos x . dx = 5\sin x + C$, etc.

Collecting the results together, we have:

$f(x)$	$\int f(x) . dx$	
x^n	$\dfrac{x^{n+1}}{n+1} + C$	$(n \neq -1)$
1	$x + C$	
a	$ax + C$	
$\sin x$	$-\cos x + C$	
$\cos x$	$\sin x + C$	
$\sec^2 x$	$\tan x + C$	
e^x	$e^x + C$	
a^x	$a^x / \ln a + C$	
$\dfrac{1}{x}$	$\ln x + C$	

At this point let us pause and summarize the main facts on integration so far

Review summary   **Unit 57**

3

1 Integration is the reverse process of differentiation. Given a derivative we have to find the expression from which it was derived.

2 Because constants have a zero derivative we find that when we reverse the process of differentiation we must introduce an integration constant into our result.

3 Such integrals are called indefinite integrals.

 # Review exercise   **Unit 57**

4

1 Determine the following integrals:

(a) $\int x^6 \, dx$ (b) $\int 3e^x \, dx$ (c) $\int \dfrac{6}{x} \, dx$

(d) $\int 5 \sin x \, dx$ (e) $\int \sec^2 x \, dx$ (f) $\int 8 \, dx$

(g) $\int x^{\frac{1}{2}} \, dx$ (h) $\int 2 \cos x \, dx$ (i) $\int x^{-3} \, dx$

(j) $\int 4^x \, dx$

2 (a) Determine $I = \int 4x^2 .dx$, given that $I = 25$ when $x = 3$.

(b) Determine $I = \int 5.dx$, given that $I = 16$ when $x = 2$.

(c) Determine $I = \int 2.\cos x.dx$, given that $I = 7$ when $x = \dfrac{\pi}{2}$ (radians).

(d) Determine $I = \int 2.e^x.dx$, given that $I = 50\cdot2$ when $x = 3$.

Complete both questions. Take your time, there is no need to rush.
If necessary, look back at the Unit.
The answers and working are in the next frame.

5

1 (a) $\int x^6 \, dx = \dfrac{x^7}{7} + C$

Check by differentiating the result – the derivative of the result will give the integrand

$$\frac{d}{dx}\left(\frac{x^7}{7} + C\right) = \frac{7x^6}{7} + 0 = x^6 \text{ the integrand}$$

(b) $\int 3e^x \, dx = 3e^x + C$

The derivative of the result is $3e^x + 0 = 3e^x$ – the integrand

▶

(c) $\int \dfrac{6}{x}\,dx = 6\ln x + C$

The derivative of the result is

$\dfrac{6}{x} + 0 = \dfrac{6}{x}$

(d) $\int 5\sin x\,dx = -5\cos x + C$

The derivative of the result is

$-5(-\sin x) + 0 = 5\sin x$

(e) $\int \sec^2 x\,dx = \tan x + C$

The derivative of the result is

$\sec^2 x + 0 = \sec^2 x$

(f) $\int 8\,dx = 8x + C$

The derivative of the result is $8 + 0 = 8$

(g) $\int x^{\frac{1}{2}}\,dx = \dfrac{x^{\frac{1}{2}+1}}{\frac{1}{2}+1} + C = \dfrac{x^{\frac{3}{2}}}{\frac{3}{2}} + C$

$= \dfrac{2x^{\frac{3}{2}}}{3} + C$

The derivative of the result is

$\dfrac{3}{2} \times \dfrac{2x^{\frac{3}{2}-1}}{3} + 0 = x^{\frac{1}{2}}$

(h) $\int 2\cos x\,dx = 2\sin x + C$

The derivative of the result is

$2\cos x + 0 = 2\cos x$

(i) $\int x^{-3}\,dx = \dfrac{x^{-3+1}}{-3+1} + C$

$= -\dfrac{x^{-2}}{2} + C$

The derivative of the result is

$-(-2)\dfrac{x^{-2-1}}{2} + 0 = x^{-3}$

(j) $\int 4^x\,dx = \dfrac{4^x}{\ln 4} + C$

Recall the standard derivative

$\dfrac{d}{dx}(a^x) = a^x \ln a$

2 (a) $I = \int 4x^2.dx = 4\dfrac{x^3}{3} + C$

$\therefore \quad 25 = 36 + C \therefore \quad C = -11$

$\therefore \quad \int 4x^2.dx = \dfrac{4x^3}{3} - 11$

(c) $I = \int 2.\cos x.dx = 2\sin x + C$

$\therefore \quad 7 = 2 + C \therefore \quad C = 5$

$\therefore \quad \int 2.\cos x.dx = 2\sin x + 5$

(b) $I = \int 5.dx = 5x + C$

$16 = 10 + C \therefore \quad C = 6$

$\therefore \quad \int 5.dx = 5x + 6$

(d) $I = \int 2.e^x.dx = 2e^x + C$

$\therefore \quad 50{\cdot}2 = 2e^3 + C = 40{\cdot}2 + C$

$\therefore \quad C = 10$

$\therefore \quad \int 2.e^x.dx = 2e^x + 10$

Now for the Review test

Review test   Unit 57

6

1 Determine the following integrals:

(a) $\int x^7\,dx$ (b) $\int 4\cos x\,dx$ (c) $\int 2e^x\,dx$ (d) $\int 12\,dx$

(e) $\int x^{-4}\,dx$ (f) $\int 6^x\,dx$ (g) $\int 9x^{\frac{1}{3}}\,dx$ (h) $\int 4\sec^2 x\,dx$

(i) $\int 9\sin x\,dx$ (j) $\int \dfrac{8}{x}\,dx$

Integration of polynomial expressions   Unit 58

1

In Unit 52 of Module 17 we differentiated a polynomial expression by dealing with the separate terms, one by one. It is not surprising, therefore, that we do much the same with the integration of polynomial expressions.

Polynomial expressions are integrated term by term with the individual constants of integration consolidated into one symbol C for the whole expression. For example:

$$\int (4x^3 + 5x^2 - 2x + 7)\,dx$$

$$= x^4 + \frac{5x^3}{3} - x^2 + 7x + C$$

So, what about this one? If $I = \int (8x^3 - 3x^2 + 4x - 5)\,dx$, determine the value of I when $x = 3$, given that at $x = 2$, $I = 26$.

First we must determine the function for I, so carrying out the integration, we get

$$I = \ldots\ldots\ldots$$

2

$$I = 2x^4 - x^3 + 2x^2 - 5x + C$$

Now we can calculate the value of C since we are told that when $x = 2$, $I = 26$. So, expressing the function for I in nested form, we have

$$I = \ldots\ldots\ldots$$

3

$$I = \{[(2x - 1)x + 2]x - 5\}x + C$$

Substituting $x = 2$:

$$26 = \ldots\ldots\ldots$$

$$\boxed{22 + C} \qquad\qquad \boxed{4}$$

We have $26 = 22 + C$ $\quad \therefore \quad C = 4$

$\therefore \quad I = \{[(2x - 1)x + 2]x - 5\}x + 4$

Finally, all we now have to do is to put $x = 3$ in this expression which gives us that, when $x = 3$, $I = \ldots\ldots\ldots$

$$\boxed{142} \qquad\qquad \boxed{5}$$

Just take one step at a time. There are no snags.

Now here is another of the same type. Determine the value of

$$I = \int (4x^3 - 6x^2 - 16x + 4)\, \mathrm{d}x \text{ when } x = -2, \text{ given that at } x = 3, I = -13.$$

As before:

(a) Perform the integration.

(b) Express the resulting function in nested form.

(c) Evaluate the constant of integration, using the fact that when $x = 3$, $I = -13$.

(d) Determine the value of I when $x = -2$.

The method is just the same as before, so work through it.

$$\therefore \quad \text{When } x = -2, \quad I = \ldots\ldots\ldots\ldots$$

$$\boxed{12} \qquad\qquad \boxed{6}$$

Here is a check on the working:

(a) $I = x^4 - 2x^3 - 8x^2 + 4x + C$

(b) In nested form, $I = \{[(x - 2)x - 8]x + 4\}x + C$

(c) At $x = 3$, $I = -13 = -33 + C$ $\therefore \quad C = 20$

$\therefore \quad I = \{[(x - 2)x - 8]x + 4\}x + 20$

(d) $\therefore \quad$ When $x = -2$, $\quad I = 12$

It is all very straightforward.

Now let us move on to something slightly different

Function of a linear function of *x*

7

It is often necessary to integrate any one of the expressions shown in our list of standard integrals when the variable x is replaced by a linear expression in x. That is, of the form $ax + b$. For example, $y = \int (3x + 2)^4 \, dx$ is of the same structure as $y = \int x^4 \, dx$ except that x is replaced by the linear expression $3x + 2$.

Let us put $u = 3x + 2$ then:

$$\int (3x + 2)^4 \, dx \text{ becomes } \int u^4 \, dx$$

We now have to change the variable x in dx before we can progress. Now, $u = 3x + 2$ so that:

$$\frac{du}{dx} = 3$$

That is: $du = 3 \, dx$ or, alternatively $dx = \dfrac{du}{3}$

We now find that our integral can be determined as:

$$y = \int u^4 \, dx = \int u^4 \frac{du}{3} = \frac{1}{3} \left(\frac{u^5}{5} \right) + C = \frac{1}{3} \left(\frac{(3x + 2)^5}{5} \right) + C$$

That is: $\quad y = \dfrac{(3x + 2)^5}{15} + C$

To integrate a 'function of a linear function of x', simply replace x in the corresponding standard result by the linear expression and divide by the coefficient of x in the linear expression.

Here are three examples:

(1) $\displaystyle\int (4x - 3)^2 . dx$ [Standard integral $\displaystyle\int x^2 . dx = \frac{x^3}{3} + C$]

$\therefore \quad \displaystyle\int (4x - 3)^2 . dx = \frac{(4x - 3)^3}{3} \times \frac{1}{4} + C = \frac{(4x - 3)^3}{12} + C$

(2) $\displaystyle\int \cos 3x . dx$ [Standard integral $\displaystyle\int \cos x . dx = \sin x + C$]

$\therefore \quad \displaystyle\int \cos 3x . dx = \sin 3x . \frac{1}{3} + C = \frac{\sin 3x}{3} + C$

(3) $\displaystyle\int e^{5x+2} . dx$ [Standard integral $\displaystyle\int e^x . dx = e^x + C$]

$\therefore \quad \displaystyle\int e^{5x+2} . dx = e^{5x+2} \frac{1}{5} + C = \frac{e^{5x+2}}{5} + C$

Just refer to the basic standard integral of the same form, replace x in the result by the linear expression and finally divide by the coefficient of x in the linear expression – and remember the constant of integration.

At this point let us pause and summarize the main facts dealing with the integration of polynomial expressions and a 'function of a linear function of x'

Review summary

Unit 58

8

1 *Integration of polynomial expressions*
Integrate term by term and combine the individual constants of integration into one symbol.

2 *Integration of a 'function of a linear function of x'*
Replace x in the corresponding standard integral by the linear expression and divide the result by the coefficient of x in the linear expression.

Review exercise

Unit 58

9

1 Determine the following integrals:

(a) $I = \int (2x^3 - 5x^2 + 6x - 9)\,dx$

(b) $I = \int (9x^3 + 11x^2 - x - 3)\,dx$, given that when $x = 1$, $I = 2$.

2 Determine the following integrals:

(a) $\int (1 - 4x)^2\,dx$ (f) $\int \cos(1 - 3x)\,dx$

(b) $\int 4e^{5x-2}\,dx$ (g) $\int 2^{3x-1}\,dx$

(c) $\int 3\sin(2x + 1)\,dx$ (h) $\int 6\sec^2(2 + 3x)\,dx$

(d) $\int (3 - 2x)^{-5}\,dx$ (i) $\int \sqrt{3 - 4x}\,dx$

(e) $\int \dfrac{7}{2x - 5}\,dx$ (j) $\int 5e^{1-3x}\,dx$

Complete both questions. Take your time, there is no need to rush.
If necessary, look back at the Unit.
The answers and working are in the next frame.

10

1 (a) $I = \int (2x^3 - 5x^2 + 6x - 9)\,dx = 2\dfrac{x^4}{4} - 5\dfrac{x^3}{3} + 6\dfrac{x^2}{2} - 9x + C$

$$= \dfrac{x^4}{2} - \dfrac{5}{3}x^3 + 3x^2 - 9x + C$$

(b) $I = \int (9x^3 + 11x^2 - x - 3)\,dx = 9\dfrac{x^4}{4} + 11\dfrac{x^3}{3} - \dfrac{x^2}{2} - 3x + C$

$$= \left(\left(\left(\dfrac{9x}{4} + \dfrac{11}{3}\right)x - \dfrac{1}{2}\right)x - 3\right)x + C$$

Given $I = 2$ when $x = 1$ we find that:

$$2 = \dfrac{29}{12} + C$$

So that $C = -\dfrac{5}{12}$ and $I = 9\dfrac{x^4}{4} + 11\dfrac{x^3}{3} - \dfrac{x^2}{2} - 3x - \dfrac{5}{12}$

2 (a) $\int (1 - 4x)^2\,dx$ [Standard integral $\int x^2\,dx = \dfrac{x^3}{3} + C$]

Therefore, $\int (1 - 4x)^2\,dx = \dfrac{(1 - 4x)^3}{3} \times \dfrac{1}{(-4)} + C = -\dfrac{(1 - 4x)^3}{12} + C$

(b) $\int 4e^{5x-2}\,dx$ [Standard integral $\int e^x\,dx = e^x + C$]

Therefore, $\int 4e^{5x-2}\,dx = 4e^{5x-2} \times \dfrac{1}{5} + C = \dfrac{4}{5}e^{5x-2} + C$

(c) $\int 3\sin(2x + 1)\,dx$ [Standard integral $\int \sin x\,dx = -\cos x + C$]

Therefore, $\int 3\sin(2x + 1)\,dx = 3(-\cos(2x + 1)) \times \dfrac{1}{2} + C$

$$= -\dfrac{3}{2}\cos(2x + 1) + C$$

(d) $\int (3 - 2x)^{-5}\,dx$ [Standard integral $\int x^{-5}\,dx = \dfrac{-x^{-4}}{4} + C$]

Therefore, $\int (3 - 2x)^{-5}\,dx = -\dfrac{(3 - 2x)^{-4}}{4} \times \dfrac{1}{(-2)} + C = \dfrac{(3 - 2x)^{-4}}{8} + C$

(e) $\int \dfrac{7}{2x - 5}\,dx$ [Standard integral $\int \dfrac{1}{x}\,dx = \ln x + C$]

Therefore, $\int \dfrac{7}{2x - 5}\,dx = 7\ln(2x - 5) \times \dfrac{1}{2} + C = \dfrac{7}{2}\ln(2x - 5) + C$

(f) $\int \cos(1 - 3x)\,dx$ [Standard integral $\int \cos x\,dx = \sin x + C$]

Therefore, $\int \cos(1 - 3x)\,dx = \sin(1 - 3x) \times \dfrac{1}{(-3)} + C$

$$= -\dfrac{\sin(1 - 3x)}{3} + C$$

(g) $\int 2^{3x-1}\,dx$ [Standard integral $\int 2^x\,dx = \dfrac{2^x}{\ln 2} + C$]

Therefore, $\int 2^{3x-1}\,dx = \dfrac{2^{3x-1}}{\ln 2} \times \dfrac{1}{3} + C = \dfrac{2^{3x-1}}{3\ln 2} + C$

(h) $\int 6\sec^2(2+3x)\,dx$ [Standard integral $\int \sec^2 x\,dx = \tan x + C$]

Therefore, $\int 6\sec^2(2+3x)\,dx = 6\tan(2+3x) \times \dfrac{1}{3} + C$

$$= 2\tan(2+3x) + C$$

(i) $\int \sqrt{3-4x}\,dx$ [Standard integral $\int \sqrt{x}\,dx = \dfrac{x^{\frac{3}{2}}}{3/2} + C$]

Therefore, $\int \sqrt{3-4x}\,dx = \dfrac{2(3-4x)^{\frac{3}{2}}}{3} \times \dfrac{1}{(-4)} + C = -\dfrac{(3-4x)^{\frac{3}{2}}}{6} + C$

(j) $\int 5e^{1-3x}\,dx$ [Standard integral $\int 5e^x\,dx = 5e^x + C$]

Therefore, $\int 5e^{1-3x}\,dx = 5e^{1-3x} \times \dfrac{1}{(-3)} + C = -\dfrac{5}{3}e^{1-3x} + C$

Now for the Review test

 # Review test

Unit 58

1 Determine the following integrals:

11

(a) $I = \int (x^3 - x^2 + x - 1)\,dx$

(b) $I = \int (4x^3 - 9x^2 + 8x - 2)\,dx$ given that $I = \dfrac{11}{16}$ when $x = \dfrac{1}{2}$.

2 Determine the following integrals:

(a) $\int (5x-1)^4\,dx$

(b) $\int \dfrac{\sin(6x-1)}{3}\,dx$

(c) $\int \sqrt{4-2x}\,dx$

(d) $\int 2e^{3x+2}\,dx$

(e) $\int 5^{1-x}\,dx$

(f) $\int \dfrac{3}{2x-3}\,dx$

(g) $\int \dfrac{\sec^2(2-5x)}{5}\,dx$

Integration by partial fractions   **Unit 59**

1

Expressions such as $\int \dfrac{7x+8}{2x^2+11x+5}\,dx$ do not appear in our list of standard integrals, but do, in fact, occur in many mathematical applications. We saw in Module 7 that such an expression as $\dfrac{7x+8}{2x^2+11x+5}$ can be expressed in partial fractions which are simpler in structure.

In fact, $\dfrac{7x+8}{2x^2+11x+5} = \dfrac{7x+8}{(x+5)(2x+1)} = \dfrac{3}{x+5} + \dfrac{1}{2x+1}$ so that

$$\int \dfrac{7x+8}{2x^2+11x+5}\,dx = \int \dfrac{3}{x+5}\,dx + \int \dfrac{1}{2x+1}\,dx$$

These partial fractions are 'functions of a linear function of x', based on the standard integral $\int \dfrac{1}{x}\,dx$, so the result is clear:

$$\int \dfrac{7x+8}{2x^2+11x+5}\,dx = \int \dfrac{7x+8}{(x+5)(2x-1)}\,dx = \int \dfrac{3}{x+5}\,dx + \int \dfrac{1}{2x+1}\,dx$$
$$= 3\ln(x+5) + \dfrac{1}{2}\ln(2x+1) + C$$

You will recall the Rules of Partial Fractions which we listed earlier and used in Module 7, so let us apply them in this example. We will only deal with simple linear denominators.

Determine $\int \dfrac{3x^2+18x+3}{3x^2+5x-2}\,dx$ by partial fractions.

The first step is ....

2

> to divide out

because the numerator is not of lower degree than that of the denominator

$$\text{So } \dfrac{3x^2+18x+3}{3x^2+5x-2} = \ldots\ldots\ldots\ldots$$

3

> $1 + \dfrac{13x+5}{3x^2+5x-2}$

The denominator factorizes into $(3x-1)(x+2)$ so the partial fractions of $\dfrac{13x+5}{(3x-1)(x+2)} = \ldots\ldots\ldots$

4

$$\frac{4}{3x-1} + \frac{3}{x+2}$$

Because

$$\frac{13x+5}{(3x-1)(x+2)} = \frac{A}{3x-1} + \frac{B}{x+2}$$

$$\therefore \quad 13x+5 = A(x+2) + B(3x-1)$$
$$= Ax + 2A + 3Bx - B$$
$$= (A+3B)x + (2A-B)$$

$[x] \quad \therefore \quad A+3B = 13 \qquad A+3B = 13$

$[CT] \quad 2A-B = 5 \qquad \underline{6A-3B = 15}$

$$7A \quad = 28 \qquad \therefore \quad A = 4$$

$$\therefore \quad 4+3B = 13 \qquad \therefore \quad 3B = 9 \qquad \therefore \quad B = 3$$

$$\therefore \quad \frac{13x+5}{(3x-1)(x+2)} = \frac{4}{3x-1} + \frac{3}{x+2}$$

$$\therefore \quad \int \frac{3x^2 + 18x + 3}{3x^2 + 5x - 2}\, dx = \int \left(1 + \frac{4}{3x-1} + \frac{3}{x+2}\right) dx$$

$$= \ldots\ldots\ldots$$

Finish it

5

$$I = x + \frac{4\ln(3x-1)}{3} + 3\ln(x+2) + C$$

Now you can do this one in like manner:

$$\int \frac{4x^2 + 26x + 5}{2x^2 + 9x + 4}\, dx = \ldots\ldots\ldots\ldots$$

Work right through it and then check the solution with the next frame

6

$$\boxed{2x + 5\ln(x + 4) - \ln(2x + 1) + C}$$

Here is the working:

$$\frac{4x^2 + 26x + 5}{2x^2 + 9x + 4} = 2 + \frac{8x - 3}{2x^2 + 9x + 4}$$

$$\frac{8x - 3}{2x^2 + 9x + 4} = \frac{8x - 3}{(x + 4)(2x + 1)} = \frac{A}{x + 4} + \frac{B}{2x + 1}$$

$$\therefore \quad 8x - 3 = A(2x + 1) + B(x + 4) = (2A + B)x + (A + 4B)$$

$$\therefore \quad 2A + B = 8 \qquad\qquad 8A + 4B = 32$$

$$A + 4B = -3 \qquad\qquad \underline{A + 4B = -3}$$

$$\therefore \quad 7A \qquad\quad = 35 \qquad \therefore \quad A = 5$$

$$10 + B = 8 \qquad\qquad\qquad\quad \therefore \quad B = -2$$

$$\therefore \quad \int \frac{4x^2 + 26x + 5}{2x^2 + 9x + 4}\,dx = \int \left(2 + \frac{5}{x + 4} - \frac{2}{2x + 1} \right) dx$$

$$= 2x + 5\ln(x + 4) - \frac{2\ln(2x + 1)}{2} + C$$

$$= 2x + 5\ln(x + 4) - \ln(2x + 1) + C$$

And finally this one:

Determine $I = \displaystyle\int \frac{16x + 7}{6x^2 + x - 12}\,dx$ by partial fractions.

$$I = \ldots\ldots\ldots$$

7

$$\boxed{\ln(2x + 3) + \frac{5}{3}\ln(3x - 4) + C}$$

$$\frac{16x + 7}{6x^2 + x - 12} = \frac{16x + 7}{(2x + 3)(3x - 4)} = \frac{A}{2x + 3} + \frac{B}{3x - 4}$$

$$\therefore \quad 16x + 7 = A(3x - 4) + B(2x + 3)$$

$$= (3A + 2B)x - (4A - 3B)$$

Equating coefficients gives $A = 2$ and $B = 5$

$$\therefore \quad \int \frac{16x + 7}{6x^2 + x - 12}\,dx = \int \left(\frac{2}{2x + 3} + \frac{5}{3x - 4} \right) dx$$

$$= \ln(2x + 3) + \frac{5}{3}\ln(3x - 4) + C$$

At this point let us pause and summarize the main facts dealing with integration by partial fractions

Review summary   Unit 59

Integration by partial fractions   **8**

1. Algebraic fractions can often be expressed in terms of partial fractions. This renders integration of such algebraic fractions possible, the integration of each partial fraction

$$\int \frac{A}{ax+b}\, dx \text{ giving } A\frac{\ln(ax+b)}{a} + C$$

 Review exercise   Unit 59

1. Integrate by partial fractions each of the following integrals: **9**

   (a) $\int \frac{5x+2}{3x^2+x-4}\, dx$ (b) $\int \frac{x+1}{4x^2-1}\, dx$ (c) $\int \frac{3x}{1+x-2x^2}\, dx$

   Complete the question.
   Look back at the Unit if necessary but don't rush.
   The answers and working are in the next frame.

1. (a) $\dfrac{5x+2}{3x^2+x-4} = \dfrac{5x+2}{(3x+4)(x-1)} = \dfrac{A}{3x+4} + \dfrac{B}{x-1}$ therefore   **10**

   $$5x+2 = A(x-1) + B(3x+4)$$
   $$= (A+3B)x + (-A+4B) \text{ so that}$$

   $$A+3B = 5$$
   $$-A+4B = 2 \qquad \text{therefore adding we find that } 4B = 3$$
   $$\text{so } B = 3/4 \text{ and } A = -1/4.$$

   Therefore: $\displaystyle\int \frac{5x+2}{3x^2+x-4}\, dx = \int \frac{2}{3x+4}\, dx + \int \frac{1}{x-1}\, dx$

   $$= \frac{2}{3}\ln(3x+4) + \ln(x-1) + C$$

(b) $\dfrac{x+1}{4x^2-1} = \dfrac{x+1}{(2x+1)(2x-1)} = \dfrac{A}{2x+1} + \dfrac{B}{2x-1}$ therefore

$$x + 1 = A(2x - 1) + B(2x + 1)$$
$$= (2A + 2B)x + (-A + B) \text{ so that}$$

$2A + 2B = 1 \qquad 2A + 2B = 1$

$-A + B = 1 \qquad -2A + 2B = 2$ therefore adding we find that $4B = 3$

so $B = 3/4$ and $A = -1/4$.

Therefore:

$$\int \dfrac{x+1}{4x^2-1}\,dx = -\int \dfrac{1/4}{2x+1}\,dx + \int \dfrac{3/4}{2x-1}\,dx$$

$$= -\dfrac{1}{8}\ln(2x+1) + \dfrac{3}{8}\ln(2x-1) + C$$

(c) $\dfrac{3x}{1+x-2x^2} = \dfrac{3x}{(1-x)(1+2x)} = \dfrac{A}{1-x} + \dfrac{B}{1+2x}$ therefore

$$3x = A(1 + 2x) + B(1 - x)$$
$$= (2A - B)x + (A + B) \text{ so that}$$

$2A - B = 3$

$A + B = 0$ therefore, adding we find that $3A = 3$ so $A = 1$

and $B = -1$.

Therefore:

$$\int \dfrac{3x}{1+x-2x^2}\,dx = \int \dfrac{1}{1-x}\,dx - \int \dfrac{1}{1+2x}\,dx$$

$$= -\ln(1-x) - \dfrac{1}{2}\ln(1+2x) + C$$

Now for the Review test

 # Review test   **Unit 59**

11   **1** Integrate by partial fractions each of the following integrals:

(a) $\displaystyle\int \dfrac{5x}{6x^2-x-1}\,dx$ (b) $\displaystyle\int \dfrac{14x+1}{2-7x-4x^2}\,dx$

(c) $\displaystyle\int \dfrac{1-9x}{1-9x^2}\,dx$

Integration by parts

Unit 60

We often need to integrate a product where either function is *not* the derivative of the other. For example, in the case of

$$\int x^2 . \ln x \, dx$$

$\ln x$ is not the derivative of x^2

x^2 is not the derivative of $\ln x$

so in situations like this, we have to find some other method of dealing with the integral. Let us establish the rule for such cases.

If u and v are functions of x, then we know that

$$\frac{d}{dx}(uv) = u\frac{dv}{dx} + v\frac{du}{dx}$$

Now integrate both sides with respect to x. On the left, we get back to the function from which we started:

$$uv = \int u\frac{dv}{dx}\,dx + \int v\frac{du}{dx}\,dx$$

and rearranging the terms, we have:

$$\int u\frac{dv}{dx}\,dx = uv - \int v\frac{du}{dx}\,dx$$

On the left-hand side, we have a product of two factors to integrate. One factor is chosen as the function u: the other is thought of as being the derivative of some function v. To find v, of course, we must integrate this particular factor separately. Then, knowing u and v we can substitute in the right-hand side and so complete the routine.

You will notice that we finish up with another product to integrate on the end of the line, but, unless we are very unfortunate, this product will be easier to tackle than the original one.

This then is the key to the routine:

$$\int u\frac{dv}{dx}\,dx = uv - \int v\frac{du}{dx}\,dx$$

For convenience, this can be memorized as:

$$\int u dv = uv - \int v du$$

In this form it is easier to remember, but the previous line gives its meaning in detail. This method is called *integration by parts*.

2

$$\text{So } \int u\frac{dv}{dx}\,dx = uv - \int v\frac{du}{dx}\,dx$$

i.e. $\int u\,dv = uv - \int v\,du$

Make a note of these results. You will soon learn them. Now for some examples involving integration by parts.

Example 1

$$\int x^2.\ln x\,dx$$

The two factors are x^2 and $\ln x$, and we have to decide which to take as u and which to link to dv. If we choose x^2 to be u and $\ln x\,dx$ to be dv, then we shall have to integrate $\ln x$ in order to find v. Unfortunately, $\int \ln x\,dx$ is not in our basic list of standard integrals, therefore we must allocate u and dv the other way round, i.e. $u = \ln x$ so that $du = \dfrac{1}{x}\,dx$ and $dv = x^2\,dx$ so that $v = \dfrac{x^3}{3}$. (We omit the integration constant here because we are in the middle of evaluating the integral. The integration constant will come out eventually.)

$$\therefore \quad \int x^2 \ln x\,dx = \int u\,dv$$

$$= uv - \int v\,du$$

$$= \ln x\left(\frac{x^3}{3}\right) - \frac{1}{3}\int x^3\frac{1}{x}\,dx$$

Notice that we can tidy up the writing of the second integral by writing the constant factors involved, outside the integral:

$$\therefore \quad \int x^2 \ln x\,dx = \ln x\left(\frac{x^3}{3}\right) - \frac{1}{3}\int x^3.\frac{1}{x}\,dx = \frac{x^3}{3}\ln x - \frac{1}{3}\int x^2\,dx$$

$$= \frac{x^3}{3}.\ln x - \frac{1}{3}.\frac{x^3}{3} + C = \frac{x^3}{3}\left\{\ln x - \frac{1}{3}\right\} + C$$

Note that if one of the factors of the product to be integrated is a log term, this must be chosen as .... (u or dv)

$$u$$

3

Example 2

$$\int x^2 e^{3x} dx \qquad \text{Let } u = x^2 \text{ so that } du = 2x\,dx \text{ and } dv = e^{3x}\,dx \text{ so that } v = \frac{e^{3x}}{3}$$

Then $\displaystyle\int x^2 e^{3x}\,dx = uv - \int v\,du$

$$= x^2\left(\frac{e^{3x}}{3}\right) - \frac{2}{3}\int e^{3x}x\,dx \qquad \text{The integral } \int e^{3x}x\,dx \text{ will also have}$$

to be integrated by parts. So that:

$$= \frac{x^2 e^{3x}}{3} - \frac{2}{3}\left\{ x\left(\frac{e^{3x}}{3}\right) - \frac{1}{3}\int e^{3x}\,dx \right\} = \frac{x^2 e^{3x}}{3} - \frac{2xe^{3x}}{9} + \frac{2}{9}\frac{e^{3x}}{3} + C$$

$$= \frac{e^{3x}}{3}\left\{ x^2 - \frac{2x}{3} + \frac{2}{9} \right\} + C$$

On to Frame 4

In Example 1 we saw that if one of the factors is a log function, that log **4** function *must* be taken as *u*.

In Example 2 we saw that, provided there is no log term present, the power of *x* is taken as *u*. (By the way, this method holds good only for positive whole-number powers of *x*. For other powers, a different method must be applied.)

So which of the two factors should we choose to be *u* in each of the following cases?

(a) $\displaystyle\int x.\ln x\,dx$

(b) $\displaystyle\int x^3.\sin x\,dx$

5

$$\text{(a)} \quad \text{In } \int x.\ln x\,dx, \quad u = \ln x$$

$$\text{(b)} \quad \text{In } \int x^3 \sin x\,dx, \quad u = x^3$$

Right. Now for a third example.

Example 3

$\int e^{3x} \sin x \, dx$. Here we have neither a log factor nor a power of x. Let us try putting $u = e^{3x}$ and $dv = \sin x \, dx$ so $v = -\cos x$.

$$\therefore \quad \int e^{3x} \sin x \, dx = e^{3x}(-\cos x) + 3 \int \cos x.e^{3x} \, dx$$

$$= -e^{3x} \cos x + 3 \int e^{3x} \cos x \, dx \quad \text{(and again by parts)}$$

$$= -e^{3x} \cos x + 3 \left\{ e^{3x}(\sin x) - 3 \int \sin x.e^{3x} \, dx \right\}$$

and it looks as though we are back where we started.

However, let us write I for the integral $\int e^{3x} \sin x \, dx$:

$$I = -e^{3x} \cos x + 3e^{3x} \sin x - 9I$$

Then, treating this as a simple equation, we get:

$$10I = e^{3x}(3 \sin x - \cos x) + C_1$$

$$I = \frac{e^{3x}}{10}(3 \sin x - \cos x) + C$$

Whenever we integrate functions of the form $e^{kx} \sin x$ or $e^{kx} \cos x$, we get similar types of results after applying the rule twice.

Move on to Frame 6

6

The three examples we have considered enable us to form a priority order for u:

   (a) $\ln x$ (b) x^n (c) e^{kx}

i.e.   If one factor is a log function, that must be taken as 'u'.
       If there is no log function but a power of x, that becomes 'u'.
       If there is neither a log function nor a power of x, then the exponential function is taken as 'u'.

Remembering the priority order will save a lot of false starts.

So which would you choose as 'u' in the following cases?

(a) $\int x^4 \cos 2x \, dx$ $u = \ldots\ldots\ldots$

(b) $\int x^4 e^{3x} \, dx$ $u = \ldots\ldots\ldots$

(c) $\int x^3 \ln(x + 4) \, dx$ $u = \ldots\ldots\ldots$

(d) $\int e^{2x} \cos 4x \, dx$ $u = \ldots\ldots\ldots$

(a) $\int x^4 \cos 2x\, dx$ $u = x^4$

(b) $\int x^4 e^{3x}\, dx$ $u = x^4$

(c) $\int x^3 \ln(x+4)\, dx$ $u = \ln(x+4)$

(d) $\int e^{2x} \cos 4x\, dx$ $u = e^{2x}$

Right. Now look at this one:

$$\int e^{5x} \sin 3x\, dx$$

Following our rule for priority for u, in this case, we should put

$u = \dots\dots\dots\dots$

$$\int e^{5x} \sin 3x\, dx \qquad \therefore\ u = e^{5x}$$

Correct. Make a note of that priority list for u.
Then go ahead and determine the integral given above.

When you have finished, check your working with that set out in the next frame

$$\int e^{5x} \sin 3x\, dx = \frac{3e^{5x}}{34}\left\{\frac{5}{3}\sin 3x - \cos 3x\right\} + C$$

Here is the working. Follow it through.

$$\int e^{5x} \sin 3x\, dx = e^{5x}\left(-\frac{\cos 3x}{3}\right) + \frac{5}{3}\int \cos 3x . e^{5x}\, dx$$

$$= -\frac{e^{5x}\cos 3x}{3} + \frac{5}{3}\left\{e^{5x}\left(\frac{\sin 3x}{3}\right) - \frac{5}{3}\int \sin 3x . e^{5x}\, dx\right\}$$

$$\therefore\ I = -\frac{e^{5x}\cos 3x}{3} + \frac{5}{9}e^{5x}\sin 3x - \frac{25}{9}I$$

$$\frac{34}{9}I = \frac{e^{5x}}{3}\left\{\frac{5}{3}\sin 3x - \cos 3x\right\} + C_1$$

$$I = \frac{3e^{5x}}{34}\left\{\frac{5}{3}\sin 3x - \cos 3x\right\} + C$$

There you are. Now do these in much the same way. Finish them both before moving on to the next frame.

(a) $\int x \ln x\, dx$ (b) $\int x^3 e^{2x}\, dx$

Solutions in Frame 10

10

(a) $\displaystyle\int x\ln x\,dx = \ln x\left(\frac{x^2}{2}\right) - \frac{1}{2}\int x^2 \cdot \frac{1}{x}\,dx$

$\displaystyle\qquad = \frac{x^2\ln x}{2} - \frac{1}{2}\int x\,dx$

$\displaystyle\qquad = \frac{x^2\ln x}{2} - \frac{1}{2}\cdot\frac{x^2}{2} + C$

$\displaystyle\qquad = \frac{x^2}{2}\left\{\ln x - \frac{1}{2}\right\} + C$

(b) $\displaystyle\int x^3 e^{2x}\,dx = x^3\left(\frac{e^{2x}}{2}\right) - \frac{3}{2}\int e^{2x}x^2\,dx$

$\displaystyle\qquad = \frac{x^3 e^{2x}}{2} - \frac{3}{2}\left\{x^2\left(\frac{e^{2x}}{2}\right) - \frac{2}{2}\int e^{2x}x\,dx\right\}$

$\displaystyle\qquad = \frac{x^3 e^{2x}}{2} - \frac{3x^2 e^{2x}}{4} + \frac{3}{2}\left\{x\left(\frac{e^{2x}}{2}\right) - \frac{1}{2}\int e^{2x}\,dx\right\}$

$\displaystyle\qquad = \frac{x^3 e^{2x}}{2} - \frac{3x^2 e^{2x}}{4} + \frac{3xe^{2x}}{4} - \frac{3}{4}\frac{e^{2x}}{2} + C$

$\displaystyle\qquad = \frac{e^{2x}}{2}\left\{x^3 - \frac{3x^2}{2} + \frac{3x}{2} - \frac{3}{4}\right\} + C$

That is it. You can now deal with the integration of products.

For now, let's pause and summarize the work on integration by parts

 # Review summary   Unit 60

11   *Parts formula:*   The integration by parts formula is often useful when trying to integrate a product of dissimilar terms. The formula is given either as:

$$\int u\,dv = uv - \int v\,du$$

or, equivalently, as:

$$\int u\frac{dv}{dx}\,dx = uv - \int v\frac{du}{dx}\,dx$$

Review exercise

Unit 60

---

1 Evaluate

$$\int (\pi - x) \cos x \, dx$$

12

> *Complete the question. Take one step at a time, there is no need to rush.*
> *If you need to, look back at the Unit.*
> *The answer and working are in the next frame.*

---

1 $\displaystyle \int (\pi - x) \cos x \, dx = \int u \, dv$ where $u = \pi - x$ so $du = -dx$ and $dv = \cos x \, dx$

13

$$\text{so } v = \sin x$$

$$= uv - \int v \, du$$

$$= (\pi - x) \sin x - \int \sin x (-dx)$$

$$= 0 + \int \sin x \, dx$$

$$= -\cos x + C$$

> *Now for the Review test*

 # Review test

Unit 60

---

1 Evaluate:

$$\int (x - \pi/2) \sin 3x \, dx$$

14

Areas under curves

Unit 61

Consider the area A of the figure bounded by the curve $y = f(x)$, the x-axis and the two vertical lines through $x = a$ and $x = b$ (where $b > a$).

1

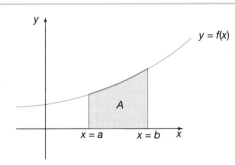

To evaluate the area A you need to consider the total area between the same curve and the x-axis from the left up to some arbitrary point P on the curve with coordinates (x, y) which we shall denote by A_x.

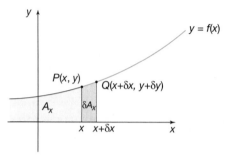

Area δA_x is the area enclosed by the strip under the arc PQ where Q has the coordinates $(x + \delta x, y + \delta y)$. If the strip is approximated by a rectangle of height y and width δx then $\delta A_x \approx y.\delta x$. This means that:

$$\frac{\delta A_x}{\delta x} \approx y$$

The error in this approximation is given by the area of PQR in the figure to the right, where the strip has been magnified.

If the width of the strip is reduced then the error is accordingly reduced. Also, if $\delta x \to 0$ then $\delta A_x \to 0$ and:

$$\frac{\delta A_x}{\delta x} \to \frac{dA_x}{dx} \quad \text{so that, in the limit,} \quad \frac{dA_x}{dx} = y$$

Consequently, because integration is the reverse process of differentiation it is seen that:

$$A_x = \int y \, dx$$

The total area between the curve and the x-axis up to the point P is given by the indefinite integral.

If $x = b$ then $A_b = \int_{(x=b)} y \, dx$ (the value of the integral and hence the area up

to b) and if $x = a$ then $A_a = \int_{(x=a)} y \, dx$ (the value of the integral and hence the

area up to a). Because $b > a$, the difference in these two areas $A_b - A_a$ gives the required area A. That is:

$$A = \int_{(x=b)} y \, dx - \int_{(x=a)} y \, dx \text{ which is written } A = \int_a^b y \, dx$$

The numbers *a* and *b* are called the *limits* of the integral where the right-hand limit is at the top of the integral sign and the left-hand limit is at the bottom. Such an integral with limits is called a *definite integral*. Notice that in the subtraction process when the integral is evaluated, the constant of integration disappears leaving the numerical value of the area.

Example 1

To determine the area bounded by the curve $y = 3x^2 + 6x + 8$, the *x*-axis and the ordinates $x = 1$ and $x = 3$.

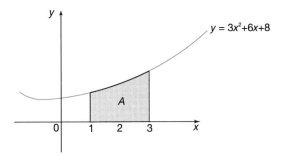

$$A = \int_1^3 y\,dx = \int_1^3 (3x^2 + 6x + 8)\,dx = \left[x^3 + 3x^2 + 8x\right]_1^3$$

Note that we enclose the expression in square brackets with the limits attached.

Now calculate the values at the upper and lower limits and subtract the second from the first which gives $A = \ldots\ldots\ldots$

$$\boxed{66 \text{ unit}^2}$$

2

Because

$$A = \left[x^3 + 3x^2 + 8x\right]_1^3$$
$$= \{27 + 27 + 24\} - \{1 + 3 + 8\} = 78 - 12 = 66 \text{ unit}^2$$

Example 2

Find the area bounded by the curve $y = 3x^2 + 14x + 15$, the *x*-axis and ordinates at $x = -1$ and $x = 2$.

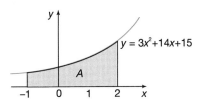

$$A = \int_{-1}^2 y\,dx = \int_{-1}^2 (3x^2 + 14x + 15)\,dx$$
$$= \left[x^3 + 7x^2 + 15x\right]_{-1}^2$$
$$\therefore \quad A = \{8 + 28 + 30\} - \{-1 + 7 - 15\}$$
$$= 66 - (-9) = 75 \text{ unit}^2$$

Example 3

Calculate the area bounded by the curve $y = -6x^2 + 24x + 10$, the x-axis and the ordinates $x = 0$ and $x = 4$.

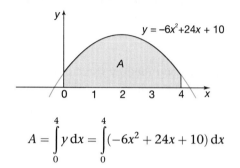

$$A = \int_0^4 y\,dx = \int_0^4 (-6x^2 + 24x + 10)\,dx$$

$$= \ldots\ldots\ldots$$

3

$$\boxed{104 \text{ unit}^2}$$

Because

$$A = \left[-2x^3 + 12x^2 + 10x\right]_0^4 = 104 - 0 \quad \therefore \quad A = 104 \text{ unit}^2$$

And now:

Example 4

Determine the area under the curve $y = e^x$ between $x = -2$ and $x = 3$.
Do this by the same method. The powers of e are available from most calculators or from tables.

$$A = \ldots\ldots\ldots$$

4

$$\boxed{19{\cdot}95 \text{ unit}^2}$$

As a check:

$$A = \int_{-2}^3 y\,dx = \int_{-2}^3 e^x\,dx = [e^x]_{-2}^3 = \{e^3\} - \{e^{-2}\}$$

$$e^3 = 20{\cdot}09 \text{ and } e^{-2} = 0{\cdot}14$$

$$\therefore \quad A = 20{\cdot}09 - 0{\cdot}14 = 19{\cdot}95 \text{ unit}^2$$

Simpson's rule   **5**

We know, of course, that integration can be used to calculate the area under a curve $y = f(x)$ between two given points $x = a$ and $x = b$.

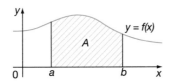

$$A = \int_a^b y \, dx = \int_a^b f(x) \, dx$$

Sometimes, however, it is not possible to evaluate the integral so, if only we could find the area A by some other means, this would give us an approximate numerical value of the integral we have to evaluate. There are various practical ways of doing this and the one we shall choose is to apply Simpson's rule.

So on to Frame 6

To find the area under the curve $y = f(x)$ between $x = a$ and $x = b$:   **6**

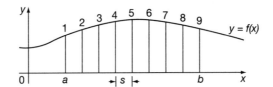

(a) Divide the figure into any even number (n) of equal-width strips (width s).

(b) Number and measure each ordinate: $y_1, y_2, y_3, \ldots, y_{n+1}$.

The number of ordinates will be one more than the number of strips.

(c) The area A of the figure is then given by:

$$A \approx \frac{s}{3} \left[(F + L) + 4E + 2R \right]$$

where $s =$ width of each strip

$F + L =$ sum of the first and last ordinates

$4E = 4 \times$ the sum of the even-numbered ordinates

$2R = 2 \times$ the sum of the remaining odd-numbered ordinates.

Note: Each ordinate is used once – and only once.

Make a note of this result for future reference

7

$$A \approx \frac{s}{3}\Big[(F+L)+4E+2R\Big]$$

The symbols themselves remind you of what they represent.

We shall now evaluate $\int_{2}^{6} y\,dx$ for the function $y = f(x)$, the graph of which is shown:

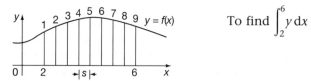

To find $\int_{2}^{6} y\,dx$

If we take 8 strips, then $s = \dfrac{6-2}{8} = \dfrac{4}{8} = \dfrac{1}{2}.$ $s = \dfrac{1}{2}$

Suppose we find lengths of the ordinates to be as follows:

Ordinate no.	1	2	3	4	5	6	7	8	9
Length	7·5	8·2	10·3	11·5	12·4	12·8	12·3	11·7	11·5

Then we have:

$$F + L = 7\cdot5 + 11\cdot5 = 19$$
$$4E = 4(8\cdot2 + 11\cdot5 + 12\cdot8 + 11\cdot7) = 4(44\cdot2) = 176\cdot8$$
$$2R = 2(10\cdot3 + 12\cdot4 + 12\cdot3) = 2(35) = 70$$

So that:

$$A \approx \frac{1/2}{3}\Big[19 + 176\cdot8 + 70\Big]$$
$$= \frac{1}{6}\Big[265\cdot8\Big] = 44\cdot3 \quad \therefore\ A = 44\cdot3 \text{ units}^2$$

$$\therefore\ \int_{2}^{6} f(x)\,dx \approx 44\cdot3$$

The accuracy of the result depends on the number of strips into which we divide the figure. A larger number of thinner strips gives a more accurate result.

Simpson's rule is important: it is well worth remembering.

Here it is again: write it out, but replace the query marks with the appropriate coefficients:

$$A \approx \frac{s}{?}\Big[(F+L)+?E+?R\Big]$$

8

$$A \approx \frac{s}{3}\Big[(F+L) + 4E + 2R\Big]$$

In practice, we do not have to plot the curve in order to measure the ordinates. We calculate them at regular intervals. Here is an example.

Example 1

To evaluate $\int_0^{\pi/3} \sqrt{\sin x}\,dx$, using six intervals.

(a) Find the value of s:

$$s = \frac{\pi/3 - 0}{6} = \frac{\pi}{18} \quad (= 10° \text{ intervals})$$

(b) Calculate the values of y (i.e. $\sqrt{\sin x}$) at intervals of $\pi/18$ between $x = 0$ (lower limit) and $x = \pi/3$ (upper limit), and set your work out in the form of the table below:

x		$\sin x$	$\sqrt{\sin x}$
0	(0°)	0·0000	0·0000
$\pi/18$	(10°)	0·1736	0·4167
$\pi/9$	(20°)	0·3420	
$\pi/6$	(30°)	0·5000	
$2\pi/9$	(40°)		
$5\pi/18$	(50°)		
$\pi/3$	(60°)		

Leave the right-hand side of your page blank for the moment.

Copy and complete the table as shown on the left-hand side above.

Here it is: check your results so far.

9

x		$\sin x$	$\sqrt{\sin x}$	(1) $F+L$	(2) E	(3) R
0	(0°)	0·0000	0·0000			
$\pi/18$	(10°)	0·1736	0·4167			
$\pi/9$	(20°)	0·3420	0·5848			
$\pi/6$	(30°)	0·5000	0·7071			
$2\pi/9$	(40°)	0·6428	0·8017			
$5\pi/18$	(50°)	0·7660	0·8752			
$\pi/3$	(60°)	0·8660	0·9306			

Now form three more columns on the right-hand side, headed as shown, and transfer the final results across as indicated. This will automatically sort out the ordinates into the correct groups.

Then on to Frame 10

10

Note that:

	(1) $F + L$	(2) E	(3) R
(a) You start in column 1	0·0000		
(b) You then zig-zag down the two right-hand columns.		0·4167	
			0·5848
(c) You finish back in column 1.		0·7071	
			0·8017
		0·8752	
	0·9306		

Now total up each of the three columns

11

Your results should be:

(1) $F + L$	(2) E	(3) R
0·9306	1·999	1·3865

Now: (a) Multiply column (2) by 4 so as to give $4E$.

(b) Multiply column (3) by 2 so as to give $2R$.

(c) Transfer the result in columns (2) and (3) to column (1) and total column (1) to obtain $(F + L) + 4E + 2R$.

Now do that

12

This gives:

	$F + L$	E	R
$F + L \longrightarrow$	0·9306	1·999	1·3865
$4E \longrightarrow$	7·996	4	2
$2R \longrightarrow$	2·773	7·996	2·773
$(F + L) + 4E + 2R \longrightarrow$	11·6996		

The formula is $A \approx \dfrac{s}{3}\left[(F + L) + 4E + 2R\right]$ so to find A we simply need to multiply our last result by $\dfrac{s}{3}$. Remember $s = \pi/18$.

So now you can finish it off.

$$\int_0^{\pi/3} \sqrt{\sin x}\, dx = \ldots\ldots\ldots\ldots$$

$$\boxed{0.681}$$

Because

$$A \approx \frac{s}{3}\Big[(F+L) + 4E + 2R\Big]$$

$$\approx \frac{\pi/18}{3}\Big[11.6996\Big]$$

$$\approx (\pi/54)\Big[11.6996\Big]$$

$$\approx 0.6807$$

$$\therefore \int_0^{\pi/3} \sqrt{\sin x}\, dx \approx 0.681$$

Before we do another example, let us see the last solution complete.

To evaluate $\int_0^{\pi/3} \sqrt{\sin x}\, dx$ by Simpson's rule, using 6 intervals.

$$s = \frac{\pi/3 - 0}{6} = \pi/18 \quad (= 10° \text{ intervals})$$

x		$\sin x$	$\sqrt{\sin x}$	$F+L$	E	R
0	(0°)	0.0000	0.0000	0.0000		
$\pi/18$	(10°)	0.1736	0.4167		0.4167	
$\pi/9$	(20°)	0.3420	0.5848			0.5848
$\pi/6$	(30°)	0.5000	0.7071		0.7071	
$2\pi/9$	(40°)	0.6428	0.8017			0.8017
$5\pi/18$	(50°)	0.7660	0.8752		0.8752	
$\pi/3$	(60°)	0.8660	0.9306	0.9306		
	$F+L$		$\longrightarrow$	0.9306	1.999	1.3865
	$4E$		$\longrightarrow$	7.996	4	2
	$2R$		$\longrightarrow$	2.773	7.996	2.773
	$(F+L)+4E+2R$		$\longrightarrow$	11.6996		

$$I \approx \frac{s}{3}\Big[(F+L) + 4E + 2R\Big]$$

$$\approx \frac{\pi}{54}\Big[11.6996\Big]$$

$$\approx 0.6807$$

$$\therefore \int_0^{\pi/3} \sqrt{\sin x}\, dx \approx 0.681$$

Now we tackle another example and set it out in much the same way.

Move to Frame 14

14

Example 2

To evaluate $\displaystyle\int_{0\cdot2}^{1\cdot0} \sqrt{1+x^3}\,\mathrm{d}x$, using 8 intervals.

First of all, find the value of s in this case.

$s = \ldots\ldots\ldots\ldots$

15

$$\boxed{0\cdot1}$$

Because

$$s = \frac{1\cdot0 - 0\cdot2}{8} = \frac{0\cdot8}{8} = 0\cdot1 \quad s = 0\cdot1$$

Now write the column headings required to build up the function values. What will they be on this occasion?

16

x	x^3	$1+x^3$	$\sqrt{1+x^3}$	$F+L$	E	R

Right. So your table will look like this, with x ranging from 0·2 to 1·0:

x	x^3	$1+x^3$	$\sqrt{1+x^3}$	$F+L$	E	R
0·2	0·008	1·008	1·0040			
0·3	0·027	1·027	1·0134			
0·4	0·064					
0·5	0·125					
0·6	0·216					
0·7	0·343					
0·8						
0·9						
1·0						
			$F+L\longrightarrow$			
			$4E\longrightarrow$		4	2
			$2R\longrightarrow$			
		$(F+L)+4E+2R\longrightarrow$				

Copy down and complete the table above and finish off the working to evaluate $\displaystyle\int_{0\cdot2}^{1\cdot0} \sqrt{1+x^3}\,\mathrm{d}x$.

Check with the next frame

17

$$\int_{0\cdot2}^{1\cdot0} \sqrt{1+x^3}\, dx = 0\cdot911$$

x	x^3	$1+x^3$	$\sqrt{1+x^3}$	$F+L$	E	R
0·2	0·008	1·008	1·0040	1·0040		
0·3	0·027	1·027	1·0134		1·0134	
0·4	0·064	1·064	1·0315			1·0315
0·5	0·125	1·125	1·0607		1·0607	
0·6	0·216	1·216	1·1027			1·1027
0·7	0·343	1·343	1·1589		1·1589	
0·8	0·512	1·512	1·2296			1·2296
0·9	0·729	1·729	1·3149		1·3149	
1·0	1·000	2·000	1·4142	1·4142		
			$F+L\longrightarrow$	2·4182	4·5479	3·3638
			$4E\longrightarrow$	18·1916	4	2
			$2R\longrightarrow$	6·7276	18·1916	6·7276
		$(F+L)+4E+2R\longrightarrow$		27·3374		

$$I = \frac{s}{3}\left[(F+L)+4E+2R\right]$$
$$= \frac{0\cdot1}{3}\left[27\cdot3374\right] = \frac{1}{3}\left[2\cdot73374\right] = 0\cdot9112$$
$$\therefore \int_{0\cdot2}^{1\cdot0}\sqrt{1+x^3}\, dx \approx 0\cdot911$$

There it is. Next frame

Here is another one: let us work through it together.

18

Example 3

Using Simpson's rule with 8 intervals, evaluate $\int_{1}^{3} y\, dx$, where the values of y at regular intervals of x are given.

x	1·0	1·25	1·50	1·75	2·00	2·25	2·50	2·75	3·00
y	2·45	2·80	3·44	4·20	4·33	3·97	3·12	2·38	1·80

If these function values are to be used as they stand, they must satisfy the requirements for Simpson's rule, which are:

(a) the function values must be spaced at .... intervals of x, and

(b) there must be an .... number of strips and therefore an .... number of ordinates.

19

> regular; even; odd

These conditions are satisfied in each case, so we can go ahead and evaluate the integral. In fact, the working will be a good deal easier for we are told the function value and there is no need to build them up as we had to do before.

In this example, $s = $ ....

20

> $s = 0.25$

Because

$$s = \frac{3-1}{8} = \frac{2}{8} = 0.25$$

Off you go, then. Set out your table and evaluate the integral defined by the values given in Frame 18. When you have finished, move on to Frame 21 to check your working.

21

> 6·62

x	y	$F + L$	E	R
1·0	2·45	2·45		
1·25	2·80		2·80	
1·50	3·44			3·44
1·75	4·20		4·20	
2·00	4·33			4·33
2·25	3·97		3·97	
2·50	3·12			3·12
2·75	2·38		2·38	
3·00	1·80	1·80		
$F + L \longrightarrow$		4·25	13·35	10·89
$4E \longrightarrow$		53·40	4	2
$2R \longrightarrow$		21·78	53·40	21·78
$(F + L) + 4E + 2R \longrightarrow$		79·43		

$$I = \frac{s}{3}\left[(F + L) + 4E + 2R\right] = \frac{0.25}{3}\left[79.43\right]$$

$$= \frac{1}{12}\left[79.43\right] = 6.62$$

$$\therefore \int_1^3 y\,dx \approx 6.62$$

Here is one further example.

22

Example 4

A pin moves along a straight guide so that its velocity v (cm/s) when it is distance x (cm) from the beginning of the guide at time t (s) is as given in the table below:

t (s)	0	0·5	1·0	1·5	2·0	2·5	3·0	3·5	4·0
v (cm/s)	0	4·00	7·94	11·68	14·97	17·39	18·25	16·08	0

Apply Simpson's rule, using 8 intervals, to find the approximate total distance travelled by the pin between $t = 0$ and $t = 4$.
 We must first interpret the problem, thus:

$$v = \frac{dx}{dt} \quad \therefore \ x = \int_0^4 v\,dt$$

and since we are given values of the function v at regular intervals of t, and there is an even number of intervals, then we are all set to apply Simpson's rule.
 Complete the problem then, entirely on your own.

When you have finished it, check with Frame 23

23

46·5 cm

t	v	$F+L$	E	R
0	0·00	0·00		
0·5	4·00		4·00	
1·0	7·94			7·94
1·5	11·68		11·68	
2·0	14·97			14·97
2·5	17·39		17·39	
3·0	18·25			18·25
3·5	16·08		16·08	
4·0	0·00	0·00		
$F+L \longrightarrow$		0·00	49·15	41·16
$4E \longrightarrow$		196·60	4	2
$2R \longrightarrow$		82·32	196·60	82·32
$(F+L)+4E+2R \longrightarrow$		278·92		

$$x = \frac{s}{3}\left[(F+L)+4E+2R\right] \text{ and } s = 0·5$$

$$\therefore \ x = \frac{1}{6}\left[278·92\right] = 46·49 \quad \therefore \ \text{Total distance} \approx 46·5 \text{ cm}$$

At this point let's pause and summarize the work so far on areas under curves

Review summary

24

1 *Area beneath a curve*

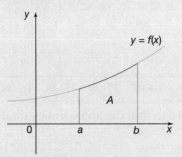

Area A, bounded by the curve $y = f(x)$, the x-axis and ordinates $x = a$ and $x = b$, is given by:

$$A = \int_a^b y \, dx$$

2 *Simpson's rule*

(a) The figure is divided into an *even* number of strips of equal width s. There will therefore be an *odd* number of ordinates or function values, including both boundary values.

(b) The value of the definite integral $\int_a^b f(x) \, dx$ is given by the numerical value of the area under the curve $y = f(x)$ between $x = a$ and $x = b$:

$$I = A \approx \frac{s}{3} \left[(F + L) + 4E + 2R \right]$$

where s = width of strip (or interval)

$F + L$ = sum of the first and last ordinates

$4E = 4 \times$ sum of the even-numbered ordinates

$2R = 2 \times$ sum of remaining odd-numbered ordinates.

(c) A practical hint to finish with:

Always set your work out in the form of a table, as we have done in the examples. It prevents your making slips in method and calculation, and enables you to check without difficulty.

Review exercise Unit 61

1 Find the area bounded by $y = 5 + 4x - x^2$, the x-axis and the ordinates $x = 1$ and $x = 4$.

2 Calculate the area under the curve $y = 2x^2 + 4x + 3$, between $x = 2$ and $x = 5$.

3 Determine the area bounded by $y = x^2 - 2x + 3$, the x-axis and ordinates $x = -1$ and $x = 3$.

4 Use Simpson's rule with 6 intervals to find the area beneath $f(x) = \sin x$ between $x = 0$ and $x = \pi$ to 3 dp.

Finish all four questions. Take your time, there is no need to rush.
If necessary, look back at the Unit.
The answers and working are in the next frame

26

$$\boxed{\textbf{1} \ 24 \ \text{unit}^2, \quad \textbf{2} \ 129 \ \text{unit}^2, \quad \textbf{3} \ 13\frac{1}{3} \ \text{unit}^2 \quad \textbf{4} \ 2\cdot001 \ \text{unit}^2}$$

Here is the working:

1 $\displaystyle A = \int_1^4 y\,dx = \int_1^4 (5 + 4x - x^2)\,dx = \left[5x + 2x^2 - \frac{x^3}{3}\right]_1^4$

$\displaystyle = \left\{20 + 32 - \frac{64}{3}\right\} - \left\{5 + 2 - \frac{1}{3}\right\}$

$\displaystyle = 30\frac{2}{3} - 6\frac{2}{3} = 24 \ \text{unit}^2$

2 $\displaystyle A = \int_2^5 y\,dx = \int_2^5 (2x^2 + 4x + 3)\,dx = \left[2\frac{x^3}{3} + 2x^2 + 3x\right]_2^5$

$\displaystyle = \left\{\frac{250}{3} + 50 + 15\right\} - \left\{\frac{16}{3} + 8 + 6\right\} = 148\frac{1}{3} - 19\frac{1}{3} = 129 \ \text{unit}^2$

3 $\displaystyle A = \int_{-1}^3 y\,dx = \int_{-1}^3 (x^2 - 2x + 3)\,dx = \left[\frac{x^3}{3} - x^2 + 3x\right]_{-1}^3$

$\displaystyle = \{9 - 9 + 9\} - \left\{-\frac{1}{3} - 1 - 3\right\}$

$\displaystyle = 9 - \left\{-4\frac{1}{3}\right\} = 13\frac{1}{3} \ \text{unit}^2.$

Notice that in all these definite integrals, we omit the constant of integration because we know it will disappear at the subtraction stage.

In an indefinite integral, however, it must always be included.

4 $f(x) = \sin x$ between $x = 0$ and $x = \pi$. The six divisions and the corresponding $f(x)$ values are, to three decimal places:

x	0	$\pi/6$	$\pi/3$	$\pi/2$	$2\pi/3$	$5\pi/6$	π
$f(x)$	0·000	0·500	0·866	1·000	0·866	0·500	0.000

Here, $a = 0$ and $b = \pi$ with $w = (b - a)/6 = \pi/6$. Therefore, the area in question is found by applying the rule:

$(w/3)[\text{sum of first and last ordinates} + 4(\text{sum of even ordinates})]$
$\quad + 2(\text{sum of remaining odd ordinates})$
$= (\pi/18)[0 + 0 + 4(0\cdot5 + 1 + 0\cdot5) + 2(0\cdot866 + 0\cdot866)]$
$= 2\cdot001 \text{ unit}^2 \text{ to 3 dp.}$

Now for the Review test

 # Review test   **Unit 61**

27 **1** Find the area enclosed between the x-axis and the curve $y = e^x$ between $x = 1$ and $x = 2$, giving your answer in terms of e.

2 Use Simpson's rule with 6 divisions of the x-axis to find the area between the x-axis and the graph of $f(x) = e^x$ between $x = -1$ and $x = +1$ to 3 dp.

Integration as a summation   **Unit 62**

1

We have identified the value of a definite integral as the area beneath a curve. However, some definite integrals have a negative value so how can we link this to an area because all areas are *positive quantities*? Before we can make this link we must consider the determination of area in a slightly different manner.

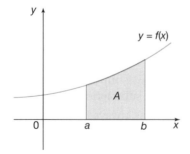

We have already seen that the area A under a curve $y = f(x)$ between $x = a$ and $x = b$ is given by the definite integral:

$$A = \int_a^b y \, dx$$

Let's look at this area a little more closely:

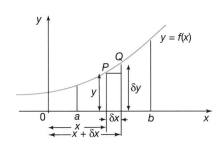

Let P be the point (x, y) on the curve and Q a similar point $(x + \delta x, y + \delta y)$. The approximate area δA of the strip under the arc PQ is given by

$$\delta A \approx y . \delta x$$

As we have indicated earlier, the error in the approximation is the area above the rectangle.

If we divide the whole figure between $x = a$ and $x = b$ into a number of such strips, the total area is approximately the sum of the areas of all rectangles $y . \delta x$.

i.e. $A \approx$ the sum of all rectangles $y . \delta x$ between $x = a$ and $x = b$. This can be written $A \approx \sum\limits_{x=a}^{x=b} y . \delta x$ where the symbol Σ represents *'the sum of all terms of the form ...'*

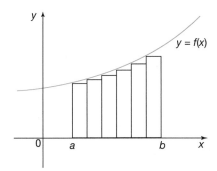

If we make the strips narrower, there will be more of them to cover the whole figure, but the total error in the approximation diminishes.

If we continue the process, we arrive at an infinite number of minutely narrow rectangles, each with an area too small to exist alone.

Then, in the limit as $\delta x \to 0$, $\quad A = \underset{\delta x \to 0}{Lim} \sum\limits_{x=a}^{x=b} y . \delta x$

But we already know that $A = \int\limits_{a}^{b} y \, dx$ $\quad \therefore \quad \underset{\delta x \to 0}{Lim} \sum\limits_{x=a}^{x=b} y . \delta x = A = \int\limits_{a}^{b} y \, dx$

Let us consider an example

2

To illustrate this, consider the area A beneath the straight line $y = x$, above the x-axis and between the values $x = 2$ and $x = 4$ as shown in the figure below.

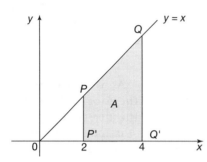

The area of a triangle is $\frac{1}{2} \times$ base $\times$ height and area A is equal to the difference in the areas of the two triangles OQQ' and OPP' so that:

$$A = \frac{1}{2} \times 4 \times 4 - \frac{1}{2} \times 2 \times 2$$

$$= 8 - 2 = 6 \text{ units}^2$$

This value will now be confirmed by dividing the area into equal strips, summing their areas and then taking the limit as the width of the strips goes to zero.

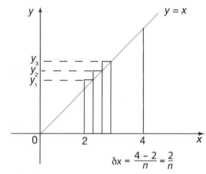

$$\delta x = \frac{4 - 2}{n} = \frac{2}{n}$$

In the figure above, the area has been subdivided into n strips each of width δx where:

$$\delta x = \frac{4 - 2}{n} = \frac{2}{n}$$

The strip heights are given as:

$$y_1 = y(2 + \delta x) = y\left(2 + \frac{2}{n}\right) = 2 + \frac{2}{n}$$

$$y_2 = y(2 + 2\delta x) = y\left(2 + 2 \times \frac{2}{n}\right) = 2 + 2 \times \frac{2}{n}$$

$$y_3 = y(2 + 3\delta x) = y\left(2 + 3 \times \frac{2}{n}\right) = 2 + 3 \times \frac{2}{n}$$

......

$$y_r = y(2 + r\delta x) = y\left(2 + r \times \frac{2}{n}\right) = 2 + r \times \frac{2}{n}$$

......

$$y_n = y(2 + n\delta x) = y\left(2 + n \times \frac{2}{n}\right) = 2 + n \times \frac{2}{n} = 4$$

Consequently:

$$\underset{\delta x \to 0}{Lim} \sum_{x=2}^{x=4} y.\delta x = \underset{n \to \infty}{Lim} \sum_{r=1}^{n} \left(2 + \frac{2r}{n}\right)\cdot\frac{2}{n}$$

Notice that as the width of each strip decreases, that is $\delta x \to 0$, so their number n increases, $n \to \infty$

$$= \underset{n \to \infty}{Lim} \sum_{r=1}^{n} \left(\frac{4}{n} + \frac{4r}{n^2}\right)$$

$$= \underset{n \to \infty}{Lim} \sum_{r=1}^{n} \frac{4}{n} + \underset{n \to \infty}{Lim} \sum_{r=1}^{n} \frac{4r}{n^2}$$

$$= \underset{n \to \infty}{Lim} \frac{4}{n} \sum_{r=1}^{n} 1 + \underset{n \to \infty}{Lim} \frac{4}{n^2} \sum_{r=1}^{n} r$$

$$= \underset{n \to \infty}{Lim} \left(\frac{4}{n} \times n\right) + \underset{n \to \infty}{Lim} \left(\frac{4}{n^2} \times \frac{n(n+1)}{2}\right)$$

$$= 4 + 2 = 6 \text{ units}^2$$

Notice the use that has been made of $\sum_{r=1}^{n} 1 = n$ and $\sum_{r=1}^{n} r = \frac{n(n+1)}{2}$ from Module 12.

Now, without following the above procedure, but by integrating normally, find the area bounded by the curve $y = x^2 - 9$ and the x-axis and between $x = -3$ and $x = 3$.

$$\boxed{A = -36 \text{ unit}^2}$$

3

Because

$$A = \int_{-3}^{3} (x^2 - 9)\,\mathrm{d}x$$

$$= \left[\frac{x^3}{3} - 9x\right]_{-3}^{3}$$

$$= (9 - 27) - (-9 + 27)$$

$$= -36$$

Simple enough, but what is meant by a negative area?

Have you any suggestions?

4

| The area that lies beneath the *x*-axis |

If we sketch the figure we get:

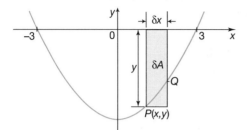

As before, the area of the strip

$$\delta A \approx y\,\delta x$$

and the total area

$$A \approx \sum_{x=-3}^{x=3} y\,\delta x$$

But, for all such strips across the figure, *y* is negative and δ*x* a positive value. Therefore, in this case, *y*.δ*x* is negative and the sum of all such quantities gives a negative total, even when δ*x* → 0.

So $\displaystyle\int_{-3}^{3} y\,dx$ has a negative value, namely *minus the value of the enclosed area*.

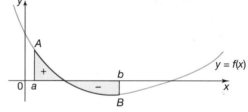

The trouble comes when part of the area to be calculated lies above the *x*-axis and part below. In that case, integration from *x* = *a* to *x* = *b* gives the algebraic sum of the two parts.

It is always wise, therefore, to sketch the figure of the problem before carrying out the integration and, if necessary, to calculate the positive and negative parts separately and to add the numerical values of each part to obtain the physical area between the limits stated.

As an example, we will determine the physical area of the figure bounded by the curve $y = x^2 - 4$, the *x*-axis and ordinates at *x* = −1 and *x* = 4.

The curve $y = x^2 - 4$ is, of course, the parabola $y = x^2$ lowered 4 units on the *y*-axis. It crosses the *x*-axis when *y* = 0, that is when $x^2 - 4 = 0$, $x^2 = 4$

∴ *x* = ±2.

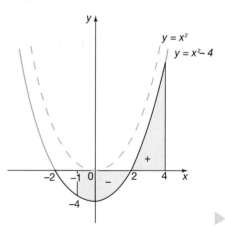

The figure of our problem extends from $x = -1$ to $x = 4$, which includes an area *beneath* the x-axis between $x = -1$ and $x = 2$ and an area *above* the x-axis between $x = 2$ and $x = 4$. Therefore, we calculate the physical area enclosed in two parts and add the results.

So let $I_1 = \displaystyle\int_{-1}^{2} y \, dx$ and $I_2 = \displaystyle\int_{2}^{4} y \, dx$. Then:

$$I_1 = \int_{-1}^{2} (x^2 - 4) \, dx = \left[\frac{x^3}{3} - 4x \right]_{-1}^{2} = \left\{ \frac{8}{3} - 8 \right\} - \left\{ -\frac{1}{3} + 4 \right\}$$

$$= -9 \text{ so } A_1 = 9 \text{ units}^2$$

and

$$I_2 = \int_{2}^{4} y \, dx = \left[\frac{x^3}{3} - 4x \right]_{2}^{4} = \left\{ \frac{64}{3} - 16 \right\} - \left\{ \frac{8}{3} - 8 \right\} = 10\frac{2}{3} \text{ units}^2$$

Consequently, $A_2 = I_2 = 10\dfrac{2}{3}$ units^2 and so:

$$A = A_1 + A_2 = 19\frac{2}{3} \text{ units}^2$$

Had we integrated right through in one calculation, we should have obtained:

$$I = \int_{-1}^{4} y \, dx = \int_{-1}^{4} (x^2 - 4) \, dx = \left[\frac{x^3}{3} - 4x \right]_{-1}^{4} = \left\{ 21\frac{1}{3} - 16 \right\} - \left\{ -\frac{1}{3} + 4 \right\}$$

$$= 1\frac{2}{3} \text{ units}^2$$

which, though it does give the correct value of the definite integral, does not give the correct value of the total area enclosed.

On to the next frame

The area between a curve and an intersecting line

5

To find the area enclosed by the curve $y = 25 - x^2$ and the straight line $y = x + 13$.

First we must develop the figure. We know that $y = x^2$ is a normal parabola:

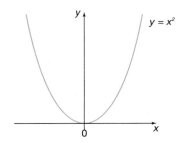

Then $y = -x^2$ is an inverted parabola:

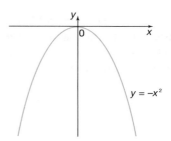

$y = 25 - x^2$ is the inverted parabola raised 25 units on the y-scale.
  Also when $y = 0$, $x^2 = 25$, so $x = \pm 5$.

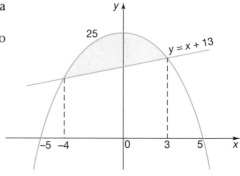

$y = x + 13$ is a straight line crossing $y = 25 - x^2$ when

$x + 13 = 25 - x^2$, i.e. $x^2 + x - 12 = 0$

$\therefore$ $(x - 3)(x + 4) = 0$ $\therefore$ $x = -4$ or 3

So the area A we need is the part shaded.

On to the next frame

6

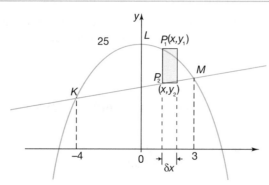

Let $P_1(x, y_1)$ be a point on $y_1 = 25 - x^2$ and $P_2(x, y_2)$ the corresponding point on $y_2 = x + 13$.
  Then area of strip

$P_1P_2 \approx (y_1 - y_2).\delta x$

Then the area of the figure

$$KLM \approx \sum_{x=-4}^{x=3} (y_1 - y_2).\delta x$$

If $\delta x \to 0$, $A = \displaystyle\int_{-4}^{3} (y_1 - y_2)\, dx$

$$\therefore \quad A = \int_{-4}^{3} (25 - x^2 - x - 13)\, dx = \int_{-4}^{3} (12 - x - x^2)\, dx$$

which you can now finish off, to give $A = \ldots\ldots\ldots$

$$\boxed{57 \cdot 2 \text{ unit}^2}$$

7

$$A = \int_{-4}^{2} (12 - x - x^2)\, dx = \left[12x - \frac{x^2}{2} - \frac{x^3}{3} \right]_{-4}^{3}$$

$$= \left\{ 36 - \frac{9}{2} - 9 \right\} - \left\{ -48 - 8 + \frac{64}{3} \right\}$$

$$= 22 \cdot 5 + 34 \cdot 67 = 57 \cdot 17 \therefore \quad A = 57 \cdot 2 \text{ unit}^2$$

At this point let us pause and summarize the main facts dealing with areas and the definite integral

---

Review summary

Unit 62

1 *Definite integral*
An integral with limits (e.g. $x = a$ and $x = b$) is called a definite integral. The constant of integration C in such cases will always disappear at the subtraction stage, since

$$\int_{a}^{b} y\, dx = \int_{x=b} y\, dx - \int_{x=a} y\, dx$$

8

2 *Integration as a summation*

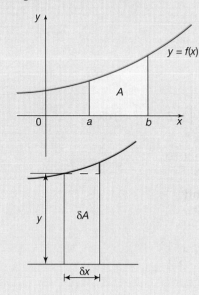

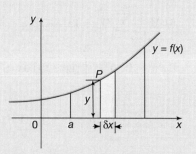

$$\delta A \approx y.\delta x \quad \therefore \quad A \approx \sum_{x=a}^{x=b} y.\delta x$$

$$\text{If } \delta x \to 0 \qquad A = \int_a^b y.\,\mathrm{d}x$$

Review exercise   **Unit 62**

9

1 Evaluate each of the following definite integrals:

(a) $\displaystyle\int_2^4 3x^5\,\mathrm{d}x$ (b) $\displaystyle\int_0^{\pi/2} (\sin x - \cos x)\,\mathrm{d}x$ (c) $\displaystyle\int_0^1 e^{2x}\,\mathrm{d}x$ (d) $\displaystyle\int_{-1}^0 x^3\,\mathrm{d}x$

2 Find the area enclosed between the x-axis and the curves:

(a) $y = x^3 + 2x^2 + x + 1$ between $x = -1$ and $x = 2$
(b) $y = x^2 - 25$ for $-5 \le x \le 5$.

3 Find the area enclosed between the curve $y = x^3$ and the straight line $y = x$.

Complete all three questions.
Look back at the Unit if necessary but don't rush.
The answers and working are in the next frame.

1 (a) $\displaystyle\int_{2}^{4} 3x^5 \, dx = \left[3\frac{x^6}{6} \right]_{2}^{4}$

$$= \left\{ 3\frac{4^6}{6} \right\} - \left\{ 3\frac{2^6}{6} \right\}$$
$$= 2048 - 32 = 2016$$

(b) $\displaystyle\int_{0}^{\pi/2} (\sin x - \cos x) \, dx = [-\cos x - \sin x]_{0}^{\pi/2}$

$$= \left\{ -\cos\frac{\pi}{2} - \sin\frac{\pi}{2} \right\} - \left\{ -\cos 0 - \sin 0 \right\}$$
$$= \{ -0 - 1 \} - \{ -1 - 0 \}$$
$$= 0$$

(c) $\displaystyle\int_{0}^{1} e^{2x} \, dx = \left[\frac{e^{2x}}{2} \right]_{0}^{1}$

$$= \left\{ \frac{e^2}{2} \right\} - \left\{ \frac{e^0}{2} \right\}$$
$$= \frac{e^2}{2} - \frac{1}{2}$$
$$= \frac{e^2 - 1}{2}$$

(d) $\displaystyle\int_{-1}^{0} x^3 \, dx = \left[\frac{x^4}{4} \right]_{-1}^{0}$

$$= \{ 0^4/4 \} - \{ (-1)^4/4 \} = -1/4$$

2 (a) The graph of $y = x^3 + 2x^2 + x + 1$ between $x = -1$ and $x = 2$ lies entirely above the x-axis

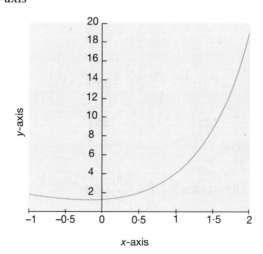

so that the area enclosed between the curve, the *x*-axis and between
$x = -1$ and $x = 2$ is given by:

$$\int_{-1}^{2} (x^3 + 2x^2 + x + 1)\, dx$$

$$= \left[\frac{x^4}{4} + 2\frac{x^3}{3} + \frac{x^2}{2} + x \right]_{-1}^{2}$$

$$= \left\{ \frac{2^4}{4} + 2.\frac{2^3}{3} + \frac{2^2}{2} + 2 \right\} - \left\{ \frac{(-1)^4}{4} - 2.\frac{(-1)^3}{3} + \frac{(-1)^2}{2} - 1 \right\}$$

$$= \left\{ 4 + \frac{16}{3} + 2 + 2 \right\} - \left\{ \frac{1}{4} - \frac{2}{3} + \frac{1}{2} - 1 \right\}$$

$$= 13\tfrac{1}{3} + \tfrac{11}{12} = 14\tfrac{1}{4} \text{ unit}^2$$

(b) The graph of $y = x^2 - 25$ for $-5 \le x \le 5$ lies entirely below the *x*-axis

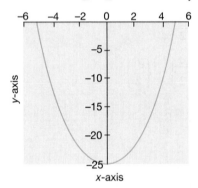

so that the area enclosed between the curve and the *x*-axis for $-5 \le x \le 5$
is given by $-I$ where:

$$I = \int_{-5}^{5} (x^2 - 25)\, dx$$

$$= \left[\frac{x^3}{3} - 25x \right]_{-5}^{5}$$

$$= \left\{ \frac{5^3}{3} - 25 \times 5 \right\} - \left\{ \frac{(-5)^3}{3} - 25 \times (-5) \right\}$$

$$= \left\{ \frac{125}{3} - 125 \right\} - \left\{ -\frac{125}{3} + 125 \right\}$$

$$= -\frac{500}{3}$$

So the enclosed area is $-I = \dfrac{500}{3}$ units^2

3   The curve $y = x^3$ and the line $y = x$ intersect when $x^3 = x$, that is when
$x^3 - x = 0$. Factorizing we see that this means $x(x^2 - 1) = 0$ which is satisfied
when $x = 0$, $x = 1$ and $x = -1$.

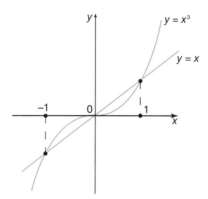

From the figure we can see that the area enclosed between the curve $y = x^3$ and the straight line $y = x$ is in two parts, one above the x-axis and the other below. From the figure it is easily seen that both parts are the same, so we only need find the area between $x = 0$ and $x = 1$ and then double it to find the total area enclosed.

The area enclosed between $x = 0$ and $x = 1$ is equal to the area beneath the line $y = x$ minus the area beneath the curve $y = x^3$. That is:

$$\int_0^1 x \, dx - \int_0^1 x^3 \, dx = \int_0^1 (x - x^3) \, dx$$

$$= \left[\frac{x^2}{2} - \frac{x^4}{4} \right]_0^1$$

$$= \left\{ \frac{1^2}{2} - \frac{1^4}{4} \right\} - \{0\}$$

$$= \frac{1}{4}$$

The total area enclosed is then twice this, namely $\frac{1}{2}$ unit^2.

Now for the Review test

Review test

Unit 62

1 Find the area enclosed between the curve $y = e^x$ and the straight line $y = 1 - x$ between $x = 1$ and $x = 2$, giving your answer to 3 dp.

11

You have now come to the end of this Module. A list of **Can You?** questions follows for you to gauge your understanding of the material in the Module. You will notice that these questions match the **Learning outcomes** listed at the beginning of the Module. Then try the **Test exercise**. *Work through the questions at your own pace, there is no need to hurry.* A set of **Further problems** provides additional valuable practice.

12

Can You?

1 Checklist: Module 19

Check this list before and after you try the end of Module test.

On a scale of 1 to 5 how confident are you that you can:

- Appreciate that integration is the reverse process of differentiation?
  Yes ☐ ☐ ☐ ☐ ☐ *No*

- Recognize the need for a constant of integration?
  Yes ☐ ☐ ☐ ☐ ☐ *No*

- Evaluate indefinite integrals of standard forms?
  Yes ☐ ☐ ☐ ☐ ☐ *No*

- Evaluate indefinite integrals of polynomials?
  Yes ☐ ☐ ☐ ☐ ☐ *No*

- Evaluate indefinite integrals of 'functions of a linear function of x'?
  Yes ☐ ☐ ☐ ☐ ☐ *No*

- Integrate by partial fractions?
  Yes ☐ ☐ ☐ ☐ ☐ *No*

- Integrate by parts
  Yes ☐ ☐ ☐ ☐ ☐ *No*

- Appreciate that a definite integral is a measure of an area under a curve?
  Yes ☐ ☐ ☐ ☐ ☐ *No*

- Evaluate definite integrals of standard forms?
  Yes ☐ ☐ ☐ ☐ ☐ *No*

- Use Simpson's rule to approximate the area beneath a curve?
  Yes ☐ ☐ ☐ ☐ ☐ *No*

- Use the definite integral to find areas between a curve and the horizontal axis?
  Yes ☐ ☐ ☐ ☐ ☐ *No*

- Use the definite integral to find areas between a curve and a given straight line?
  Yes ☐ ☐ ☐ ☐ ☐ *No*

Test exercise 19

2

1 Determine the following integrals:

(a) $\int x^4 \, dx$ (f) $\int 5^x \, dx$

(b) $\int 3 \sin x \, dx$ (g) $\int x^{-4} \, dx$

(c) $\int 4.e^x \, dx$ (h) $\int \frac{4}{x} \, dx$

(d) $\int 6 \, dx$ (i) $\int 3 \cos x \, dx$

(e) $\int 3x^{\frac{1}{2}} \, dx$ (j) $\int 2 \sec^2 x \, dx$

2 (a) Determine $\int (8x^3 + 6x^2 - 5x + 4) \, dx$.

   (b) If $I = \int (4x^3 - 3x^2 + 6x - 2) \, dx$, determine the value of I

   when $x = 4$, given that when $x = 2$, $I = 20$.

3 Determine:

(a) $\int 2 \sin(3x + 1) \, dx$ (e) $\int 4^{2x-3} \, dx$

(b) $\int \sqrt{5 - 2x} \, dx$ (f) $\int 6 \cos(1 - 2x) \, dx$

(c) $\int 6.e^{1-3x} \, dx$ (g) $\int \frac{5}{3x - 2} \, dx$

(d) $\int (4x + 1)^3 \, dx$ (h) $\int 3 \sec^2(1 + 4x) \, dx$

4 Express $\dfrac{2x^2 + 14x + 10}{2x^2 + 9x + 4}$ in partial fractions and hence
   determine

   $\int \frac{2x^2 + 14x + 10}{2x^2 + 9x + 4} \, dx$

5 Evaluate $\int e^{-3x} \cos 2x \, dx$.

6 (a) Evaluate the area under the curve $y = 5.e^x$ between $x = 0$ and $x = 3$.

   (b) Show that the straight line $y = -2x + 28$ crosses the curve $y = 36 - x^2$ at
   $x = -2$ and $x = 4$. Hence, determine the area enclosed by the curve
   $y = 36 - x^2$ and the line $y = -2x + 28$

7 Evaluate $\int_0^{\pi/2} \sqrt{\cos \theta} \, d\theta$, using 6 intervals.

Further problems 19

3

1 Determine the following:

(a) $\int (5 - 6x)^2 \, dx$ (e) $\int \sqrt{5 - 3x} \, dx$ (h) $\int 4\cos(1 - 2x) \, dx$

(b) $\int 4\sin(3x + 2) \, dx$ (f) $\int 6.e^{3x+1} \, dx$ (i) $\int 3\sec^2(4 - 3x) \, dx$

(c) $\int \dfrac{5}{2x + 3} \, dx$ (g) $\int (4 - 3x)^{-2} \, dx$ (j) $\int 8.e^{3 - 4x} \, dx$

(d) $\int 3^{2x-1} \, dx$

2 Determine $\displaystyle \int \left(6.e^{3x-5} + \dfrac{4}{3x - 2} - 5^{2x+1} \right) dx$.

3 If $I = \int (8x^3 + 3x^2 - 6x + 7) \, dx$, determine the value of I when $x = -3$, given that when $x = 2$, $I = 50$.

4 Determine the following using partial fractions:

(a) $\int \dfrac{6x + 1}{4x^2 + 4x - 3} \, dx$ (e) $\int \dfrac{6x^2 - 2x - 2}{6x^2 - 7x + 2} \, dx$

(b) $\int \dfrac{x + 11}{x^2 - 3x - 4} \, dx$ (f) $\int \dfrac{4x^2 - 9x - 19}{2x^2 - 7x - 4} \, dx$

(c) $\int \dfrac{3x - 17}{12x^2 - 19x + 4} \, dx$ (g) $\int \dfrac{6x^2 + 2x + 1}{2x^2 + x - 6} \, dx$

(d) $\int \dfrac{20x + 2}{8x^2 - 14x - 15} \, dx$ (h) $\int \dfrac{4x^2 - 2x - 11}{2x^2 - 7x - 4} \, dx$

5 Evaluate:

(a) $\displaystyle \int_0^\pi x^2 \sin x \, dx$ (b) $\displaystyle \int_0^\pi x^2 \sin^2 x \, dx$

(c) $\displaystyle \int_0^1 x \tan^{-1} x \, dx$ (d) $\displaystyle \int_0^\pi e^{2x} \cos 4x \, dx$

(e) $\displaystyle \int_0^{\pi/6} e^{2\theta} \cos 3\theta \, d\theta$

6 Determine the area bounded by the curve $y = f(x)$, the x-axis and the stated ordinates in the following cases:

(a) $y = x^2 - 3x + 4$, $x = 0$ and $x = 4$

(b) $y = 3x^2 + 5$, $x = -2$ and $x = 3$

(c) $y = 5 + 6x - 3x^2$, $x = 0$ and $x = 2$

(d) $y = -3x^2 + 12x + 10$, $x = 1$ and $x = 4$

(e) $y = x^3 + 10$, $x = -1$ and $x = 2$

7 In each of the following, determine the area enclosed by the given boundaries:

(a) $y = 10 - x^2$ and $y = x^2 + 2$

(b) $y = x^2 - 4x + 20$, $y = 3x$, $x = 0$ and $x = 4$

(c) $y = 4e^{2x}$, $y = 4e^{-x}$, $x = 1$ and $x = 3$

(d) $y = 5e^x$, $y = x^3$, $x = 1$ and $x = 4$

(e) $y = 20 + 2x - x^2$, $y = \dfrac{e^x}{2}$, $x = 1$ and $x = 3$

8 In each of the following cases, apply Simpson's rule (6 intervals) to obtain an approximate value of the integral:

(a) $\displaystyle\int_0^{\pi/2} \frac{dx}{1 + 3\cos x}$

(b) $\displaystyle\int_0^{\pi} (5 - 4\cos\theta)^{\frac{1}{2}}\, d\theta$

(c) $\displaystyle\int_0^{\pi/2} \frac{d\theta}{\sqrt{1 - \frac{1}{2}\sin^2\theta}}$

9 The coordinates of a point on a curve are given below:

x	0	1	2	3	4	5	6	7	8
y	4	5·9	7·0	6·4	4·8	3·4	2·5	1·7	1

The plane figure bounded by the curve, the x-axis and the ordinates at $x = 0$ and $x = 8$, rotates through a complete revolution about the x-axis. Use Simpson's rule (8 intervals) to obtain an approximate value of the volume generated.

10 The perimeter of an ellipse with parametric equations $x = 3\cos\theta$, $y = 2\sin\theta$, is $2\sqrt{2}\displaystyle\int_0^{\pi/2} (13 - 5\cos 2\theta)^{\frac{1}{2}}\, d\theta$. Evaluate this integral using Simpson's rule with 6 intervals.

11 Calculate the area bounded by the curve $y = e^{-x^2}$, the x-axis and the ordinates at $x = 0$ and $x = 1$. Use Simpson's rule with 6 intervals.

12 The voltage of a supply at regular intervals of 0·01 s, over a half-cycle, is found to be: 0, 19·5, 35, 45, 40·5, 25, 20·5, 29, 27, 12·5, 0. By Simpson's rule (10 intervals) find the rms value of the voltage over the half-cycle.

13 Show that the length of arc of the curve $x = 3\theta - 4\sin\theta$, $y = 3 - 4\cos\theta$, between $\theta = 0$ and $\theta = 2\pi$, is given by the integral $\displaystyle\int_0^{2\pi} \sqrt{25 - 24\cos\theta}\, d\theta$. Evaluate the integral, using Simpson's rule with 8 intervals.

14   Obtain the first four terms of the expansion of $(1+x^3)^{\frac{1}{2}}$ and use them to determine the approximate value of $\int_0^{\frac{1}{2}} \sqrt{1+x^3}\,dx$, correct to three decimal places.

15   Establish the integral in its simplest form representing the length of the curve $y = \frac{1}{2}\sin\theta$ between $\theta = 0$ and $\theta = \frac{\pi}{2}$. Apply Simpson's rule, using 6 intervals, to find an approximate value of this integral.

Answers

Unit 1 Review test

1 (a) $-3 < -2$ (b) $8 > -13$ (c) $-25 < 0$ **2** (a) 6 (b) 110
3 (a) 1350, 1400, 1000 (b) 2500, 2500, 3000 (c) $-2450, -2500, -2000$
  (d) $-23\,630, -23\,600, -24\,000$

Unit 2 Review test

1 (a) $2 \times 5 \times 17$ (b) $5 \times 7 \times 13$ (c) $3 \times 5 \times 5 \times 11 \times 11$ (d) $2 \times 2 \times 3 \times 5 \times 19$
2 (a) HCF $= 4$, LCM $= 1848$ (b) HCF $= 3$, LCM $= 2310$

Unit 3 Review test

1 (a) $\dfrac{2}{3}$ (b) $\dfrac{48}{7}$ (c) $-\dfrac{7}{2}$ (d) 16 **2** (a) $\dfrac{2}{7}$ (b) $\dfrac{11}{25}$ (c) $\dfrac{67}{91}$ (d) $-\dfrac{49}{48}$
3 (a) $5 : 2 : 3$ (b) $15 : 20 : 12 : 13$ **4** (a) 80 per cent (b) 24
  (c) 64·3 per cent (to 1 dp) (d) 3·75

Unit 4 Review test

1 (a) 21·4, 21·36 (b) 0·0246, 0·02 (c) 0·311, 0·31 (d) 5130, 5134·56
2 (a) 0·267 (b) $-0·538$ (c) 1·800 (d) $-2·154$ **3** (a) $\dfrac{4}{5}$ (b) $\dfrac{14}{5}$ (c) $\dfrac{329}{99}$
  (d) $-\dfrac{11}{2}$ **4** (a) $1·\dot{0}\dot{1}$ (b) $9·24\dot{5}\dot{6}$

Unit 5 Review test

1 (a) 3^9 (b) $2^1 = 2$ (c) 9^6 (d) 1 **2** (a) 2·466 (b) 1·380 (c) -3
  (d) not possible **3** (a) 3204·4 (b) 1611·05 **4** (a) $1·3465 \times 10^2$
  (b) $2·401 \times 10^{-3}$ **5** (a) $1·61105 \times 10^3$ (b) 9·3304 **6** (a) 0·024 (b) 5·21

Unit 6 Review test

1 (a) $11·25_{10}$ (b) $302·908\ldots$ (c) $171·9357\ldots$ (d) $3233·6958\ldots$
2 $17·465_8\ldots$, $1111·100_2\ldots$, $13·731_{12}\ldots$, $F·9AE_{16}\ldots$

Test exercise 1

1 (a) $-12 > -15$ (b) $9 > -17$ (c) $-11 < 10$ **2** (a) 14 (b) 48
3 (a) $2 \times 2 \times 3 \times 13$ (b) $2 \times 3 \times 7 \times 13$ (c) $5 \times 17 \times 17$ (d) $3 \times 3 \times 3 \times 5 \times 11$
4 (a) 5050, 5000, 5000 (b) 1100, 1100, 1000 (c) $-1550, -1600, -2000$
  (d) $-5000, -5000, -5000$ **5** (a) HCF $= 13$, LCM $= 19\,110$

(b) HCF $= 2$, LCM $= 576$ **6**   (a) $\dfrac{4}{7}$ (b) $\dfrac{9}{2}$ (c) $-\dfrac{31}{3}$ (d) -27 **7**   (a) $\dfrac{14}{15}$

(b) $\dfrac{11}{63}$ (c) $\dfrac{16}{5}$ (d) $\dfrac{8}{75}$ (e) 3 (f) $\dfrac{9}{4}$ **8**   (a) $3:1$ (b) $10:1:4$ (c) $6:18:5:1$

9   (a) 60 per cent   (b) $\dfrac{4}{25}$ (c) £2·19   **10**   (a) $83\cdot54$, $83\cdot543$ (b) $83\cdot54$, $83\cdot543$

(c) -2692, $-2692\cdot228$ (d) $-550\cdot3$, $-550\cdot341$ **11**   (a) $0\cdot176$ (b) $-0\cdot133$

(c) $5\cdot667$ (d) $-2\cdot182$ **12**   (a) $6\cdot\dot{7}$ (b) $0\cdot\dot{0}1\dot{0}$ **13**   (a) $\dfrac{2}{5}$ (b) $\dfrac{92}{25}$ (c) $\dfrac{13}{9}$

(d) $-\dfrac{61}{10}$ **14**   (a) 2^{11} (b) $\left(\dfrac{6}{5}\right)^2 = (1\cdot2)^2$ (c) (-4) (d) 1 **15**   (a) $1\cdot821$

(b) $1\cdot170$ (c) $-2\cdot408$ (d) not possible   **16**   (a) $5\cdot376 \times 10^2$ (b) $3\cdot64 \times 10^{-1}$

(c) $4\cdot902 \times 10^3$ (d) $1\cdot25 \times 10^{-4}$ **17**   (a) $61\cdot47 \times 10^6$ (b) $243\cdot9 \times 10^{-6}$
(c) $528\cdot6 \times 10^3$ (d) $437\cdot1 \times 10^{-9}$ **18** $4\cdot72605 \times 10^8$, $472\cdot605 \times 10^6$
19   (a) 1 (b) $0\cdot55$ **20**   (a) $15\cdot75$ (b) $511\cdot876\ldots$ (c) $567\cdot826\ldots$
    (d) $3586\cdot9792\ldots$ **21** $23\cdot676_8\ldots$, $10011\cdot110_2\ldots$, $17\cdot X56_{12}\ldots$, $13\cdot DF3_{16}\ldots$

Further problems 1

1   (a) $-4 > -11$ (b) $7 > -13$ (c) $-15 < 13$ **2**   (a) 2 (b) 12
3   (a) 3510, 3500, 4000 (b) 500, 500, 0 (c) -2470, -2500, -2000
    (d) $-9010, -9000, -9000$ **4**   (a) (i) $-19\cdot691$ (ii) $-19\cdot6915$ (iii) $-19\cdot69$
    (b) (i) $94\cdot541$ (ii) $94\cdot5414$ (iii) $94\cdot54$ (c) (i) $0\cdot48058$ (ii) $0\cdot4806$ (iii) $0\cdot4806$
    (d) (i) $1\cdot5692$ (ii) $1\cdot5692$ (iii) $1\cdot569$ (e) (i) $0\cdot86276$ (ii) $0\cdot8628$ (iii) $0\cdot863$
    (f) (i) $0\cdot87927$ (ii) $0\cdot8793$ (iii) $0\cdot8793$ **5**   (a) $2 \times 2 \times 3 \times 7 \times 11$
    (b) $3 \times 5 \times 5 \times 11$ (c) $2 \times 3 \times 5 \times 7 \times 11$ (d) $2 \times 5 \times 11 \times 17 \times 19$
6   (a) HCF $= 3$, LCM $= 63$ (b) HCF $= 5$, LCM $= 255$ (c) HCF $= 3$, LCM $= 462$
    (d) HCF $= 2$, LCM $= 2880$ **7**   (a) $\dfrac{1}{4}$ (b) $\dfrac{13}{6}$ (c) -8 (d) $-\dfrac{15}{7}$ **8**   (a) $\dfrac{441}{110}$
    (b) $\dfrac{513}{104}$ (c) $\dfrac{16\,641}{64}$ (d) $\dfrac{32\,725}{8168}$ **9**   (a) $\dfrac{9}{25}$ (b) $\dfrac{7}{40}$ (c) $\dfrac{87}{1000}$ (d) $\dfrac{18}{25}$
10   (a) 80% (b) $27\cdot3\%$ (c) $22\cdot2\%$ (d) $14\cdot3\%$ (e) $47\cdot4\%$ (f) $48\cdot1\%$ (g) $6\cdot9\%$
    (h) $99\cdot5\%$ **11**   (a) 20 (b) $0\cdot54$ to 2 sig fig   (c) $-1\cdot8225$ (d) $0\cdot12$ to 2 sig fig
12   (a) $3:10:2$ (b) $9:5:10$ (c) $4:10:11$ (d) $13:91:63$ **13**   (a) $8\cdot\dot{7}\dot{6}$
    (b) $212\cdot\dot{2}1\dot{1}$ **14**   (a) $\dfrac{3}{25}$ (b) $\dfrac{21}{4}$ (c) $\dfrac{589}{111}$ (d) $-\dfrac{93}{10}$ **15**   (a) 8^7 (b) 2^3
    (c) 5^{15} (d) 1 **16**   (a) $3\cdot106$ (b) $1\cdot899$ (c) $-2\cdot924$ (d) 25 **17**   (a) $0\cdot238$
    (b) $-0\cdot118$ (c) $2\cdot667$ (d) $-1\cdot684$ **18**   (a) $5\cdot2876 \times 10^1$ (b) $1\cdot5243 \times 10^4$
    (c) $8\cdot765 \times 10^{-2}$ (d) $4\cdot92 \times 10^{-5}$ (e) $4\cdot362 \times 10^2$ (f) $5\cdot728 \times 10^{-1}$
19   (a) $4\cdot285 \times 10^3$ (b) $16\cdot9 \times 10^{-3}$ (c) $852\cdot6 \times 10^{-6}$ (d) $362\cdot9 \times 10^3$
    (e) $10\cdot073 \times 10^6$ (f) $569\cdot4 \times 10^6$ **20** $1\cdot257 \times 10^8$, $125\cdot7 \times 10^6$
21   (a) $1\cdot110_2\ldots$, $1\cdot650_8\ldots$, $1\cdot9 \Lambda 6_{12}\ldots$, $1\cdot D47_{16}\ldots$
    (b) $101010110\cdot1_2$, $526\cdot4_8$, $246\cdot6_{12}$, $156\cdot8_{16}$
22   (a) $0\cdot101$, $0\cdot748$, $0\cdot8 \Lambda 8\ldots$, $0\cdot BF7\ldots$ (b) 101110011, 371, 26Λ, 173
23   (a) $0\cdot111\ldots$, $0\cdot770\ldots$, $0\cdot986\ldots$, $0\cdot FC8$ (b) 10110010101, 2625, 1429, 595
24   (a) $0\cdot59375$, $0\cdot46$, $0\cdot716$, $0\cdot98$ (b) 460, 714, 324, $1CC$
25   (a) $0\cdot111\ldots$, $0\cdot751\ldots$, $0\cdot \Lambda 57\ldots$, $0\cdot955\ldots$ (b) 1110100101, 1645, 659, 933

Unit 7 Review test

1 (a) $4pq - pr - 2qr$ (b) $mn(8l^2 + ln - m)$ (c) w^{4+a-b} (d) $s^{\frac{1}{2}}t^{\frac{7}{4}}$
2 (a) $8xy^2 - 4x^2y - 24x^3$ (b) $6a^3 - 29a^2b + 46ab^2 - 24b^3$ (c) $2x - 6y + 5z + 54$

Unit 8 Review test

1 (a) $-1{\cdot}996$ (b) $1{\cdot}244$ (c) $1{\cdot}660$ **2** $F = \dfrac{GmM}{r^2}$

3 $\log T = \log 2 + \log \pi + \dfrac{1}{2}(\log l - \log g)$

Unit 9 Review test

1 (a) $4n^3 + 13n^2 - 2n - 15$ (b) $2v^5 - 5v^4 + 4v^3 - 5v^2 + 6v - 2$ **2** (a) $\left(\dfrac{p}{q}\right)^4$

(b) $\dfrac{a^4}{2b^2}$ **3** (a) $2y - 5$ (b) $q^2 + 2q + 4$ (c) $2r + 1$

Unit 10 Review test

1 (a) $6xy(3x - 2y)$ (b) $(x^2 - 3y^2)(x + 4y)$ (c) $(3x - 5y)(x + y)$ (d) $(4x - 3)(3x - 4)$

Test exercise 2

1 (a) $6ab - 4ac - 2cb$ (b) $yz(7xz - 3y)$ (c) c^{p-q+2} (d) $x^{-\frac{13}{12}}y^{-\frac{7}{2}}$
2 (a) $6fg^2 + 4fgh - 16fh^2$ (b) $50x^3 + 80x^2y - 198xy^2 + 36y^3$
   (c) $12p - 24q - 12r + 96$ **3** (a) $-1{\cdot}5686$ (b) $3{\cdot}8689$ (c) $2{\cdot}4870$
4 (a) $\log V = \log \pi + \log h + \log(D - h) + \log(D + h) - \log 4$

(b) $\log P = 2\log(2d - 1) + \log N + \dfrac{1}{2}\log S - \log 16$

5 (a) $x = \dfrac{PQ^2}{1000K}$ (b) $R = 10S^3 . \sqrt[3]{M}$ (c) $P = \dfrac{e^2\sqrt{Q+1}}{R^3}$
6 (a) $3x^3 - 4x^2 - 2x + 1$ (b) $3a^4 + 10a^3 + 18a^2 + 16a + 8$
7 (a) x^2y (b) $\dfrac{b^3}{ac}$ **8** (a) $x + 3$ (b) $n^2 - 3n + 9$ (c) $3a^2 - a + 1$
9 (a) $4x^2y(9xy - 2)$ (b) $(x + 3y)(x^2 + 2y^2)$ (c) $(2x + 3)^2$ (d) $(x + 5y)(5x + 3y)$
   (e) $(x + 6)(x + 4)$ (f) $(x - 2)(x - 8)$ (g) $(x - 9)(x + 4)$ (h) $(2x + 3)(3x - 2)$

Further problems 2

1 (a) $8x^3 + 6x^2 - 47x + 21$ (b) $20x^3 - 11x^2 - 27x + 18$ (c) $15x^3 - 34x^2 + 3x + 20$

(d) $3x^2 + 2x - 4$ (e) $3x^2 + 5x - 7$ (f) $6x^2 - 4x + 7$ **2** $(x + 1)^2\sqrt{x^2 - 1}$

3 (a) $ab^{\frac{3}{2}}c^{\frac{2}{3}}$ (b) 16 (c) $\dfrac{12x^3y^3}{z}$ (d) $(x - y)^2$ **4** (a) $-2{\cdot}0720$ (b) $3{\cdot}2375$

(c) $2{\cdot}8660$ **5** (a) $\log f = -\left(\log \pi + \log d + \dfrac{1}{2}\log L + \dfrac{1}{2}\log C\right)$

(b) $\log K = 3\log a + \dfrac{1}{2}\log b - \dfrac{1}{6}\log c - \dfrac{2}{5}\log d$

6 (a) $W = \dfrac{A^2w^2}{32\pi^2 r^2 c}$ (b) $S = \dfrac{K}{2}.\dfrac{\pi^2 n^2 yrL^2}{h^2 g}$ (c) $I = \dfrac{2Ve^{KL}}{K(KR + r)}$ **7** (a) $5xy^2(3x + 4y)$
   (b) $2a^2b(7a - 6b)$ (c) $(2x + 3y)(x - 5)$ (d) $(4x - 7y)(y - 3)$ (e) $(5x + 2)(3x + 4y)$

(f) $(3x-4)(2y+5)$ (g) $(3x+4y)^2$ (h) $(4x-5y)^2$ (i) $xy^2(5xy-4)(5xy+4)$
(j) $(x+8y)(3x+2y)$ **8**   (a) $(x+2)(5x+3)$ (b) $(2x-3)(x-4)$
(c) $(2x-3)(3x+2)$ (d) no factors   (e) $(5x-4)(x-3)$ (f) no factors
(g) no factors   (h) $(3x-2)(3x-4)$ (i) $(5x-2)(2x+3)$ (j) $(5x-3)(3x-2)$
(k) $(2x+3)(4x-5)$

Unit 11 Review test

1 $16\,739 \cdot 41$ **2**   (a) $l = \sqrt{\dfrac{T^2 3g(r-t)}{4\pi^2} - 4t^2}$ (b) $r = t + \dfrac{4\pi^2}{3gT^2}(l^2 + 4t^2)$

Unit 12 Review test

1 $f(x) = ((7x-6)x+4)x+1$ so $f(-2) = ((7(-2)-6)(-2)+4)(-2)+1 = -87$
2 remainder 235 **3** $(2x+1)(x-2)(3x-2)(x+3)$

Test exercise 3

1 (a) $1 \cdot 74$ (b) $0 \cdot 401$ **2** $p = \sqrt{\dfrac{\sqrt{q^2+2}+1}{5}}$ **3** (a) 38 (b) 24 (c) 136 (d) 250

4 81 **5** $(x-2)(2x^2+6x-5)$ **6** $(x+2)(x+4)(2x-5)$
7 $(x+1)(x-2)(x-3)(2x+1)$

Further problems 3

1 (a) $K = 8 \cdot 755$ (b) $P = 130 \cdot 6$ (c) $Q = 0 \cdot 8628$ **2**   (a) $I = \dfrac{V}{R}$ (b) $u = v - at$

(c) $u = \dfrac{2s - at^2}{2}$ (d) $L = \dfrac{1}{4\pi^2 f^2 C}$ (e) $C = \dfrac{SF}{S-P}$ (f) $L = \dfrac{8S^2 + 3D^2}{3D}$

(g) $M = \dfrac{T+m}{1-Tm}$ (h) $h = \dfrac{\sqrt{A^2 - \pi^2 r^4}}{\pi r}$ (i) $R = \sqrt{\dfrac{6V - \pi h^3}{3\pi h}}$

3 (a) 18 (b) 143 (c) -79 (d) 69 (e) 226 **4**   (a) 363 (b) 261 (c) -76
(d) -59 (e) 595 **5**   (a) $(x-1)(x+3)(x+4)$ (b) $(x-2)(x+5)(2x+3)$
(c) $(x+2)(x-4)(3x+2)$ (d) $(x+1)(3x^2-4x+5)$ (e) $(x-2)(2x+5)(3x-4)$
(f) $(x-1)(2x-5)(2x+7)$ (g) $(x-2)(2x^2+3x-4)$ (h) $(x+3)(3x+4)(5x-4)$
(i) $(x-2)(x^2+3x+4)$ (j) $(x+3)(2x+5)(3x+2)$
6 (a) $(x-1)(x+2)(x-4)(2x+1)$ (b) $(x-2)(x-3)(x+3)(3x-1)$
(c) $(x-2)(x+3)(4x^2-8x-3)$ (d) $(x-1)(x+1)(x^2+2x-5)$
(e) $(x+2)(x-3)(2x+1)(3x-4)$ (f) $(x-2)(x+2)(x+4)(2x-3)$

Unit 13 Review test

1 (a) $y = \pm\sqrt{1+x^2}$ (b) $(-10, \pm 10 \cdot 0)$, $(-8, \pm 8 \cdot 1)$, $(-6, \pm 6 \cdot 1)$, $(-4, \pm 4 \cdot 1)$, $(-2, \pm 2 \cdot 2)$,
$(0, \pm 1 \cdot 0)$, $(2, \pm 2 \cdot 2)$, $(4, \pm 4 \cdot 1)$, $(6, \pm 6 \cdot 1)$, $(8, \pm 8 \cdot 1)$, $(10, \pm 10 \cdot 0)$

(c)

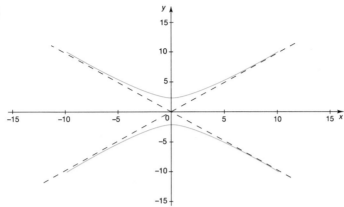

2 (a)

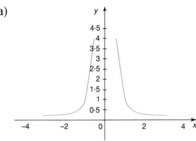

(b)

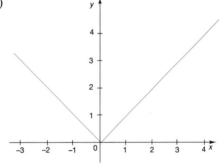

Unit 14 Review test

1 (a)

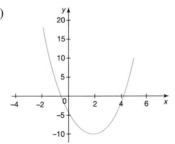

(b)

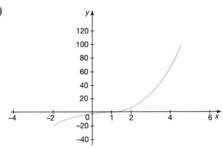

Unit 15 Review test

1 (a) region above the line $y = -x$ (b) region below and on the curve $y = 3x^3$
(c) inside and on the circle centred on the origin and of radius 1

Unit 16 Review test

1 (a) $-7 < x < 3$ (b) $x < -7$ or $x > 3$ (c) $-1/2 < x < 7/2$ (d) $x < -4/5$ or $x > 4$

Test exercise 4

1   (a) $y = \pm\sqrt{1 - x^2}$ (b) $(-1\cdot0, 0)$, $(-0\cdot8, \pm0\cdot6)$, $(-0\cdot6, \pm0\cdot8)$, $(-0\cdot4, \pm0\cdot9)$,

$(-0\cdot2, \pm1\cdot0)$, $(0\cdot0, \pm1\cdot0)$, $(0\cdot2, \pm1\cdot0)$, $(0\cdot4, \pm0\cdot9)$, $(0\cdot6, \pm0\cdot8)$, $(0\cdot8, \pm0\cdot6)$, $(1\cdot0, 0)$
(c)

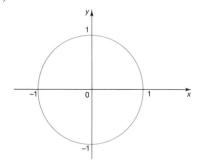

2   (a)

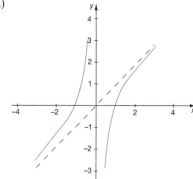

(b)

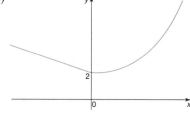

3   (a)

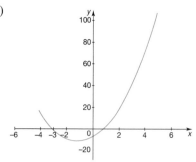

(b)

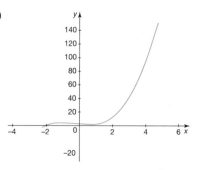

4   (a) region below the line $y = 2 - 3x$ (b) region above the line $y = x - \dfrac{2}{x}$

(c) region on and below the line $y = 1 - x$

5   (a) $-21 < x < 5$ (b) $x < -21$ or $x > 5$ (c) $-3/4 < x < 13/4$

(d) $x < -8/3$ or $x > 20/3$

Further problems 4

1

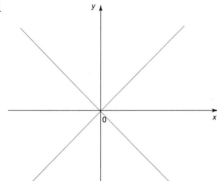

2

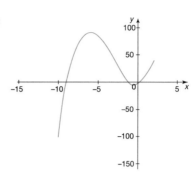

3

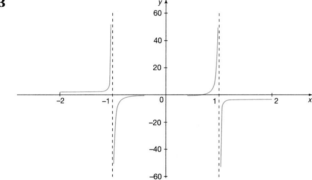

4 $y = \pm\sqrt{4 - x^2}$ circle centre $(0,0)$ radius 2   **5** $y = \pm 4\sqrt{1 - \dfrac{x^2}{3}}$ ellipse

6 $(x + 1)^2 + (y + 1)^2 = 1$ so that $y = \pm\sqrt{1 - (1 + x)^2} - 1$; circle centre $(-1, -1)$ radius 1

7 (a) region on and above the line $y = 2x + 4$ (b) region below the line $y = 3 - x$

(c) region on and below the line $y = \dfrac{3x}{4} - \dfrac{1}{4}$

(d) region outside the circle centred on the origin and radius $\sqrt{2}$

8

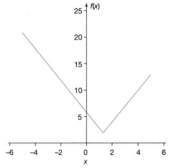

No difference since $|-x| = |x|$

9 (a) $0 \le x < \pi/6$, $5\pi/6 < x < 7\pi/6$ and $11\pi/6 < x \le 2\pi$

(b) $0 \le x < \pi/3$, $2\pi/3 < x < 4\pi/3$ and $5\pi/3 < x \le 2\pi$

Unit 17 Review test

1 (a) $x = -2$ (b) $x = 2$ (c) $x = -3$ 2 $x = 1$, $y = -1$ 3 $x = 1$, $y = 3$, $z = 5$

Test exercise 5

1 (a) $x = -3$ (b) $x = 25$ (c) $x = \dfrac{3}{2}$ (d) $x = 1$ 2 (a) $x = 2$, $y = 1$

(b) $x = 6{\cdot}5$, $y = -10{\cdot}5$ 3 $x = 3$, $y = -2$, $z = 5$ 4 $x = 1$, $y = 3$, $z = -5$

Further problems 5

1 $x = -\dfrac{3}{4}$ 2 $x = \dfrac{10}{3}$ 3 $x = -3$ 4 $x = \dfrac{3}{4}$ 5 $x = \dfrac{2}{3}$ 6 $x = \dfrac{11}{4}$

7 $x = -7{\cdot}5$, $y = -14$ 8 $x = -\dfrac{3}{2}$ 9 $x = 2$ 10 $x = 1$, $y = 2$, $z = 3$

11 $x = 6$, $y = 7$, $z = 5$ 12 $x = 2$, $y = -2$, $z = 0$ 13 $x = 3$, $y = 5$, $z = 2$

14 $x = -\dfrac{3}{2}$, $y = \dfrac{1}{2}$ 15 $x = -\dfrac{35}{6}$ 16 $x = 4{\cdot}5$ 17 $x = 5$, $y = 4$ 18 $x = \dfrac{1}{15}$, $y = \dfrac{1}{30}$

19 $x = -14$, $y = -1$ 20 $x = -0{\cdot}5$, $y = 0{\cdot}25$

Unit 18 Review test

1 $x = \dfrac{2}{3}$ or $x = \dfrac{3}{2}$ 2 $x = 0{\cdot}215$ or $x = -1{\cdot}549$ 3 $x = -3{\cdot}303$ or $x = 0{\cdot}303$

4 $x = 3$ or $x = 0{\cdot}697$ or $x = 4{\cdot}303$ 5 $x = 1$ or $x = 2$ or $x = -0{\cdot}281$ or $x = 1{\cdot}781$

Test exercise 6

1 (a) $x = -2$ or $x = -9$ (b) $x = 6$ or $x = 7$ (c) $x = 3$ or $x = -7$

(d) $x = -4$ or $x = -2{\cdot}5$ (e) $x = -3$ or $x = \dfrac{4}{3}$ (f) $x = 4$ or $x = 1{\cdot}2$

2 (a) $x = -0{\cdot}581$ or $2{\cdot}581$ (b) $x = -1{\cdot}775$ or $-0{\cdot}225$ (c) $x = -5{\cdot}162$ or $1{\cdot}162$
3 (a) $x = -2{\cdot}554$ or $-0{\cdot}196$ (b) $x = -0{\cdot}696$ or $1{\cdot}196$ (c) $x = -0{\cdot}369$ or $2{\cdot}169$
4 $x = -1$, $-4{\cdot}812$ or $0{\cdot}312$ 5 $x = 1$, $x = 2$, $x = -0{\cdot}396$ or $1{\cdot}896$

Further problems 6

1 (a) $x = -8$ or 5 (b) $x = 4$ or 7 (c) $x = -4$ or -6 (d) $x = 9$ or -5
2 (a) $x = 2$ or $-0{\cdot}75$ (b) $x = -1{\cdot}768$ or $-0{\cdot}566$ (c) $x = -1{\cdot}290$ or $-0{\cdot}310$
(d) $x = -1{\cdot}178$ or $0{\cdot}606$ (e) $x = 0{\cdot}621$ or $1{\cdot}879$ (f) $x = -1{\cdot}608$ or $0{\cdot}233$
3 (a) $x = -1$, -2, $0{\cdot}2$ (b) $x = 1$, $-2{\cdot}175$, $-0{\cdot}575$ (c) $x = -2$, $-0{\cdot}523$, $3{\cdot}189$
(d) $x = 1$, $-2{\cdot}366$, $-0{\cdot}634$ (e) $x = 2$, $-2{\cdot}151$, $-0{\cdot}349$ (f) $x = 3$, $0{\cdot}319$, $-1{\cdot}569$
4 (a) $x = -1$, -2, $0{\cdot}321$ or $4{\cdot}679$ (b) $x = 1$, -1, $-1{\cdot}914$ or $0{\cdot}314$
(c) $x = 1$, -3, $0{\cdot}431$ or $2{\cdot}319$ (d) $x = -1$, -2, $-3{\cdot}225$ or $-0{\cdot}775$
(e) $x = 2$, 3, $-2{\cdot}758$ or $-0{\cdot}242$ (f) $x = -1$, 2, $-0{\cdot}344$ or $1{\cdot}744$
(g) $x = 1$, 2, $0{\cdot}321$ or $4{\cdot}679$

Unit 19 Review test

1 $\dfrac{1}{4}\left(\dfrac{3}{x+2}+\dfrac{9}{x+6}\right)$ 2 $1 - \dfrac{3}{x+2}+\dfrac{1}{x+1}$

Unit 20 Review test

1 $\dfrac{6}{x+1}+\dfrac{x-1}{x^2+x+1}$ **2** $\dfrac{5}{x-1}+\dfrac{11}{(x-1)^2}$ **3** $2x-16+\dfrac{91}{x+4}-\dfrac{95}{(x+4)^2}$

Test exercise 7

1 $\dfrac{5}{x-4}-\dfrac{4}{x-6}$ **2** $\dfrac{4}{5x-3}+\dfrac{1}{2x-1}$ **3** $4+\dfrac{3}{x-4}+\dfrac{2}{x+5}$

4 $\dfrac{4}{x+1}+\dfrac{2x-3}{x^2+5x-2}$ **5** $\dfrac{5}{2x-3}+\dfrac{2}{(2x-3)^2}$ **6** $\dfrac{3}{x-5}-\dfrac{4}{(x-5)^2}+\dfrac{2}{(x-5)^3}$

7 $\dfrac{2}{x+4}+\dfrac{1}{2x-3}-\dfrac{3}{3x+1}$ **8** $\dfrac{4}{x-1}+\dfrac{3}{x+2}+\dfrac{1}{x+3}$

Further problems 7

1 $\dfrac{2}{x+4}+\dfrac{5}{x+8}$ **2** $\dfrac{3}{x-7}+\dfrac{2}{x+4}$ **3** $\dfrac{4}{x-5}-\dfrac{3}{x-2}$ **4** $\dfrac{1}{x-6}+\dfrac{2}{x+3}$

5 $\dfrac{2}{x-5}+\dfrac{3}{2x+3}$ **6** $\dfrac{4}{3x-2}+\dfrac{2}{2x+1}$ **7** $\dfrac{4}{5x-3}+\dfrac{1}{2x-1}$ **8** $\dfrac{5}{2x+3}-\dfrac{4}{3x+1}$

9 $\dfrac{6}{3x+4}-\dfrac{4}{(3x+4)^2}$ **10** $\dfrac{7}{5x+2}+\dfrac{3}{(5x+2)^2}$ **11** $\dfrac{3}{4x-5}-\dfrac{1}{(4x-5)^2}$

12 $\dfrac{5}{x-2}+\dfrac{7}{(x-2)^2}-\dfrac{1}{(x-2)^3}$ **13** $\dfrac{3}{5x+2}-\dfrac{5}{(5x+2)^2}-\dfrac{6}{(5x+2)^3}$

14 $\dfrac{4}{4x-5}+\dfrac{3}{(4x-5)^2}-\dfrac{7}{(4x-5)^3}$ **15** $\dfrac{3}{x+2}+\dfrac{5}{x-1}+\dfrac{2}{(x-1)^2}$

16 $\dfrac{3}{x+2}+\dfrac{1}{x-3}-\dfrac{5}{(x-3)^2}$ **17** $\dfrac{2}{x-1}+\dfrac{3}{2x+3}+\dfrac{4}{(2x+3)^2}$ **18** $4-\dfrac{2}{x-5}+\dfrac{7}{x-8}$

19 $5+\dfrac{4}{x+3}+\dfrac{6}{x-5}$ **20** $\dfrac{6}{x-2}+\dfrac{2x-3}{x^2-2x-5}$ **21** $\dfrac{1}{2x+3}+\dfrac{2x-1}{x^2+5x+2}$

22 $\dfrac{2}{3x-1}+\dfrac{x-3}{2x^2-4x-5}$ **23** $\dfrac{6}{x+1}-\dfrac{1}{x+2}-\dfrac{5}{x+3}$ **24** $\dfrac{4}{x-1}-\dfrac{2}{x+3}+\dfrac{3}{x+4}$

25 $\dfrac{4}{4x+1}+\dfrac{2}{2x+3}-\dfrac{5}{3x-2}$ **26** $\dfrac{2}{2x-1}+\dfrac{4}{3x+2}+\dfrac{3}{4x-3}$

27 $\dfrac{3}{x-1}+\dfrac{2}{2x+1}-\dfrac{4}{3x-1}$ **28** $\dfrac{4}{x+1}-\dfrac{5}{2x+3}+\dfrac{1}{3x-5}$ **29** $\dfrac{3}{x-2}-\dfrac{2}{2x-3}-\dfrac{4}{3x-4}$

30 $\dfrac{3}{5x+3}+\dfrac{6}{4x-1}-\dfrac{2}{x+2}$

Unit 21 Review test

1 $253\cdot3116°$ **2** $73°24'54''$ **3** (a) $0\cdot82$ rad (b) $0\cdot22$ rad (c) $3\pi/4$ rad
4 (a) $264\cdot76°$ (b) $405°$ (c) $468°$ **5** (a) $0\cdot9135$ (b) $0\cdot9659$ (c) $0\cdot5774$
(d) $3\cdot2535$ (e) $1\cdot0045$ (f) $0\cdot0987$ **6** $11\cdot0$ cm **8** $18\cdot4$ km, $8\cdot8$ km

Test exercise 8

1 $39\cdot951°$ to 3 dp **2** $52°30'18''$ **3** (a) $1\cdot47$ rad (b) $1\cdot21$ rad (c) $4\pi/3$ rad
4 (a) $122\cdot56°$ (b) $300\cdot00°$ (c) $162\cdot00°$ **5** (a) $0\cdot9511$ (b) $0\cdot2817$ (c) $0\cdot6235$
(d) $0\cdot8785$ (e) $1\cdot1595$ (f) $1\cdot5152$ **6** $10\cdot3$ cm to 1 dp
8 yes, it is a 3, 4, 5 triangle

Further problems 8

1 81·306° to 3 dp **2** 63°12′58″ to the nearest second **3** (a) 0·54 rad
(b) 0·84 rad   (c) 5π/4 rad   **4**   (a) 102·22°   (b) 135·00°   (c) 144·00°
5 (a) 0·5095 (b) 0·5878 (c) 5·6713 (d) 71·6221 (e) 1432·3946 (f) 0·4120
6 6·5 cm to 1 dp **8** 2

Unit 23 Review test

1 (a) and (b) do, (c) does not

2 (a) $h(x) = \dfrac{1}{4 - x} + 3x - 9$; domain $0 < x < 4$, range $-35/4 < h(x) < \infty$

(b) $k(x) = \dfrac{1}{2(4 - x)(x - 3)}$; domain $0 < x < 4$, $x \neq 3$, range $-\infty < k(x) < \infty$

3 (a) inverse is a function

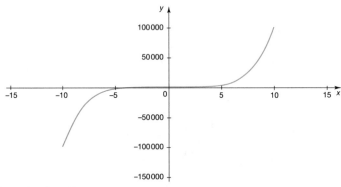

(b) inverse is not a function

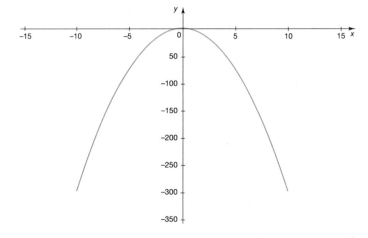

(c) inverse is not a function

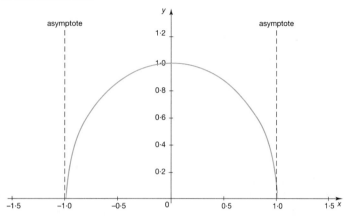

Unit 24 Review test

1 (a) $f(x) = -2(\sqrt{x} - 1)^3$ (b) $f(x) = 4\sqrt{x}$ (c) $f(x) = ((x-1)^3 - 1)^3$
2 $f(x) = b[a(b[c(x)])]$ where $a(x) = x^2$, $b(x) = x + 4$ and $c(x) = 3x$, inverse is not a
function

Unit 25 Review test

1 (a) 0·5   (b) −2·3662   (c) −0·2309   **2** (a) period $\pi/3$, amplitude 2, phase 0
(b) period $\pi/2$, amplitude ∞, phase −1   (c) period 2π, amplitude 1, phase $-\pi$
3 (a) 2   (b) 2   (c) 0

Unit 26 Review test

1 (a) $1·4573 \pm 2n\pi$ radians or $83·4944° \pm n \times 360°$,
 $4·8295 \pm 2n\pi$ radians or $276·5056° \pm n \times 360°$

(b) $\dfrac{\pi}{8} \pm n\pi$ radians or $22·5° \pm n \times 360°$, $\dfrac{3\pi}{8} \pm n\pi$ radians or $67·5° \pm n \times 360°$

(c) $0·0415 \pm \dfrac{n\pi}{3}$ radians or $2·3750° \pm n \times 60°$

Unit 27 Review test

1 (a) 0   (b) −0·75   (c) 1·1763   (d) 1·3195   (e) 7·3891   **2** (a) 2·6021   (b) 1, 3
(c) 0·5848   (d) ±1·732

Unit 28 Review test

2 $f_e(x) = x^2$ and $f_o(x) = x(x^2 + 1)$

Unit 29 Review test

1 (a) 6   (b) 2/3
(c) $-1/2e$

Test exercise 9

1 (a) does not, (b) and (c) do

2 (a) $h(x) = \dfrac{2}{x-2} - 3x + 3$; domain $2 < x < 3$, range $-4 < x < \infty$

(b) $k(x) = -\dfrac{3}{5(x-2)(x-1)}$; domain $2 < x < 3$, range $-\infty < x < -0 \cdot 3$

4 (a) $f(x) = 5(\sqrt{x}+3)^4$ (b) $f(x) = 25\sqrt{x}$ (c) $f(x) = ((x+3)^4 + 3)^4$

5 $f(x) = a[c(a[b(x)])]$ where $a(x) = x - 4$, $b(x) = 5x$ and $c(x) = x^3$
$f^{-1}(x) = b^{-1}[a^{-1}(c^{-1}[a^{-1}(x)])]$: $a^{-1}(x) = x + 4$, $b^{-1}(x) = x/5$, $c^{-1}(x) = x^{\frac{1}{3}}$
$f^{-1}(x) = \dfrac{(x+4)^{\frac{1}{3}} + 4}{5}$

6 (a) $0 \cdot 6428$ (b) $3 \cdot 5495$ (c) 0   **7** (a) period $2\pi/7$, amplitude 4, phase 0
(b) period π, amplitude 2, phase 0 relative to $\cos 2\theta$
(c) period $2\pi/3$, amplitude ∞, phase 4/3   **8**   (a) 3   (b) 4·5   (c) 0

9 (a) $-6 \cdot 45° \pm 360n°$, $173 \cdot 55° \pm 360n°$ (b) $\pi/28 \pm n\pi$ radians   (c) $\pm 9° \pm 360n°$

10 (a) 0 (b) 1/3 (c) $3 \cdot 9949$ (d) 625 (e) $4 \cdot 2618$ (f) $4 \cdot 6052$ **11**   (a) 2

(b) $0 \cdot 6931$, $1 \cdot 0986$ (c) 6 (d) $\pm 32/81$ (e) 1/9   **12** $f_e(x) = \dfrac{a^x + a^{-x}}{2}$ and

$f_o(x) = \dfrac{a^x - a^{-x}}{2}$ **13** (a) 0 (b) 6 (c) $-5/28$ (d) $-1/9$ (e) $\pi/4$

Further problems 9

1 $x = \sqrt{3}$ **2** (a) domain $-1 < x < 1$, range $-\infty < f(x) < \infty$
3 the two straight lines $y = x/3$ and $y = -x/3$

6 yes; consider $a = \sqrt{2}$ and $x = 2$ as an example   **7** (a) $f^{-1}(x) = \left(\dfrac{x}{6}+2\right)^{\frac{1}{3}}$

(b) $f^{-1}(x) = \left(x^{\frac{1}{3}}+2\right)^{\frac{1}{3}}$ (c) $f^{-1}(x) = \left(\dfrac{(x+2)^{\frac{1}{3}}}{6}+2\right)^{\frac{1}{3}}$

10 (b) $f'(x) = \begin{cases} -1 & \text{if } x < 0 \\ 1 & \text{if } x > 0 \end{cases}$ (c) $f'(0)$ does not exist

13 (a) $-5 \cdot 702$ (b) $0 \cdot 8594$ (c) $-2 \cdot 792$ **14** (a) 0 (b) -4 (c) 1/2

Unit 30 Review test

1 (a) $\begin{pmatrix} -20 & 24 \\ 9 & -7 \end{pmatrix}$ (b) $\begin{pmatrix} -6 & -22 \\ 9 & 3 \end{pmatrix}$ (c) $\begin{pmatrix} -14 & 46 \\ 0 & -10 \end{pmatrix}$ (d) $\begin{pmatrix} 298 & -53 \\ -45 & 10 \end{pmatrix}$

(e) $\begin{pmatrix} 91 & -304 \\ -63 & 217 \end{pmatrix}$ (f) $\begin{pmatrix} -538 \\ 85 \end{pmatrix}$ (g) $(-21 \quad 16)$

Unit 31 Review test

1 (a) 14 (b) -48 (c) 0 (d) does not exist

2 (a) $\begin{pmatrix} 5 & -2 \\ -2 & 1 \end{pmatrix}$ (b) $\begin{pmatrix} -9 & -14 \\ 2 & 3 \end{pmatrix}$ (c) does not exist

Unit 32 Review test

1 $x = -2$, $y = 3$ **2** $x = 2$, $y = 4$, $z = -2$

Test exercise 10

1 (a) $\begin{pmatrix} 6 & 2 \\ -5 & 12 \end{pmatrix}$ (b) $\begin{pmatrix} 4 & -2 \\ 1 & -4 \end{pmatrix}$ (c) $\begin{pmatrix} 2 & 4 \\ -6 & 16 \end{pmatrix}$ (d) $\begin{pmatrix} 1 & 8 \\ -31 & 32 \end{pmatrix}$

(e) $\begin{pmatrix} 5 & 10 \\ -14 & 28 \end{pmatrix}$ (f) $\begin{pmatrix} 13 \\ 45 \end{pmatrix}$ (g) $\begin{pmatrix} 19 & 12 \end{pmatrix}$

2 (a) -10 (b) -44 (c) 0 (d) does not exist

3 (a) $\begin{pmatrix} -7/2 & -3 \\ -1 & 1 \end{pmatrix}$ (b) $\begin{pmatrix} 7/44 & -3/22 \\ 1/22 & -2/11 \end{pmatrix}$ (c) does not exist

4 (a) $x = 7$, $y = -1$ (b) $x = -2$, $y = -4$
5 (a) $x = 2$, $y = -2$, $z = 3$ (b) $a = 2$, $b = -2$, $c = 1$ (c) $s = 5$, $t = 7$, $u = 9$

Further problems 10

1 (a) $\begin{pmatrix} 3 & 8 \\ 23 & -4 \end{pmatrix}$ (b) $\begin{pmatrix} -19 & -43 \\ 0 & 10 \end{pmatrix}$ (c) $\begin{pmatrix} 32 & -6 \\ -14 & -8 \end{pmatrix}$ (d) $\begin{pmatrix} -11 & -3 \\ 15 & 35 \end{pmatrix}$

(e) $\begin{pmatrix} 6 \\ -26 \end{pmatrix}$ (f) $\begin{pmatrix} -23 & -6 \end{pmatrix}$ (g) $\begin{pmatrix} -27 \end{pmatrix}$ (h) $\begin{pmatrix} -15 & 20 \\ 9 & -12 \end{pmatrix}$

6 (a) $x = -1$, $y = 3$ (b) $a = 1/2$, $b = -7/2$ (c) $p = 15/4$, $q = 3/2$
(d) $m = -2$, $n = -1$ **7** (a) $x = 1$, $y = -1$ (b) $m = 2$, $n = 3$ (c) $a = 5$, $b = -3$
(d) $s = 2$, $t = 4$ **8** (a) -2 (b) 198 (c) 115 (d) 0
9 (a) $x = 14$, $y = 1$, $z = 8$ (b) $p = -5$, $q = 7$, $r = -9$ (c) $l = 1$, $m = 1$, $n = 1$
(d) $a = -3$, $b = 2$, $c = -1$

Unit 33 Review test

1 $\overline{AB} = 3\mathbf{i} + 4\mathbf{j}$, $\overline{BC} = 2\mathbf{i} - 3\mathbf{j}$, $\overline{CD} = 3\mathbf{i} + 5\mathbf{j}$ and $\overline{DA} = -8\mathbf{i} - 6\mathbf{j}$
$|\overline{AB}| = 5$, $|\overline{BC}| = \sqrt{13}$, $|\overline{CD}| = \sqrt{34}$ and $|\overline{DA}| = 10$

Unit 34 Review test

1 $l = -\dfrac{1}{\sqrt{3}}$; $m = \dfrac{1}{\sqrt{3}}$; $n = -\dfrac{1}{\sqrt{3}}$ **2** $\theta = 90°$

Unit 35 Review test

1 (a) 5 (b) $\mathbf{i} + 6\mathbf{j} + 5\mathbf{k}$ **2** (a) $\mathbf{a.b} = 48$, $\theta = 69\cdot5°$
(b) $|\mathbf{a} \times \mathbf{b}| = 128\cdot05$, $l = \dfrac{6}{128\cdot05}$, $m = \dfrac{114}{128\cdot05}$, $n = \dfrac{58}{128\cdot05}$, $\theta = 58\cdot4°$

Test exercise 11

1 $\overline{AB} = 2\mathbf{i} - 5\mathbf{j}$, $\overline{BC} = -4\mathbf{i} + \mathbf{j}$, $\overline{CA} = 2\mathbf{i} + 4\mathbf{j}$, $AB = \sqrt{29}$, $BC = \sqrt{17}$, $CA = \sqrt{20}$
2 $l = 3/13$, $m = 4/13$, $n = 12/13$ **3** (a) -8 (b) $-2\mathbf{i} - 7\mathbf{j} - 18\mathbf{k}$
4 (a) 6, $\theta = 82°44'$ (b) $47\cdot05$, $\theta = 19°31'$

Further problems 11

1 $\overline{OG} = \frac{1}{3}(10\mathbf{i} + 2\mathbf{j})$ **2** $\frac{1}{\sqrt{50}}(3, 4, 5)$; $\frac{1}{\sqrt{14}}(1, 2, -3)$; $\theta = 98°42'$

3 moduli: $\sqrt{74}$, $3\sqrt{10}$, $2\sqrt{46}$; direction cosines: $\frac{1}{\sqrt{74}}(3, 7, -4)$, $\frac{1}{3\sqrt{10}}(1, -5, -8)$,

$\frac{1}{\sqrt{46}}(3, -1, 6)$; sum $= 10\mathbf{i}$, direction cosines: $(1, 0, 0)$ **4** 8, $17\mathbf{i} - 7\mathbf{j} + 2\mathbf{k}$,

$\theta = 66°36'$ **5** (a) -7 (b) $7(\mathbf{i} - \mathbf{j} - \mathbf{k})$ (c) $\cos \theta = -0\cdot5$ **6** $\cos \theta = -0\cdot4768$

7 (a) 7, $5\mathbf{i} - 3\mathbf{j} - \mathbf{k}$ (b) 8, $11\mathbf{i} + 18\mathbf{j} - 19\mathbf{k}$

8 $-\frac{3}{\sqrt{155}}\mathbf{i} + \frac{5}{\sqrt{155}}\mathbf{j} + \frac{11}{\sqrt{155}}\mathbf{k}$; $\sin \theta = 0\cdot997$ **9** $\frac{2}{\sqrt{13}}, \frac{-3}{\sqrt{13}}, 0$; $\frac{5}{\sqrt{30}}, \frac{1}{\sqrt{30}}, \frac{-2}{\sqrt{30}}$

10 $6\sqrt{5}$; $\frac{-2}{3\sqrt{5}}, \frac{4}{3\sqrt{5}}, \frac{5}{3\sqrt{5}}$ **11** (a) 0, $\theta = 90°$ (b) $68\cdot53$, $(-0\cdot1459, -0\cdot5982, -0\cdot7879)$

12 $4\mathbf{i} - 5\mathbf{j} + 11\mathbf{k}$; $\frac{1}{9\sqrt{2}}(4, -5, 11)$ **13** (a) $\mathbf{i} + 3\mathbf{j} - 7\mathbf{k}$ (b) $-4\mathbf{i} + \mathbf{j} + 2\mathbf{k}$

(c) $13(\mathbf{i} + 2\mathbf{j} + \mathbf{k})$ (d) $\frac{\sqrt{6}}{6}(\mathbf{i} + 2\mathbf{j} + \mathbf{k})$ **14** (a) $-2\mathbf{i} + 2\mathbf{j} - 6\mathbf{k}$ (b) -40

(c) $-30\mathbf{i} - 10\mathbf{j} + 30\mathbf{k}$ **15** $2\mathbf{i} + 12\mathbf{j} + 34\mathbf{k}$ **20** $2x - 3y + z = 0$

Unit 36 Review test

1 $^{12}C_7 = \frac{12!}{(12-7)!7!} = 792$ **2** (a) 720 (b) $39\,916\,800$ (c) 2730 (d) 720 (e) $1/6$

3 (a) 6 (b) 210 (c) 1 (d) 36

Unit 37 Review test

1 $81a^4 + 432a^3b + 864a^2b^2 + 768ab^3 + 256b^4$ **2** (a) $-\frac{21}{2}x^3$ (b) $-\frac{63}{16}$

Unit 38 Review test

1 (a) 210 (b) $n(n+4)$ **2** 12, 110

Test exercise 12

1 $13\,983\,816$ **2** (a) $40\,320$ (b) $3\,628\,800$ (c) 4080 (d) 24 (e) 24 **3** (a) 56

(b) 455 (c) 159 (d) 1

4 $128a^7 - 2240a^6b + 16\,800a^5b^2 - 70\,000a^4b^3 + 175\,000a^3b^4 - 262\,500a^2b^5 +$
$218\,750ab^6 - 78\,125b^7$

5 (a) $\frac{1\cdot2 \times 10^9}{x^7}$ (b) $4\cdot5 \times 10^8$ **6** (a) 1035 (b) $\frac{3n(5-n)}{2}$ **7** 9, 400

Further problems 12

1 (a) $^{12}C_5 = 792$ (b) $^{12}C_5 \times {}^7C_4 = 27\,720$ (c) $^{12}C_5 \times {}^7C_4 \times {}^3C_2 = 83\,160$

3 (a) $1 - 12x + 54x^2 - 108x^3 + 81x^4$ (b) $32 + 40x + 20x^2 + 5x^3 + \frac{5}{8}x^4 + \frac{1}{32}x^5$

(c) $1 - 5x^{-1} + 10x^{-2} - 10x^{-3} + 5x^{-4} - x^{-5}$

(d) $x^6 + 6x^4 + 15x^2 + 20 + 15x^{-2} + 6x^{-4} + x^{-6}$ **4** (a) 568 (b) $\frac{n(3-5n)}{2}$

Unit 39 Review test

1 (c), (d), (f), (g) **2** 8 **3** (a) {Head, Tail} (b) {$x : x$ is a playing card}

Unit 40 Review test

1 (a) {15, 16, 17} (b) B
(c) {$x : x$ is a black, even-numbered card from a deck of playing cards}
(d) {$x : x$ is a paperback edition of *A Christmas Carol* by Charles Dickens}
2 (a) {$x : x$ is a colour of the rainbow} (b) {$l, n, p, r, s, t, u, v, w$}
(c) {$x : x$ is a black car with automatic transmission} (d) B
3 (a) {red, yellow, indigo} (b) {11, 13, 17, 19} (c) {$-3, -2, 0, 2, 3$}
(d) {$x : x$ is a sheet of unlined A4 paper}

Unit 41 Review test

1 (a) $(A \cap \bar{B}) \cup B$ $A \cup B$

 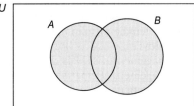

$$(A \cap \bar{B}) \cup B = A \cup B$$

(b) $(A \cap \bar{B}) \cup (\bar{A} \cap B)$ $(A \cup B) \cap (\bar{A} \cup \bar{B})$

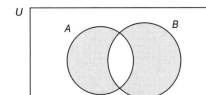 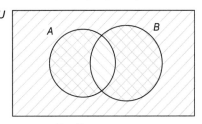

$$(A \cap \bar{B}) \cup (\bar{A} \cap B) = (A \cup B) \cap (\bar{A} \cup \bar{B})$$

3 83 **4** 14 **5** 21

Test exercise 13

1 (b), (c), (d) **2** 16 **3** (a) {q, r} (b) B
(c) {$x : x$ is a red ace or 2 of hearts from a deck of playing cards}
(d) {$x : x$ is a white cotton bed sheet} **4** (a) {u, v, w, x, y, z}
(b) {$-8, -6, -4, -2, 0, 2, 4, 6$} (c) B (a circle is a special kind of ellipse)
(d) {$x : x$ is a court playing card or the 3 of diamonds}
5 (a) {$-8, -4, -2, 0, 2, 4, 8$} (b) {1, 3, 5, 7} (c) {$-3\pi, -2\pi, -\pi$}
(d) {$x : x$ is a computer printer that is not a laser printer}

6 (a) $(\bar{A} \cap B) \cap \bar{C}$

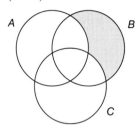

(b) $(\overline{A \cap B}) \cap (A \cup B)$

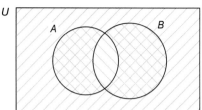

8 6   **9** 26   **10** 2

Further problems 13

1 (a) $\{x : x(x - 1) = 0\}$ (b) $\{2, 3\}$

2 $F \cap M$ is the set of female management employees, $\bar{F} \cap M$ is the set of male management employees, $F \cap \bar{M}$ is the set of female employees who are not in management, $\bar{F} \cap \bar{M}$ is the set of male employees who are not in management, $\overline{F \cap M}$ is the set of male employees or those who are not in management, $F \cup M$ is the set of female or management employees, $\bar{F} \cup M$ is the set of male or management employees, $F \cup \bar{M}$ is the set of female employees or those who are not in management, $\bar{F} \cup \bar{M}$ is the set of male employees or those who are not in management, $\overline{F \cup M}$ is the set of male employees who are not in management   **3** (a), (c)   **4** (a), (b), (c), (d)

5 Ø, {Ø}, {{Ø}}, {Ø, {Ø}}   **6** {Ø, {1}, {{1}}, {1, {1}}}

7 (a) $\{1, 3, 7\}$ (b) $\{2, 4, 8\}$ (c) $\{5\}$ (d) $\{6, 9\}$ (e) $\{1, 2, 3, 4, 5, 7, 8\}$
(f) $\{1, 2, 3, 4, 6, 7, 8, 9, 10\}$ (g) $\{1, 3, 5, 6, 7, 9, 10\}$ (h) $\{2, 4, 5, 6, 8, 9, 10\}$
(i) $\{2, 4, 5, 6, 8, 9, 10\}$ (j) $\{6, 9, 10\}$

8 (a) $(\bar{A} \cap B) \cup (B \cap \bar{C})$

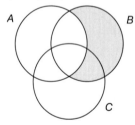

(b) $(\bar{A} \cup \bar{B}) \cap C$

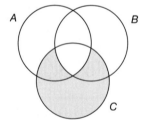

9 (a) $(A \cap B) \cup (\bar{A} \cap \bar{B})$

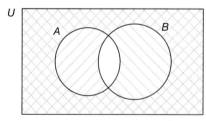

$(\bar{A} \cup B) \cap (A \cup \bar{B})$

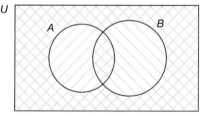

$$(A \cap B) \cup (\bar{A} \cap \bar{B}) = (\bar{A} \cup B) \cap (A \cup \bar{B})$$

(b) $(A \cap \bar{B}) \cup (\bar{A} \cap C) \cup (A \cap B)$ $A \cup C$

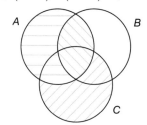 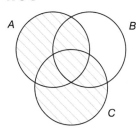

$$(A \cap \bar{B}) \cup (\bar{A} \cap C) \cup (A \cap B) = A \cup C$$

(c) $(A \cap B \cap C) \cup (A \cap B \cap \bar{C})$ $A \cap B$

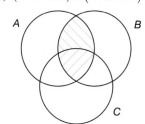 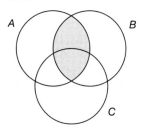

$$(A \cap B \cap C) \cup (A \cap B \cap \bar{C}) = A \cap B$$

11 (a) Difference $B - A$ Symmetric difference

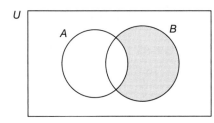 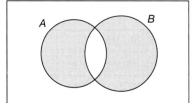

12 290

13

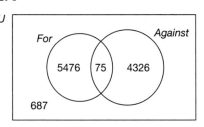

14 10   **15** 4500   **16** 18   **17** 79

Unit 42 Review test

1 (a) (i) 0·92   (ii) 0·08   (b) 22

Unit 43 Review test

1 (a) 5/13   (b) 51/65   (c) 39/65

Unit 44 Review test

1 (a) 0·1   (b) 0·095

Unit 45 Review test

1 0·06   **2** (a) 5/18   (b) 5/18

Test exercise 14

1 (a)(i) 0·12   (ii) 0·88   (b) 3520   **2** (a) 0·2400   (b) 0·1600   (c) 0·3600
3 (a) 0·1576   (b) 0·0873   (c) 0·1939

Further problems 14

1 3/8   **2** (a) 7/16   (b) 3/8   (c) 9/16   **3** 121/125   **4** (a) 1/6   (b) 1/18
(c) 2/9   **5** 8/16575   **6** 2/16575   **7** (a) 1/5   (b) 1/2   **8** 3/8   **9** $1/2^n$
10 0·94   **11** 3/7   **12** (a) 3%   (b) 2%   (c) 93%   **13** (a)(i) 1/10   (ii) 1/90
(b)(i) 8/90   (ii) 0   (c) 3/445   **14** 1/20, 50   **15** (a) 25/198   (b) 12/99

Unit 46 Review test

1

Number	1	2	3	4	5
Frequency	9	11	14	17	9
Relative frequency	0·150	0·183	0·233	0·283	0·150

2

Weight (gm)	94·9–95·0	95·1–95·2	95·3–95·4
Frequency	22	19	9
Relative frequency	0·44	0·38	0·18

Unit 47 Review test

1 Mode = 122 gm, Median = 123 gm, Mean = 122·6 gm
2 Mode = 17·333 m, Median = 17·336 m, Mean = 17·318 m

Unit 48 Review test

1 2·5   **2** 0·1

Unit 49 Review test

1 (a) 0·44   (b) −0·28   (c) $-1·01 \le z \le 0·61$, area 0·5729

Test exercise 15

1 (a), (b)

Mass (kg) x	Frequency f
4·2	1
4·3	3
4·4	7
4·5	10
4·6	12
4·7	10
4·8	5
4·9	2
	$n = \sum f = 50$

(c)

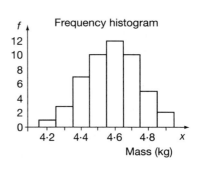

2 (a) (i), (ii)

Diam (mm) x	x_m	Rel f (%)
8·82–8·86	8·84	1·33
8·87–8·91	8·89	10·67
8·92–8·96	8·94	21·33
8·97–9·01	8·99	24·00
9·02–9·06	9·04	20·00
9·07–9·11	9·09	13·33
9·12–9·16	9·14	6·67
9·17–9·21	9·19	2·67
		100·00%

(b)

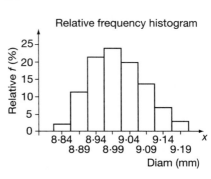

(c) (i) 8·915 mm   (ii) 9·115 mm   (iii) 0·05 mm   **3** (a) $\bar{x} = 10·62$ mm
(b) $\sigma = 0·437$ mm   (c) mode $= 10·51$ mm   (d) median $= 10·59$ mm
4 (a) $\bar{x} = 15·05$ mm; $\sigma = 0·4314$ mm   (b) (i) 13·76 mm to 16·34 mm   (ii) 1200
5 212

Further problems 15

1 (a), (b) (i)

x	f
60–64	3
65–69	5
70–74	7
75–79	11
80–84	8
85–89	4
90–94	2
	$n = \sum f = 40$

(b) (ii) $\bar{x} = 76 \cdot 5$ (iii) $\sigma = 7 \cdot 73$

2 (a), (b)

Length (mm) x	Frequency f
13·5–14·4	2
14·5–15·4	5
15·5–16·4	9
16·5–17·4	13
17·5–18·4	7
18·5–19·4	4
	$n = \sum f = 40$

(c) (i) $\bar{x} = 16 \cdot 77$ mm (ii) $\sigma = 1 \cdot 343$ mm **3** (a) (i) $\bar{x} = 4 \cdot 283$ kg (ii) $\sigma = 0 \cdot 078$ kg
(b) (i) 4·050 kg to 4·516 kg (ii) 288 **4** (a) $\bar{x} = 2 \cdot 382$ MΩ (b) $\sigma = 0 \cdot 0155$ MΩ
(c) mode = 2·378 MΩ (d) median = 2·382 MΩ **5** (a) $\bar{x} = 31 \cdot 06$ kN
(b) $\sigma = 0 \cdot 237$ kN (c) mode = 31·02 kN (d) median = 31·04 kN
6 (a) (i) $\bar{x} = 11 \cdot 538$ min (ii) $\sigma = 0 \cdot 724$ min (iii) mode = 11·6 min
(iv) median = 11·57 min (b) (i) 0·5 min (ii) 10·75 min (iii) 13·25 min
7 (a) 32·60 to 32·72 mm (b) 1956 **8** (a) (i) $\bar{x} = 7 \cdot 556$ kg (ii) $\sigma = 0 \cdot 126$ kg
(b) (i) 7·18 kg to 7·94 kg (ii) 1680 **9** (a) (i) $\bar{x} = 29 \cdot 64$ mm (ii) $\sigma = 0 \cdot 123$ mm
(b) (i) 29·27 mm to 30·01 mm (ii) 2100 **10** (a) (i) $\bar{x} = 25 \cdot 14$ mm
(ii) $\sigma = 0 \cdot 690$ mm (b) (i) 23·07 mm to 27·21 mm (ii) 480 **11** (a) 0·0078
(b) 0·9028 **12** (a) 74·9% (b) 0·14% **13** 333 **14** (a) 0·0618 (b) 0·2206
15 (a) 7 (b) 127 **16** (a) 0·4835 (b) 1·46% **17** (a) 215 (b) 89·3% (c) 0·0022

Unit 50 Review test

1 $y = 84 \cdot 97 - 7 \cdot 91x$

Unit 51 Review test

1 $r = -0 \cdot 857$ **2** $r_S = 0 \cdot 2$

Test exercise 16

1 $y = -0.85 + 0.54x$ **2** $r = 0.92$ **3** $r_S = 0.71$

Further problems 16

1 $y = -0.06 + 4.44x$ **2** $a = 2.26, n = 1.22$ **3** $W = 1.3 \times 10^9 e^{-0.66T}$ **4** $r = 0.998$
5 $r = 0.998$ **6** $r = -0.998$ **7** $r_S = 0.05$ **8** $r_S = 0.61$ **9** $r_S = 0.95$

Unit 52 Review test

1 (a) 2 (b) $-11/9$ (c) 2 **2** 14·4 **3** $12x^3 - 15x^2 + 2x + 2$, 538

Unit 53 Review test

1 (a) (i) $12x^2 + 2x - 6$, $24x + 2$ (ii) 0·629 or -0.795 to 3 dp
(b) $-18\sin(9x - 1) - 3\cos(3x + 7)$, $-162\cos(9x - 1) + 9\sin(3x + 7)$
2 (a) $x^3(4\cos x - x\sin x)$ (b) $x(2\ln x + 1)$ (c) $\dfrac{x\cos x - \sin x}{x^2}$ (d) $\dfrac{3}{x^5}(1 - 4\ln x)$
3 (a) $15(3x - 2)^4$ (b) $6\cos(3x - 1)$ (c) $-2e^{(2x+3)}$ (d) $\dfrac{2}{2x - 3}$
(e) $e^{2x}(2\cos 3x - 3\sin 3x)$ (f) $\dfrac{\sin(2 - 5x)}{x} - 5(\ln x)\cos(2 - 5x)$ **4** 1·442250

Test exercise 17

1 (a) $\dfrac{4}{3}$ (b) $-\dfrac{2}{5}$ (c) $-\dfrac{6}{7}$ (d) $-\dfrac{13}{5}$ **2** 7·2 **3** $\dfrac{dy}{dx} = 4x^3 + 15x^2 - 12x + 7$, $\dfrac{dy}{dx} = 59$

4 (a) (i) $6x^2 - 22x + 12$, $12x - 22$ (ii) $\dfrac{2}{3}$, 3

(b) $\dfrac{dy}{dx} = 6\cos(2x + 1) - 12\sin(3x - 1)$, $\dfrac{d^2y}{dx^2} = -12\sin(2x + 1) - 36\cos(3x - 1)$

5 (a) $x(x\cos x + 2\sin x)$ (b) $x^2 e^x(x + 3)$ (c) $\dfrac{-x\sin x - 2\cos x}{x^3}$ (d) $\dfrac{2e^x(\tan x - \sec^2 x)}{\tan^2 x}$
6 (a) $20(5x + 2)^3$ (b) $3\cos(3x + 2)$ (c) $4e^{4x-1}$ (d) $-10\sin(2x + 3)$
(e) $-3\cos^2 x . \sin x$ (f) $\dfrac{4}{4x - 5}$ **7** 1·179509

Further problems 17

1 (a) $m = -4$ (b) $m = 25$ (c) $m = 14$ **2** (a) $6x^2 + 8x - 2$, 6
(b) $12x^3 - 15x^2 + 8x - 1$, 212 (c) $20x^4 + 8x^3 - 9x^2 + 14x - 2$, 31
3 (a) $x^4(x\cos x + 5\sin x)$ (b) $e^x(\cos x - \sin x)$ (c) $x^2(x\sec^2 x + 3\tan x)$
(d) $x^3(4\cos x - x\sin x)$ (e) $5x(x\cos x + 2\sin x)$ (f) $2e^x\left(\dfrac{1}{x} + \ln x\right)$
4 (a) $\dfrac{-x\sin x - 3\cos x}{x^4}$ (b) $\dfrac{\cos x - \sin x}{2e^x}$ (c) $\dfrac{-\sin x\tan x - \cos x\sec^2 x}{\tan^2 x}$
(d) $\dfrac{4x^2(3\cos x + x\sin x)}{\cos^2 x}$ (e) $\dfrac{\sec^2 x - \tan x}{e^x}$ (f) $\dfrac{1 - 3\ln x}{x^4}$ **5** (a) $6(2x - 3)^2$

(b) $3e^{3x+2}$ (c) $-12\sin(3x + 1)$ (d) $\dfrac{2x}{x^2 + 4}$ (e) $2\cos(2x - 3)$ (f) $2x\sec^2(x^2 - 3)$

(g) $40(4x + 5)$ (h) $12x.e^{x^2+2}$ (i) $\dfrac{2}{x}$ (j) $-15\cos(4 - 5x)$ **6** (a) -2.456
(b) 1·765 (c) 0·739 (d) 1·812 (e) 1·8175 (f) 0·5170 (g) 0·8449 (h) 0·8806

Unit 54 Review test

1 $\dfrac{\partial z}{\partial x} = 6x^2 - 7y, \quad \dfrac{\partial z}{\partial y} = -7x + 27y^2$ **2** $\dfrac{\partial z}{\partial x} = 48x + 14y, \quad \dfrac{\partial z}{\partial y} = 14x - 48y$

3 $\dfrac{\partial z}{\partial x} = 6\cos(6x + 2y), \quad \dfrac{\partial z}{\partial y} = 2\cos(6x + 2y)$

4 $\dfrac{\partial z}{\partial x} = -\dfrac{x\sin(x - 4y) + \cos(x - 4y)}{x^2 y}, \quad \dfrac{\partial z}{\partial y} = \dfrac{4y\sin(x - 4y) - \cos(x - 4y)}{xy^2}$

Unit 55 Review test

1 (a) $\dfrac{\partial z}{\partial x} = 15x^2 + 3y, \quad \dfrac{\partial z}{\partial y} = 3x - 6y^2,$

$\dfrac{\partial^2 z}{\partial x^2} = 30x, \quad \dfrac{\partial^2 z}{\partial y^2} = -12y, \quad \dfrac{\partial^2 z}{\partial x \partial y} = 3$

(b) $\dfrac{\partial z}{\partial x} = y\sec^2 xy, \quad \dfrac{\partial z}{\partial y} = x\sec^2 xy,$

$\dfrac{\partial^2 z}{\partial x^2} = 2y^2 \sec^2 xy \tan xy, \quad \dfrac{\partial^2 z}{\partial y^2} = 2x^2 \sec^2 xy \tan xy,$

$\dfrac{\partial^2 z}{\partial x \partial y} = \sec^2 xy + 2xy \sec^2 xy \tan xy$

(c) $\dfrac{\partial z}{\partial x} = -\dfrac{13y}{(3x - 2y)^2}, \quad \dfrac{\partial z}{\partial y} = \dfrac{13x}{(3x - 2y)^2},$

$\dfrac{\partial^2 z}{\partial x^2} = \dfrac{78y}{(3x - 2y)^3}, \quad \dfrac{\partial^2 z}{\partial y^2} = \dfrac{52x}{(3x - 2y)^3}, \quad \dfrac{\partial^2 z}{\partial x \partial y} = -\dfrac{13(3x + 2y)}{(3x - 2y)^3}$

Test exercise 18

1 (a) $\dfrac{\partial z}{\partial x} = 12x^2 - 5y^2, \quad \dfrac{\partial z}{\partial y} = -10xy + 9y^2, \quad \dfrac{\partial^2 z}{\partial x^2} = 24x, \quad \dfrac{\partial^2 z}{\partial y^2} = -10x + 18y$

$\dfrac{\partial^2 z}{\partial y.\partial x} = -10y, \quad \dfrac{\partial^2 z}{\partial x.\partial y} = -10y$ (b) $\dfrac{\partial z}{\partial x} = -2\sin(2x + 3y), \quad \dfrac{\partial z}{\partial y} = -3\sin(2x + 3y)$

$\dfrac{\partial^2 z}{\partial x^2} = -4\cos(2x + 3y), \quad \dfrac{\partial^2 z}{\partial y^2} = -9\cos(2x + 3y), \quad \dfrac{\partial^2 z}{\partial y.\partial x} = -6\cos(2x + 3y)$

$\dfrac{\partial^2 z}{\partial x.\partial y} = -6\cos(2x + 3y)$ (c) $\dfrac{\partial z}{\partial x} = 2xe^{x^2 - y^2}, \quad \dfrac{\partial z}{\partial y} = -2ye^{x^2 - y^2}$

$\dfrac{\partial^2 z}{\partial x^2} = 2e^{x^2 - y^2}(2x^2 + 1), \quad \dfrac{\partial^2 z}{\partial y^2} = 2e^{x^2 - y^2}(2y^2 - 1), \quad \dfrac{\partial^2 z}{\partial y.\partial x} = -4xye^{x^2 - y^2}$

$\dfrac{\partial^2 z}{\partial x.\partial y} = -4xye^{x^2 - y^2}$ (d) $\dfrac{\partial z}{\partial x} = 2x^2 \cos(2x + 3y) + 2x\sin(2x + 3y)$

$\dfrac{\partial^2 z}{\partial x^2} = (2 - 4x^2)\sin(2x + 3y) + 8x\cos(2x + 3y)$

$\dfrac{\partial^2 z}{\partial y.\partial x} = -6x^2 \sin(2x + 3y) + 6x\cos(2x + 3y), \quad \dfrac{\partial z}{\partial y} = 3x^2 \cos(2x + 3y)$

$\dfrac{\partial^2 z}{\partial y^2} = -9x^2 \sin(2x + 3y), \quad \dfrac{\partial^2 z}{\partial x.\partial y} = -6x^2 \sin(2x + 3y) + 6x\cos(2x + 3y)$

2 (a) $2V$ **3** P decreases 375 W **4** $\pm 2{\cdot}5\%$

Further problems 18

10 $\pm 0 \cdot 67E \times 10^{-5}$ approx. **12** $\pm(x + y + z)\%$ **13** y decreases by 19% approx.
14 $\pm 4 \cdot 25\%$ **16** 19% **18** $\delta y = y\{\delta x.p \cot(px + a) - \delta t.q \tan(qt + b)\}$

Unit 57 Review test

1 (a) $\dfrac{x^8}{8} + C$ (b) $4 \sin x + C$ (c) $2e^x + C$ (d) $12x + C$ (e) $-\dfrac{x^{-3}}{3} + C$ (f) $\dfrac{6^x}{\ln 6} + C$

  (g) $\dfrac{27}{4} x^{\frac{4}{3}} + C$ (h) $4 \tan x + C$ (i) $-9 \cos x + C$ (j) $8 \ln x + C$

Unit 58 Review test

1 (a) $I = \dfrac{x^4}{4} - \dfrac{x^3}{3} + \dfrac{x^2}{2} - x + C$ (b) $I = x^4 - 3x^3 + 4x^2 - 2x + 1$ **2** (a) $\dfrac{(5x - 1)^5}{25} + C$

  (b) $\dfrac{\cos(6x - 1)}{18} + C$ (c) $-\dfrac{(4 - 2x)^{\frac{3}{2}}}{3} + C$ (d) $\dfrac{2e^{3x+2}}{3} + C$ (e) $-\dfrac{5^{1-x}}{\ln 5} + C$

  (f) $\dfrac{3 \ln(2x - 3)}{2} + C$ (g) $-\dfrac{\tan(2 - 5x)}{25} + C$

Unit 59 Review test

1 (a) $\dfrac{\ln(3x + 1)}{3} + \dfrac{\ln(2x - 1)}{2} + C$ (b) $-\dfrac{\ln(1 - 4x)}{2} - 3 \ln(x + 2) + C$

  (c) $2\dfrac{\ln(1 + 3x)}{3} + \dfrac{\ln(1 - 3x)}{3} + C$

Unit 60 Review test

1 $-(x - \pi/2)\dfrac{\cos 3x}{3} + \dfrac{\sin 3x}{9} + C$

Unit 61 Review test

1 $e(e - 1) \text{unit}^2$ **2** $2 \cdot 351 \text{unit}^2$

Unit 62 Review test

1 $10 \cdot 478 \text{units}^2$

Test exercise 19

1 (a) $\dfrac{x^5}{5} + C$ (b) $-3 \cos x + C$ (c) $4e^x + C$ (d) $6x + C$ (e) $2x^{\frac{3}{2}} + C$ (f) $5^x / \ln 5 + C$

  (g) $-\dfrac{1}{3x^3} + C$ (h) $4 \ln x + C$ (i) $3 \sin x + C$ (j) $2 \tan x + C$

2 (a) $2x^4 + 2x^3 - \dfrac{5}{2} x^2 + 4x + C$ (b) $I = 236$

3 (a) $\dfrac{-2 \cos(3x + 1)}{3} + C$ (b) $\dfrac{-(5 - 2x)^{\frac{3}{2}}}{3} + C$ (c) $-2e^{1-3x} + C$ (d) $\dfrac{(4x + 1)^4}{16} + C$

(e) $\dfrac{4^{2x-3}}{2\ln 4}+C$ (f) $-3\sin(1-2x)+C$ (g) $\dfrac{5\ln(3x-2)}{3}+C$ (h) $\dfrac{3\tan(1+4x)}{4}+C$

4 $x+\dfrac{1}{2}\ln(2x+1)+2\ln(x+4)+C$

5 $\dfrac{2e^{-3x}}{13}\left\{\sin 2x-\dfrac{3}{2}\cos 2x\right\}+C$ **6** (a) 95·43 unit^2 (b) 36 unit^2 **7** 1·188

Further problems 19

1 (a) $-\dfrac{(5-6x)^3}{18}+C$ (b) $-\dfrac{4\cos(3x+2)}{3}+C$ (c) $\dfrac{5\ln(2x+3)}{2}+C$

(d) $\dfrac{1}{2}3^{2x-1}/\ln 3+C$ (e) $-\dfrac{2}{9}(5-3x)^{\frac{3}{2}}+C$ (f) $2e^{3x+1}+C$ (g) $\dfrac{1}{3}(4-3x)^{-1}+C$

(h) $-2\sin(1-2x)+C$ (i) $-\tan(4-3x)+C$ (j) $-2e^{3-4x}+C$

2 $2e^{3x-5}+\dfrac{4}{3}\ln(3x-2)-\dfrac{1}{2}5^{2x+1}/\ln 5+C$ **3** $I=95$

4 (a) $\dfrac{1}{2}\ln(2x-1)+\ln(2x+3)+C$ (b) $-2\ln(x+1)+3\ln(x-4)+C$

(c) $-\ln(3x-4)+\dfrac{5}{4}\ln(4x-1)+C$ (d) $2\ln(2x-5)+\dfrac{1}{2}\ln(4x+3)+C$

(e) $x+\dfrac{3}{2}\ln(2x-1)-\dfrac{2}{3}\ln(3x-2)+C$ (f) $2x+\ln(x-4)+\dfrac{3}{2}\ln(2x+1)+C$

(g) $3x-3\ln(x+2)+\dfrac{5}{2}\ln(2x-3)+C$ (h) $2x+5\ln(x-4)+\ln(2x+1)+C$

5 (a) π^2-4 (b) $\dfrac{\pi^3}{6}-\dfrac{\pi}{4}$ (c) $\dfrac{\pi}{4}-\dfrac{1}{2}$ (d) $\dfrac{1}{10}(e^{2\pi}-1)=53\cdot45$ (e) $\dfrac{1}{13}\left\{3e^{\pi/3}-2\right\}$

6 (a) $13\frac{1}{3}$ unit^2 (b) 60 unit^2 (c) 14 unit^2 (d) 57 unit^2 (e) 33·75 unit^2

7 (a) $21\frac{1}{3}$ unit^2 (b) $45\frac{1}{3}$ unit^2 (c) 791 unit^2 (d) 196 unit^2 (e) 30·6 unit^2

8 (a) 0·625 (b) 6·682 (c) 1·854 **9** 560 unit^3 **10** 15·86 **11** 0·747 unit^2

12 28·4 **13** 28·92 **14** 0·508 **15** $\dfrac{\sqrt{2}}{4}\displaystyle\int_0^{\pi/2}\sqrt{9+\cos 2\theta}\,.d\theta$; 1·66

Index

FOUNDATION
MATHEMATICS

K. A. Stroud
Formerly Principal Lecturer
Department of Mathematics
Coventry University

with

Dexter J. Booth
Formerly Principal Lecturer
School of Computing and Engineering
University of Huddersfield

First published 2009 by
PALGRAVE MACMILLAN

Palgrave Macmillan in the UK is an imprint of Macmillan Publishers Limited, registered in England, company number 785998, of Houndmills, Basingstoke, Hampshire RG21 6XS.

Palgrave Macmillan in the US is a division of St Martin's Press LLC, 175 Fifth Avenue, New York, NY 10010.

Palgrave Macmillan is the global academic imprint of the above companies and has companies and representatives throughout the world.

Palgrave® and Macmillan® are registered trademarks in the United States, the United Kingdom, Europe and other countries.

ISBN-13 978–0–230–57907–1   paperback
ISBN-10 0–230–57907–8   paperback

This book is printed on paper suitable for recycling and made from fully managed and sustained forest sources. Logging, pulping and manufacturing processes are expected to conform to the environmental regulations of the country of origin.

A catalogue record for this book is available from the British Library.

A catalog record for this book is available from the Library of Congress.

10   9   8   7   6   5   4   3   2   1
18   17   16   15   14   13   12   11   10   09

Printed and bound in Great Britain by
CPI Antony Rowe, Chippenham and Eastbourne